The Mathematics of The Standard Model of Physics

Volume 1 of 2

Edited by Paul F. Kisak

Contents

Chapter 1

Standard Model (mathematical formulation)

For a less mathematical description, see Standard Model.

This article describes the mathematics of the **Standard Model** of particle physics, a gauge quantum field theory con-

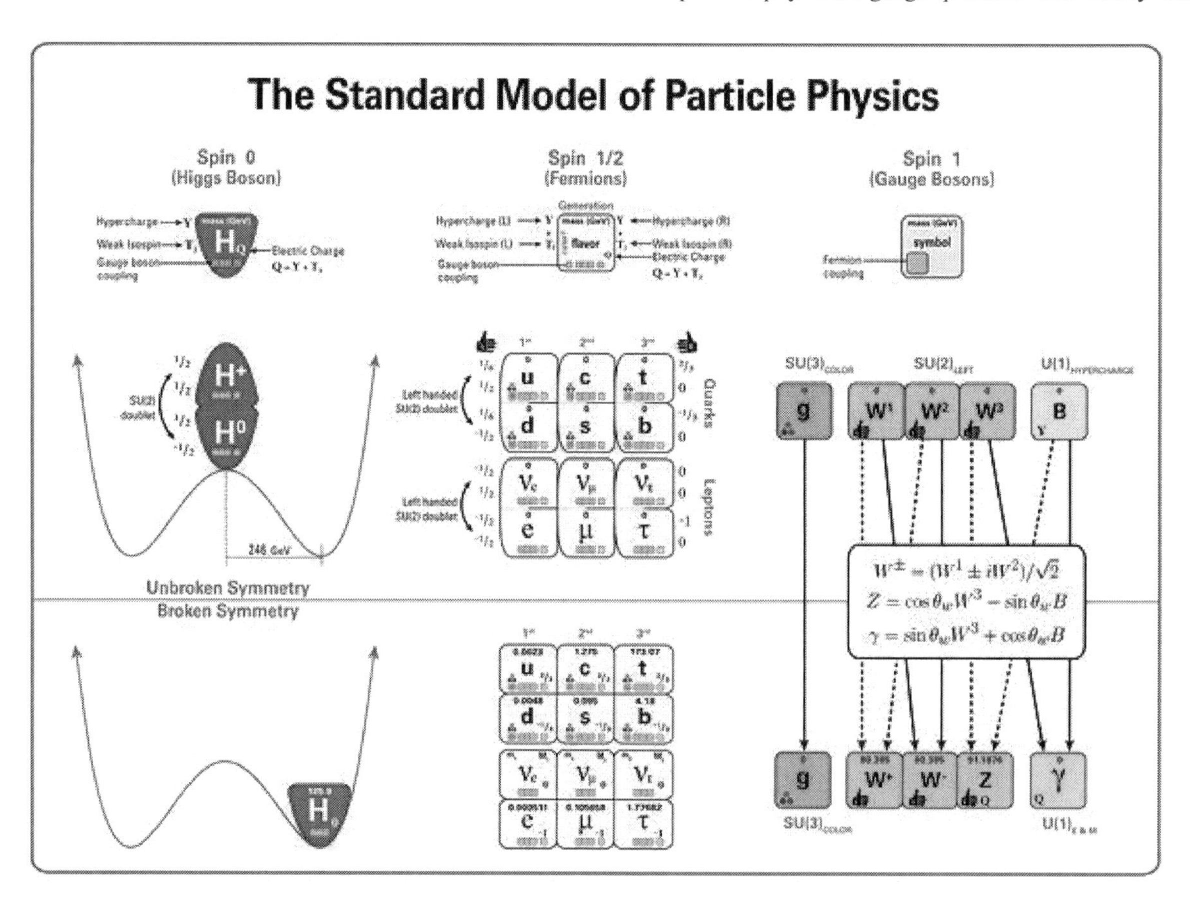

Standard Model of Particle Physics. The diagram shows the elementary particles of the Standard Model (the Higgs boson, the three generations of quarks and leptons, and the gauge bosons), including their names, masses, spins, charges, chiralities, and interactions with the strong, weak and electromagnetic forces. It also depicts the crucial role of the Higgs boson in electroweak symmetry breaking, and shows how the properties of the various particles differ in the (high-energy) symmetric phase (top) and the (low-energy) broken-symmetry phase (bottom).

taining the internal symmetries of the unitary product group SU(3) × SU(2) × U(1). The theory is commonly viewed as containing the fundamental set of particles – the leptons, quarks, gauge bosons and the Higgs particle.

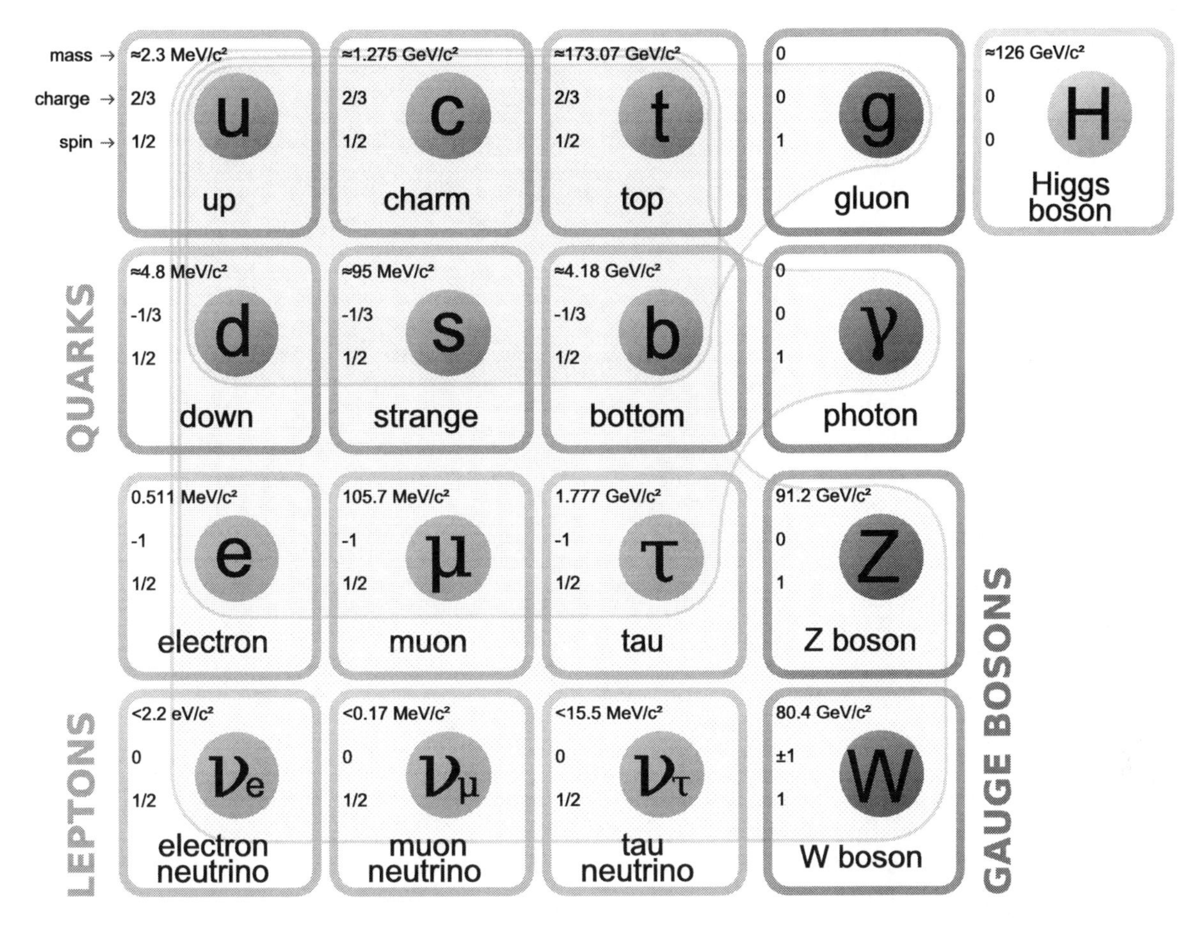

The Standard Model of Particle Physics: More Schematic Depiction

The Standard Model is renormalizable and mathematically self-consistent,[1] however despite having huge and continued successes in providing experimental predictions it does leave some unexplained phenomena. In particular, although the physics of special relativity is incorporated, general relativity is not, and the Standard Model will fail at energies or distances where the graviton is expected to emerge. Therefore in a modern field theory context, it is seen as an effective field theory.

This article requires some background in physics and mathematics, but is designed as both an introduction and a reference.

1.1 Quantum field theory

The standard model is a quantum field theory, meaning its fundamental objects are *quantum fields* which are defined at all points in spacetime. These fields are

- the fermion field, ψ, which accounts for "matter particles";
- the electroweak boson fields W_1, W_2, W_3 , and B;
- the gluon field, G_a; and
- the Higgs field, φ.

That these are *quantum* rather than *classical* fields has the mathematical consequence that they are operator-valued. In particular, values of the fields generally do not commute. As operators, they act upon the quantum state (ket vector).

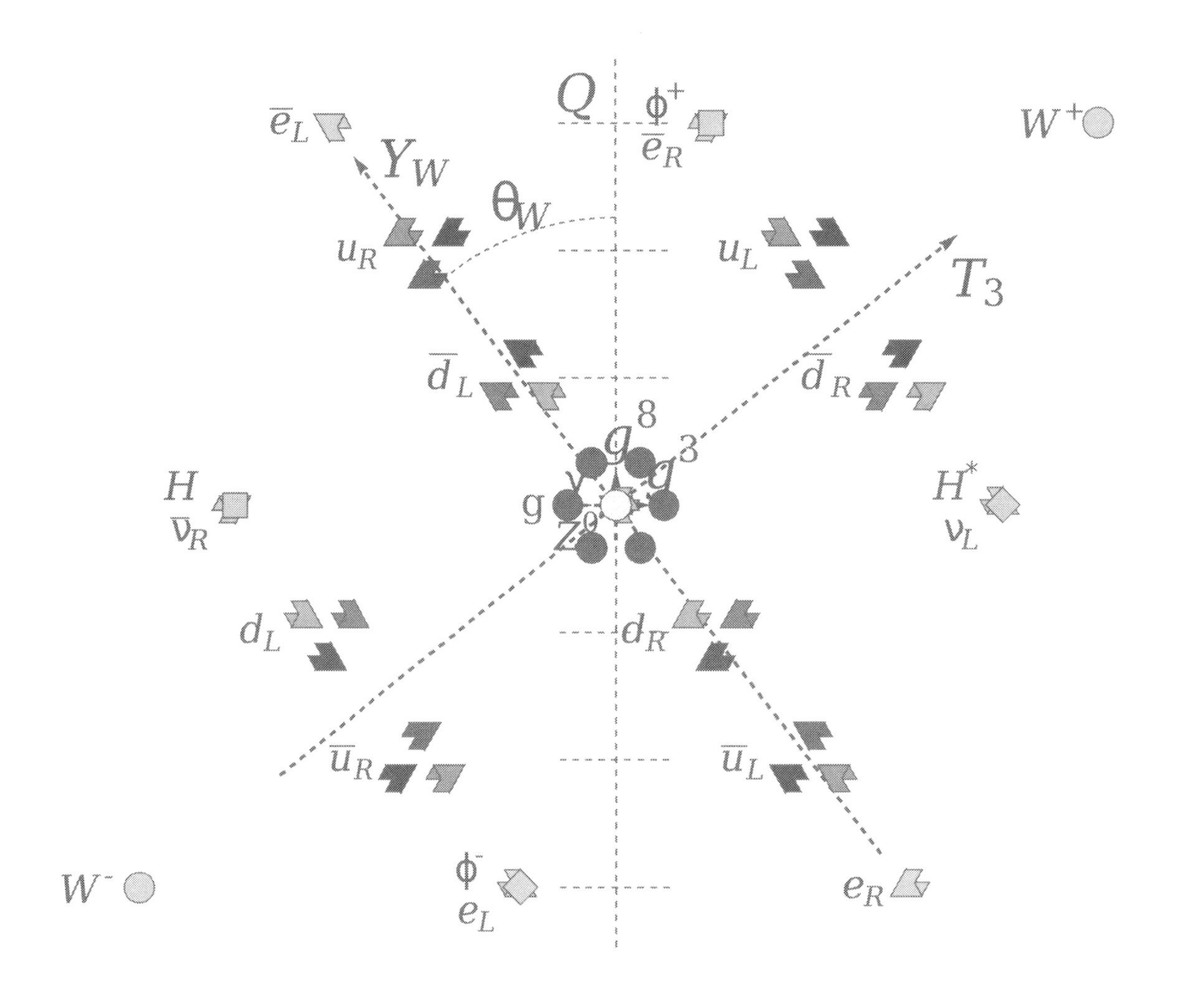

The pattern of weak isospin T_3, *weak hypercharge YW, and color charge of all known elementary particles, rotated by the weak mixing angle to show electric charge Q, roughly along the vertical. The neutral Higgs field (gray square) breaks the electroweak symmetry and interacts with other particles to give them mass.*

The dynamics of the quantum state and the fundamental fields are determined by the Lagrangian density $\mathcal{L}$ (usually for short just called the Lagrangian). This plays a role similar to that of the Schrödinger equation in non-relativistic quantum mechanics, but a Lagrangian is not an equation – rather, it is a polynomial function of the fields and their derivatives, and used with the principle of least action. While it would be possible to derive a system of differential equations governing the fields from the Langrangian, it is more common to use other techniques to compute with quantum field theories.

The standard model is furthermore a gauge theory, which means there are degrees of freedom in the mathematical formalism which do not correspond to changes in the physical state. The gauge group of the standard model is SU(3) × SU(2) × U(1), where U(1) acts on B and φ, SU(2) acts on W and φ, and SU(3) acts on G. The fermion field ψ also transforms under these symmetries, although all of them leave some parts of it unchanged.

1.1.1 The role of the quantum fields

In classical mechanics, the state of a system can usually be captured by a small set of variables, and the dynamics of the system is thus determined by the time evolution of these variables. In classical field theory, the *field* is part of the state of the system, so in order to describe it completely one effectively introduces separate variables for every point in spacetime (even though there are many restrictions on how the values of the field "variables" may vary from point to point, for example in the form of field equations involving partial derivatives of the fields).

In quantum mechanics, the classical variables are turned into operators, but these do not capture the state of the system, which is instead encoded into a wavefunction ψ or more abstract ket vector. If ψ is an eigenstate with respect to an operator P, then $P\psi = \lambda\psi$ for the corresponding eigenvalue λ, and hence letting an operator P act on ψ is analogous to multiplying ψ by the value of the classical variable to which P corresponds. By extension, a classical formula where all variables have been replaced by the corresponding operators will behave like an operator which, when it acts upon the state of the system, multiplies it by the analogue of the quantity that the classical formula would compute. The formula as such does however not contain any information about the state of the system; it would evaluate to the same operator regardless of what state the system is in.

Quantum fields relate to quantum mechanics as classical fields do to classical mechanics, i.e., there is a separate operator for every point in spacetime, and these operators do not carry any information about the state of the system; they are merely used to exhibit some aspect of the state, at the point to which they belong. In particular, the quantum fields are *not* wavefunctions, even though the equations which govern their time evolution may be deceptively similar to those of the corresponding wavefunction in a semiclassical formulation. There is no variation in strength of the fields between different points in spacetime; the variation that happens is rather one of phase factors.

1.1.2 Vectors, scalars, and spinors

Mathematically it may look as though all of the fields are vector-valued (in addition to being operator-valued), since they all have several components, can be multiplied by matrices, etc., but physicists assign a more specific physical meaning to the word: a **vector** is something which transforms like a four-vector under Lorentz transformations, and a **scalar** is something which is invariant under Lorentz transformations. The B, W_j, and G_a fields are all vectors in this sense, so the corresponding particles are said to be vector bosons. The Higgs field φ is a scalar.

The fermion field ψ does transform under Lorentz transformations, but not like a vector should; rotations will only turn it by half the angle a proper vector should. Therefore these constitute a third kind of quantity, which is known as a spinor.

It is common to make use of abstract index notation for the vector fields, in which case the vector fields all come with a Lorentzian index μ, like so: B^μ, W_j^μ , and G_a^μ . If abstract index notation is used also for spinors then these will carry a spinorial index and the Dirac gamma will carry one Lorentzian and two spinorian indices, but it is more common to regard spinors as column matrices and the Dirac gamma $\gamma\mu$ as a matrix which additionally carries a Lorentzian index. The Feynman slash notation can be used to turn a vector field into a linear operator on spinors, like so: $\not{B} = \gamma^\mu B_\mu$; this may involve raising and lowering indices.

1.2 Alternative presentations of the fields

As is common in quantum theory, there is more than one way to look at things. At first the basic fields given above may not seem to correspond well with the "fundamental particles" in the chart above, but there are several alternative presentations which, in particular contexts, may be more appropriate than those that are given above.

1.2.1 Fermions

Rather than having one fermion field ψ, it can be split up into separate components for each type of particle. This mirrors the historical evolution of quantum field theory, since the electron component ψ_e (describing the electron and its antiparticle the positron) is then the original ψ field of quantum electrodynamics, which was later accompanied by $\psi\mu$

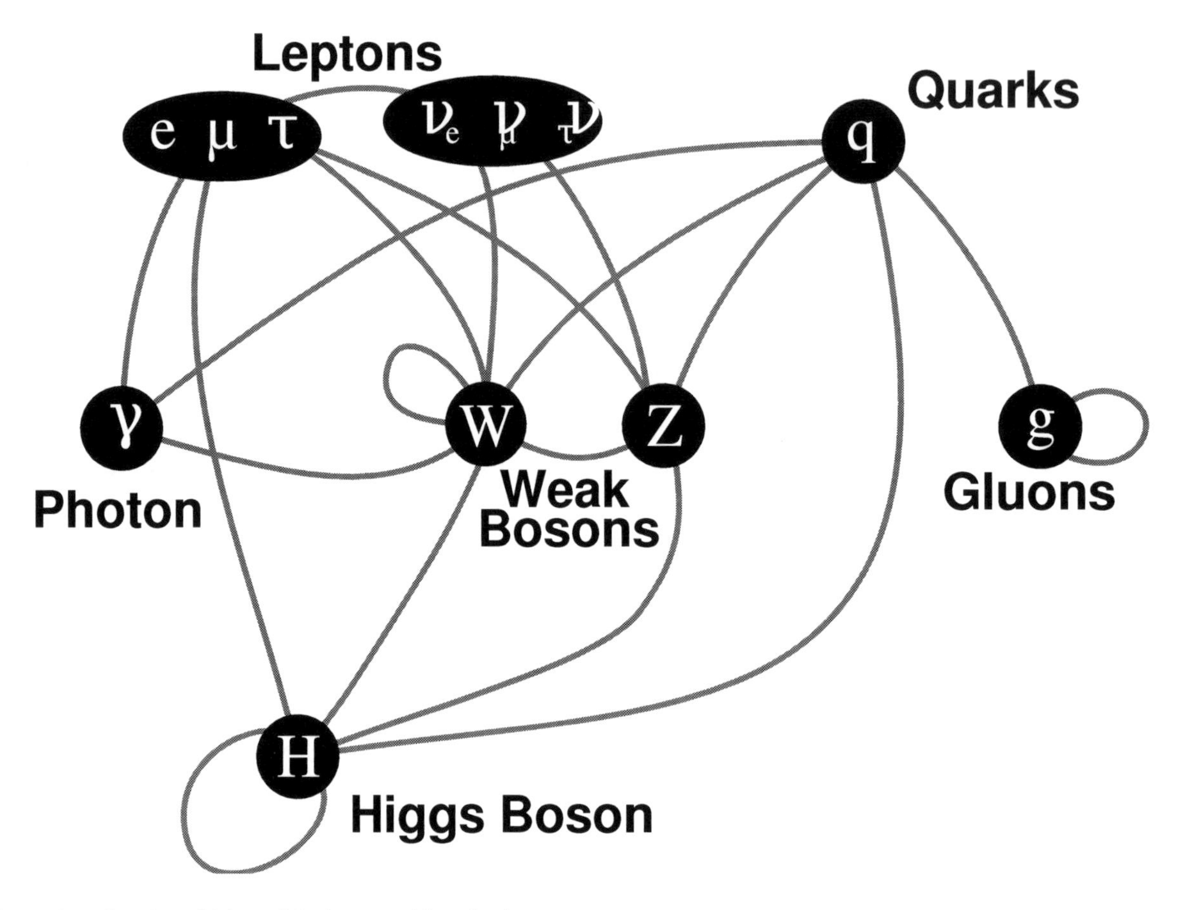

Connections denoting which particles interact with each other.

and ψτ fields for the muon and tauon respectively (and their antiparticles). Electroweak theory added ψ_{ν_e}, ψ_{ν_μ} , and ψ_{ν_τ} for the corresponding neutrinos, and the quarks add still further components. In order to be four-spinors like the electron and other lepton components, there must be one quark component for every combination of flavour and colour, bringing the total to 24 (3 for charged leptons, 3 for neutrinos, and 2·3·3 = 18 for quarks).

An important definition is the barred fermion field $\bar{\psi}$ is defined to be $\psi^\dagger\gamma^0$, where † denotes the Hermitian adjoint and γ^0 is the zeroth gamma matrix. If ψ is thought of as an $n \times 1$ matrix then $\bar{\psi}$ should be thought of as a $1 \times n$ matrix.

A chiral theory

An independent decomposition of ψ is that into chirality components:

"Left" chirality: $\psi^L = \frac{1}{2}(1 - \gamma_5)\psi$

"Right" chirality: $\psi^R = \frac{1}{2}(1 + \gamma_5)\psi$

where γ_5 is the fifth gamma matrix. This is very important in the Standard Model because *left and right chirality components are treated differently by the gauge interactions.*

In particular, under weak isospin SU(2) transformations the left-handed particles are weak-isospin doublets, whereas the right-handed are singlets – i.e. the weak isospin of ψR is zero. Put more simply, the weak interaction could rotate e.g. a left-handed electron into a left-handed neutrino (with emission of a W^-), but could not do so with the same right-handed particles. As an aside, the right-handed neutrino originally did not exist in the standard model – but the discovery of neutrino oscillation implies that neutrinos must have mass, and since chirality can change during the propagation of a massive particle, right-handed neutrinos must exist in reality. This does not however change the (experimentally-proven) chiral nature of the weak interaction.

Furthermore, U(1) acts differently on ψ_e^L than on ψ_e^R (because they have different weak hypercharges).

Mass and interaction eigenstates

A distinction can thus be made between, for example, the mass and interaction eigenstates of the neutrino. The former is the state which propagates in free space, whereas the latter is the *different* state that participates in interactions. Which is the "fundamental" particle? For the neutrino, it is conventional to define the "flavour" (ν
e, ν
μ, or ν
τ) by the interaction eigenstate, whereas for the quarks we define the flavour (up, down, etc.) by the mass state. We can switch between these states using the CKM matrix for the quarks, or the PMNS matrix for the neutrinos (the charged leptons on the other hand are eigenstates of both mass and flavour).

As an aside, if a complex phase term exists within either of these matrices, it will give rise to direct CP violation, which could explain the dominance of matter over antimatter in our current universe. This has been proven for the CKM matrix, and is expected for the PMNS matrix.

Positive and negative energies

Finally, the quantum fields are sometimes decomposed into "positive" and "negative" energy parts: $\psi = \psi^+ + \psi^-$. This is not so common when a quantum field theory has been set up, but often features prominently in the process of quantizing a field theory.

1.2.2 Bosons

Due to the Higgs mechanism, the electroweak boson fields W_1, W_2, W_3 , and B "mix" to create the states which are physically observable. To retain gauge invariance, the underlying fields must be massless, but the observable states can *gain masses* in the process. These states are:

The massive neutral boson:

$$Z = \cos\theta_W W_3 - \sin\theta_W B$$

The massless neutral boson:

$$A = \sin\theta_W W_3 + \cos\theta_W B$$

The massive charged W bosons:

$$W^{\pm} = \frac{1}{\sqrt{2}}(W_1 \mp iW_2)$$

where θW is the Weinberg angle.

The A field is the photon, which corresponds classically to the well-known electromagnetic four-potential – i.e. the electric and magnetic fields. The Z field actually contributes in every process the photon does, but due to its large mass, the contribution is usually negligible.

1.3 Perturbative QFT and the interaction picture

Much of the qualitative descriptions of the standard model in terms of "particles" and "forces" comes from the perturbative quantum field theory view of the model. In this, the Langrangian is decomposed as $\mathcal{L} = \mathcal{L}_0 + \mathcal{L}_\text{I}$ into separate *free field* and *interaction* Langrangians. The free fields care for particles in isolation, whereas processes involving several particles arise through interactions. The idea is that the state vector should only change when particles interact, meaning a free particle is one whose quantum state is constant. This corresponds to the interaction picture in quantum mechanics.

In the more common Schrödinger picture, even the states of free particles change over time: typically the phase changes at a rate which depends on their energy. In the alternative Heisenberg picture, state vectors are kept constant, at the price of having the operators (in particular the observables) be time-dependent. The interaction picture constitutes an intermediate between the two, where some time dependence is placed in the operators (the quantum fields) and some in the state vector. In QFT, the former is called the free field part of the model, and the latter is called the interaction part. The free field model can be solved exactly, and then the solutions to the full model can be expressed as perturbations of the free field solutions, for example using the Dyson series.

It should be observed that the decomposition into free fields and interactions is in principle arbitrary. For example renormalization in QED modifies the mass of the free field electron to match that of a physical electron (with an electromagnetic field), and will in doing so add a term to the free field Lagrangian which must be cancelled by a counterterm in the interaction Lagrangian, that then shows up as a two-line vertex in the Feynman diagrams. This is also how the Higgs field is thought to give particles mass: the part of the interaction term which corresponds to the (nonzero) vacuum expectation value of the Higgs field is moved from the interaction to the free field Lagrangian, where it looks just like a mass term having nothing to do with Higgs.

1.3.1 Free fields

Under the usual free/interaction decomposition, which is suitable for low energies, the free fields obey the following equations:

- The fermion field ψ satisfies the Dirac equation; $(i\hbar\partial\!\!\!/ - m_f c)\psi_f = 0$ for each type f of fermion.
- The photon field A satisfies the wave equation $\partial_\mu \partial^\mu A^\nu = 0$.
- The Higgs field φ satisfies the Klein–Gordon equation.
- The weak interaction fields Z, $W^\pm$ also satisfy the Klein–Gordon equation.

These equations can be solved exactly. One usually does so by considering first solutions that are periodic with some period L along each spatial axis; later taking the limit: $L \to \infty$ will lift this periodicity restriction.

In the periodic case, the solution for a field F (any of the above) can be expressed as a Fourier series of the form

$$F(x) = \beta \sum_{\mathbf{p}} \sum_{r} E_{\mathbf{p}}^{-\frac{1}{2}} \left(a_r(\mathbf{p}) u_r(\mathbf{p}) e^{-\frac{ipx}{\hbar}} + b_r^\dagger(\mathbf{p}) v_r(\mathbf{p}) e^{\frac{ipx}{\hbar}} \right)$$

where:

- β is a normalization factor; for the fermion field ψ_f it is $\sqrt{m_f c^2/V}$, where $V = L^3$ is the volume of the fundamental cell considered; for the photon field A^μ it is $\hbar c/\sqrt{2V}$.
- The sum over $\mathbf{p}$ is over all momenta consistent with the period L, i.e., over all vectors $\frac{2\pi\hbar}{L}(n_1, n_2, n_3)$ where n_1, n_2, n_3 are integers.
- The sum over r covers other degrees of freedom specific for the field, such as polarization or spin; it usually comes out as a sum from 1 to 2 or from 1 to 3.

- $E_{\mathbf{p}}$ is the relativistic energy for a momentum **p** quantum of the field, $= \sqrt{m^2c^4 + c^2\mathbf{p}^2}$ when the rest mass is m.
- *ar*(**p**) and $b_r^\dagger(\mathbf{p})$ are annihilation and creation respectively operators for "a-particles" and "b-particles" respectively of momentum **p**; "b-particles" are the antiparticles of "a-particles". Different fields have different "a-" and "b-particles". For some fields, a and b are the same.
- *ur*(**p**) and *vr*(**p**) are non-operators which carry the vector or spinor aspects of the field (where relevant).
- $p = (E_{\mathbf{p}}/c, \mathbf{p})$ is the four-momentum for a quanta with momentum **p**. $px = p_\mu x^\mu$ denotes an inner product of four-vectors.

In the limit $L \to \infty$, the sum would turn into an integral with help from the V hidden inside β. The numeric value of β also depends on the normalization chosen for $u_r(\mathbf{p})$ and $v_r(\mathbf{p})$.

Technically, $a_r^\dagger(\mathbf{p})$ is the Hermitian adjoint of the operator *ar*(**p**) in the inner product space of ket vectors. The identification of $a_r^\dagger(\mathbf{p})$ and *ar*(**p**) as creation and annihilation operators comes from comparing conserved quantities for a state before and after one of these have acted upon it. $a_r^\dagger(\mathbf{p})$ can for example be seen to add one particle, because it will add 1 to the eigenvalue of the a-particle number operator, and the momentum of that particle ought to be **p** since the eigenvalue of the vector-valued momentum operator increases by that much. For these derivations, one starts out with expressions for the operators in terms of the quantum fields. That the operators with † are creation operators and the one without annihilation operators is a convention, imposed by the sign of the commutation relations postulated for them.

An important step in preparation for calculating in perturbative quantum field theory is to separate the "operator" factors a and b above from their corresponding vector or spinor factors u and v. The vertices of Feynman graphs come from the way that u and v from different factors in the interaction Lagrangian fit together, whereas the edges come from the way that the as and bs must be moved around in order to put terms in the Dyson series on normal form.

1.3.2 Interaction terms and the path integral approach

The Lagrangian can also be derived without using creation and annihilation operators (the "canonical" formalism), by using a "path integral" approach, pioneered by Feynman building on the earlier work of Dirac. See e.g. Path integral formulation on Wikipedia or A. Zee's QFT in a nutshell. This is one possible way that the Feynman diagrams, which are pictorial representations of interaction terms, can be derived relatively easily. A quick derivation is indeed presented at the article on Feynman diagrams.

1.4 Lagrangian formalism

We can now give some more detail about the aforementioned free and interaction terms appearing in the Standard Model Lagrangian density. Any such term must be both gauge and reference-frame invariant, otherwise the laws of physics would depend on an arbitrary choice or the frame of an observer. Therefore the global Poincaré symmetry, consisting of translational symmetry, rotational symmetry and the inertial reference frame invariance central to the theory of special relativity must apply. The local SU(3) × SU(2) × U(1) gauge symmetry is the internal symmetry. The three factors of the gauge symmetry together give rise to the three fundamental interactions, after some appropriate relations have been defined, as we shall see.

A complete formulation of the Standard Model Lagrangian with all the terms written together can be found e.g. here.

1.4.1 Kinetic terms

A free particle can be represented by a mass term, and a *kinetic* term which relates to the "motion" of the fields.

Standard Model Interactions
(Forces Mediated by Gauge Bosons)

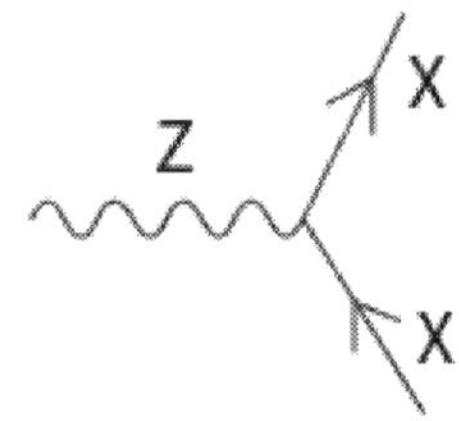

X is any fermion in the Standard Model.

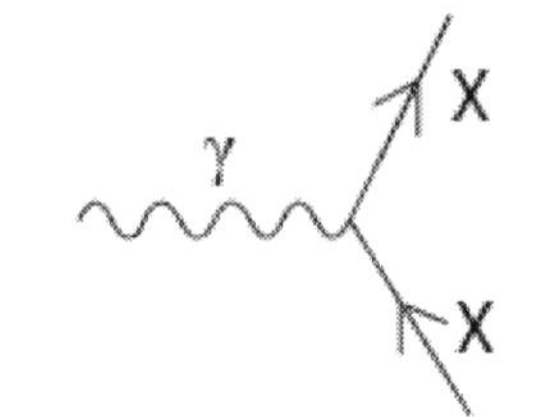

X is electrically charged.

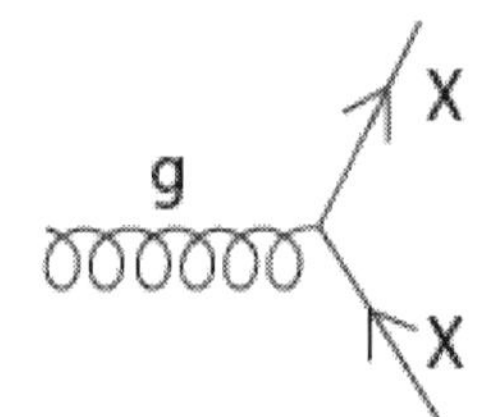

X is any quark.

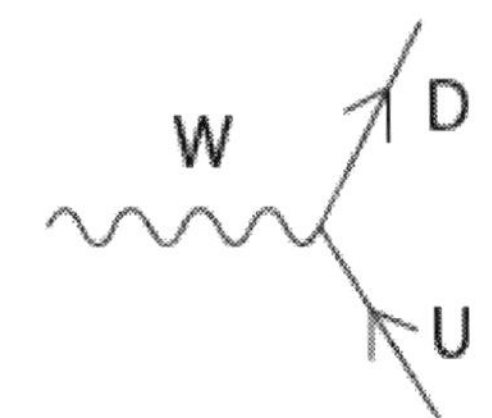

U is a up-type quark;
D is a down-type quark.

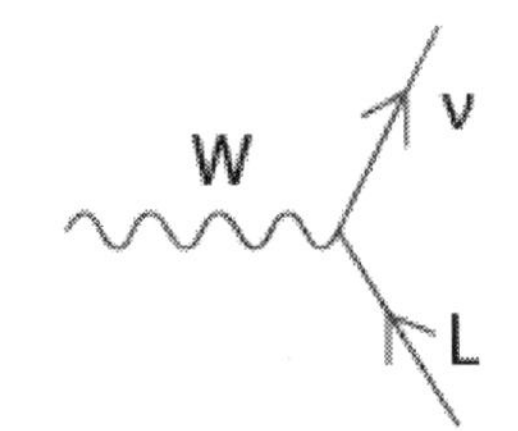

L is a lepton and v is the corresponding neutrino.

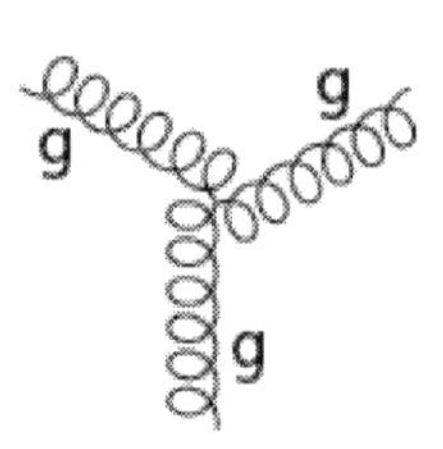

X is a photon or Z-boson.

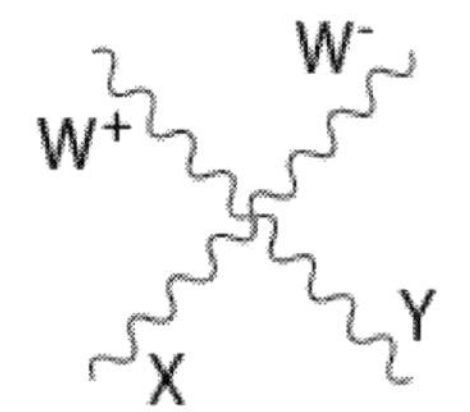

X and Y are any two electroweak bosons such that charge is conserved.

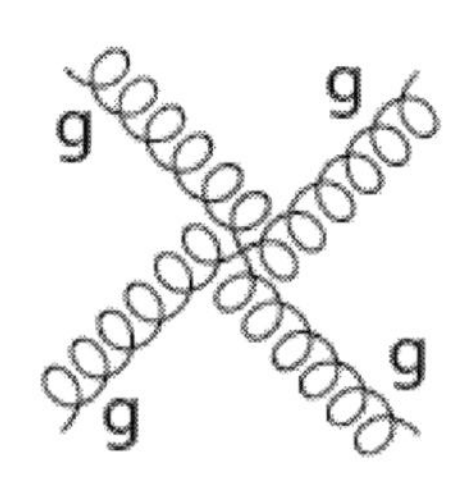

The above interactions show some basic interaction vertices – Feynman diagrams in the standard model are built from these vertices. Higgs boson interactions are however not shown, and neutrino oscillations are commonly added. The charge of the W bosons are dictated by the fermions they interact with.

Fermion fields

The kinetic term for a Dirac fermion is

$$i\bar{\psi}\gamma^{\mu}\partial_{\mu}\psi$$

where the notations are carried from earlier in the article. ψ can represent any, or all, Dirac fermions in the standard model. Generally, as below, this term is included within the couplings (creating an overall "dynamical" term).

Gauge fields

For the spin-1 fields, first define the field strength tensor

$$F^a_{\mu\nu} = \partial_\mu A^a_\nu - \partial_\nu A^a_\mu + g f^{abc} A^b_\mu A^c_\nu$$

for a given gauge field (here we use A), with gauge coupling constant g. The quantity f^{abc} is the structure constant of the particular gauge group, defined by the commutator

$$[t_a, t_b] = i f^{abc} t_c,$$

where t_i are the generators of the group. In an Abelian (commutative) group (such as the U(1) we use here), since the generators t_a all commute with each other, the structure constants vanish. Of course, this is not the case in general – the standard model includes the non-Abelian SU(2) and SU(3) groups (such groups lead to what is called a Yang–Mills gauge theory).

We need to introduce three gauge fields corresponding to each of the subgroups SU(3) × SU(2) × U(1).

- The gluon field tensor will be denoted by $G^a_{\mu\nu}$, where the index a labels elements of the **8** representation of colour SU(3). The strong coupling constant is conventionally labelled g_s (or simply g where there is no ambiguity). *The observations leading to the discovery of this part of the Standard Model are discussed in the article in quantum chromodynamics.*
- The notation $W^a_{\mu\nu}$ will be used for the gauge field tensor of SU(2) where a runs over the 3 generators of this group. The coupling can be denoted g_w or again simply g. The gauge field will be denoted by W^a_μ .
- The gauge field tensor for the U(1) of weak hypercharge will be denoted by Bμν, the coupling by g′, and the gauge field by Bμ.

The kinetic term can now be written simply as

$$\mathcal{L}_{\text{kin}} = -\frac{1}{4} B_{\mu\nu} B^{\mu\nu} - \frac{1}{2} \text{tr} W_{\mu\nu} W^{\mu\nu} - \frac{1}{2} \text{tr} G_{\mu\nu} G^{\mu\nu}$$

where the traces are over the SU(2) and SU(3) indices hidden in W and G respectively. The two-index objects are the field strengths derived from W and G the vector fields. There are also two extra hidden parameters: the theta angles for SU(2) and SU(3).

1.4.2 Coupling terms

The next step is to "couple" the gauge fields to the fermions, allowing for interactions.

Electroweak sector

Main article: Electroweak interaction

The electroweak sector interacts with the symmetry group U(1) × SU(2)L, where the subscript L indicates coupling only to left-handed fermions.

$$\mathcal{L}_{\text{EW}} = \sum_\psi \bar{\psi} \gamma^\mu \left(i\partial_\mu - g' \frac{1}{2} Y_{\text{W}} B_\mu - g \frac{1}{2} \boldsymbol{\tau} \mathbf{W}_\mu \right) \psi$$

Where Bμ is the U(1) gauge field; YW is the weak hypercharge (the generator of the U(1) group); **W**μ is the three-component SU(2) gauge field; and the components of $\boldsymbol{\tau}$ are the Pauli matrices (infinitesimal generators of the SU(2) group) whose eigenvalues give the weak isospin. Note that we have to redefine a new U(1) symmetry of *weak hypercharge*, different from QED, in order to achieve the unification with the weak force. The electric charge Q, third component of weak isospin T_3 (also called Tz, I_3 or I_z) and weak hypercharge YW are related by

$$Q = T_3 + \tfrac{1}{2}Y_W,$$

or by the alternate convention Q = T_3 + YW. The first convention (used in this article) is equivalent to the earlier Gell-Mann–Nishijima formula. We can then define the conserved current for weak isospin as

$$\mathbf{j}_\mu = \frac{1}{2}\bar{\psi}_L \gamma_\mu \boldsymbol{\tau} \psi_L$$

and for weak hypercharge as

$$j_\mu^Y = 2(j_\mu^{em} - j_\mu^3)$$

where j_μ^{em} is the electric current and j_μ^3 the third weak isospin current. As explained above, *these currents mix* to create the physically observed bosons, which also leads to testable relations between the coupling constants.

To explain in a simpler way, we can see the effect of the electroweak interaction by picking out terms from the Lagrangian. We see that the SU(2) symmetry acts on each (left-handed) fermion doublet contained in ψ, for example

$$-\frac{g}{2}(\bar{\nu}_e \; \bar{e})\tau^+ \gamma_\mu (W^-)^\mu \begin{pmatrix} \nu_e \\ e \end{pmatrix} = -\frac{g}{2}\bar{\nu}_e \gamma_\mu (W^-)^\mu e$$

where the particles are understood to be left-handed, and where

$$\tau^\pm \equiv \frac{1}{2}(\tau^1 \pm i\tau^2) = \begin{pmatrix} 0 & 1 \\ 0 & 0 \end{pmatrix}$$

This is an interaction corresponding to a "rotation in weak isospin space" or in other words, a *transformation between eL and veL via emission of a* W^- *boson.* The U(1) symmetry, on the other hand, is similar to electromagnetism, but acts on all "*weak hypercharged*" fermions (both left and right handed) via the neutral Z^0, as well as the *charged* fermions via the photon.

Quantum chromodynamics sector

Main article: Quantum chromodynamics

The quantum chromodynamics (QCD) sector defines the interactions between quarks and gluons, with SU(3) symmetry, generated by T_a. Since leptons do not interact with gluons, they are not affected by this sector. The Dirac Lagrangian of the quarks coupled to the gluon fields is given by

$$\mathcal{L}_{\rm QCD} = i\overline{U}\left(\partial_\mu - ig_s G_\mu^a T^a\right)\gamma^\mu U + i\overline{D}\left(\partial_\mu - ig_s G_\mu^a T^a\right)\gamma^\mu D.$$

where D and U are the Dirac spinors associated with up- and down-type quarks, and other notations are continued from the previous section.

1.4.3 Mass terms and the Higgs mechanism

Mass terms

The mass term arising from the Dirac Lagrangian (for any fermion ψ) is $-m\bar{\psi}\psi$ which is *not* invariant under the electroweak symmetry. This can be seen by writing ψ in terms of left and right handed components (skipping the actual calculation):

$$-m\bar{\psi}\psi = -m(\bar{\psi}_L\psi_R + \bar{\psi}_R\psi_L)$$

i.e. contribution from $\bar{\psi}_L\psi_L$ and $\bar{\psi}_R\psi_R$ terms do not appear. We see that the mass-generating interaction is achieved by constant flipping of particle chirality. The spin-half particles have no right/left chirality pair with the same SU(2) representations and equal and opposite weak hypercharges, so assuming these gauge charges are conserved in the vacuum, none of the spin-half particles could ever swap chirality, and must remain massless. Additionally, we know experimentally that the W and Z bosons are massive, but a boson mass term contains the combination e.g. $A^{\mu}A_{\mu}$, which clearly depends on the choice of gauge. Therefore, none of the standard model fermions *or* bosons can "begin" with mass, but must acquire it by some other mechanism.

The Higgs mechanism

Main article: Higgs mechanism

The solution to both these problems comes from the Higgs mechanism, which involves scalar fields (the number of which depend on the exact form of Higgs mechanism) which (to give the briefest possible description) are "absorbed" by the massive bosons as degrees of freedom, and which couple to the fermions via Yukawa coupling to create what looks like mass terms.

In the Standard Model, the Higgs field is a complex scalar of the group SU(2)L:

$$\phi = \frac{1}{\sqrt{2}}\begin{pmatrix}\phi^+\\ \phi^0\end{pmatrix},$$

where the superscripts + and 0 indicate the electric charge (Q) of the components. The weak hypercharge (YW) of both components is 1.

The Higgs part of the Lagrangian is

$$\mathcal{L}_H = \left[\left(\partial_\mu - igW_\mu^a t^a - ig'Y_\phi B_\mu\right)\phi\right]^2 + \mu^2\phi^\dagger\phi - \lambda(\phi^\dagger\phi)^2,$$

where $\lambda > 0$ and $\mu^2 > 0$, so that the mechanism of spontaneous symmetry breaking can be used. There is a parameter here, at first hidden within the shape of the potential, that is very important. In a unitarity gauge one can set $\varphi^+ = 0$ and make φ^0 real. Then $\langle\phi^0\rangle = v$ is the non-vanishing vacuum expectation value of the Higgs field. v has units of mass, and it is the only parameter in the Standard Model which is not dimensionless. It is also much smaller than the Planck scale; it is approximately equal to the Higgs mass, and sets the scale for the mass of everything else. This is the only real fine-tuning to a small nonzero value in the Standard Model, and it is called the Hierarchy problem. Quadratic terms in W_μ and B_μ arise, which give masses to the W and Z bosons:

$$M_W = \tfrac{1}{2}v|g|$$
$$M_Z = \tfrac{1}{2}v\sqrt{g^2 + g'^2}$$

The Yukawa interaction terms are

$$\mathcal{L}_{YU} = \overline{U}_L G_u U_R \phi^0 - \overline{D}_L G_u U_R \phi^- + \overline{U}_L G_d D_R \phi^+ + \overline{D}_L G_d D_R \phi^0 + hc$$

where $G_{u,d}$ are 3 × 3 matrices of Yukawa couplings, with the ij term giving the coupling of the generations i and j.

Neutrino masses

As previously mentioned, evidence shows neutrinos must have mass. But within the standard model, the right-handed neutrino does not exist, so even with a Yukawa coupling neutrinos remain massless. An obvious solution[2] is to simply *add a right-handed neutrino* νR resulting in a **Dirac mass** term as usual. This field however must be a sterile neutrino, since being right-handed it experimentally belongs to an isospin singlet ($T_3 = 0$) and also has charge $Q = 0$, implying $YW = 0$ (see above) i.e. it does not even participate in the weak interaction. Current experimental status is that evidence for observation of sterile neutrinos is not convincing.[3]

Another possibility to consider is that the neutrino satisfies the **Majorana equation**, which at first seems possible due to its zero electric charge. In this case the mass term is

$$-\frac{m}{2}\left(\overline{\nu}^C \nu + \overline{\nu} \nu^C\right)$$

where C denotes a charge conjugated (i.e. anti-) particle, and the terms are consistently all left (or all right) chirality (note that a left-chirality projection of an antiparticle is a right-handed field; care must be taken here due to different notations sometimes used). Here we are essentially flipping between LH neutrinos and RH anti-neutrinos (it is furthermore possible but *not* necessary that neutrinos are their own antiparticle, so these particles are the same). However for the left-chirality neutrinos, this term changes weak hypercharge by 2 units - not possible with the standard Higgs interation, requiring the Higgs field to be extended to include an extra triplet with weak hypercharge 2[4] - whereas for right-chirality neutrinos, no Higgs extensions are necessary. For both left and right chirality cases, Majorana terms violate lepton number, but possibly at a level beyond the current sensitivity of experiments to detect such violations.

It is possible to include **both** Dirac and Majorana mass terms in the same theory, which (in contrast to the Dirac-mass-only approach) can provide a “natural” explanation for the smallness of the observed neutrino masses, by linking the RH neutrinos to yet-unknown physics around the GUT scale[5] (see seesaw mechanism).

Since in any case new fields must be postulated to explain the experimental results, neutrinos are an obvious gateway to searching physics beyond the Standard Model.

1.5 Detailed Information

This section provides more detail on some aspects, and some reference material.

1.5.1 Field content in detail

The Standard Model has the following fields. These describe one *generation* of leptons and quarks, and there are three generations, so there are three copies of each field. By CPT symmetry, there is a set of right-handed fermions with the opposite quantum numbers. The column "**representation**" indicates under which representations of the gauge groups that each field transforms, in the order (SU(3), SU(2), U(1)). Symbols used are common but not universal; superscript C denotes an antiparticle; and for the U(1) group, the value of the weak hypercharge is listed. Note that there are twice as many left-handed lepton field components as left-handed antilepton field components in each generation, but an equal number of left-handed quark and antiquark fields.

1.5.2 Fermion content

This table is based in part on data gathered by the Particle Data Group.[6]

[1] These are not ordinary abelian charges, which can be added together, but are labels of group representations of Lie groups.

[2] Mass is really a coupling between a left-handed fermion and a right-handed fermion. For example, the mass of an electron is really a coupling between a left-handed electron and a right-handed electron, which is the antiparticle of a left-handed positron. Also neutrinos show large mixings in their mass coupling, so it's not accurate to talk about neutrino masses in the flavor basis or to suggest a left-handed electron antineutrino.

[3] The Standard Model assumes that neutrinos are massless. However, several contemporary experiments prove that neutrinos oscillate between their flavour states, which could not happen if all were massless. It is straightforward to extend the model to fit these data but there are many possibilities, so the mass eigenstates are still open. See neutrino mass.

[4] W.-M. Yao *et al.* (Particle Data Group) (2006). "Review of Particle Physics: Neutrino mass, mixing, and flavor change" (PDF). *Journal of Physics G* **33**: 1. arXiv:astro-ph/0601168. Bibcode:2006JPhG...33....1Y. doi:10.1088/0954-3899/33/1/001.

[5] The masses of baryons and hadrons and various cross-sections are the experimentally measured quantities. Since quarks can't be isolated because of QCD confinement, the quantity here is supposed to be the mass of the quark at the renormalization scale of the QCD scale.

1.5.3 Free parameters

Upon writing the most general Lagrangian without neutrinos, one finds that the dynamics depend on 19 parameters, whose numerical values are established by experiment. With neutrinos 7 more parameters are needed, 3 masses and 4 PMNS matrix parameters, for a total of 26 parameters.[7] The neutrino parameter values are still uncertain. The 19 certain parameters are summarized here (note: with the Higgs mass is at 125 GeV, the Higgs self-coupling strength $\lambda \sim 1/8$).

1.5.4 Additional symmetries of the Standard Model

From the theoretical point of view, the Standard Model exhibits four additional global symmetries, not postulated at the outset of its construction, collectively denoted **accidental symmetries**, which are continuous U(1) global symmetries. The transformations leaving the Lagrangian invariant are:

$$\psi_\text{q}(x) \to e^{i\alpha/3}\psi_\text{q}$$

$$E_L \to e^{i\beta}E_L \text{ and } (e_R)^c \to e^{i\beta}(e_R)^c$$

$$M_L \to e^{i\beta}M_L \text{ and } (\mu_R)^c \to e^{i\beta}(\mu_R)^c$$

$$T_L \to e^{i\beta}T_L \text{ and } (\tau_R)^c \to e^{i\beta}(\tau_R)^c$$

The first transformation rule is shorthand meaning that all quark fields for all generations must be rotated by an identical phase simultaneously. The fields ML, TL and $(\mu_R)^c$, $(\tau_R)^c$ are the 2nd (muon) and 3rd (tau) generation analogs of EL and $(e_R)^c$ fields.

By Noether's theorem, each symmetry above has an associated conservation law: the conservation of baryon number, electron number, muon number, and tau number. Each quark is assigned a baryon number of $\frac{1}{3}$, while each antiquark is assigned a baryon number of $-\frac{1}{3}$. Conservation of baryon number implies that the number of quarks minus the number of antiquarks is a constant. Within experimental limits, no violation of this conservation law has been found.

Similarly, each electron and its associated neutrino is assigned an electron number of +1, while the anti-electron and the associated anti-neutrino carry a −1 electron number. Similarly, the muons and their neutrinos are assigned a muon number of +1 and the tau leptons are assigned a tau lepton number of +1. The Standard Model predicts that each of these three numbers should be conserved separately in a manner similar to the way baryon number is conserved. These numbers are collectively known as lepton family numbers (LF).

In addition to the accidental (but exact) symmetries described above, the Standard Model exhibits several **approximate symmetries**. These are the "SU(2) custodial symmetry" and the "SU(2) or SU(3) quark flavor symmetry."

1.5.5 The U(1) symmetry

For the leptons, the gauge group can be written $SU(2)_l \times U(1)L \times U(1)R$. The two U(1) factors can be combined into $U(1)Y \times U(1)_l$ where l is the lepton number. Gauging of the lepton number is ruled out by experiment, leaving only the possible gauge group SU(2)L × U(1)Y. A similar argument in the quark sector also gives the same result for the electroweak theory.

1.5.6 The charged and neutral current couplings and Fermi theory

The charged currents $j^{\pm} = j^1 \pm ij^2$ are

$$j_\mu^+ = \overline{U}_{iL}\gamma_\mu D_{iL} + \overline{\nu}_{iL}\gamma_\mu l_{iL}.$$

These charged currents are precisely those that entered the Fermi theory of beta decay. The action contains the charge current piece

$$\mathcal{L}_{CC} = \frac{g}{\sqrt{2}}(j_\mu^+ W^{-\mu} + j_\mu^- W^{+\mu}).$$

For energy much less than the mass of the W-boson, the effective theory becomes the current–current interaction of the Fermi theory.

However, gauge invariance now requires that the component W^3 of the gauge field also be coupled to a current that lies in the triplet of SU(2). However, this mixes with the U(1), and another current in that sector is needed. These currents must be uncharged in order to conserve charge. So we require the **neutral currents**

$$j_\mu^3 = \frac{1}{2}(\overline{U}_{iL}\gamma_\mu U_{iL} - \overline{D}_{iL}\gamma_\mu D_{iL} + \overline{\nu}_{iL}\gamma_\mu \nu_{iL} - \overline{l}_{iL}\gamma_\mu l_{iL})$$

$$j_\mu^{em} = \frac{2}{3}\overline{U}_i\gamma_\mu U_i - \frac{1}{3}\overline{D}_i\gamma_\mu D_i - \overline{l}_i\gamma_\mu l_i.$$

The neutral current piece in the Lagrangian is then

$$\mathcal{L}_{NC} = ej_\mu^{em}A^\mu + \frac{g}{\cos\theta_W}(J_\mu^3 - \sin^2\theta_W J_\mu^{em})Z^\mu.$$

1.6 See also

- Overview of Standard Model of particle physics
- Fundamental interaction
- Noncommutative standard model
- Open questions: CP violation, Neutrino masses, Quark matter
- Physics beyond the Standard Model
- Strong interactions: Flavour, Quantum chromodynamics, Quark model
- Weak interactions: Electroweak interaction, Fermi's interaction
- Weinberg angle
- Symmetry in quantum mechanics

1.7 References and external links

[1] In fact, there are mathematical issues regarding quantum field theories still under debate (see e.g. Landau pole), but the predictions extracted from the Standard Model by current methods are all self-consistent. For a further discussion see e.g. R. Mann, chapter 25.

[2] https://fas.org/sgp/othergov/doe/lanl/pubs/00326607.pdf

[3] http://t2k-experiment.org/neutrinos/oscillations-today/

[4] https://fas.org/sgp/othergov/doe/lanl/pubs/00326607.pdf

[5] http://www.mpi-hd.mpg.de/personalhomes/schwetz/tueb-2.pdf

[6] W.-M. Yao *et al.* (Particle Data Group) (2006). "Review of Particle Physics: Quarks" (PDF). *Journal of Physics G* **33**: 1. arXiv:astro-ph/0601168. Bibcode:2006JPhG...33....1Y. doi:10.1088/0954-3899/33/1/001.

[7] Mark Thomson (5 September 2013). *Modern Particle Physics*. Cambridge University Press. pp. 499–500. ISBN 978-1-107-29254-3.

- *An introduction to quantum field theory*, by M.E. Peskin and D.V. Schroeder (HarperCollins, 1995) ISBN 0-201-50397-2.
- *Gauge theory of elementary particle physics*, by T.P. Cheng and L.F. Li (Oxford University Press, 1982) ISBN 0-19-851961-3.
- Standard Model Lagrangian with explicit Higgs terms (T.D. Gutierrez, ca 1999) (PDF, PostScript, and LaTeX version)
- *The quantum theory of fields* (vol 2), by S. Weinberg (Cambridge University Press, 1996) ISBN 0-521-55002-5.
- *Quantum Field Theory in a Nutshell* (Second Edition), by A. Zee (Princeton University Press, 2010) ISBN 978-1-4008-3532-4.
- *An Introduction to Particle Physics and the Standard Model*, by R. Mann (CRC Press, 2010) ISBN 978-1420082982

Chapter 2

Particle physics

For other uses of the word “particle” in physics and elsewhere, see particle (disambiguation).

Particle physics is the branch of physics that studies the nature of the particles that constitute *matter* (particles with mass) and *radiation* (massless particles). Although the word "particle" can refer to various types of very small objects (e.g. protons, gas particles, or even household dust), “particle physics” usually investigates the irreducibly smallest detectable particles and the irreducibly fundamental force fields necessary to explain them. By our current understanding, these elementary particles are excitations of the quantum fields that also govern their interactions. The currently dominant theory explaining these fundamental particles and fields, along with their dynamics, is called the Standard Model. Thus, modern particle physics generally investigates the Standard Model and its various possible extensions, e.g. to the newest “known” particle, the Higgs boson, or even to the oldest known force field, gravity.[1][2]

2.1 Subatomic particles

Modern particle physics research is focused on subatomic particles, including atomic constituents such as electrons, protons, and neutrons (protons and neutrons are composite particles called baryons, made of quarks), produced by radioactive and scattering processes, such as photons, neutrinos, and muons, as well as a wide range of exotic particles. Dynamics of particles is also governed by quantum mechanics; they exhibit wave–particle duality, displaying particle-like behaviour under certain experimental conditions and wave-like behaviour in others. In more technical terms, they are described by quantum state vectors in a Hilbert space, which is also treated in quantum field theory. Following the convention of particle physicists, the term *elementary particles* is applied to those particles that are, according to current understanding, presumed to be indivisible and not composed of other particles.[3]

All particles, and their interactions observed to date, can be described almost entirely by a quantum field theory called the Standard Model.[4] The Standard Model, as currently formulated, has 61 elementary particles.[3] Those elementary particles can combine to form composite particles, accounting for the hundreds of other species of particles that have been discovered since the 1960s. The Standard Model has been found to agree with almost all the experimental tests conducted to date. However, most particle physicists believe that it is an incomplete description of nature, and that a more fundamental theory awaits discovery (See Theory of Everything). In recent years, measurements of neutrino mass have provided the first experimental deviations from the Standard Model.

2.2 History

Main article: History of subatomic physics

The idea that all matter is composed of elementary particles dates to at least the 6th century BC.[5] In the 19th century,

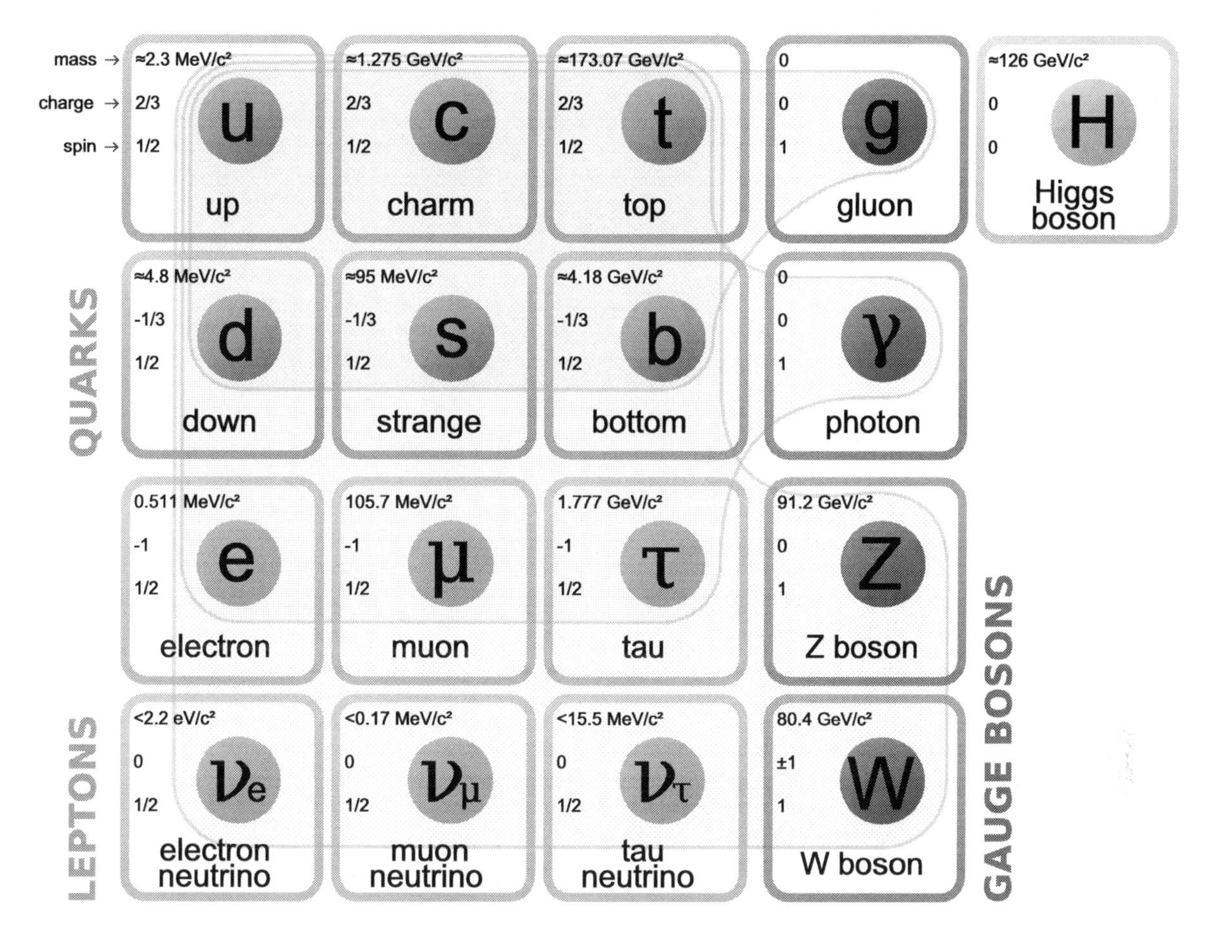

The particle content of the Standard Model of Physics

John Dalton, through his work on stoichiometry, concluded that each element of nature was composed of a single, unique type of particle.[6] The word *atom*, after the Greek word *atomos* meaning “indivisible”, denotes the smallest particle of a chemical element since then, but physicists soon discovered that *atoms* are not, in fact, the fundamental particles of nature, but conglomerates of even smaller particles, such as the electron. The early 20th-century explorations of nuclear physics and quantum physics culminated in proofs of nuclear fission in 1939 by Lise Meitner (based on experiments by Otto Hahn), and nuclear fusion by Hans Bethe in that same year; both discoveries also led to the development of nuclear weapons. Throughout the 1950s and 1960s, a bewildering variety of particles were found in scattering experiments. It was referred to as the "particle zoo". That term was deprecated after the formulation of the Standard Model during the 1970s in which the large number of particles was explained as combinations of a (relatively) small number of fundamental particles.

2.3 Standard Model

Main article: Standard Model

The current state of the classification of all elementary particles is explained by the Standard Model. It describes the strong, weak, and electromagnetic fundamental interactions, using mediating gauge bosons. The species of gauge bosons are the gluons, W−, W+ and Z bosons, and the photons.[4] The Standard Model also contains 24 fundamental particles, (12 particles and their associated anti-particles), which are the constituents of all matter.[7] Finally, the Standard Model also predicted the existence of a type of boson known as the Higgs boson. Early in the morning on 4 July 2012, physicists with the Large Hadron Collider at CERN announced they had found a new particle that behaves similarly to what is expected from the Higgs boson.[8]

2.4 Experimental laboratories

In particle physics, the major international laboratories are located at the:

- Brookhaven National Laboratory (Long Island, United States). Its main facility is the Relativistic Heavy Ion Collider (RHIC), which collides heavy ions such as gold ions and polarized protons. It is the world's first heavy ion collider, and the world's only polarized proton collider.[9]
- Budker Institute of Nuclear Physics (Novosibirsk, Russia). Its main projects are now the electron-positron colliders VEPP-2000,[10] operated since 2006, and VEPP-4,[11] started experiments in 1994. Earlier facilities include the first electron-electron beam-beam collider VEP-1, which conducted experiments from 1964 to 1968; the electron-positron colliders VEPP-2, operated from 1965 to 1974; and, its successor VEPP-2M,[12] performed experiments from 1974 to 2000.[13]
- CERN, (Conseil Européen pour la Recherche Nucléaire) (Franco-Swiss border, near Geneva). Its main project is now the Large Hadron Collider (LHC), which had its first beam circulation on 10 September 2008, and is now the world's most energetic collider of protons. It also became the most energetic collider of heavy ions after it began colliding lead ions. Earlier facilities include the Large Electron–Positron Collider (LEP), which was stopped on 2 November 2000 and then dismantled to give way for LHC; and the Super Proton Synchrotron, which is being reused as a pre-accelerator for the LHC.[14]
- DESY (Deutsches Elektronen-Synchrotron) (Hamburg, Germany). Its main facility is the Hadron Elektron Ring Anlage (HERA), which collides electrons and positrons with protons.[15]
- Fermi National Accelerator Laboratory (Fermilab), (Batavia, United States). Its main facility until 2011 was the Tevatron, which collided protons and antiprotons and was the highest-energy particle collider on earth until the Large Hadron Collider surpassed it on 29 November 2009.[16]
- KEK, (Tsukuba, Japan). It is the home of a number of experiments such as the K2K experiment, a neutrino oscillation experiment and Belle, an experiment measuring the CP violation of B mesons.[17]
- SLAC National Accelerator Laboratory, (Menlo Park, United States). Its 2-mile-long linear particle accelerator began operating in 1962 and was the basis for numerous electron and positron collision experiments until 2008. Since then the linear accelerator is being used for the Linac Coherent Light Source X-ray laser as well as advanced accelerator design research. SLAC staff continue to participate in developing and building many particle physics experiments around the world.[18]

Many other particle accelerators do exist.

The techniques required to do modern, experimental, particle physics are quite varied and complex, constituting a sub-specialty nearly completely distinct from the theoretical side of the field.

2.5 Theory

Theoretical particle physics attempts to develop the models, theoretical framework, and mathematical tools to understand current experiments and make predictions for future experiments. See also theoretical physics. There are several major interrelated efforts being made in theoretical particle physics today. One important branch attempts to better understand the Standard Model and its tests. By extracting the parameters of the Standard Model, from experiments with less uncertainty, this work probes the limits of the Standard Model and therefore expands our understanding of nature's building blocks. Those efforts are made challenging by the difficulty of calculating quantities in quantum chromodynamics. Some theorists working in this area refer to themselves as **phenomenologists** and they may use the tools of quantum field theory and effective field theory. Others make use of lattice field theory and call themselves *lattice theorists*.

Another major effort is in model building where model builders develop ideas for what physics may lie beyond the Standard Model (at higher energies or smaller distances). This work is often motivated by the hierarchy problem and is constrained by existing experimental data. It may involve work on supersymmetry, alternatives to the Higgs mechanism, extra spatial dimensions (such as the Randall-Sundrum models), Preon theory, combinations of these, or other ideas.

A third major effort in theoretical particle physics is string theory. *String theorists* attempt to construct a unified description of quantum mechanics and general relativity by building a theory based on small strings, and branes rather than particles. If the theory is successful, it may be considered a "Theory of Everything", or "TOE".

There are also other areas of work in theoretical particle physics ranging from particle cosmology to loop quantum gravity.

This division of efforts in particle physics is reflected in the names of categories on the arXiv, a preprint archive:[19] hep-th (theory), hep-ph (phenomenology), hep-ex (experiments), hep-lat (lattice gauge theory).

2.6 Practical applications

In principle, all physics (and practical applications developed therefrom) can be derived from the study of fundamental particles. In practice, even if "particle physics" is taken to mean only "high-energy atom smashers", many technologies have been developed during these pioneering investigations that later find wide uses in society. Cyclotrons are used to produce medical isotopes for research and treatment (for example, isotopes used in PET imaging), or used directly for certain cancer treatments. The development of Superconductors has been pushed forward by their use in particle physics. The World Wide Web and touchscreen technology were initially developed at CERN.

Additional applications are found in medicine, national security, industry, computing, science, and workforce development, illustrating a long and growing list of beneficial practical applications with contributions from particle physics.[20]

2.7 Future

The primary goal, which is pursued in several distinct ways, is to find and understand what physics may lie beyond the standard model. There are several powerful experimental reasons to expect new physics, including dark matter and neutrino mass. There are also theoretical hints that this new physics should be found at accessible energy scales. Furthermore, there may be surprises that will give us opportunities to learn about nature.

Much of the effort to find this new physics are focused on new collider experiments. The Large Hadron Collider (LHC) was completed in 2008 to help continue the search for the Higgs boson, supersymmetric particles, and other new physics. An intermediate goal is the construction of the International Linear Collider (ILC), which will complement the LHC by allowing more precise measurements of the properties of newly found particles. In August 2004, a decision for the technology of the ILC was taken but the site has still to be agreed upon.

In addition, there are important non-collider experiments that also attempt to find and understand physics beyond the Standard Model. One important non-collider effort is the determination of the neutrino masses, since these masses may arise from neutrinos mixing with very heavy particles. In addition, cosmological observations provide many useful constraints on the dark matter, although it may be impossible to determine the exact nature of the dark matter without the colliders. Finally, lower bounds on the very long lifetime of the proton put constraints on Grand Unified Theories at energy scales much higher than collider experiments will be able to probe any time soon.

In May 2014, the Particle Physics Project Prioritization Panel released its report on particle physics funding priorities for the United States over the next decade. This report emphasized continued U.S. participation in the LHC and ILC, and expansion of the Long Baseline Neutrino Experiment, among other recommendations.

2.8 See also

- Atomic physics

- High pressure
- International Conference on High Energy Physics
- Introduction to quantum mechanics
- List of accelerators in particle physics
- List of particles
- Magnetic monopole
- Micro black hole
- Number theory
- Resonance (particle physics)
- Self-consistency principle in high energy Physics
- Non-extensive self-consistent thermodynamical theory
- Standard Model (mathematical formulation)
- Stanford Physics Information Retrieval System
- Timeline of particle physics
- Unparticle physics
- Tetraquark

2.9 References

[1] http://home.web.cern.ch/topics/higgs-boson

[2] http://www.nobelprize.org/nobel_prizes/physics/laureates/2013/advanced-physicsprize2013.pdf

[3] Braibant, S.; Giacomelli, G.; Spurio, M. (2009). *Particles and Fundamental Interactions: An Introduction to Particle Physics*. Springer. pp. 313–314. ISBN 978-94-007-2463-1.

[4] "Particle Physics and Astrophysics Research". The Henryk Niewodniczanski Institute of Nuclear Physics. Retrieved 31 May 2012.

[5] "Fundamentals of Physics and Nuclear Physics" (PDF). Retrieved 21 July 2012.

[6] "Scientific Explorer: Quasiparticles". Sciexplorer.blogspot.com. 22 May 2012. Retrieved 21 July 2012.

[7] Nakamura, K (1 July 2010). "Review of Particle Physics". *Journal of Physics G: Nuclear and Particle Physics* **37** (7A): 075021. Bibcode:2010JPhG...37g5021N. doi:10.1088/0954-3899/37/7A/075021.

[8] Mann, Adam (28 March 2013). "Newly Discovered Particle Appears to Be Long-Awaited Higgs Boson - Wired Science". Wired.com. Retrieved 6 February 2014.

[9] "Brookhaven National Laboratory – A Passion for Discovery". Bnl.gov. Retrieved 23 June 2012.

[10] "index". Vepp2k.inp.nsk.su. Retrieved 21 July 2012.

[11] "The VEPP-4 accelerating-storage complex". V4.inp.nsk.su. Retrieved 21 July 2012.

[12] "VEPP-2M collider complex" (in Russian). Inp.nsk.su. Retrieved 21 July 2012.

[13] "The Budker Institute Of Nuclear Physics". English Russia. 21 January 2012. Retrieved 23 June 2012.

[14] "Welcome to". Info.cern.ch. Retrieved 23 June 2012.

[15] "Germany's largest accelerator centre – Deutsches Elektronen-Synchrotron DESY". Desy.de. Retrieved 23 June 2012.

[16] "Fermilab | Home". Fnal.gov. Retrieved 23 June 2012.

[17] "Kek | High Energy Accelerator Research Organization". Legacy.kek.jp. Retrieved 23 June 2012.

[18] "SLAC National Accelerator Laboratory Home Page". Retrieved 19 February 2015.

[19] arxiv.org

[20] "Fermilab | Science at Fermilab | Benefits to Society". Fnal.gov. Retrieved 23 June 2012.

2.10 Further reading

Introductory reading

- Close, Frank (2004). *Particle Physics: A Very Short Introduction*. Oxford University Press. ISBN 0-19-280434-0.
- Close, Frank; Marten, Michael; Sutton, Christine (2004). *The Particle Odyssey: A Journey to the Heart of the Matter*. Oxford University Press. ISBN 9780198609438.
- Ford, Kenneth W. (2005). *The Quantum World*. Harvard University Press.
- Oerter, Robert (2006). *The Theory of Almost Everything: The Standard Model, the Unsung Triumph of Modern Physics*. Plume.
- Schumm, Bruce A. (2004). *Deep Down Things: The Breathtaking Beauty of Particle Physics*. Johns Hopkins University Press. ISBN 0-8018-7971-X.
- Close, Frank (2006). *The New Cosmic Onion*. Taylor & Francis. ISBN 1-58488-798-2.

Advanced reading

- Robinson, Matthew B.; Bland, Karen R.; Cleaver, Gerald. B.; Dittmann, Jay R. (2008). "A Simple Introduction to Particle Physics". arXiv:0810.3328 [hep-th].
- Robinson, Matthew B.; Ali, Tibra; Cleaver, Gerald B. (2009). "A Simple Introduction to Particle Physics Part II". arXiv:0908.1395 [hep-th].
- Griffiths, David J. (1987). *Introduction to Elementary Particles*. Wiley, John & Sons, Inc. ISBN 0-471-60386-4.
- Kane, Gordon L. (1987). *Modern Elementary Particle Physics*. Perseus Books. ISBN 0-201-11749-5.
- Perkins, Donald H. (1999). *Introduction to High Energy Physics*. Cambridge University Press. ISBN 0-521-62196-8.
- Povh, Bogdan (1995). *Particles and Nuclei: An Introduction to the Physical Concepts*. Springer-Verlag. ISBN 0-387-59439-6.
- Boyarkin, Oleg (2011). *Advanced Particle Physics Two-Volume Set*. CRC Press. ISBN 978-1-4398-0412-4.

2.11 External links

- *Symmetry* magazine
- Fermilab
- Particle physics – it matters – the Institute of Physics
- Nobes, Matthew (2002) "Introduction to the Standard Model of Particle Physics" on Kuro5hin: Part 1, Part 2, Part 3a, Part 3b.
- CERN – European Organization for Nuclear Research
- The Particle Adventure – educational project sponsored by the Particle Data Group of the Lawrence Berkeley National Laboratory (LBNL)

Chapter 3

Quantum field theory

“Relativistic quantum field theory” redirects here. For other uses, see Relativity.

In theoretical physics, **quantum field theory (QFT)** is a theoretical framework for constructing quantum mechanical models of subatomic particles in particle physics and quasiparticles in condensed matter physics. A QFT treats particles as excited states of an underlying physical field, so these are called field quanta.

In quantum field theory, quantum mechanical interactions between particles and are described by interaction terms between the corresponding underlying fields.

3.1 Definition

Quantum electrodynamics (QED) has one electron field and one photon field; quantum chromodynamics (QCD) has one field for each type of quark; and, in condensed matter, there is an atomic displacement field that gives rise to phonon particles. Edward Witten describes QFT as “by far” the most difficult theory in modern physics.[1]

3.1.1 Dynamics

See also: Relativistic dynamics

Ordinary quantum mechanical systems have a fixed number of particles, with each particle having a finite number of degrees of freedom. In contrast, the excited states of a QFT can represent any number of particles. This makes quantum field theories especially useful for describing systems where the particle count/number may change over time, a crucial feature of relativistic dynamics.

3.1.2 States

QFT interaction terms are similar in spirit to those between charges with electric and magnetic fields in Maxwell’s equations. However, unlike the classical fields of Maxwell’s theory, fields in QFT generally exist in quantum superpositions of states and are subject to the laws of quantum mechanics.

Because the fields are continuous quantities over space, there exist excited states with arbitrarily large numbers of particles in them, providing QFT systems with an effectively infinite number of degrees of freedom. Infinite degrees of freedom can easily lead to divergences of calculated quantities (i.e., the quantities become infinite). Techniques such as renormalization of QFT parameters or discretization of spacetime, as in lattice QCD, are often used to avoid such infinities so as to yield physically meaningful results.

3.1.3 Fields and radiation

The gravitational field and the electromagnetic field are the only two fundamental fields in nature that have infinite range and a corresponding classical low-energy limit, which greatly diminishes and hides their "particle-like" excitations. Albert Einstein in 1905, attributed "particle-like" and discrete exchanges of momenta and energy, characteristic of "field quanta", to the electromagnetic field. Originally, his principal motivation was to explain the thermodynamics of radiation. Although the photoelectric effect and Compton scattering strongly suggest the existence of the photon, it might alternately be explained by a mere quantization of emission; more definitive evidence of the quantum nature of radiation is now taken up into modern quantum optics as in the antibunching effect.[2]

3.2 Theories

There is currently no complete quantum theory of the remaining fundamental force, gravity. Many of the proposed theories to describe gravity as a QFT postulate the existence of a graviton particle that mediates the gravitational force. Presumably, the as yet unknown correct quantum field-theoretic treatment of the gravitational field will behave like Einstein's general theory of relativity in the low-energy limit. Quantum field theory of the fundamental forces itself has been postulated to be the low-energy effective field theory limit of a more fundamental theory such as superstring theory.

Most theories in standard particle physics are formulated as **relativistic quantum field theories**, such as QED, QCD, and the Standard Model. QED, the quantum field-theoretic description of the electromagnetic field, approximately reproduces Maxwell's theory of electrodynamics in the low-energy limit, with small non-linear corrections to the Maxwell equations required due to virtual electron–positron pairs.

In the perturbative approach to quantum field theory, the full field interaction terms are approximated as a perturbative expansion in the number of particles involved. Each term in the expansion can be thought of as forces between particles being mediated by other particles. In QED, the electromagnetic force between two electrons is caused by an exchange of photons. Similarly, intermediate vector bosons mediate the weak force and gluons mediate the strong force in QCD. The notion of a force-mediating particle comes from perturbation theory, and does not make sense in the context of non-perturbative approaches to QFT, such as with bound states.

3.3 History

Main article: History of quantum field theory

3.3.1 Foundations

The early development of the field involved Dirac, Fock, Pauli, Heisenberg and Bogolyubov. This phase of development culminated with the construction of the theory of quantum electrodynamics in the 1950s.

3.3.2 Gauge theory

Gauge theory was formulated and quantized, leading to the **unification of forces** embodied in the standard model of particle physics. This effort started in the 1950s with the work of Yang and Mills, was carried on by Martinus Veltman and a host of others during the 1960s and completed by the 1970s through the work of Gerard 't Hooft, Frank Wilczek, David Gross and David Politzer.

3.3.3 Grand synthesis

Parallel developments in the understanding of phase transitions in condensed matter physics led to the study of the renormalization group. This in turn led to the **grand synthesis** of theoretical physics, which unified theories of particle and condensed matter physics through quantum field theory. This involved the work of Michael Fisher and Leo Kadanoff in the 1970s, which led to the seminal reformulation of quantum field theory by Kenneth G. Wilson in 1975.

3.4 Principles

3.4.1 Classical and quantum fields

Main article: Classical field theory

A classical field is a function defined over some region of space and time.[3] Two physical phenomena which are described by classical fields are Newtonian gravitation, described by Newtonian gravitational field $\mathbf{g}(\mathbf{x}, t)$, and classical electromagnetism, described by the electric and magnetic fields $\mathbf{E}(\mathbf{x}, t)$ and $\mathbf{B}(\mathbf{x}, t)$. Because such fields can in principle take on distinct values at each point in space, they are said to have infinite degrees of freedom.[3]

Classical field theory does not, however, account for the quantum-mechanical aspects of such physical phenomena. For instance, it is known from quantum mechanics that certain aspects of electromagnetism involve discrete particles—photons—rather than continuous fields. The business of *quantum* field theory is to write down a field that is, like a classical field, a function defined over space and time, but which also accommodates the observations of quantum mechanics. This is a *quantum field*.

It is not immediately clear *how* to write down such a quantum field, since quantum mechanics has a structure very unlike a field theory. In its most general formulation, quantum mechanics is a theory of abstract operators (observables) acting on an abstract state space (Hilbert space), where the observables represent physically observable quantities and the state space represents the possible states of the system under study.[4] For instance, the fundamental observables associated with the motion of a single quantum mechanical particle are the position and momentum operators $\hat{x}$ and $\hat{p}$. Field theory, in contrast, treats x as a way to index the field rather than as an operator.[5]

There are two common ways of developing a quantum field: the path integral formalism and canonical quantization.[6] The latter of these is pursued in this article.

Lagrangian formalism

Quantum field theory frequently makes use of the Lagrangian formalism from classical field theory. This formalism is analogous to the Lagrangian formalism used in classical mechanics to solve for the motion of a particle under the influence of a field. In classical field theory, one writes down a Lagrangian density, $\mathcal{L}$, involving a field, $\varphi(\mathbf{x},t)$, and possibly its first derivatives ($\partial\varphi/\partial t$ and $\nabla\varphi$), and then applies a field-theoretic form of the Euler–Lagrange equation. Writing coordinates $(t, \mathbf{x}) = (x^0, x^1, x^2, x^3) = x^\mu$, this form of the Euler–Lagrange equation is[3]

$$\frac{\partial}{\partial x^\mu}\left[\frac{\partial\mathcal{L}}{\partial(\partial\phi/\partial x^\mu)}\right]-\frac{\partial\mathcal{L}}{\partial\phi}=0,$$

where a sum over μ is performed according to the rules of Einstein notation.

By solving this equation, one arrives at the “equations of motion” of the field.[3] For example, if one begins with the Lagrangian density

$$\mathcal{L}(\phi,\nabla\phi)=-\rho(t,\mathbf{x})\,\phi(t,\mathbf{x})-\frac{1}{8\pi G}|\nabla\phi|^2,$$

and then applies the Euler–Lagrange equation, one obtains the equation of motion

$$4\pi G\rho(t, \mathbf{x}) = \nabla^2\phi.$$

This equation is Newton's law of universal gravitation, expressed in differential form in terms of the gravitational potential $\varphi(t, \mathbf{x})$ and the mass density $\rho(t, \mathbf{x})$. Despite the nomenclature, the "field" under study is the gravitational potential, φ, rather than the gravitational field, **g**. Similarly, when classical field theory is used to study electromagnetism, the "field" of interest is the electromagnetic four-potential (V/c, **A**), rather than the electric and magnetic fields **E** and **B**.

Quantum field theory uses this same Lagrangian procedure to determine the equations of motion for quantum fields. These equations of motion are then supplemented by commutation relations derived from the canonical quantization procedure described below, thereby incorporating quantum mechanical effects into the behavior of the field.

3.4.2 Single- and many-particle quantum mechanics

Main articles: Quantum mechanics and First quantization

In quantum mechanics, a particle (such as an electron or proton) is described by a complex wavefunction, $\psi(x, t)$, whose time-evolution is governed by the Schrödinger equation:

$$-\frac{\hbar^2}{2m}\frac{\partial^2}{\partial x^2}\psi(x,t) + V(x)\psi(x,t) = i\hbar\frac{\partial}{\partial t}\psi(x,t).$$

Here m is the particle's mass and $V(x)$ is the applied potential. Physical information about the behavior of the particle is extracted from the wavefunction by constructing expected values for various quantities; for example, the expected value of the particle's position is given by integrating $\psi^*(x)\ x\ \psi(x)$ over all space, and the expected value of the particle's momentum is found by integrating $-i\hbar\psi^*(x)\mathrm{d}\psi/\mathrm{d}x$. The quantity $\psi^*(x)\psi(x)$ is itself in the Copenhagen interpretation of quantum mechanics interpreted as a probability density function. This treatment of quantum mechanics, where a particle's wavefunction evolves against a classical background potential $V(x)$, is sometimes called *first quantization*.

This description of quantum mechanics can be extended to describe the behavior of multiple particles, so long as the number and the type of particles remain fixed. The particles are described by a wavefunction $\psi(x_1, x_2, \ldots, xN, t)$, which is governed by an extended version of the Schrödinger equation.

Often one is interested in the case where N particles are all of the same type (for example, the 18 electrons orbiting a neutral argon nucleus). As described in the article on identical particles, this implies that the state of the entire system must be either symmetric (bosons) or antisymmetric (fermions) when the coordinates of its constituent particles are exchanged. This is achieved by using a Slater determinant as the wavefunction of a fermionic system (and a Slater permanent for a bosonic system), which is equivalent to an element of the symmetric or antisymmetric subspace of a tensor product.

For example, the general quantum state of a system of N bosons is written as

$$|\phi_1 \cdots \phi_N\rangle = \sqrt{\frac{\prod_j N_j!}{N!}} \sum_{p\in S_N} |\phi_{p(1)}\rangle \otimes \cdots \otimes |\phi_{p(N)}\rangle,$$

where $|\phi_i\rangle$ are the single-particle states, Nj is the number of particles occupying state j, and the sum is taken over all possible permutations p acting on N elements. In general, this is a sum of $N!$ (N factorial) distinct terms. $\sqrt{\frac{\prod_j N_j!}{N!}}$ is a normalizing factor.

There are several shortcomings to the above description of quantum mechanics, which are addressed by quantum field theory. First, it is unclear how to extend quantum mechanics to include the effects of special relativity.[7] Attempted replacements for the Schrödinger equation, such as the Klein–Gordon equation or the Dirac equation, have many unsatisfactory qualities; for instance, they possess energy eigenvalues that extend to $-\infty$, so that there seems to be no easy definition of a ground state. It turns out that such inconsistencies arise from relativistic wavefunctions not having a well-defined probabilistic interpretation in position space, as probability conservation is not a relativistically covariant concept. The second shortcoming, related to the first, is that in quantum mechanics there is no mechanism to describe particle creation and annihilation;[8] this is crucial for describing phenomena such as pair production, which result from the conversion between mass and energy according to the relativistic relation $E = mc^2$.

3.4.3 Second quantization

Main article: Second quantization

In this section, we will describe a method for constructing a quantum field theory called **second quantization**. This basically involves choosing a way to index the quantum mechanical degrees of freedom in the space of multiple identical-particle states. It is based on the Hamiltonian formulation of quantum mechanics.

Several other approaches exist, such as the Feynman path integral,[9] which uses a Lagrangian formulation. For an overview of some of these approaches, see the article on quantization.

Bosons

For simplicity, we will first discuss second quantization for bosons, which form perfectly symmetric quantum states. Let us denote the mutually orthogonal single-particle states which are possible in the system by $|\phi_1\rangle, |\phi_2\rangle, |\phi_3\rangle$, and so on. For example, the 3-particle state with one particle in state $|\phi_1\rangle$ and two in state $|\phi_2\rangle$ is

$$\frac{1}{\sqrt{3}}\left[|\phi_1\rangle|\phi_2\rangle|\phi_2\rangle + |\phi_2\rangle|\phi_1\rangle|\phi_2\rangle + |\phi_2\rangle|\phi_2\rangle|\phi_1\rangle\right].$$

The first step in second quantization is to express such quantum states in terms of **occupation numbers**, by listing the number of particles occupying each of the single-particle states $|\phi_1\rangle, |\phi_2\rangle$, etc. This is simply another way of labelling the states. For instance, the above 3-particle state is denoted as

$$|1, 2, 0, 0, 0, \dots\rangle.$$

An N-particle state belongs to a space of states describing systems of N particles. The next step is to combine the individual N-particle state spaces into an extended state space, known as Fock space, which can describe systems of any number of particles. This is composed of the state space of a system with no particles (the so-called vacuum state, written as $|0\rangle$), plus the state space of a 1-particle system, plus the state space of a 2-particle system, and so forth. States describing a definite number of particles are known as Fock states: a general element of Fock space will be a linear combination of Fock states. There is a one-to-one correspondence between the occupation number representation and valid boson states in the Fock space.

At this point, the quantum mechanical system has become a quantum field in the sense we described above. The field's elementary degrees of freedom are the occupation numbers, and each occupation number is indexed by a number j indicating which of the single-particle states $|\phi_1\rangle, |\phi_2\rangle, \dots, |\phi_j\rangle, \dots$ it refers to:

$$|N_1, N_2, N_3, \dots, N_j, \dots\rangle.$$

The properties of this quantum field can be explored by defining creation and annihilation operators, which add and subtract particles. They are analogous to ladder operators in the quantum harmonic oscillator problem, which added and subtracted energy quanta. However, these operators literally create and annihilate particles of a given quantum state. The bosonic annihilation operator a_2 and creation operator $a_2^\dagger$ are easily defined in the occupation number representation as having the following effects:

$$a_2|N_1, N_2, N_3, \ldots\rangle = \sqrt{N_2} \mid N_1, (N_2 - 1), N_3, \ldots\rangle,$$

$$a_2^\dagger|N_1, N_2, N_3, \ldots\rangle = \sqrt{N_2 + 1} \mid N_1, (N_2 + 1), N_3, \ldots\rangle.$$

It can be shown that these are operators in the usual quantum mechanical sense, i.e. linear operators acting on the Fock space. Furthermore, they are indeed Hermitian conjugates, which justifies the way we have written them. They can be shown to obey the commutation relation

$$[a_i, a_j] = 0 \quad , \quad \left[a_i^\dagger, a_j^\dagger\right] = 0 \quad , \quad \left[a_i, a_j^\dagger\right] = \delta_{ij},$$

where δ stands for the Kronecker delta. These are precisely the relations obeyed by the ladder operators for an infinite set of independent quantum harmonic oscillators, one for each single-particle state. Adding or removing bosons from each state is therefore analogous to exciting or de-exciting a quantum of energy in a harmonic oscillator.

Applying an annihilation operator a_k followed by its corresponding creation operator $a_k^\dagger$ returns the number N_k of particles in the kth single-particle eigenstate:

$$a_k^\dagger\, a_k|\ldots, N_k, \ldots\rangle = N_k|\ldots, N_k, \ldots\rangle.$$

The combination of operators $a_k^\dagger a_k$ is known as the number operator for the kth eigenstate.

The Hamiltonian operator of the quantum field (which, through the Schrödinger equation, determines its dynamics) can be written in terms of creation and annihilation operators. For instance, for a field of free (non-interacting) bosons, the total energy of the field is found by summing the energies of the bosons in each energy eigenstate. If the kth single-particle energy eigenstate has energy E_k and there are N_k bosons in this state, then the total energy of these bosons is $E_k N_k$. The energy in the *entire* field is then a sum over k :

$$E_{\text{tot}} = \sum_k E_k N_k$$

This can be turned into the Hamiltonian operator of the field by replacing N_k with the corresponding number operator, $a_k^\dagger a_k$. This yields

$$H = \sum_k E_k\, a_k^\dagger\, a_k.$$

Fermions

It turns out that a different definition of creation and annihilation must be used for describing fermions. According to the Pauli exclusion principle, fermions cannot share quantum states, so their occupation numbers *Ni* can only take on the value 0 or 1. The fermionic annihilation operators *c* and creation operators $c^\dagger$ are defined by their actions on a Fock state thus

$$c_j|N_1, N_2, \ldots, N_j = 0, \ldots\rangle = 0$$

$$c_j|N_1, N_2, \ldots, N_j = 1, \ldots\rangle = (-1)^{(N_1 + \cdots + N_{j-1})}|N_1, N_2, \ldots, N_j = 0, \ldots\rangle$$

$$c_j^\dagger|N_1, N_2, \ldots, N_j = 0, \ldots\rangle = (-1)^{(N_1 + \cdots + N_{j-1})}|N_1, N_2, \ldots, N_j = 1, \ldots\rangle$$

$c_j^\dagger|N_1, N_2, \ldots, N_j = 1, \ldots\rangle = 0.$

These obey an anticommutation relation:

$$\{c_i, c_j\} = 0 \quad , \quad \left\{c_i^\dagger, c_j^\dagger\right\} = 0 \quad , \quad \left\{c_i, c_j^\dagger\right\} = \delta_{ij}.$$

One may notice from this that applying a fermionic creation operator twice gives zero, so it is impossible for the particles to share single-particle states, in accordance with the exclusion principle.

Field operators

We have previously mentioned that there can be more than one way of indexing the degrees of freedom in a quantum field. Second quantization indexes the field by enumerating the single-particle quantum states. However, as we have discussed, it is more natural to think about a "field", such as the electromagnetic field, as a set of degrees of freedom indexed by position.

To this end, we can define *field operators* that create or destroy a particle at a particular point in space. In particle physics, these operators turn out to be more convenient to work with, because they make it easier to formulate theories that satisfy the demands of relativity.

Single-particle states are usually enumerated in terms of their momenta (as in the particle in a box problem.) We can construct field operators by applying the Fourier transform to the creation and annihilation operators for these states. For example, the bosonic field annihilation operator $\phi(\mathbf{r})$ is

$$\phi(\mathbf{r}) \stackrel{\text{def}}{=} \sum_j e^{i\mathbf{k}_j \cdot \mathbf{r}} a_j.$$

The bosonic field operators obey the commutation relation

$$[\phi(\mathbf{r}), \phi(\mathbf{r}')] = 0 \quad , \quad [\phi^\dagger(\mathbf{r}), \phi^\dagger(\mathbf{r}')] = 0 \quad , \quad [\phi(\mathbf{r}), \phi^\dagger(\mathbf{r}')] = \delta^3(\mathbf{r} - \mathbf{r}')$$

where $\delta(x)$ stands for the Dirac delta function. As before, the fermionic relations are the same, with the commutators replaced by anticommutators.

The field operator is not the same thing as a single-particle wavefunction. The former is an operator acting on the Fock space, and the latter is a quantum-mechanical amplitude for finding a particle in some position. However, they are closely related, and are indeed commonly denoted with the same symbol. If we have a Hamiltonian with a space representation, say

$$H = -\frac{\hbar^2}{2m} \sum_i \nabla_i^2 + \sum_{i<j} U(|\mathbf{r}_i - \mathbf{r}_j|)$$

where the indices i and j run over all particles, then the field theory Hamiltonian (in the non-relativistic limit and for negligible self-interactions) is

$$H = -\frac{\hbar^2}{2m} \int d^3r \; \phi^\dagger(\mathbf{r}) \nabla^2 \phi(\mathbf{r}) + \frac{1}{2} \int d^3r \int d^3r' \; \phi^\dagger(\mathbf{r}) \phi^\dagger(\mathbf{r}') U(|\mathbf{r} - \mathbf{r}'|) \phi(\mathbf{r}') \phi(\mathbf{r}).$$

This looks remarkably like an expression for the expectation value of the energy, with ϕ playing the role of the wavefunction. This relationship between the field operators and wavefunctions makes it very easy to formulate field theories starting from space-projected Hamiltonians.

3.4.4 Dynamics

Once the Hamiltonian operator is obtained as part of the canonical quantization process, the time dependence of the state is described with the Schrödinger equation, just as with other quantum theories. Alternatively, the Heisenberg picture can be used where the time dependence is in the operators rather than in the states.

3.4.5 Implications

Unification of fields and particles

The “second quantization” procedure that we have outlined in the previous section takes a set of single-particle quantum states as a starting point. Sometimes, it is impossible to define such single-particle states, and one must proceed directly to quantum field theory. For example, a quantum theory of the electromagnetic field *must* be a quantum field theory, because it is impossible (for various reasons) to define a wavefunction for a single photon.[10] In such situations, the quantum field theory can be constructed by examining the mechanical properties of the classical field and guessing the corresponding quantum theory. For free (non-interacting) quantum fields, the quantum field theories obtained in this way have the same properties as those obtained using second quantization, such as well-defined creation and annihilation operators obeying commutation or anticommutation relations.

Quantum field theory thus provides a unified framework for describing “field-like” objects (such as the electromagnetic field, whose excitations are photons) and “particle-like” objects (such as electrons, which are treated as excitations of an underlying electron field), so long as one can treat interactions as “perturbations” of free fields. There are still unsolved problems relating to the more general case of interacting fields that may or may not be adequately described by perturbation theory. For more on this topic, see Haag’s theorem.

Physical meaning of particle indistinguishability

The second quantization procedure relies crucially on the particles being identical. We would not have been able to construct a quantum field theory from a distinguishable many-particle system, because there would have been no way of separating and indexing the degrees of freedom.

Many physicists prefer to take the converse interpretation, which is that *quantum field theory explains what identical particles are*. In ordinary quantum mechanics, there is not much theoretical motivation for using symmetric (bosonic) or antisymmetric (fermionic) states, and the need for such states is simply regarded as an empirical fact. From the point of view of quantum field theory, particles are identical if and only if they are excitations of the same underlying quantum field. Thus, the question “why are all electrons identical?" arises from mistakenly regarding individual electrons as fundamental objects, when in fact it is only the electron field that is fundamental.

Particle conservation and non-conservation

During second quantization, we started with a Hamiltonian and state space describing a fixed number of particles (N), and ended with a Hamiltonian and state space for an arbitrary number of particles. Of course, in many common situations N is an important and perfectly well-defined quantity, e.g. if we are describing a gas of atoms sealed in a box. From the point of view of quantum field theory, such situations are described by quantum states that are eigenstates of the number operator $\hat{N}$, which measures the total number of particles present. As with any quantum mechanical observable, $\hat{N}$ is conserved if it commutes with the Hamiltonian. In that case, the quantum state is trapped in the N-particle subspace of the total Fock space, and the situation could equally well be described by ordinary N-particle quantum mechanics. (Strictly speaking, this is only true in the noninteracting case or in the low energy density limit of renormalized quantum field theories)

For example, we can see that the free-boson Hamiltonian described above conserves particle number. Whenever the Hamiltonian operates on a state, each particle destroyed by an annihilation operator a_k is immediately put back by the creation operator $a_k^\dagger$.

On the other hand, it is possible, and indeed common, to encounter quantum states that are *not* eigenstates of $\hat{N}$, which do not have well-defined particle numbers. Such states are difficult or impossible to handle using ordinary quantum mechanics, but they can be easily described in quantum field theory as quantum superpositions of states having different values of N. For example, suppose we have a bosonic field whose particles can be created or destroyed by interactions with a fermionic field. The Hamiltonian of the combined system would be given by the Hamiltonians of the free boson and free fermion fields, plus a "potential energy" term such as

$$H_I = \sum_{k,q} V_q(a_q + a_{-q}^\dagger)c_{k+q}^\dagger c_k,$$

where $a_k^\dagger$ and a_k denotes the bosonic creation and annihilation operators, $c_k^\dagger$ and c_k denotes the fermionic creation and annihilation operators, and V_q is a parameter that describes the strength of the interaction. This "interaction term" describes processes in which a fermion in state k either absorbs or emits a boson, thereby being kicked into a different eigenstate $k + q$. (In fact, this type of Hamiltonian is used to describe interaction between conduction electrons and phonons in metals. The interaction between electrons and photons is treated in a similar way, but is a little more complicated because the role of spin must be taken into account.) One thing to notice here is that even if we start out with a fixed number of bosons, we will typically end up with a superposition of states with different numbers of bosons at later times. The number of fermions, however, is conserved in this case.

In condensed matter physics, states with ill-defined particle numbers are particularly important for describing the various superfluids. Many of the defining characteristics of a superfluid arise from the notion that its quantum state is a superposition of states with different particle numbers. In addition, the concept of a coherent state (used to model the laser and the BCS ground state) refers to a state with an ill-defined particle number but a well-defined phase.

3.4.6 Axiomatic approaches

The preceding description of quantum field theory follows the spirit in which most physicists approach the subject. However, it is not mathematically rigorous. Over the past several decades, there have been many attempts to put quantum field theory on a firm mathematical footing by formulating a set of axioms for it. These attempts fall into two broad classes.

The first class of axioms, first proposed during the 1950s, include the Wightman, Osterwalder–Schrader, and Haag–Kastler systems. They attempted to formalize the physicists' notion of an "operator-valued field" within the context of functional analysis, and enjoyed limited success. It was possible to prove that any quantum field theory satisfying these axioms satisfied certain general theorems, such as the spin-statistics theorem and the CPT theorem. Unfortunately, it proved extraordinarily difficult to show that any realistic field theory, including the Standard Model, satisfied these axioms. Most of the theories that could be treated with these analytic axioms were physically trivial, being restricted to low-dimensions and lacking interesting dynamics. The construction of theories satisfying one of these sets of axioms falls in the field of constructive quantum field theory. Important work was done in this area in the 1970s by Segal, Glimm, Jaffe and others.

During the 1980s, a second set of axioms based on geometric ideas was proposed. This line of investigation, which restricts its attention to a particular class of quantum field theories known as topological quantum field theories, is associated most closely with Michael Atiyah and Graeme Segal, and was notably expanded upon by Edward Witten, Richard Borcherds, and Maxim Kontsevich. However, most of the physically relevant quantum field theories, such as the Standard Model, are not topological quantum field theories; the quantum field theory of the fractional quantum Hall effect is a notable exception. The main impact of axiomatic topological quantum field theory has been on mathematics, with important applications in representation theory, algebraic topology, and differential geometry.

Finding the proper axioms for quantum field theory is still an open and difficult problem in mathematics. One of the Millennium Prize Problems—proving the existence of a mass gap in Yang–Mills theory—is linked to this issue.

3.5 Associated phenomena

In the previous part of the article, we described the most general features of quantum field theories. Some of the quantum field theories studied in various fields of theoretical physics involve additional special ideas, such as renormalizability, gauge symmetry, and supersymmetry. These are described in the following sections.

3.5.1 Renormalization

Main article: Renormalization

Early in the history of quantum field theory, it was found that many seemingly innocuous calculations, such as the perturbative shift in the energy of an electron due to the presence of the electromagnetic field, give infinite results. The reason is that the perturbation theory for the shift in an energy involves a sum over all other energy levels, and there are infinitely many levels at short distances that each give a finite contribution which results in a divergent series.

Many of these problems are related to failures in classical electrodynamics that were identified but unsolved in the 19th century, and they basically stem from the fact that many of the supposedly "intrinsic" properties of an electron are tied to the electromagnetic field that it carries around with it. The energy carried by a single electron—its self energy—is not simply the bare value, but also includes the energy contained in its electromagnetic field, its attendant cloud of photons. The energy in a field of a spherical source diverges in both classical and quantum mechanics, but as discovered by Weisskopf with help from Furry, in quantum mechanics the divergence is much milder, going only as the logarithm of the radius of the sphere.

The solution to the problem, presciently suggested by Stueckelberg, independently by Bethe after the crucial experiment by Lamb, implemented at one loop by Schwinger, and systematically extended to all loops by Feynman and Dyson, with converging work by Tomonaga in isolated postwar Japan, comes from recognizing that all the infinities in the interactions of photons and electrons can be isolated into redefining a finite number of quantities in the equations by replacing them with the observed values: specifically the electron's mass and charge: this is called renormalization. The technique of renormalization recognizes that the problem is essentially purely mathematical, that extremely short distances are at fault. In order to define a theory on a continuum, first place a cutoff on the fields, by postulating that quanta cannot have energies above some extremely high value. This has the effect of replacing continuous space by a structure where very short wavelengths do not exist, as on a lattice. Lattices break rotational symmetry, and one of the crucial contributions made by Feynman, Pauli and Villars, and modernized by 't Hooft and Veltman, is a symmetry-preserving cutoff for perturbation theory (this process is called regularization). There is no known symmetrical cutoff outside of perturbation theory, so for rigorous or numerical work people often use an actual lattice.

On a lattice, every quantity is finite but depends on the spacing. When taking the limit of zero spacing, we make sure that the physically observable quantities like the observed electron mass stay fixed, which means that the constants in the Lagrangian defining the theory depend on the spacing. Hopefully, by allowing the constants to vary with the lattice spacing, all the results at long distances become insensitive to the lattice, defining a continuum limit.

The renormalization procedure only works for a certain class of quantum field theories, called **renormalizable quantum field theories**. A theory is **perturbatively renormalizable** when the constants in the Lagrangian only diverge at worst as logarithms of the lattice spacing for very short spacings. The continuum limit is then well defined in perturbation theory, and even if it is not fully well defined non-perturbatively, the problems only show up at distance scales that are exponentially small in the inverse coupling for weak couplings. The Standard Model of particle physics is perturbatively renormalizable, and so are its component theories (quantum electrodynamics/electroweak theory and quantum chromodynamics). Of the three components, quantum electrodynamics is believed to not have a continuum limit, while the asymptotically free SU(2) and SU(3) weak hypercharge and strong color interactions are nonperturbatively well defined.

The renormalization group describes how renormalizable theories emerge as the long distance low-energy effective field theory for any given high-energy theory. Because of this, renormalizable theories are insensitive to the precise nature of the underlying high-energy short-distance phenomena. This is a blessing because it allows physicists to formulate low energy theories without knowing the details of high energy phenomenon. It is also a curse, because once a renormalizable theory like the standard model is found to work, it gives very few clues to higher energy processes. The only way high energy processes can be seen in the standard model is when they allow otherwise forbidden events, or if they predict quantitative relations between the coupling constants.

3.5.2 Haag's theorem

See also: Haag's theorem

From a mathematically rigorous perspective, there exists no interaction picture in a Lorentz-covariant quantum field theory. This implies that the perturbative approach of Feynman diagrams in QFT is not strictly justified, despite producing vastly precise predictions validated by experiment. This is called Haag's theorem, but most particle physicists relying on QFT largely shrug it off.

3.5.3 Gauge freedom

A gauge theory is a theory that admits a symmetry with a local parameter. For example, in every quantum theory the global phase of the wave function is arbitrary and does not represent something physical. Consequently, the theory is invariant under a global change of phases (adding a constant to the phase of all wave functions, everywhere); this is a global symmetry. In quantum electrodynamics, the theory is also invariant under a *local* change of phase, that is – one may shift the phase of all wave functions so that the shift may be different at every point in space-time. This is a *local* symmetry. However, in order for a well-defined derivative operator to exist, one must introduce a new field, the gauge field, which also transforms in order for the local change of variables (the phase in our example) not to affect the derivative. In quantum electrodynamics this gauge field is the electromagnetic field. The change of local gauge of variables is termed gauge transformation. It is worth noting that by Noether's theorem, for every such symmetry there exists an associated conserved current. The aforementioned symmetry of the wavefunction under global phase changes implies the conservation of electric charge.

In quantum field theory the excitations of fields represent particles. The particle associated with excitations of the gauge field is the gauge boson, which is the photon in the case of quantum electrodynamics.

The degrees of freedom in quantum field theory are local fluctuations of the fields. The existence of a gauge symmetry reduces the number of degrees of freedom, simply because some fluctuations of the fields can be transformed to zero by gauge transformations, so they are equivalent to having no fluctuations at all, and they therefore have no physical meaning. Such fluctuations are usually called "non-physical degrees of freedom" or *gauge artifacts*; usually some of them have a negative norm, making them inadequate for a consistent theory. Therefore, if a classical field theory has a gauge symmetry, then its quantized version (i.e. the corresponding quantum field theory) will have this symmetry as well. In other words, a gauge symmetry cannot have a quantum anomaly. If a gauge symmetry is anomalous (i.e. not kept in the quantum theory) then the theory is non-consistent: for example, in quantum electrodynamics, had there been a gauge anomaly, this would require the appearance of photons with longitudinal polarization and polarization in the time direction, the latter having a negative norm, rendering the theory inconsistent; another possibility would be for these photons to appear only in intermediate processes but not in the final products of any interaction, making the theory non-unitary and again inconsistent (see optical theorem).

In general, the gauge transformations of a theory consist of several different transformations, which may not be commutative. These transformations are together described by a mathematical object known as a gauge group. Infinitesimal gauge transformations are the gauge group generators. Therefore the number of gauge bosons is the group dimension (i.e. number of generators forming a basis).

All the fundamental interactions in nature are described by gauge theories. These are:

- Quantum chromodynamics, whose gauge group is **SU**(3). The gauge bosons are eight gluons.
- The electroweak theory, whose gauge group is **U**(1) × **SU**(2), (a direct product of **U**(1) and **SU**(2)).
- Gravity, whose classical theory is general relativity, admits the equivalence principle, which is a form of gauge symmetry. However, it is explicitly non-renormalizable.

3.5.4 Multivalued gauge transformations

The gauge transformations which leave the theory invariant involve, by definition, only single-valued gauge functions $\Lambda(x_i)$ which satisfy the Schwarz integrability criterion

$$\partial_{x_i x_j}\Lambda = \partial_{x_j x_i}\Lambda.$$

An interesting extension of gauge transformations arises if the gauge functions $\Lambda(x_i)$ are allowed to be multivalued functions which violate the integrability criterion. These are capable of changing the physical field strengths and are therefore not proper symmetry transformations. Nevertheless, the transformed field equations describe correctly the physical laws in the presence of the newly generated field strengths. See the textbook by H. Kleinert cited below for the applications to phenomena in physics.

3.5.5 Supersymmetry

Main article: Supersymmetry

Supersymmetry assumes that every fundamental fermion has a superpartner that is a boson and vice versa. It was introduced in order to solve the so-called Hierarchy Problem, that is, to explain why particles not protected by any symmetry (like the Higgs boson) do not receive radiative corrections to its mass driving it to the larger scales (GUT, Planck...). It was soon realized that supersymmetry has other interesting properties: its gauged version is an extension of general relativity (Supergravity), and it is a key ingredient for the consistency of string theory.

The way supersymmetry protects the hierarchies is the following: since for every particle there is a superpartner with the same mass, any loop in a radiative correction is cancelled by the loop corresponding to its superpartner, rendering the theory UV finite.

Since no superpartners have yet been observed, if supersymmetry exists it must be broken (through a so-called soft term, which breaks supersymmetry without ruining its helpful features). The simplest models of this breaking require that the energy of the superpartners not be too high; in these cases, supersymmetry is expected to be observed by experiments at the Large Hadron Collider. The Higgs particle has been detected at the LHC, and no such superparticles have been discovered.

3.6 See also

- Abraham–Lorentz force
- Basic concepts of quantum mechanics
- Common integrals in quantum field theory
- Einstein–Maxwell–Dirac equations
- Form factor (quantum field theory)
- Green–Kubo relations
- Green's function (many-body theory)
- Invariance mechanics
- List of quantum field theories
- Quantum field theory in curved spacetime

- Quantum flavordynamics
- Quantum hydrodynamics
- Quantum magnetodynamics
- Quantum triviality
- Relation between Schrödinger's equation and the path integral formulation of quantum mechanics
- Relationship between string theory and quantum field theory
- Schwinger–Dyson equation
- Static forces and virtual-particle exchange
- Symmetry in quantum mechanics
- Theoretical and experimental justification for the Schrödinger equation
- Ward–Takahashi identity
- Wheeler–Feynman absorber theory
- Wigner's classification
- Wigner's theorem

3.7 Notes

3.8 References

[1] "Beautiful Minds, Vol. 20: Ed Witten". la Repubblica. 2010. Retrieved 22 June 2012. See here.

[2] J. J. Thorn et al. (2004). Observing the quantum behavior of light in an undergraduate laboratory. . J. J. Thorn, M. S. Neel, V. W. Donato, G. S. Bergreen, R. E. Davies, and M. Beck. American Association of Physics Teachers, 2004.DOI: 10.1119/1.1737397.

[3] David Tong, *Lectures on Quantum Field Theory*, chapter 1.

[4] Srednicki, Mark. *Quantum Field Theory* (1st ed.). p. 19.

[5] Srednicki, Mark. *Quantum Field Theory* (1st ed.). pp. 25–6.

[6] Zee, Anthony. *Quantum Field Theory in a Nutshell* (2nd ed.). p. 61.

[7] David Tong, *Lectures on Quantum Field Theory*, Introduction.

[8] Zee, Anthony. *Quantum Field Theory in a Nutshell* (2nd ed.). p. 3.

[9] Abraham Pais, *Inward Bound: Of Matter and Forces in the Physical World* ISBN 0-19-851997-4. Pais recounts how his astonishment at the rapidity with which Feynman could calculate using his method. Feynman's method is now part of the standard methods for physicists.

[10] Newton, T.D.; Wigner, E.P. (1949). "Localized states for elementary systems". *Reviews of Modern Physics* **21** (3): 400–406. Bibcode:1949RvMP...21..400N. doi:10.1103/RevModPhys.21.400.

3.9 Further reading

General readers

- Feynman, R.P. (2001) [1964]. *The Character of Physical Law*. MIT Press. ISBN 0-262-56003-8.
- Feynman, R.P. (2006) [1985]. *QED: The Strange Theory of Light and Matter*. Princeton University Press. ISBN 0-691-12575-9.
- Gribbin, J. (1998). *Q is for Quantum: Particle Physics from A to Z*. Weidenfeld & Nicolson. ISBN 0-297-81752-3.
- Schumm, Bruce A. (2004) *Deep Down Things*. Johns Hopkins Univ. Press. Chpt. 4.

Introductory texts

- McMahon, D. (2008). *Quantum Field Theory*. McGraw-Hill. ISBN 978-0-07-154382-8.
- Bogoliubov, N.; Shirkov, D. (1982). *Quantum Fields*. Benjamin-Cummings. ISBN 0-8053-0983-7.
- Frampton, P.H. (2000). *Gauge Field Theories. Frontiers in Physics (2nd ed.). Wiley.*
- Greiner, W; Müller, B. (2000). *Gauge Theory of Weak Interactions*. Springer. ISBN 3-540-67672-4.
- Itzykson, C.; Zuber, J.-B. (1980). *Quantum Field Theory*. McGraw-Hill. ISBN 0-07-032071-3.
- Kane, G.L. (1987). *Modern Elementary Particle Physics*. Perseus Books. ISBN 0-201-11749-5.
- Kleinert, H.; Schulte-Frohlinde, Verena (2001). *Critical Properties of φ^4-Theories*. World Scientific. ISBN 981-02-4658-7.
- Kleinert, H. (2008). *Multivalued Fields in Condensed Matter, Electrodynamics, and Gravitation* (PDF). World Scientific. ISBN 978-981-279-170-2.
- Loudon, R (1983). *The Quantum Theory of Light*. Oxford University Press. ISBN 0-19-851155-8.
- Mandl, F.; Shaw, G. (1993). *Quantum Field Theory*. John Wiley & Sons. ISBN 978-0-471-94186-6.
- Peskin, M.; Schroeder, D. (1995). *An Introduction to Quantum Field Theory*. Westview Press. ISBN 0-201-50397-2.
- Ryder, L.H. (1985). *Quantum Field Theory*. Cambridge University Press. ISBN 0-521-33859-X.
- Schwartz, M.D. (2014). *Quantum Field Theory and the Standard Model*. Cambridge University Press. ISBN 978-1107034730.
- Srednicki, Mark (2007) *Quantum Field Theory*. Cambridge Univ. Press.
- Ynduráin, F.J. (1996). *Relativistic Quantum Mechanics and Introduction to Field Theory* (1st ed.). Springer. ISBN 978-3-540-60453-2.
- Zee, A. (2003). *Quantum Field Theory in a Nutshell*. Princeton University Press. ISBN 0-691-01019-6.

Advanced texts

- Brown, Lowell S. (1994). *Quantum Field Theory*. Cambridge University Press. ISBN 978-0-521-46946-3.
- Bogoliubov, N.; Logunov, A.A.; Oksak, A.I.; Todorov, I.T. (1990). *General Principles of Quantum Field Theory*. Kluwer Academic Publishers. ISBN 978-0-7923-0540-8.
- Weinberg, S. (1995). *The Quantum Theory of Fields* **1–3**. Cambridge University Press.

Articles:

- Gerard 't Hooft (2007) "The Conceptual Basis of Quantum Field Theory" in Butterfield, J., and John Earman, eds., *Philosophy of Physics, Part A*. Elsevier: 661–730.
- Frank Wilczek (1999) "Quantum field theory", *Reviews of Modern Physics* 71: S83–S95. Also doi=10.1103/Rev. Mod. Phys. 71.

3.10 External links

- Hazewinkel, Michiel, ed. (2001), “Quantum field theory”, *Encyclopedia of Mathematics*, Springer, ISBN 978-1-55608-010-4
- Stanford Encyclopedia of Philosophy: "Quantum Field Theory", by Meinard Kuhlmann.
- Siegel, Warren, 2005. *Fields*. A free text, also available from arXiv:hep-th/9912205.
- Quantum Field Theory by P. J. Mulders

Chapter 4

Field (physics)

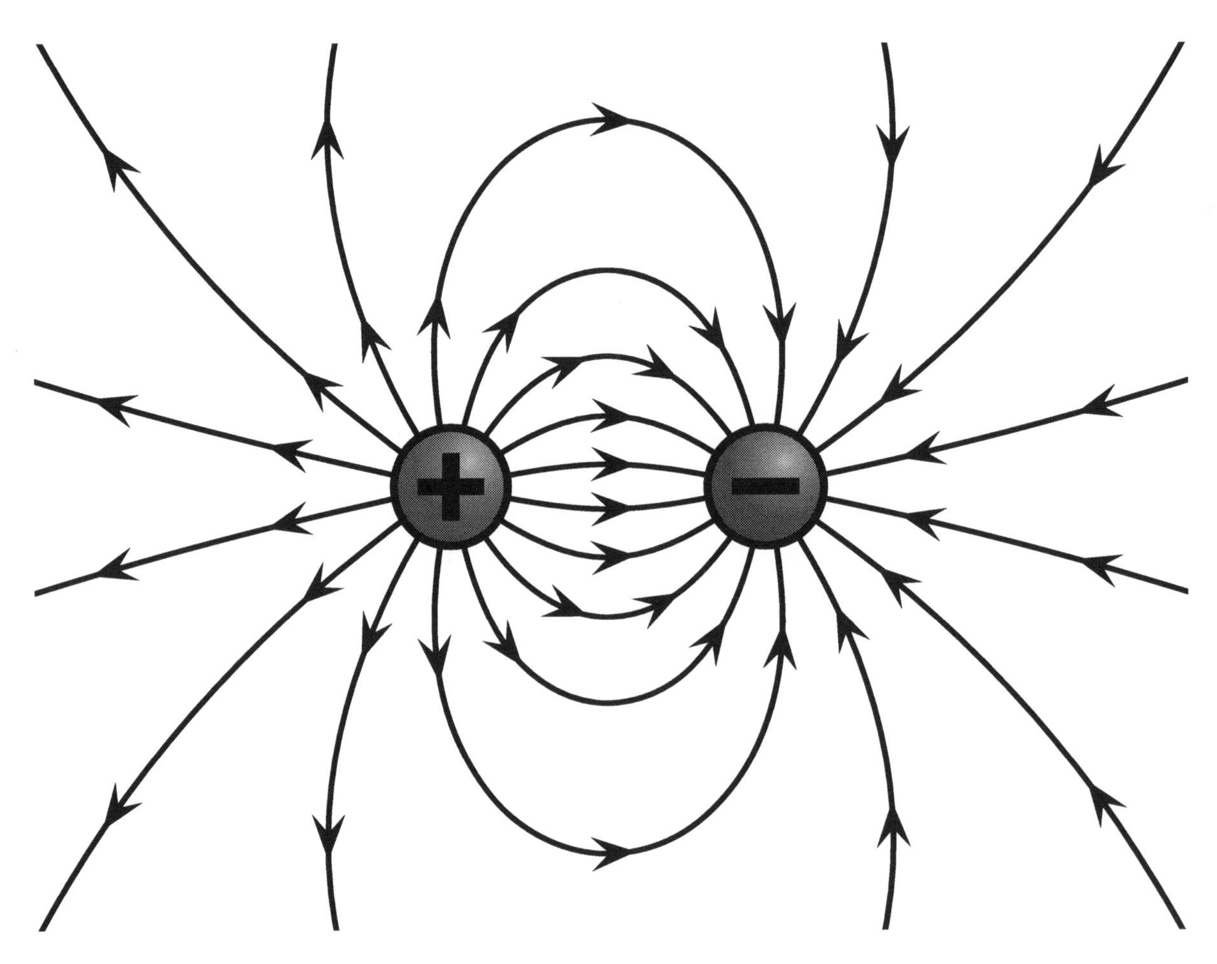

Illustration of the electric field surrounding a positive (red) and a negative (blue) charge.

In physics, a **field** is a physical quantity that has a value for each point in space and time.[1][2][3] For example, on a weather map, the surface wind velocity is described by assigning a vector to each point on a map. Each vector represents the speed and direction of the movement of air at that point. As another example, an electric field can be thought of as a "condition in space"[4] emanating from an electric charge and extending throughout the whole of space. When a test electric charge is placed in this electric field, the particle accelerates due to a force. Physicists have found the notion of a field to be of such practical utility for the analysis of forces that they have come to think of a force as due to a field.[5]

In the modern framework of the quantum theory of fields, even without referring to a test particle, a field occupies space, contains energy, and its presence eliminates a true vacuum.[6] This lead physicists to consider electromagnetic fields to be a physical entity, making the field concept a supporting paradigm of the edifice of modern physics. "The fact that the electromagnetic field can possess momentum and energy makes it very real... a particle makes a field, and a field acts on another particle, and the field has such familiar properties as energy content and momentum, just as particles can have".[7] In practice, the strength of most fields has been found to diminish with distance to the point of being undetectable. For instance the strength of many relevant classical fields, such as the gravitational field in Newton's theory of gravity or the electrostatic field in classical electromagnetism, is inversely proportional to the square of the distance from the source (i.e. they follow the Gauss's law). One consequence is that the Earth's gravitational field quickly becomes undetectable on cosmic scales.

A field can be classified as a scalar field, a vector field, a spinor field or a tensor field according to whether the represented physical quantity is a scalar, a vector, a spinor or a tensor, respectively. A field has a unique tensorial character in every point where it is defined: i.e. a field cannot be a scalar field somewhere and a vector field somewhere else. For example, the Newtonian gravitational field is a vector field: specifying its value at a point in spacetime requires three numbers, the components of the gravitational field vector at that point. Moreover, within each category (scalar, vector, tensor), a field can be either a *classical field* or a *quantum field*, depending on whether it is characterized by numbers or quantum operators respectively. In fact in this theory an equivalent representation of field is a field particle, namely a boson.[8]

4.1 History

To Isaac Newton his law of universal gravitation simply expressed the gravitational force that acted between any pair of massive objects. When looking at the motion of many bodies all interacting with each other, such as the planets in the Solar System, dealing with the force between each pair of bodies separately rapidly becomes computationally inconvenient. In the eighteenth century, a new quantity was devised to simplify the bookkeeping of all these gravitational forces. This quantity, the gravitational field, gave at each point in space the total gravitational force which would be felt by an object with unit mass at that point. This did not change the physics in any way: it did not matter if you calculated all the gravitational forces on an object individually and then added them together, or if you first added all the contributions together as a gravitational field and then applied it to an object.[9]

The development of the independent concept of a field truly began in the nineteenth century with the development of the theory of electromagnetism. In the early stages, André-Marie Ampère and Charles-Augustin de Coulomb could manage with Newton-style laws that expressed the forces between pairs of electric charges or electric currents. However, it became much more natural to take the field approach and express these laws in terms of electric and magnetic fields; in 1849 Michael Faraday became the first to coin the term "field".[9]

The independent nature of the field became more apparent with James Clerk Maxwell's discovery that waves in these fields propagated at a finite speed. Consequently, the forces on charges and currents no longer just depended on the positions and velocities of other charges and currents at the same time, but also on their positions and velocities in the past.[9]

Maxwell, at first, did not adopt the modern concept of a field as fundamental quantity that could independently exist. Instead, he supposed that the electromagnetic field expressed the deformation of some underlying medium—the luminiferous aether—much like the tension in a rubber membrane. If that were the case, the observed velocity of the electromagnetic waves should depend upon the velocity of the observer with respect to the aether. Despite much effort, no experimental evidence of such an effect was ever found; the situation was resolved by the introduction of the special theory of relativity by Albert Einstein in 1905. This theory changed the way the viewpoints of moving observers should be related to each other in such a way that velocity of electromagnetic waves in Maxwell's theory would be the same for all observers. By doing away with the need for a background medium, this development opened the way for physicists to start thinking about fields as truly independent entities.[9]

In the late 1920s, the new rules of quantum mechanics were first applied to the electromagnetic fields. In 1927, Paul Dirac used quantum fields to successfully explain how the decay of an atom to lower quantum state lead to the spontaneous emission of a photon, the quantum of the electromagnetic field. This was soon followed by the realization (following the work of Pascual Jordan, Eugene Wigner, Werner Heisenberg, and Wolfgang Pauli) that all particles, including electrons and protons, could be understood as the quanta of some quantum field, elevating fields to the status of the most fundamental objects in nature.[9] That said, John Wheeler and Richard Feynman seriously considered Newton's pre-field concept of action at a distance (although they set it aside because of the ongoing utility of the field concept for research in general relativity and quantum electrodynamics).

4.2 Classical fields

Main article: Classical field theory

There are several examples of classical fields. Classical field theories remain useful wherever quantum properties do not arise, and can be active areas of research. Elasticity of materials, fluid dynamics and Maxwell's equations are cases in point.

Some of the simplest physical fields are vector force fields. Historically, the first time that fields were taken seriously was with Faraday's lines of force when describing the electric field. The gravitational field was then similarly described.

4.2.1 Newtonian gravitation

A classical field theory describing gravity is Newtonian gravitation, which describes the gravitational force as a mutual interaction between two masses.

Any body with mass M is associated with a gravitational field $\mathbf{g}$ which describes its influence on other bodies with mass. The gravitational field of M at a point $\mathbf{r}$ in space corresponds to the ratio between force $\mathbf{F}$ that M exerts on a small or negligible test mass m located at $\mathbf{r}$ and the test mass itself:[10]

$$\mathbf{g}(\mathbf{r}) = \frac{\mathbf{F}(\mathbf{r})}{m}.$$

Stipulating that m is much smaller than M ensures that the presence of m has a negligible influence on the behavior of M.

According to Newton's law of universal gravitation, $\mathbf{F}(\mathbf{r})$ is given by[10]

$$\mathbf{F}(\mathbf{r}) = -\frac{GMm}{r^2}\hat{\mathbf{r}},$$

where $\hat{\mathbf{r}}$ is a unit vector lying along the line joining M and m and pointing from m to M. Therefore, the gravitational field of $\mathbf{M}$ is[10]

$$\mathbf{g}(\mathbf{r}) = \frac{\mathbf{F}(\mathbf{r})}{m} = -\frac{GM}{r^2}\hat{\mathbf{r}}.$$

The experimental observation that inertial mass and gravitational mass are equal to an unprecedented level of accuracy leads to the identity that gravitational field strength is identical to the acceleration experienced by a particle. This is the starting point of the equivalence principle, which leads to general relativity.

Because the gravitational force $\mathbf{F}$ is conservative, the gravitational field $\mathbf{g}$ can be rewritten in terms of the gradient of a scalar function, the gravitational potential $\Phi(\mathbf{r})$:

$$\mathbf{g}(\mathbf{r}) = -\nabla\Phi(\mathbf{r}).$$

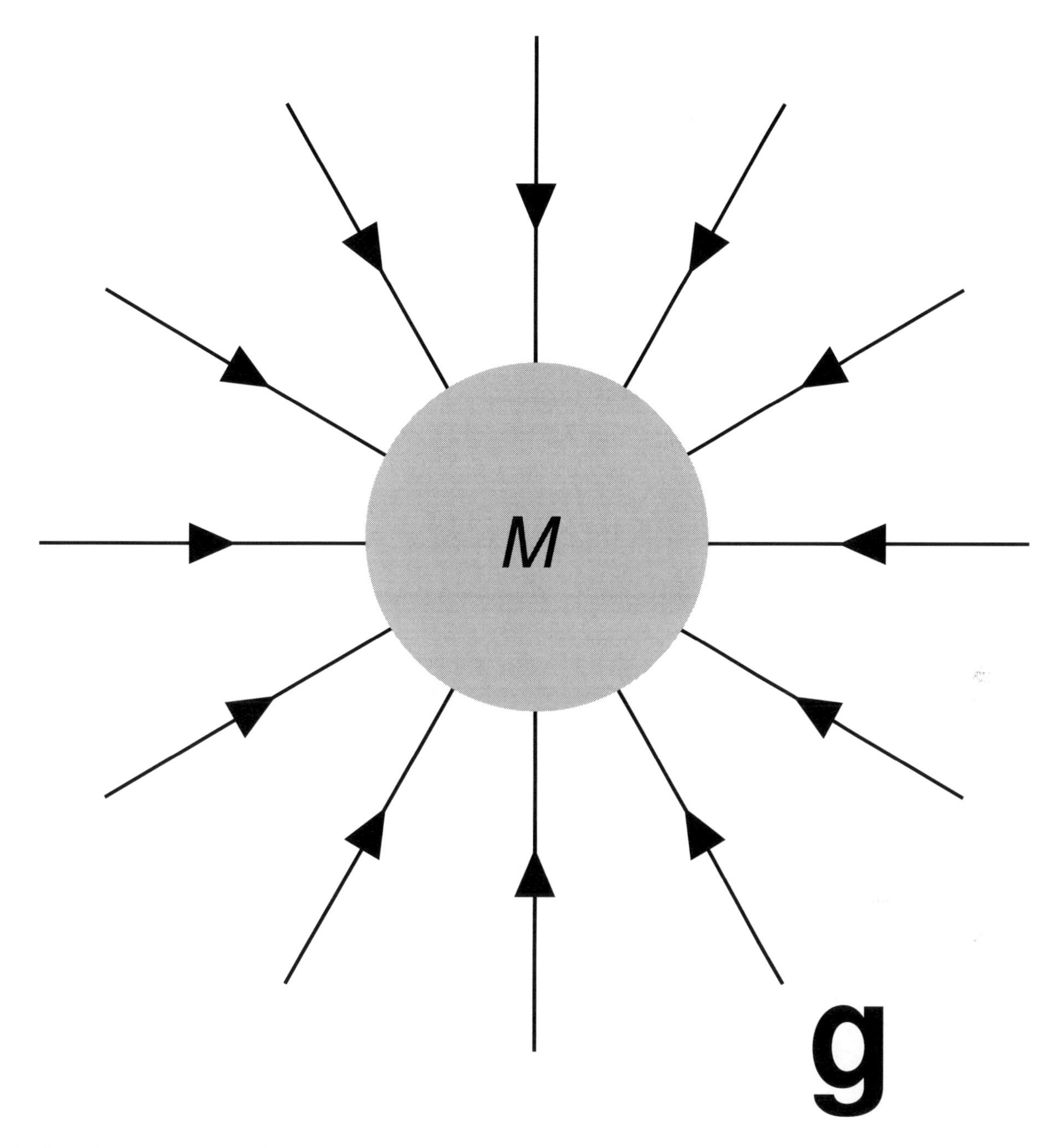

In classical gravitation, mass is the source of an attractive gravitational field **g**.

4.2.2 Electromagnetism

Main article: Electromagnetism

Michael Faraday first realized the importance of a field as a physical quantity, during his investigations into magnetism. He realized that electric and magnetic fields are not only fields of force which dictate the motion of particles, but also have an independent physical reality because they carry energy.

These ideas eventually led to the creation, by James Clerk Maxwell, of the first unified field theory in physics with the introduction of equations for the electromagnetic field. The modern version of these equations is called Maxwell's equations.

Electrostatics

Main article: Electrostatics

A charged test particle with charge q experiences a force **F** based solely on its charge. We can similarly describe the electric field **E** so that $\mathbf{F} = q\mathbf{E}$. Using this and Coulomb's law tells us that the electric field due to a single charged particle as

$$\mathbf{E} = \frac{1}{4\pi\epsilon_0}\frac{q}{r^2}\hat{\mathbf{r}}.$$

The electric field is conservative, and hence can be described by a scalar potential, $V(\mathbf{r})$:

$$\mathbf{E}(\mathbf{r}) = -\nabla V(\mathbf{r}).$$

Magnetostatics

Main article: Magnetostatics

A steady current I flowing along a path ℓ will exert a force on nearby moving charged particles that is quantitatively different from the electric field force described above. The force exerted by I on a nearby charge q with velocity **v** is

$$\mathbf{F}(\mathbf{r}) = q\mathbf{v} \times \mathbf{B}(\mathbf{r}),$$

where $\mathbf{B}(\mathbf{r})$ is the magnetic field, which is determined from I by the Biot–Savart law:

$$\mathbf{B}(\mathbf{r}) = \frac{\mu_0 I}{4\pi}\int \frac{d\boldsymbol{\ell} \times d\hat{\mathbf{r}}}{r^2}.$$

The magnetic field is not conservative in general, and hence cannot usually be written in terms of a scalar potential. However, it can be written in terms of a vector potential, $\mathbf{A}(\mathbf{r})$:

$$\mathbf{B}(\mathbf{r}) = \boldsymbol{\nabla} \times \mathbf{A}(\mathbf{r})$$

Electrodynamics

Main article: Electrodynamics

In general, in the presence of both a charge density $\rho(\mathbf{r}, t)$ and current density $\mathbf{J}(\mathbf{r}, t)$, there will be both an electric and a magnetic field, and both will vary in time. They are determined by Maxwell's equations, a set of differential equations which directly relate **E** and **B** to ρ and **J**.[13]

Alternatively, one can describe the system in terms of its scalar and vector potentials V and **A**. A set of integral equations known as *retarded potentials* allow one to calculate V and **A** from ρ and **J**,[note 1] and from there the electric and magnetic fields are determined via the relations[14]

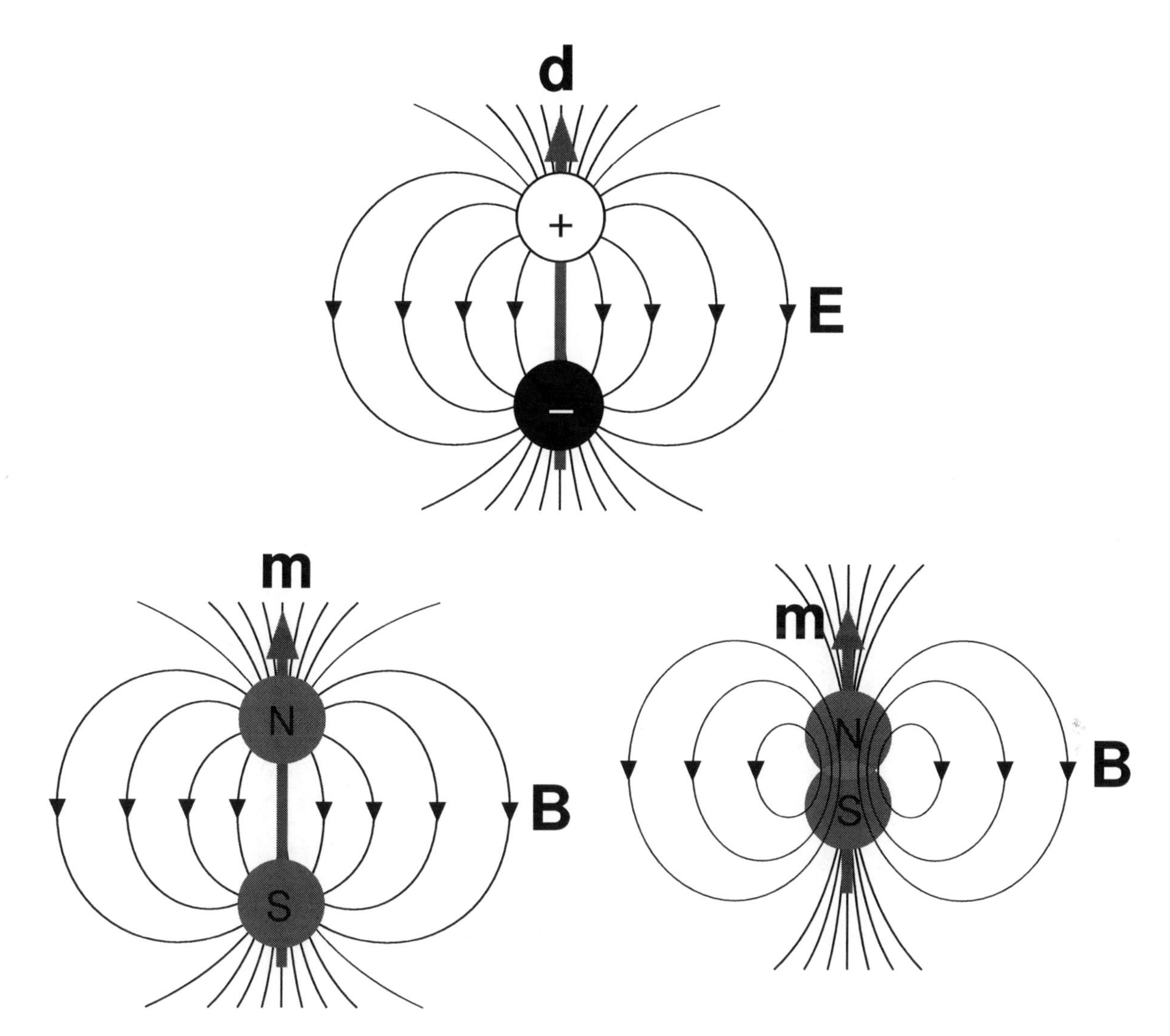

*The **E** fields and **B** fields due to electric charges (black/white) and magnetic poles (red/blue).*[11][12] ***Top:** **E** field due to an electric dipole moment **d**. **Bottom left:** **B** field due to a* mathematical *magnetic dipole **m** formed by two magnetic monopoles. **Bottom right:** **B** field due to a pure magnetic dipole moment **m** found in ordinary matter (*not *from monopoles).*

$$\mathbf{E} = -\boldsymbol{\nabla} V - \frac{\partial \mathbf{A}}{\partial t}$$

$$\mathbf{B} = \boldsymbol{\nabla} \times \mathbf{A}.$$

At the end of the 19th century, the electromagnetic field was understood as a collection of two vector fields in space. Nowadays, one recognizes this as a single antisymmetric 2nd-rank tensor field in spacetime.

4.2.3 Gravitation in general relativity

Einstein's theory of gravity, called general relativity, is another example of a field theory. Here the principal field is the metric tensor, a symmetric 2nd-rank tensor field in spacetime. This replaces Newton's law of universal gravitation.

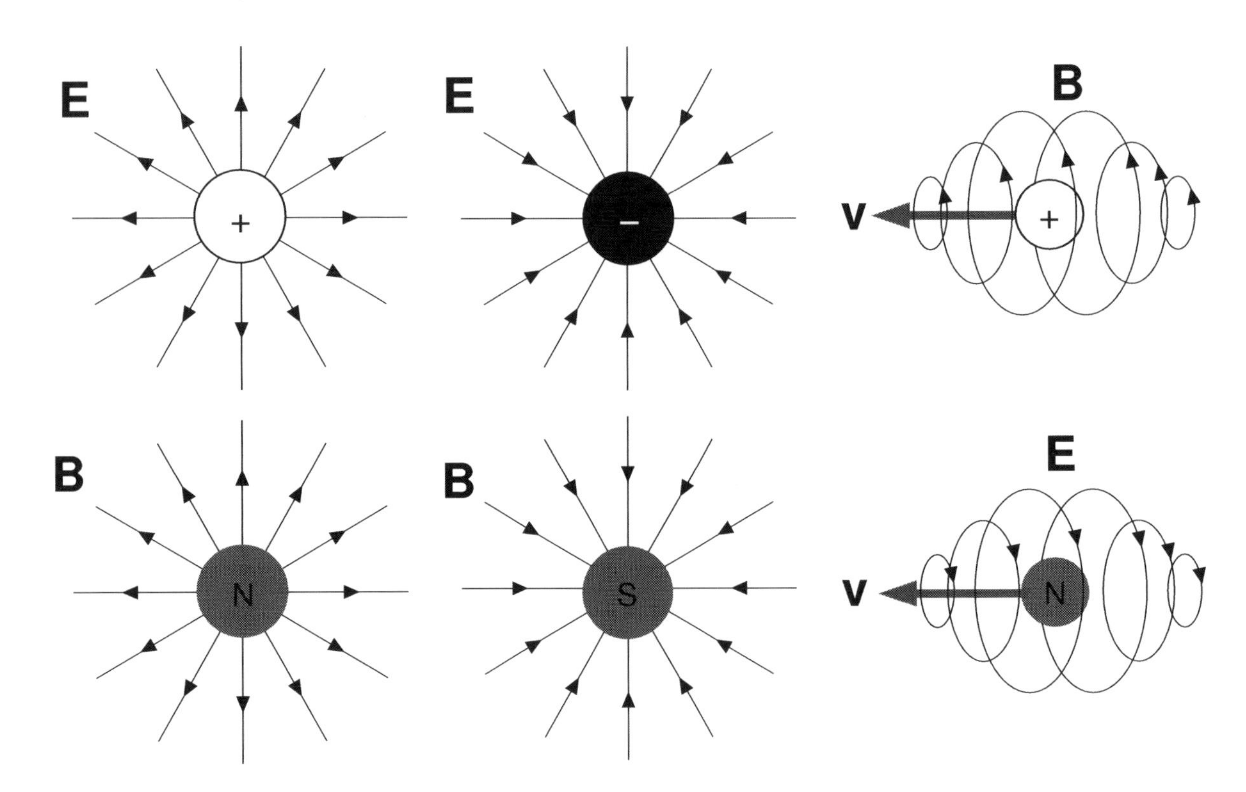

*The **E** fields and **B** fields due to electric charges (black/white) and magnetic poles (red/blue).*[11][12] ***E** fields due to stationary electric charges and **B** fields due to stationary magnetic charges (note in nature N and S monopoles do not exist). In motion (velocity **v**), an* electric *charge induces a **B** field while a* magnetic *charge (not found in nature) would induce an **E** field. Conventional current is used.*

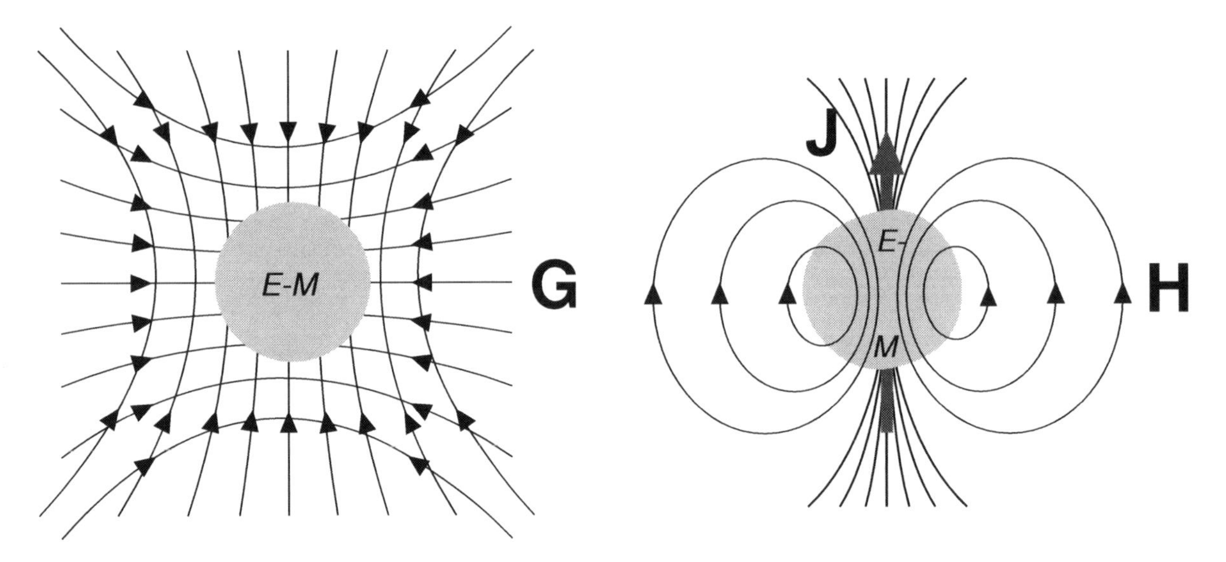

*In general relativity, mass-energy warps space time (Einstein tensor **G**),*[15] *and rotating asymmetric mass-energy distributions with angular momentum **J** generate GEM fields **H***[16]

4.2.4 Waves as fields

Waves can be constructed as physical fields, due to their finite propagation speed and causal nature when a simplified physical model of an isolated closed system is set . They are also subject to the inverse-square law.

For electromagnetic waves, there are optical fields, and terms such as near- and far-field limits for diffraction. In practice, though the field theories of optics are superseded by the electromagnetic field theory of Maxwell.

4.3 Quantum fields

Main article: Quantum field theory

It is now believed that quantum mechanics should underlie all physical phenomena, so that a classical field theory should, at least in principle, permit a recasting in quantum mechanical terms; success yields the corresponding quantum field theory. For example, quantizing classical electrodynamics gives quantum electrodynamics. Quantum electrodynamics is arguably the most successful scientific theory; experimental data confirm its predictions to a higher precision (to more significant digits) than any other theory.[17] The two other fundamental quantum field theories are quantum chromodynamics and the electroweak theory.

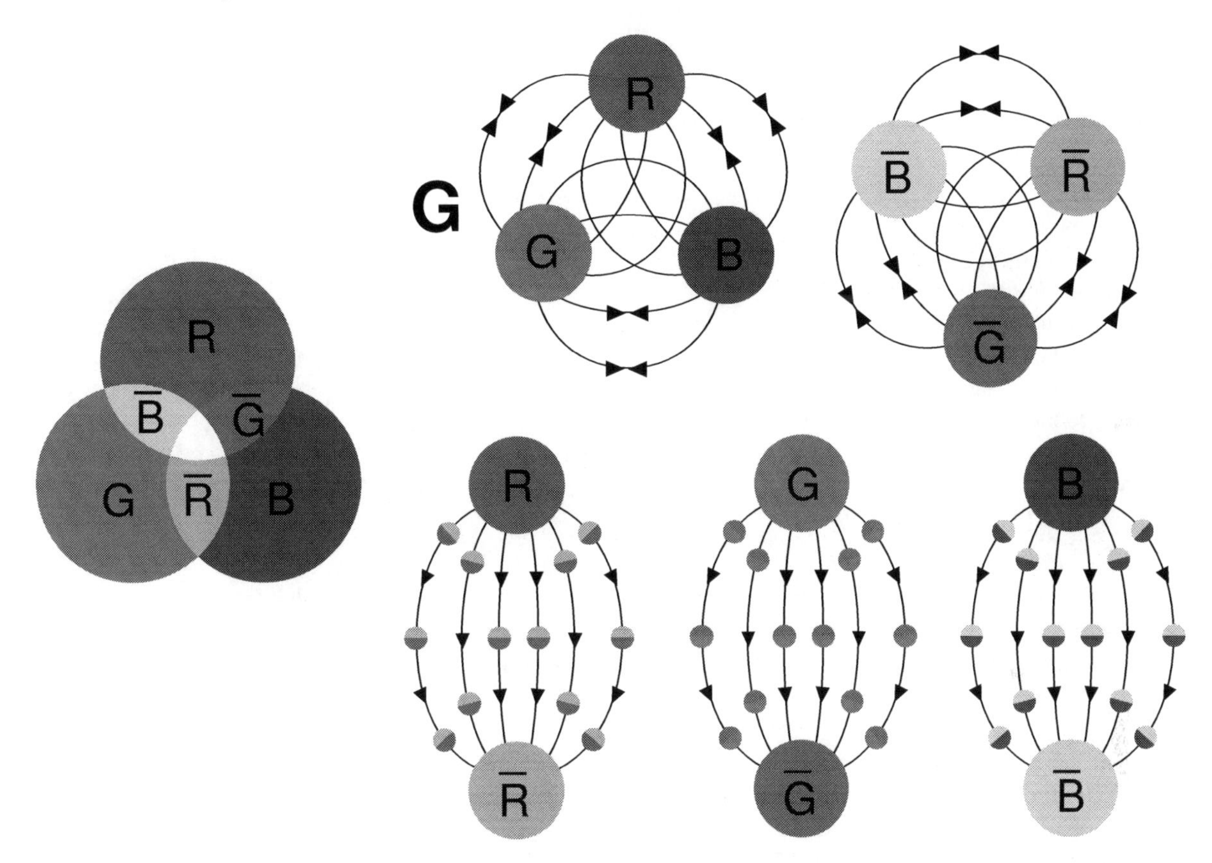

*Fields due to color charges, like in quarks (**G** is the gluon field strength tensor). These are "colorless" combinations. **Top:** Color charge has "ternary neutral states" as well as binary neutrality (analogous to electric charge). **Bottom:** The quark/antiquark combinations.*[11][12]

In quantum chromodynamics, the color field lines are coupled at short distances by gluons, which are polarized by the field and line up with it. This effect increases within a short distance (around 1 fm from the vicinity of the quarks) making the color force increase within a short distance, confining the quarks within hadrons. As the field lines are pulled together tightly by gluons, they do not "bow" outwards as much as an electric field between electric charges.[18]

These three quantum field theories can all be derived as special cases of the so-called standard model of particle physics. General relativity, the Einsteinian field theory of gravity, has yet to be successfully quantized. However an extension, thermal field theory, deals with quantum field theory at *finite temperatures*, something seldom considered in quantum field theory.

In BRST theory one deals with odd fields, e.g. Faddeev–Popov ghosts. There are different descriptions of odd classical fields both on graded manifolds and supermanifolds.

As above with classical fields, it is possible to approach their quantum counterparts from a purely mathematical view using similar techniques as before. The equations governing the quantum fields are in fact PDEs (specifically, relativistic wave equations (RWEs)). Thus one can speak of Yang–Mills, Dirac, Klein–Gordon and Schrödinger fields as being solutions to their respective equations. A possible problem is that these RWEs can deal with complicated mathematical objects with exotic algebraic properties (e.g. spinors are not tensors, so may need calculus over spinor fields), but these in theory can still be subjected to analytical methods given appropriate mathematical generalization.

4.4 Field theory

Field theory usually refers to a construction of the dynamics of a field, i.e. a specification of how a field changes with time or with respect to other independent physical variables on which the field depends. Usually this is done by writing a Lagrangian or a Hamiltonian of the field, and treating it as the classical mechanics (or quantum mechanics) of a system with an infinite number of degrees of freedom. The resulting field theories are referred to as classical or quantum field theories.

The dynamics of a classical field are usually specified by the Lagrangian density in terms of the field components; the dynamics can be obtained by using the action principle.

It is possible to construct simple fields without any a priori knowledge of physics using only mathematics from several variable calculus, potential theory and partial differential equations (PDEs). For example, scalar PDEs might consider quantities such as amplitude, density and pressure fields for the wave equation and fluid dynamics; temperature/concentration fields for the heat/diffusion equations. Outside of physics proper (e.g., radiometry and computer graphics), there are even light fields. All these previous examples are scalar fields. Similarly for vectors, there are vector PDEs for displacement, velocity and vorticity fields in (applied mathematical) fluid dynamics, but vector calculus may now be needed in addition, being calculus over vector fields (as are these three quantities, and those for vector PDEs in general). More generally problems in continuum mechanics may involve for example, directional elasticity (from which comes the term *tensor*, derived from the Latin word for stretch), complex fluid flows or anisotropic diffusion, which are framed as matrix-tensor PDEs, and then require matrices or tensor fields, hence matrix or tensor calculus. It should be noted that the scalars (and hence the vectors, matrices and tensors) can be real or complex as both are fields in the abstract-algebraic/ring-theoretic sense.

In a general setting, classical fields are described by sections of fiber bundles and their dynamics is formulated in the terms of jet manifolds (covariant classical field theory).[19]

In modern physics, the most often studied fields are those that model the four fundamental forces which one day may lead to the Unified Field Theory.

4.4.1 Symmetries of fields

A convenient way of classifying a field (classical or quantum) is by the symmetries it possesses. Physical symmetries are usually of two types:

Spacetime symmetries

Main articles: Global symmetry and Spacetime symmetries

Fields are often classified by their behaviour under transformations of spacetime. The terms used in this classification are:

- scalar fields (such as temperature) whose values are given by a single variable at each point of space. This value does not change under transformations of space.
- vector fields (such as the magnitude and direction of the force at each point in a magnetic field) which are specified by attaching a vector to each point of space. The components of this vector transform between themselves contravariantly under rotations in space. Similarly, a dual (or co-) vector field attaches a dual vector to each point of space, and the components of each dual vector transform covariantly.

- tensor fields, (such as the stress tensor of a crystal) specified by a tensor at each point of space. Under rotations in space, the components of the tensor transform in a more general way which depends on the number of covariant indices and contravariant indices.
- spinor fields (such as the Dirac spinor) arise in quantum field theory to describe particles with spin which transform like vectors except for the one of their component; in other words, when one rotates a vector field 360 degrees around a specific axis, the vector field turns to itself; however, spinors in same case turn to their negatives.

Internal symmetries

Main article: Local symmetry

Fields may have internal symmetries in addition to spacetime symmetries. For example, in many situations one needs fields which are a list of space-time scalars: (φ_1, φ_2, ... φN). For example, in weather prediction these may be temperature, pressure, humidity, etc. In particle physics, the color symmetry of the interaction of quarks is an example of an internal symmetry of the strong interaction, as is the isospin or flavour symmetry.

If there is a symmetry of the problem, not involving spacetime, under which these components transform into each other, then this set of symmetries is called an *internal symmetry*. One may also make a classification of the charges of the fields under internal symmetries.

4.4.2 Statistical field theory

Main article: Statistical field theory

Statistical field theory attempts to extend the field-theoretic paradigm toward many-body systems and statistical mechanics. As above, it can be approached by the usual infinite number of degrees of freedom argument.

Much like statistical mechanics has some overlap between quantum and classical mechanics, statistical field theory has links to both quantum and classical field theories, especially the former with which it shares many methods. One important example is mean field theory.

4.4.3 Continuous random fields

Classical fields as above, such as the electromagnetic field, are usually infinitely differentiable functions, but they are in any case almost always twice differentiable. In contrast, generalized functions are not continuous. When dealing carefully with classical fields at finite temperature, the mathematical methods of continuous random fields are used, because thermally fluctuating classical fields are nowhere differentiable. Random fields are indexed sets of random variables; a continuous random field is a random field that has a set of functions as its index set. In particular, it is often mathematically convenient to take a continuous random field to have a Schwartz space of functions as its index set, in which case the continuous random field is a tempered distribution.

We can think about a continuous random field, in a (very) rough way, as an ordinary function that is $\pm\infty$ almost everywhere, but such that when we take a weighted average of all the infinities over any finite region, we get a finite result. The infinities are not well-defined; but the finite values can be associated with the functions used as the weight functions to get the finite values, and that can be well-defined. We can define a continuous random field well enough as a linear map from a space of functions into the real numbers.

4.5 See also

- Field strength

- Lagrangian and Eulerian specification of a field
- Covariant Hamiltonian field theory
- Scalar field theory

4.6 Notes

[1] This is contingent on the correct choice of gauge. V and $\mathbf{A}$ are not completely determined by ρ and $\mathbf{J}$; rather, they are only determined up to some scalar function $f(\mathbf{r}, t)$ known as the gauge. The retarded potential formalism requires one to choose the Lorenz gauge.

4.7 References

[1] John Gribbin (1998). *Q is for Quantum: Particle Physics from A to Z*. London: Weidenfeld & Nicolson. p. 138. ISBN 0-297-81752-3.

[2] Richard Feynman (1970). *The Feynman Lectures on Physics Vol II*. Addison Wesley Longman. ISBN 978-0-201-02115-8. A "field" is any physical quantity which takes on different values at different points in space.

[3] Ernan McMullin (2002). "The Origins of the Field Concept in Physics" (PDF). *Phys. Perspect.* **4**: 13–39.

[4] Richard P. Feynman (1970). *The Feynman Lectures on Physics Vol II*. Addison Wesley Longman.

[5] Richard P. Feynman (1970). *The Feynman Lectures on Physics Vol I*. Addison Wesley Longman.

[6] John Archibald Wheeler (1998). *Geons, Black Holes, and Quantum Foam: A Life in Physics*. London: Norton. p. 163.

[7] Richard P. Feynman (1970). *The Feynman Lectures on Physics Vol I*. Addison Wesley Longman.

[8] Steven Weinberg (November 7, 2013). "Physics: What We Do and Don't Know". *New York Review of Books*.

[9] Weinberg, Steven (1977). "The Search for Unity: Notes for a History of Quantum Field Theory". *Daedalus* **106** (4): 17–35. JSTOR 20024506.

[10] Kleppner, David; Kolenkow, Robert. *An Introduction to Mechanics*. p. 85.

[11] Parker, C.B. (1994). *McGraw Hill Encyclopaedia of Physics* (2nd ed.). Mc Graw Hill. ISBN 0-07-051400-3.

[12] M. Mansfield, C. O'Sullivan (2011). *Understanding Physics* (4th ed.). John Wiley & Sons. ISBN 978-0-47-0746370.

[13] Griffiths, David. *Introduction to Electrodynamics* (3rd ed.). p. 326.

[14] Wangsness, Roald. *Electromagnetic Fields* (2nd ed.). p. 469.

[15] J.A. Wheeler, C. Misner, K.S. Thorne (1973). *Gravitation*. W.H. Freeman & Co. ISBN 0-7167-0344-0.

[16] I. Ciufolini and J.A. Wheeler (1995). *Gravitation and Inertia*. Princeton Physics Series. ISBN 0-691-03323-4.

[17] Peskin, Michael E.; Schroeder, Daniel V. (1995). *An Introduction to Quantum Fields*. Westview Press. p. 198. ISBN 0-201-50397-2.. Also see precision tests of QED.

[18] R. Resnick, R. Eisberg (1985). *Quantum Physics of Atoms, Molecules, Solids, Nuclei and Particles* (2nd ed.). John Wiley & Sons. p. 684. ISBN 978-0-471-87373-0.

[19] Giachetta, G., Mangiarotti, L., Sardanashvily, G. (2009) *Advanced Classical Field Theory*. Singapore: World Scientific, ISBN 978-981-283-895-7 (arXiv: 0811.0331v2)

4.8 Further reading

- "Fields". *Principles of Physical Science. Encyclopaedia Britannica (Macropaedia)* **25** (fifteenth ed.). 1994. p. 815.
- Landau, Lev D. and Lifshitz, Evgeny M. (1971). *Classical Theory of Fields* (3rd ed.). London: Pergamon. ISBN 0-08-016019-0. Vol. 2 of the Course of Theoretical Physics.
- Jepsen, Kathryn (July 18, 2013). "Real talk: Everything is made of fields" (PDF). *Symmetry Magazine*.

4.9 External links

- Particle and Polymer Field Theories

Chapter 5

Special unitary group

"SU(5)" redirects here. For the specific grand unification theory, see Georgi–Glashow model.

In mathematics, the **special unitary group** of degree n, denoted SU(n), is the Lie group of n×n unitary matrices with determinant 1 (i.e., real-valued determinant, not complex as for general unitary matrices). The group operation is that of matrix multiplication. The special unitary group is a subgroup of the unitary group U(n), consisting of all n×n unitary matrices. As a compact classical group, U(n) is the group that preserves the standard inner product on $\mathbf{C}^n$.[nb 1] It is itself a subgroup of the general linear group, SU(n) ⊂ U(n) ⊂ GL(n, **C**).

The SU(n) groups find wide application in the Standard Model of particle physics, especially SU(2) in the electroweak interaction and SU(3) in quantum chromodynamics.[1]

The simplest case, SU(1), is the trivial group, having only a single element. The group SU(2) is isomorphic to the group of quaternions of norm 1, and is thus diffeomorphic to the 3-sphere. Since unit quaternions can be used to represent rotations in 3-dimensional space (up to sign), there is a surjective homomorphism from SU(2) to the rotation group SO(3) whose kernel is $\{+I, -I\}$.[nb 2] SU(2) is also identical to one of the symmetry groups of spinors, Spin(3), that enables a spinor presentation of rotations.

5.1 Properties

The special unitary group SU(n) is a real Lie group (though not a complex Lie group). Its dimension as a real manifold is $n^2 - 1$. Topologically, it is compact and simply connected. Algebraically, it is a simple Lie group (meaning its Lie algebra is simple; see below). [2]

The center of SU(n) is isomorphic to the cyclic group Zn, and is composed of the diagonal matrices $\zeta\, I$ for ζ an nth root of unity and I the n×n identity matrix.

Its outer automorphism group, for $n \geq 3$, is Z_2, while the outer automorphism group of SU(2) is the trivial group.

A maximal torus, of rank $n - 1$, is given by the set of diagonal matrices with determinant 1. The Weyl group is the symmetric group Sn, which is represented by signed permutation matrices (the signs being necessary to ensure the determinant is 1).

The Lie algebra of SU(n), denoted by **su**(n), can be identified with the set of traceless antihermitian n×n complex matrices, with the regular commutator as Lie bracket. Particle physicists often use a different, equivalent representation: the set of traceless hermitian n×n complex matrices with Lie bracket given by $-i$ times the commutator.

5.2 Infinitesimal generators

The Lie algebra **su**(n) can be generated by n^2 operators $\hat{O}_{ij}$, i, j= 1, 2, ..., n, which satisfy the commutator relationships

$$\left[\hat{O}_{ij}, \hat{O}_{k\ell}\right] = \delta_{jk}\hat{O}_{i\ell} - \delta_{i\ell}\hat{O}_{kj}$$

for $i, j, k, \ell = 1, 2, ..., n$, where δ_{jk} denotes the Kronecker delta. Additionally, the operator

$$\hat{N} = \sum_{i=1}^{n} \hat{O}_{ii}$$

satisfies

$$\left[\hat{N}, \hat{O}_{ij}\right] = 0,$$

which implies that the number of *independent* generators of the Lie algebra is $n^2 - 1$.[3]

5.2.1 Fundamental representation

In the defining, or fundamental, representation of **su**(n) the generators Ta are represented by traceless hermitian matrices complex nxn matrices, where:

$$T_a T_b = \frac{1}{2n}\delta_{ab} I_n + \frac{1}{2}\sum_{c=1}^{n^2-1} (i f_{abc} + d_{abc}) T_c$$

where the f are the structure constants and are antisymmetric in all indices, while the d-coefficients are symmetric in all indices. As a consequence:

$$[T_a, T_b]_+ = \frac{1}{n}\delta_{ab} I_n + \sum_{c=1}^{n^2-1} d_{abc} T_c$$

$$[T_a, T_b]_- = i\sum_{c=1}^{n^2-1} f_{abc} T_c .$$

We also take

$$\sum_{c,e=1}^{n^2-1} d_{ace} d_{bce} = \frac{n^2-4}{n}\delta_{ab}$$

as a normalization convention.

5.2.2 Adjoint representation

In the $(n^2 - 1)$ -dimensional adjoint representation, the generators are represented by $(n^2 - 1) \times (n^2 - 1)$ matrices, whose elements are defined by the structure constants themselves:

$$(T_a)_{jk} = -i f_{ajk}.$$

5.3 $n = 2$

See also: Versor and Pauli matrices

SU(2) is the following group,

$$\mathrm{SU}(2) = \left\{ \begin{pmatrix} \alpha & -\overline{\beta} \\ \beta & \overline{\alpha} \end{pmatrix} : \ \alpha, \beta \in \mathbf{C}, |\alpha|^2 + |\beta|^2 = 1 \right\} ,$$

where the overline denotes complex conjugation.

Now, consider the following map,

$$\varphi : \mathbf{C}^2 \to \mathrm{M}(2, \mathbf{C})$$
$$\varphi(\alpha, \beta) = \begin{pmatrix} \alpha & -\overline{\beta} \\ \beta & \overline{\alpha} \end{pmatrix},$$

where M(2, **C**) denotes the set of 2 by 2 complex matrices. By considering $\mathbf{C}^2$ diffeomorphic to $\mathbf{R}^4$ and M(2, **C**) diffeomorphic to $\mathbf{R}^8$, we can see that φ is an injective real linear map and hence an embedding. Now, considering the restriction of φ to the 3-sphere (since modulus is 1), denoted S^3, we can see that this is an embedding of the 3-sphere onto a compact submanifold of M(2, **C**). However, it is also clear that $\varphi(S^3)$ = SU(2).

Therefore, as a manifold S^3 is diffeomorphic to SU(2) and so SU(2) is a compact, connected Lie group.

The Lie algebra of SU(2) is

$$\mathfrak{su}(2) = \left\{ \begin{pmatrix} ia & -\overline{z} \\ z & -ia \end{pmatrix} : \ a \in \mathbf{R}, z \in \mathbf{C} \right\} .$$

It is easily verified that matrices of this form have trace zero and are antihermitian. The Lie algebra is then generated by the following matrices,

$$u_1 = \begin{pmatrix} 0 & i \\ i & 0 \end{pmatrix} \quad u_2 = \begin{pmatrix} 0 & -1 \\ 1 & 0 \end{pmatrix} \quad u_3 = \begin{pmatrix} i & 0 \\ 0 & -i \end{pmatrix},$$

which are easily seen to have the form of the general element specified above.

These satisfy $u_3u_2 = -u_2u_3 = -u_1$ and $u_2u_1 = -u_1u_2 = -u_3$. The commutator bracket is therefore specified by

$$[u_3, u_1] = 2u_2, \qquad [u_1, u_2] = 2u_3, \qquad [u_2, u_3] = 2u_1 .$$

The above generators are related to the Pauli matrices by $u_1 = i\,\sigma_1$,$u_2 = -i\,\sigma_2$ and $u_3 = i\,\sigma_3$. This representation is routinely used in quantum mechanics to represent the spin of fundamental particles such as electrons. They also serve as unit vectors for the description of our 3 spatial dimensions in loop quantum gravity.

The Lie algebra serves to work out the representations of SU(2).

See also: Rotation group SO(3) § A note on representations

5.4 $n = 3$

The generators of **su**(3), T, in the defining representation, are:

$$T_a = \frac{\lambda_a}{2}.$$

where λ the Gell-Mann matrices, are the SU(3) analog of the Pauli matrices for SU(2):

These λ_a span all traceless Hermitian matrices H of the Lie algebra, as required.

They obey the relations

$$[T_a, T_b] = i \sum_{c=1}^{8} f_{abc} T_c$$

$$\{T_a, T_b\} = \frac{1}{3}\delta_{ab} + \sum_{c=1}^{8} d_{abc} T_c$$

$$\{\lambda_a, \lambda_b\} = \frac{4}{3}\delta_{ab} + 2\sum_{c=1}^{8} d_{abc} \lambda_c$$

The f are the structure constants of the Lie algebra, given by:

$$f_{123} = 1$$

$$f_{147} = -f_{156} = f_{246} = f_{257} = f_{345} = -f_{367} = \frac{1}{2}$$

$$f_{458} = f_{678} = \frac{\sqrt{3}}{2},$$

while all other f_{abc} not related to these by permutation are zero.

The symmetric coefficients d take the values:

$$d_{118} = d_{228} = d_{338} = -d_{888} = \frac{1}{\sqrt{3}}$$

$$d_{448} = d_{558} = d_{668} = d_{778} = -\frac{1}{2\sqrt{3}}$$

$$d_{146} = d_{157} = -d_{247} = d_{256} = d_{344} = d_{355} = -d_{366} = -d_{377} = \frac{1}{2}.$$

As a topological space, *SU(3)* is a direct product of a 3-sphere and a 5-sphere, S^3 ⍰ S^5.

A generic *SU(3)* group element generated by a traceless 3×3 hermitian matrix H, normalized as tr(H^2) = 2 , is given by[4]

$$\exp(i\theta H) =$$

$$\left[-\tfrac{1}{3}\, I \sin(\phi + 2\pi/3)\sin(\phi - 2\pi/3) - \tfrac{1}{2\sqrt{3}}\, H \sin(\phi) - \tfrac{1}{4}\, H^2\right] \frac{\exp\left(\tfrac{2}{\sqrt{3}}\, i\theta \sin\phi\right)}{\cos(\phi + 2\pi/3)\cos(\phi - 2\pi/3)}$$

$$+\left[-\tfrac{1}{3}\, I \sin(\phi)\sin(\phi-2\pi/3) - \tfrac{1}{2\sqrt{3}}\, H \sin(\phi+2\pi/3) - \tfrac{1}{4}\, H^2\right] \frac{\exp\left(\frac{2}{\sqrt{3}}\, i\theta \sin(\phi+2\pi/3)\right)}{\cos(\phi)\cos(\phi-2\pi/3)}$$

$$+\left[-\tfrac{1}{3}\, I \sin(\phi)\sin(\phi+2\pi/3) - \tfrac{1}{2\sqrt{3}}\, H \sin(\phi-2\pi/3) - \tfrac{1}{4}\, H^2\right] \frac{\exp\left(\frac{2}{\sqrt{3}}\, i\theta \sin(\phi-2\pi/3)\right)}{\cos(\phi)\cos(\phi+2\pi/3)}$$

where

$$\phi \equiv \tfrac{1}{3}\left(\arccos\left(\tfrac{3}{2}\sqrt{3}\det H\right) - \tfrac{\pi}{2}\right)$$

See also: Clebsch–Gordan coefficients for SU(3)

for elementary representation theory facts.

5.5 Lie algebra structure

The above representation bases generalize to $n > 3$, using generalized Pauli matrices.

If we choose an (arbitrary) particular basis, then the subspace of traceless diagonal $n \times n$ matrices with imaginary entries forms an $(n-1)$-dimensional Cartan subalgebra.

Complexify the Lie algebra, so that any traceless $n \times n$ matrix is now allowed. The weight eigenvectors are the Cartan subalgebra itself, as well as the matrices with only one nonzero entry which is off diagonal. Even though the Cartan subalgebra **h** is only $(n-1)$-dimensional, to simplify calculations, it is often convenient to introduce an auxiliary element, the unit matrix which commutes with everything else (which is not an element of the Lie algebra!) for the purpose of computing weights—and that only. So, we have a basis where the i-th basis vector is the matrix with 1 on the i-th diagonal entry and zero elsewhere. Weights would then be given by n coordinates and the sum over all n coordinates has to be zero (because the unit matrix is only auxiliary).

So, SU(n) is of rank $n-1$ and its Dynkin diagram is given by A_{n-1}, a chain of $n-1$ vertices, o–o–o–o---o. Its root system consists of $n(n-1)$ roots spanning a $n-1$ Euclidean space. Here, we use n redundant coordinates instead of $n-1$ to emphasize the symmetries of the root system (the n coordinates have to add up to zero).

In other words, we are embedding this $n-1$ dimensional vector space in an n-dimensional one. Thus, the roots consists of all the $n(n-1)$ permutations of $(1, -1, 0, \ldots, 0)$. The construction given above explains why. A choice of simple roots is

$$(1,-1,0,\ldots,0),$$
$$(0,1,-1,\ldots,0),$$
$$\ldots$$
$$(0,0,0,\ldots,1,-1).$$

Its Cartan matrix is

$$\begin{pmatrix} 2 & -1 & 0 & \ldots & 0 \\ -1 & 2 & -1 & \ldots & 0 \\ 0 & -1 & 2 & \ldots & 0 \\ \vdots & \vdots & \vdots & \ddots & \vdots \\ 0 & 0 & 0 & \ldots & 2 \end{pmatrix}.$$

Its Weyl group or Coxeter group is the symmetric group S_n, the symmetry group of the $(n-1)$-simplex.

5.6 Generalized special unitary group

For a field F, the **generalized special unitary group over F**, SU(p, q; F), is the group of all linear transformations of determinant 1 of a vector space of rank $n = p + q$ over F which leave invariant a nondegenerate, Hermitian form of signature (p, q). This group is often referred to as the **special unitary group of signature p q over F**. The field F can be replaced by a commutative ring, in which case the vector space is replaced by a free module.

Specifically, fix a Hermitian matrix A of signature p q in GL(n, **R**), then all

$$M \in \mathrm{SU}(p, q, R)$$

satisfy

$$M^* AM = A$$
$$\det M = 1.$$

Often one will see the notation SU(p, q) without reference to a ring or field; in this case, the ring or field being referred to is **C** and this gives one of the classical Lie groups. The standard choice for A when F = **C** is

$$A = \begin{bmatrix} 0 & 0 & i \\ 0 & I_{n-2} & 0 \\ -i & 0 & 0 \end{bmatrix}.$$

However there may be better choices for A for certain dimensions which exhibit more behaviour under restriction to subrings of **C**.

5.6.1 Example

An important example of this type of group is the Picard modular group SU(2, 1; **Z**[i]) which acts (projectively) on complex hyperbolic space of degree two, in the same way that SL(2,9;**Z**) acts (projectively) on real hyperbolic space of dimension two. In 2005 Gábor Francsics and Peter Lax computed an explicit fundamental domain for the action of this group on HC^2.[5]

A further example is SU(1, 1; **C**), which is isomorphic to SL(2,**R**).

5.7 Important subgroups

In physics the special unitary group is used to represent bosonic symmetries. In theories of symmetry breaking it is important to be able to find the subgroups of the special unitary group. Subgroups of SU(n) that are important in GUT physics are, for $p > 1$, $n - p > 1$,

$$\mathrm{SU}(n) \supset \mathrm{SU}(p) \times \mathrm{SU}(n-p) \times \mathrm{U}(1)$$

where × denotes the direct product and U(1), known as the circle group, is the multiplicative group of all complex numbers with absolute value 1.

For completeness, there are also the orthogonal and symplectic subgroups,

$$\mathrm{SU}(n) \supset \mathrm{SO}(n),$$
$$\mathrm{SU}(2n) \supset \mathrm{Sp}(n).$$

Since the rank of SU(n) is $n - 1$ and of U(1) is 1, a useful check is that the sum of the ranks of the subgroups is less than or equal to the rank of the original group. SU(n) is a subgroup of various other Lie groups,

$$\begin{aligned} \mathrm{SO}(2n) &\supset \mathrm{SU}(n) \\ \mathrm{Sp}(n) &\supset \mathrm{SU}(n) \\ \mathrm{Spin}(4) &= \mathrm{SU}(2)\times\mathrm{SU}(2) \\ \mathrm{E}_6 &\supset \mathrm{SU}(6) \\ \mathrm{E}_7 &\supset \mathrm{SU}(8) \\ \mathrm{G}_2 &\supset \mathrm{SU}(3) \end{aligned}$$

See spin group, and simple Lie groups for E_6, E_7, and G_2.

There are also the accidental isomorphisms: SU(4) = Spin(6) , SU(2) = Spin(3) = Sp(1) ,[nb 3] and U(1) = Spin(2) = SO(2) .

One may finally mention that SU(2) is the double covering group of SO(3), a relation that plays an important role in the theory of rotations of 2-spinors in non-relativistic quantum mechanics.

5.8 See also

- Projective special unitary group, PSU(n)
- Generalizations of Pauli matrices

5.9 Remarks

[1] For a characterization of U(n) and hence SU(n) in terms of preservation of the standard inner product on $\mathbb{C}^n$, see Classical group.

[2] For an explicit description of the homomorphism SU(2) → SO(3), see Connection between SO(3) and SU(2).

[3] Sp(n) is the compact real form of Sp($2n$, **C**). It is sometimes denoted USp($2n$). The dimension of the Sp(n)-matrices is $2n \times 2n$.

5.10 References

[1] Halzen, Francis; Martin, Alan (1984). *Quarks & Leptons: An Introductory Course in Modern Particle Physics*. John Wiley & Sons. ISBN 0-471-88741-2.

[2] Wybourne, B G (1974). *Classical Groups for Physicists*, Wiley-Interscience. ISBN 0471965057 .

[3] R.R. Puri, *Mathematical Methods of Quantum Optics*, Springer, 2001.

[4] Rosen, S P (1971). "Finite Transformations in Various Representations of SU(3)". *Journal of Mathematical Physics* **12** (4): 673. doi:10.1063/1.1665634. ISSN 0022-2488.; Curtright, T L; Zachos, C K (2015). "Elementary results for the fundamental representation of SU(3)". *Researchgate*. doi:10.13140/RG.2.1.1743.2163.

[5] Francsics, Gabor; Lax, Peter D. "An Explicit Fundamental Domain For The Picard Modular Group In Two Complex Dimensions". arXiv:math/0509708v1.

Chapter 6

Circle group

For the jazz group, see Circle (jazz band).

In mathematics, the **circle group**, denoted by **T**, is the multiplicative group of all complex numbers with absolute value 1, i.e., the unit circle in the complex plane or simply the **unit complex numbers**[1]

$\mathbb{T} = \{z \in \mathbb{C} : |z| = 1\}.$

The circle group forms a subgroup of $\mathbf{C}^\times$, the multiplicative group of all nonzero complex numbers. Since $\mathbf{C}^\times$ is abelian, it follows that **T** is as well. The circle group is also the group **U(1)** of 1×1 unitary matrices; these act on the complex plane by rotation about the origin. The circle group can be parametrized by the angle θ of rotation by

$\theta \mapsto z = e^{i\theta} = \cos\theta + i\sin\theta.$

This is the exponential map for the circle group.

The circle group plays a central role in Pontryagin duality, and in the theory of Lie groups.

The notation **T** for the circle group stems from the fact that, with the standard topology (see below), the circle group is a 1-torus. More generally $\mathbf{T}^n$ (the direct product of **T** with itself *n* times) is geometrically an *n*-torus.

6.1 Elementary introduction

One way to think about the circle group is that it describes how to add *angles*, where only angles between 0° and 360° are permitted. For example, the diagram illustrates how to add 150° to 270°. The answer should be 150° + 270° = 420°, but when thinking in terms of the circle group, we need to “forget” the fact that we have wrapped once around the circle. Therefore we adjust our answer by 360° which gives 420° = 60° (mod 360°).

Another description is in terms of ordinary addition, where only numbers between 0 and 1 are allowed (with 1 corresponding to a full rotation). To achieve this, we might need to throw away digits occurring before the decimal point. For example, when we work out 0.784 + 0.925 + 0.446, the answer should be 2.155, but we throw away the leading 2, so the answer (in the circle group) is just 0.155.

6.2 Topological and analytic structure

The circle group is more than just an abstract algebraic object. It has a natural topology when regarded as a subspace of the complex plane. Since multiplication and inversion are continuous functions on $\mathbf{C}^\times$, the circle group has the structure

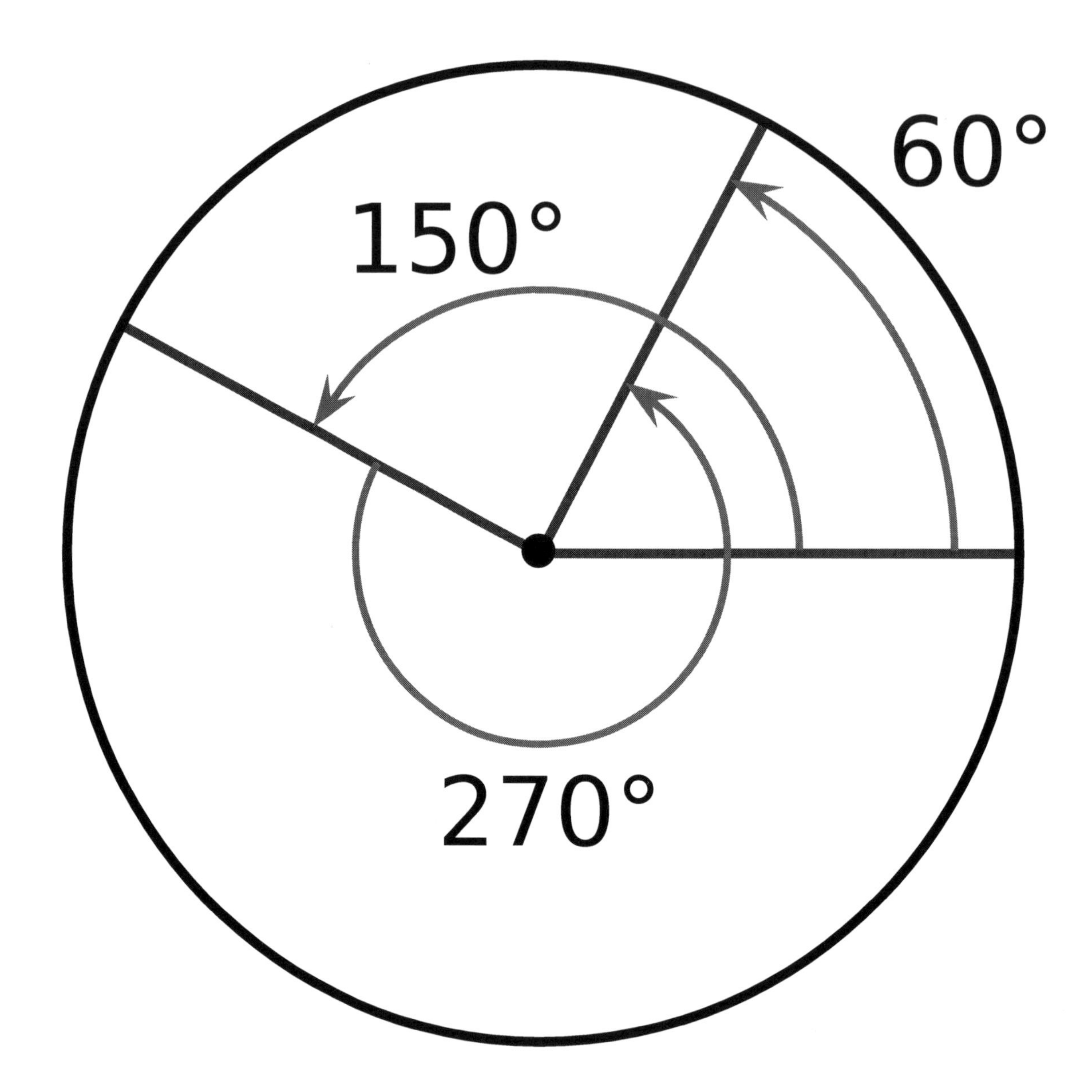

Multiplication on the circle group is equivalent to addition of angles

of a topological group. Moreover, since the unit circle is a closed subset of the complex plane, the circle group is a closed subgroup of $\mathbf{C}^{\times}$ (itself regarded as a topological group).

One can say even more. The circle is a 1-dimensional real manifold and multiplication and inversion are real-analytic maps on the circle. This gives the circle group the structure of a one-parameter group, an instance of a Lie group. In fact, up to isomorphism, it is the unique 1-dimensional compact, connected Lie group. Moreover, every n-dimensional compact, connected, abelian Lie group is isomorphic to $\mathbf{T}^n$.

6.3 Isomorphisms

The circle group shows up in a variety of forms in mathematics. We list some of the more common forms here. Specifically, we show that

$\mathbb{T} \cong \mathrm{U}(1) \cong \mathbb{R}/\mathbb{Z} \cong \mathrm{SO}(2).$

Note that the slash (/) denotes here quotient group.

The set of all 1×1 unitary matrices clearly coincides with the circle group; the unitary condition is equivalent to the condition that its element have absolute value 1. Therefore, the circle group is canonically isomorphic to U(1), the first unitary group.

The exponential function gives rise to a group homomorphism exp : **R** → **T** from the additive real numbers **R** to the circle group **T** via the map

$$\theta \mapsto e^{i\theta} = \cos\theta + i\sin\theta.$$

The last equality is Euler's formula or the complex exponential. The real number θ corresponds to the angle on the unit circle as measured from the positive *x*-axis. That this map is a homomorphism follows from the fact that the multiplication of unit complex numbers corresponds to addition of angles:

$$e^{i\theta_1}e^{i\theta_2} = e^{i(\theta_1+\theta_2)}.$$

This exponential map is clearly a surjective function from **R** to **T**. It is not, however, injective. The kernel of this map is the set of all integer multiples of 2π. By the first isomorphism theorem we then have that

$$\mathbb{T} \cong \mathbb{R}/2\pi\mathbb{Z}.$$

After rescaling we can also say that **T** is isomorphic to **R/Z**.

If complex numbers are realized as 2×2 real matrices (see complex number), the unit complex numbers correspond to 2×2 orthogonal matrices with unit determinant. Specifically, we have

$$e^{i\theta} \leftrightarrow \begin{bmatrix} \cos\theta & -\sin\theta \\ \sin\theta & \cos\theta \end{bmatrix}.$$

The circle group is therefore isomorphic to the special orthogonal group SO(2). This has the geometric interpretation that multiplication by a unit complex number is a proper rotation in the complex plane, and every such rotation is of this form.

6.4 Properties

Every compact Lie group *G* of dimension > 0 has a subgroup isomorphic to the circle group. That means that, thinking in terms of symmetry, a compact symmetry group acting *continuously* can be expected to have one-parameter circle subgroups acting; the consequences in physical systems are seen for example at rotational invariance, and spontaneous symmetry breaking.

The circle group has many subgroups, but its only proper closed subgroups consist of roots of unity: For each integer *n* > 0, the *n*th roots of unity form a cyclic group of order *n*, which is unique up to isomorphism.

6.5 Representations

The representations of the circle group are easy to describe. It follows from Schur's lemma that the irreducible complex representations of an abelian group are all 1-dimensional. Since the circle group is compact, any representation ρ : **T** →

$GL(1, \mathbf{C}) \cong \mathbf{C}^\times$, must take values in $U(1) \cong \mathbf{T}$. Therefore, the irreducible representations of the circle group are just the homomorphisms from the circle group to itself.

These representations are all inequivalent. The representation φ_{-n} is conjugate to φ_n,

$$\phi_{-n} = \overline{\phi_n}.$$

These representations are just the characters of the circle group. The character group of **T** is clearly an infinite cyclic group generated by φ_1:

$$\mathrm{Hom}(\mathbb{T}, \mathbb{T}) \cong \mathbb{Z}.$$

The irreducible real representations of the circle group are the trivial representation (which is 1-dimensional) and the representations

$$\rho_n(e^{i\theta}) = \begin{bmatrix} \cos n\theta & -\sin n\theta \\ \sin n\theta & \cos n\theta \end{bmatrix}, \quad n \in \mathbb{Z}^+,$$

taking values in SO(2). Here we only have positive integers n since the representation ρ_{-n} is equivalent to ρ_n .

6.6 Group structure

In this section we will forget about the topological structure of the circle group and look only at its structure as an abstract group.

The circle group **T** is a divisible group. Its torsion subgroup is given by the set of all nth roots of unity for all n, and is isomorphic to **Q/Z**. The structure theorem for divisible groups and the axiom of choice together tell us that **T** is isomorphic to the direct sum of **Q/Z** with a number of copies of **Q**. The number of copies of **Q** must be c (the cardinality of the continuum) in order for the cardinality of the direct sum to be correct. But the direct sum of c copies of **Q** is isomorphic to **R**, as **R** is a vector space of dimension c over **Q**. Thus

$$\mathbb{T} \cong \mathbb{R} \oplus (\mathbb{Q}/\mathbb{Z}).$$

The isomorphism

$$\mathbb{C}^\times \cong \mathbb{R} \oplus (\mathbb{Q}/\mathbb{Z})$$

can be proved in the same way, as $\mathbf{C}^\times$ is also a divisible abelian group whose torsion subgroup is the same as the torsion subgroup of **T**.

6.7 See also

- Rotation number
- Torus
- One-parameter subgroup
- Unitary group

- Orthogonal group
- Group of rational points on the unit circle
- Phase factor (application in quantum-mechanics)

6.8 Notes

[1] "a **unit complex number** is a complex number of unit absolute value" (James & James 1992, p. 436)

6.9 References

- James, Robert C.; James, Glenn (1992), *Mathematics Dictionary* (Fifth ed.), Chapman & Hall

6.10 Further reading

- Hua Luogeng (1981) *Starting with the unit circle*, Springer Verlag, ISBN 0-387-90589-8 .

6.11 External links

- Homeomorphism and the Group Structure on a Circle

Chapter 7

Lepton

For other uses, see Lepton (disambiguation).

A **lepton** is an elementary, half-integer spin (spin ½) particle that does not undergo strong interactions, but is subject to the Pauli exclusion principle.[1] The best known of all leptons is the electron, which is directly tied to all chemical properties. Two main classes of leptons exist: charged leptons (also known as the *electron-like* leptons), and neutral leptons (better known as neutrinos). Charged leptons can combine with other particles to form various composite particles such as atoms and positronium, while neutrinos rarely interact with anything, and are consequently rarely observed.

There are six types of leptons, known as *flavours*, forming three *generations*.[2] The first generation is the *electronic leptons*, comprising the electron (e−) and electron neutrino (ν
e); the second is the *muonic leptons*, comprising the muon (μ−) and muon neutrino (ν
μ); and the third is the *tauonic leptons*, comprising the tau (τ−) and the tau neutrino (ν
τ). Electrons have the least mass of all the charged leptons. The heavier muons and taus will rapidly change into electrons through a process of particle decay: the transformation from a higher mass state to a lower mass state. Thus electrons are stable and the most common charged lepton in the universe, whereas muons and taus can only be produced in high energy collisions (such as those involving cosmic rays and those carried out in particle accelerators).

Leptons have various intrinsic properties, including electric charge, spin, and mass. Unlike quarks however, leptons are not subject to the strong interaction, but they are subject to the other three fundamental interactions: gravitation, electromagnetism (excluding neutrinos, which are electrically neutral), and the weak interaction. For every lepton flavor there is a corresponding type of antiparticle, known as antilepton, that differs from the lepton only in that some of its properties have equal magnitude but opposite sign. However, according to certain theories, neutrinos may be their own antiparticle, but it is not currently known whether this is the case or not.

The first charged lepton, the electron, was theorized in the mid-19th century by several scientists[3][4][5] and was discovered in 1897 by J. J. Thomson.[6] The next lepton to be observed was the muon, discovered by Carl D. Anderson in 1936, which was classified as a meson at the time.[7] After investigation, it was realized that the muon did not have the expected properties of a meson, but rather behaved like an electron, only with higher mass. It took until 1947 for the concept of "leptons" as a family of particle to be proposed.[8] The first neutrino, the electron neutrino, was proposed by Wolfgang Pauli in 1930 to explain certain characteristics of beta decay.[8] It was first observed in the Cowan–Reines neutrino experiment conducted by Clyde Cowan and Frederick Reines in 1956.[8][9] The muon neutrino was discovered in 1962 by Leon M. Lederman, Melvin Schwartz and Jack Steinberger,[10] and the tau discovered between 1974 and 1977 by Martin Lewis Perl and his colleagues from the Stanford Linear Accelerator Center and Lawrence Berkeley National Laboratory.[11] The tau neutrino remained elusive until July 2000, when the DONUT collaboration from Fermilab announced its discovery.[12][13]

Leptons are an important part of the Standard Model. Electrons are one of the components of atoms, alongside protons and neutrons. Exotic atoms with muons and taus instead of electrons can also be synthesized, as well as lepton–antilepton particles such as positronium.

7.1 Etymology

The name *lepton* comes from the Greek λεπτός *leptós*, "fine, small, thin" (neuter form: λεπτόν *leptón*);[14][15] the earliest attested form of the word is the Mycenaean Greek ???, *re-po-to*, written in Linear B syllabic script.[16] *Lepton* was first used by physicist Léon Rosenfeld in 1948:[17]

> Following a suggestion of Prof. C. Møller, I adopt — as a pendant to "nucleon" — the denomination "lepton" (from λεπτός, small, thin, delicate) to denote a particle of small mass.

The etymology incorrectly implies that all the leptons are of small mass. When Rosenfeld named them, the only known leptons were electrons and muons, which are in fact of small mass — the mass of an electron (0.511 MeV/c^2)[18] and the mass of a muon (with a value of 105.7 MeV/c^2)[19] are fractions of the mass of the "heavy" proton (938.3 MeV/c^2).[20] However, the mass of the tau (discovered in the mid 1970s) (1777 MeV/c^2)[21] is nearly twice that of the proton, and about 3,500 times that of the electron.

7.2 History

See also: Electron § Discovery, Muon § History and Tau (particle) § History

The first lepton identified was the electron, discovered by J.J. Thomson and his team of British physicists in 1897.[22][23]

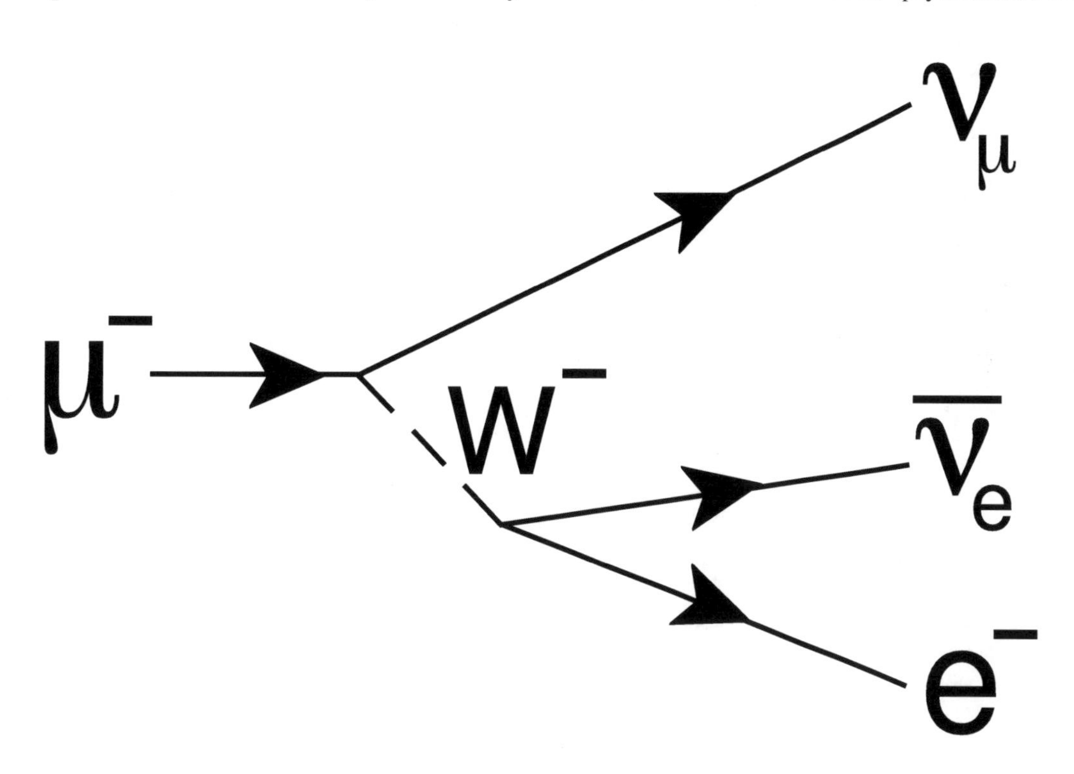

A muon transmutes into a muon neutrino by emitting a W− boson. The W− boson subsequently decays into an electron and an electron antineutrino.

Then in 1930 Wolfgang Pauli postulated the electron neutrino to preserve conservation of energy, conservation of momentum, and conservation of angular momentum in beta decay.[24] Pauli theorized that an undetected particle was carrying away the difference between the energy, momentum, and angular momentum of the initial and observed final particles. The electron neutrino was simply called the neutrino, as it was not yet known that neutrinos came in different flavours (or different "generations").

Nearly 40 years after the discovery of the electron, the muon was discovered by Carl D. Anderson in 1936. Due to its mass, it was initially categorized as a meson rather than a lepton.[25] It later became clear that the muon was much more similar to the electron than to mesons, as muons do not undergo the strong interaction, and thus the muon was reclassified: electrons, muons, and the (electron) neutrino were grouped into a new group of particles – the leptons. In 1962 Leon M. Lederman, Melvin Schwartz and Jack Steinberger showed that more than one type of neutrino exists by first detecting interactions of the muon neutrino, which earned them the 1988 Nobel Prize, although by then the different flavours of neutrino had already been theorized.[26]

The tau was first detected in a series of experiments between 1974 and 1977 by Martin Lewis Perl with his colleagues at the SLAC LBL group.[27] Like the electron and the muon, it too was expected to have an associated neutrino. The first evidence for tau neutrinos came from the observation of "missing" energy and momentum in tau decay, analogous to the "missing" energy and momentum in beta decay leading to the discovery of the electron neutrino. The first detection of tau neutrino interactions was announced in 2000 by the DONUT collaboration at Fermilab, making it the latest particle of the Standard Model to have been directly observed,[28] apart from the Higgs boson, which probably has been discovered in 2012.

Although all present data is consistent with three generations of leptons, some particle physicists are searching for a fourth generation. The current lower limit on the mass of such a fourth charged lepton is 100.8 GeV/c^2,[29] while its associated neutrino would have a mass of at least 45.0 GeV/c^2.[30]

7.3 Properties

7.3.1 Spin and chirality

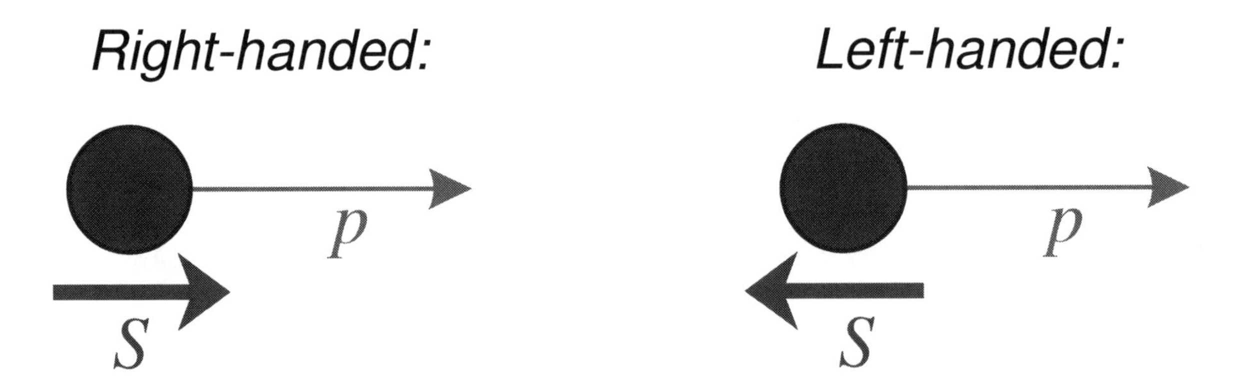

Left-handed and right-handed helicities

Leptons are spin-½ particles. The spin-statistics theorem thus implies that they are fermions and thus that they are subject to the Pauli exclusion principle; no two leptons of the same species can be in exactly the same state at the same time. Furthermore, it means that a lepton can have only two possible spin states, namely up or down.

A closely related property is chirality, which in turn is closely related to a more easily visualized property called helicity. The helicity of a particle is the direction of its spin relative to its momentum; particles with spin in the same direction as their momentum are called *right-handed* and otherwise they are called *left-handed*. When a particle is mass-less, the direction of its momentum relative to its spin is frame independent, while for massive particles it is possible to 'overtake' the particle by a Lorentz transformation flipping the helicity. Chirality is a technical property (defined through the transformation behaviour under the Poincaré group) that agrees with helicity for (approximately) massless particles and is still well defined for massive particles.

In many quantum field theories—such as quantum electrodynamics and quantum chromodynamics—left and right-handed fermions are identical. However in the Standard Model left-handed and right-handed fermions are treated asymmetrically. Only left-handed fermions participate in the weak interaction, while there are no right-handed neutrinos. This is an example of parity violation. In the literature left-handed fields are often denoted by a capital *L* subscript (e.g. e–L) and right-handed fields are denoted by a capital *R* subscript.

7.3.2 Electromagnetic interaction

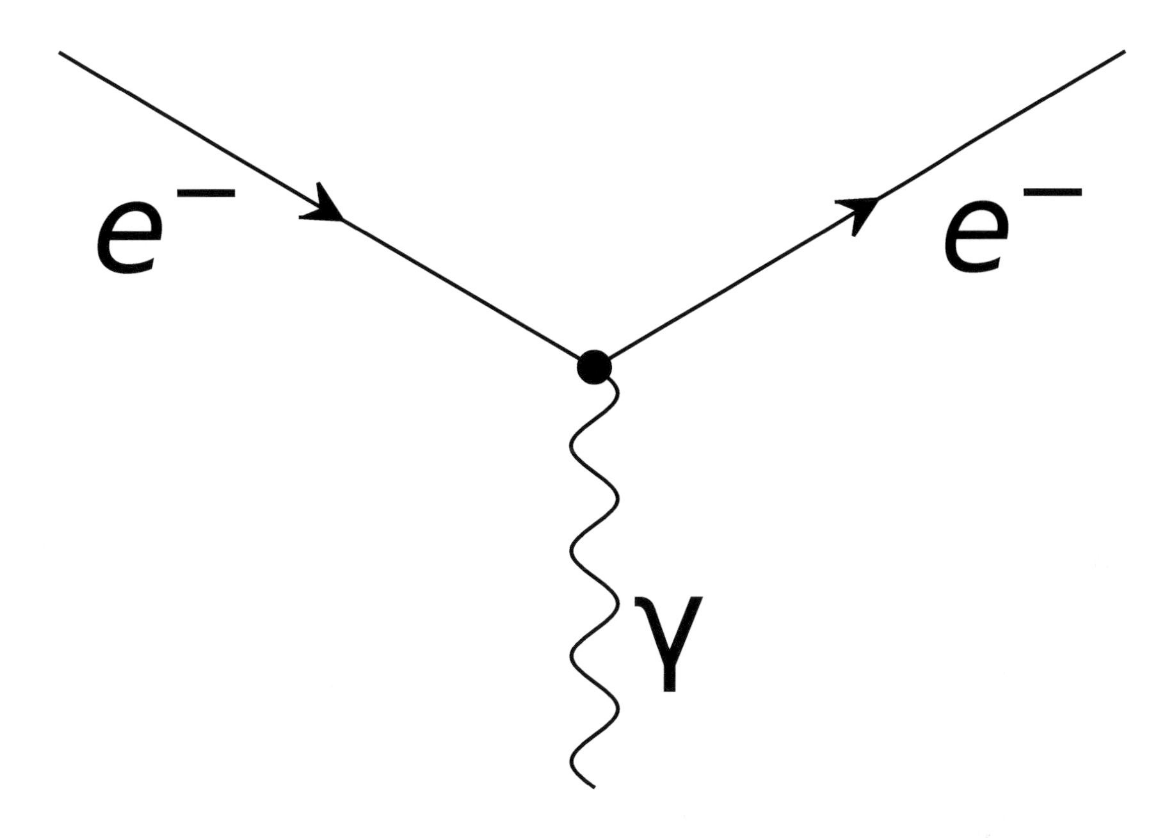

Lepton–photon interaction

One of the most prominent properties of leptons is their electric charge, *Q*. The electric charge determines the strength of their electromagnetic interactions. It determines the strength of the electric field generated by the particle (see Coulomb's law) and how strongly the particle reacts to an external electric or magnetic field (see Lorentz force). Each generation contains one lepton with $Q = -e$ (conventionally the charge of a particle is expressed in units of the elementary charge) and one lepton with zero electric charge. The lepton with electric charge is commonly simply referred to as a 'charged lepton' while the neutral lepton is called a neutrino. For example the first generation consists of the electron e– with a negative electric charge and the electrically neutral electron neutrino ν
e.

In the language of quantum field theory the electromagnetic interaction of the charged leptons is expressed by the fact that the particles interact with the quantum of the electromagnetic field, the photon. The Feynman diagram of the electron-photon interaction is shown on the right.

Because leptons possess an intrinsic rotation in the form of their spin, charged leptons generate a magnetic field. The size of their magnetic dipole moment μ is given by,

$$\mu = g\frac{Q\hbar}{4m},$$

where *m* is the mass of the lepton and *g* is the so-called g-factor for the lepton. First order approximation quantum mechanics predicts that the g-factor is 2 for all leptons. However, higher order quantum effects caused by loops in Feynman diagrams introduce corrections to this value. These corrections, referred to as the anomalous magnetic dipole moment, are very sensitive to the details of a quantum field theory model and thus provide the opportunity for precision tests of the standard model. The theoretical and measured values for the electron anomalous magnetic dipole moment are within agreement within eight significant figures.[31]

7.3.3 Weak Interaction

In the Standard Model the left-handed charged lepton and the left-handed neutrino are arranged in doublet (ν
eL, e−L) that transforms in the spinor representation (T = ½) of the weak isospin SU(2) gauge symmetry. This means that these particles are eigenstates of the isospin projection T_3 with eigenvalues ½ and −½ respectively. In the meantime, the right-handed charged lepton transforms as a weak isospin scalar (T = 0) and thus does not participate in the weak interaction, while there is no right-handed neutrino at all.

The Higgs mechanism recombines the gauge fields of the weak isospin SU(2) and the weak hypercharge U(1) symmetries to three massive vector bosons (W+, W−, Z0) mediating the weak interaction, and one massless vector boson, the photon, responsible for the electromagnetic interaction. The electric charge Q can be calculated from the isospin projection T_3 and weak hypercharge YW through the Gell-Mann–Nishijima formula,

$$Q = T_3 + YW/2$$

To recover the observed electric charges for all particles the left-handed weak isospin doublet (ν
eL, e−L) must thus have $YW = -1$, while the right-handed isospin scalar e−
R must have $YW = -2$. The interaction of the leptons with the massive weak interaction vector bosons is shown in the figure on the left.

7.3.4 Mass

In the Standard Model each lepton starts out with no intrinsic mass. The charged leptons (i.e. the electron, muon, and tau) obtain an effective mass through interaction with the Higgs field, but the neutrinos remain massless. For technical reasons the masslessness of the neutrinos implies that there is no mixing of the different generations of charged leptons as there is for quarks. This is in close agreement with current experimental observations.[32]

However, it is known from experiments – most prominently from observed neutrino oscillations[33] – that neutrinos do in fact have some very small mass, probably less than 2 eV/c^2.[34] This implies the existence of physics beyond the Standard Model. The currently most favoured extension is the so-called seesaw mechanism, which would explain both why the left-handed neutrinos are so light compared to the corresponding charged leptons, and why we have not yet seen any right-handed neutrinos.

7.3.5 Leptonic numbers

Main article: Lepton number

The members of each generation's weak isospin doublet are assigned leptonic numbers that are conserved under the Standard Model.[35] Electrons and electron neutrinos have an *electronic number* of $L_e = 1$, while muons and muon neutrinos have a *muonic number* of $L\mu = 1$, while tau particles and tau neutrinos have a *tauonic number* of $L\tau = 1$. The antileptons have their respective generation's leptonic numbers of −1.

Conservation of the leptonic numbers means that the number of leptons of the same type remains the same, when particles interact. This implies that leptons and antileptons must be created in pairs of a single generation. For example, the following processes are allowed under conservation of leptonic numbers:

e− + e+ → γ + γ,

$$\begin{pmatrix} \nu_e \\ e^- \end{pmatrix}, \begin{pmatrix} \nu_\mu \\ \mu^- \end{pmatrix}, \begin{pmatrix} \nu_\tau \\ \tau^- \end{pmatrix}$$

Each generation forms a weak isospin doublet.

τ− + τ+ → Z0 + Z0,

but not these:

γ → e− + μ+,

W− → e− + ν

τ,

Z0 → μ− + τ+.

However, neutrino oscillations are known to violate the conservation of the individual leptonic numbers. Such a violation is considered to be smoking gun evidence for physics beyond the Standard Model. A much stronger conservation law is the conservation of the total number of leptons (L), conserved even in the case of neutrino oscillations, but even it is still violated by a tiny amount by the chiral anomaly.

7.4 Universality

The coupling of the leptons to gauge bosons are flavour-independent (i.e., the interactions between leptons and gauge bosons are the same for all leptons).[35] This property is called *lepton universality* and has been tested in measurements of the tau and muon lifetimes and of Z boson partial decay widths, particularly at the Stanford Linear Collider (SLC) and Large Electron-Positron Collider (LEP) experiments.[36]:241–243[37]:138

The decay rate (Γ) of muons through the process μ− → e− + ν
e + ν
μ is approximately given by an expression of the form (see muon decay for more details)[35]

$$\Gamma\left(\mu^- \to e^- + \bar{\nu_e} + \nu_\mu\right) = K_1 G_F^2 m_\mu^5,$$

where K_1 is some constant, and GF is the Fermi coupling constant. The decay rate of tau particles through the process
τ− → e− + ν
e + ν
τ is given by an expression of the same form[35]

$$\Gamma\left(\tau^- \to e^- + \bar{\nu_e} + \nu_\tau\right) = K_2 G_F^2 m_\tau^5,$$

where K_2 is some constant. Muon–Tauon universality implies that $K_1 = K_2$. On the other hand, electron–muon universality implies[35]

$$\Gamma\left(\tau^- \to e^- + \bar{\nu_e} + \nu_\tau\right) = \Gamma\left(\tau^- \to \mu^- + \bar{\nu_\mu} + \nu_\tau\right).$$

This explains why the branching ratios for the electronic mode (17.85%) and muonic (17.36%) mode of tau decay are equal (within error).[21]

Universality also accounts for the ratio of muon and tau lifetimes. The lifetime of a lepton (τ_l) is related to the decay rate by[35]

$$\tau_l = \frac{B\left(l^- \to e^- + \bar{\nu_e} + \nu_l\right)}{\Gamma\left(l^- \to e^- + \bar{\nu_e} + \nu_l\right)},$$

where B(x → y) and Γ(x → y) denotes the branching ratios and the resonance width of the process x → y.

The ratio of tau and muon lifetime is thus given by[35]

$$\frac{\tau_\tau}{\tau_\mu} = \frac{B\left(\tau^- \to e^- + \bar{\nu_e} + \nu_\tau\right)}{B\left(\mu^- \to e^- + \bar{\nu_e} + \nu_\mu\right)} \left(\frac{m_\mu}{m_\tau}\right)^5.$$

Using the values of the 2008 *Review of Particle Physics* for the branching ratios of muons[19] and tau[21] yields a lifetime ratio of ~1.29×10^{-7}, comparable to the measured lifetime ratio of ~1.32×10^{-7}. The difference is due to K_1 and K_2 not actually being constants; they depend on the mass of leptons.

7.5 Table of leptons

7.6 See also

- Koide formula
- List of particles
- Preons – hypothetical particles which were once postulated to be subcomponents of quarks and leptons

7.7 Notes

[1] "Lepton (physics)". *Encyclopædia Britannica*. Retrieved 2010-09-29.

[2] R. Nave. "Leptons". *HyperPhysics*. Georgia State University, Department of Physics and Astronomy. Retrieved 2010-09-29.

[3] W.V. Farrar (1969). "Richard Laming and the Coal-Gas Industry, with His Views on the Structure of Matter". *Annals of Science* **25** (3): 243–254. doi:10.1080/00033796900200141.

[4] T. Arabatzis (2006). *Representing Electrons: A Biographical Approach to Theoretical Entities*. University of Chicago Press. pp. 70–74. ISBN 0-226-02421-0.

[5] J.Z. Buchwald, A. Warwick (2001). *Histories of the Electron: The Birth of Microphysics*. MIT Press. pp. 195–203. ISBN 0-262-52424-4.

[6] J.J. Thomson (1897). "Cathode Rays". *Philosophical Magazine* **44** (269): 293. doi:10.1080/14786449708621070.

[7] S.H. Neddermeyer, C.D. Anderson; Anderson (1937). "Note on the Nature of Cosmic-Ray Particles". *Physical Review* **51** (10): 884–886. Bibcode:1937PhRv...51..884N. doi:10.1103/PhysRev.51.884.

[8] "The Reines-Cowan Experiments: Detecting the Poltergeist" (PDF). *Los Alamos Science* **25**: 3. 1997. Retrieved 2010-02-10.

[9] F. Reines, C.L. Cowan, Jr.; Cowan (1956). "The Neutrino". *Nature* **178** (4531): 446. Bibcode:1956Natur.178..446R. doi:10.1038/178446a0.

[10] G. Danby; Gaillard, J-M.; Goulianos, K.; Lederman, L.; Mistry, N.; Schwartz, M.; Steinberger, J. et al. (1962). "Observation of high-energy neutrino reactions and the existence of two kinds of neutrinos". *Physical Review Letters* **9**: 36. Bibcode:1962PhRvL...9...36D. doi:10.1103/PhysRevLett.9.36.

[11] M.L. Perl; Abrams, G.; Boyarski, A.; Breidenbach, M.; Briggs, D.; Bulos, F.; Chinowsky, W.; Dakin, J.; Feldman, G.; Friedberg, C.; Fryberger, D.; Goldhaber, G.; Hanson, G.; Heile, F.; Jean-Marie, B.; Kadyk, J.; Larsen, R.; Litke, A.; Lüke, D.; Lulu, B.; Lüth, V.; Lyon, D.; Morehouse, C.; Paterson, J.; Pierre, F.; Pun, T.; Rapidis, P.; Richter, B.; Sadoulet, B. et al. (1975). "Evidence for Anomalous Lepton Production in e+e− Annihilation". *Physical Review Letters* **35** (22): 1489. Bibcode:1975PhRvL..35.1489P. doi:10.1103/PhysRevLett.35.1489.

[12] "Physicists Find First Direct Evidence for Tau Neutrino at Fermilab" (Press release). Fermilab. 20 July 2000.

[13] K. Kodama *et al.* (DONUT Collaboration); Kodama; Ushida; Andreopoulos; Saoulidou; Tzanakos; Yager; Baller; Boehnlein; Freeman; Lundberg; Morfin; Rameika; Yun; Song; Yoon; Chung; Berghaus; Kubantsev; Reay; Sidwell; Stanton; Yoshida; Aoki; Hara; Rhee; Ciampa; Erickson; Graham et al. (2001). "Observation of tau neutrino interactions". *Physics Letters B* **504** (3): 218. arXiv:hep-ex/0012035. Bibcode:2001PhLB..504..218D. doi:10.1016/S0370-2693(01)00307-0.

[14] "lepton". *Online Etymology Dictionary*.

[15] λεπτός. Liddell, Henry George; Scott, Robert; *A Greek–English Lexicon* at the Perseus Project.

[16] Found on the KN L 693 and PY Un 1322 tablets. "The Linear B word re-po-to". Palaeolexicon. Word study tool of ancient languages. Raymoure, K.A. "re-po-to". *Minoan Linear A & Mycenaean Linear B*. Deaditerranean. "KN 693 L (103)". "PY 1322 Un + fr. (Cii)". *DĀMOS: Database of Mycenaean at Oslo*. University of Oslo.

[17] L. Rosenfeld (1948)

[18] C. Amsler *et al.* (2008): Particle listings – e−

[19] C. Amsler *et al.* (2008): Particle listings – μ−

[20] C. Amsler *et al.* (2008): Particle listings – p+

[21] C. Amsler *et al.* (2008): Particle listings – τ−

[22] S. Weinberg (2003)

[23] R. Wilson (1997)

[24] K. Riesselmann (2007)

[25] S.H. Neddermeyer, C.D. Anderson (1937)

[26] I.V. Anicin (2005)

[27] M.L. Perl et al. (1975)

[28] K. Kodama (2001)

[29] C. Amsler *et al.* (2008) Heavy Charged Leptons Searches

[30] C. Amsler *et al.* (2008) Searches for Heavy Neutral Leptons

[31] M.E. Peskin, D.V. Schroeder (1995), p. 197

[32] M.E. Peskin, D.V. Schroeder (1995), p. 27

[33] Y. Fukuda *et al.* (1998)

[34] C.Amsler et al. (2008): Particle listings – Neutrino properties

[35] B.R. Martin, G. Shaw (1992)

[36] J. P. Cumalat (1993). *Physics in Collision 12*. Atlantica Séguier Frontières. ISBN 978-2-86332-129-4.

[37] G Fraser (1 January 1998). *The Particle Century*. CRC Press. ISBN 978-1-4200-5033-2.

[38] J. Peltoniemi, J. Sarkamo (2005)

7.8 References

- C. Amsler *et al.* (Particle Data Group); Amsler; Doser; Antonelli; Asner; Babu; Baer; Band; Barnett; Bergren; Beringer; Bernardi; Bertl; Bichsel; Biebel; Bloch; Blucher; Blusk; Cahn; Carena; Caso; Ceccucci; Chakraborty; Chen; Chivukula; Cowan; Dahl; d'Ambrosio; Damour et al. (2008). "Review of Particle Physics". *Physics Letters B* **667**: 1. Bibcode:2008PhLB..667....1P. doi:10.1016/j.physletb.2008.07.018.
- I.V. Anicin (2005). "The Neutrino – Its Past, Present and Future". *SFIN (Institute of Physics, Belgrade) year XV, Series A: Conferences, No. A2 (2002) 3–59*: 3172. arXiv:physics/0503172. Bibcode:2005physics...3172A.
- Y.Fukuda; Hayakawa, T.; Ichihara, E.; Inoue, K.; Ishihara, K.; Ishino, H.; Itow, Y.; Kajita, T. et al. (1998). "Evidence for Oscillation of Atmospheric Neutrinos". *Physical Review Letters* **81** (8): 1562–1567. arXiv:hep-ex/9807003. Bibcode:1998PhRvL..81.1562F. doi:10.1103/PhysRevLett.81.1562.
- K. Kodama; Ushida, N.; Andreopoulos, C.; Saoulidou, N.; Tzanakos, G.; Yager, P.; Baller, B.; Boehnlein, D.; Freeman, W.; Lundberg, B.; Morfin, J.; Rameika, R.; Yun, J.C.; Song, J.S.; Yoon, C.S.; Chung, S.H.; Berghaus, P.; Kubantsev, M.; Reay, N.W.; Sidwell, R.; Stanton, N.; Yoshida, S.; Aoki, S.; Hara, T.; Rhee, J.T.; Ciampa, D.; Erickson, C.; Graham, M.; Heller, K. et al. (2001). "Observation of tau neutrino interactions". *Physics Letters B* **504** (3): 218. arXiv:hep-ex/0012035. Bibcode:2001PhLB..504..218D. doi:10.1016/S0370-2693(01)00307-0.
- B.R. Martin, G. Shaw (1992). "Chapter 2 – Leptons, quarks and hadrons". *Particle Physics*. John Wiley & Sons. pp. 23–47. ISBN 0-471-92358-3.
- S.H. Neddermeyer, C.D. Anderson; Anderson (1937). "Note on the Nature of Cosmic-Ray Particles". *Physical Review* **51** (10): 884–886. Bibcode:1937PhRv...51..884N. doi:10.1103/PhysRev.51.884.
- J. Peltoniemi, J. Sarkamo (2005). "Laboratory measurements and limits for neutrino properties". *The Ultimate Neutrino Page*. Retrieved 2008-11-07.
- M.L. Perl; Abrams, G.; Boyarski, A.; Breidenbach, M.; Briggs, D.; Bulos, F.; Chinowsky, W.; Dakin, J. et al. (1975). "Evidence for Anomalous Lepton Production in e^+–e^- Annihilation". *Physical Review Letters* **35** (22): 1489–1492. Bibcode:1975PhRvL..35.1489P. doi:10.1103/PhysRevLett.35.1489.
- M.E. Peskin, D.V. Schroeder (1995). *Introduction to Quantum Field Theory*. Westview Press. ISBN 0-201-50397-2.
- K. Riesselmann (2007). "Logbook: Neutrino Invention". *Symmetry Magazine* **4** (2).
- L. Rosenfeld (1948). *Nuclear Forces*. Interscience Publishers. p. xvii.
- R. Shankar (1994). "Chapter 2 – Rotational Invariance and Angular Momentum". *Principles of Quantum Mechanics* (2nd ed.). Springer. pp. 305–352. ISBN 978-0-306-44790-7.
- S. Weinberg (2003). *The Discovery of Subatomic Particles*. Cambridge University Press. ISBN 0-521-82351-X.
- R. Wilson (1997). *Astronomy Through the Ages: The Story of the Human Attempt to Understand the Universe*. CRC Press. p. 138. ISBN 0-7484-0748-0.

7.9 External links

- Particle Data Group homepage. The PDG compiles authoritative information on particle properties.
- Leptons, a summary of leptons from *Hyperphysics*.

Chapter 8

Quark

This article is about the particle. For other uses, see Quark (disambiguation).

A **quark** (/ˈkwɔrk/ or /ˈkwɑrk/) is an elementary particle and a fundamental constituent of matter. Quarks combine to form composite particles called hadrons, the most stable of which are protons and neutrons, the components of atomic nuclei.[1] Due to a phenomenon known as *color confinement*, quarks are never directly observed or found in isolation; they can be found only within hadrons, such as baryons (of which protons and neutrons are examples), and mesons.[2][3] For this reason, much of what is known about quarks has been drawn from observations of the hadrons themselves.

Quarks have various intrinsic properties, including electric charge, mass, color charge and spin. Quarks are the only elementary particles in the Standard Model of particle physics to experience all four fundamental interactions, also known as *fundamental forces* (electromagnetism, gravitation, strong interaction, and weak interaction), as well as the only known particles whose electric charges are not integer multiples of the elementary charge.

There are six types of quarks, known as *flavors*: up, down, strange, charm, top, and bottom.[4] Up and down quarks have the lowest masses of all quarks. The heavier quarks rapidly change into up and down quarks through a process of particle decay: the transformation from a higher mass state to a lower mass state. Because of this, up and down quarks are generally stable and the most common in the universe, whereas strange, charm, bottom, and top quarks can only be produced in high energy collisions (such as those involving cosmic rays and in particle accelerators). For every quark flavor there is a corresponding type of antiparticle, known as an *antiquark*, that differs from the quark only in that some of its properties have equal magnitude but opposite sign.

The quark model was independently proposed by physicists Murray Gell-Mann and George Zweig in 1964.[5] Quarks were introduced as parts of an ordering scheme for hadrons, and there was little evidence for their physical existence until deep inelastic scattering experiments at the Stanford Linear Accelerator Center in 1968.[6][7] Accelerator experiments have provided evidence for all six flavors. The top quark was the last to be discovered at Fermilab in 1995.[5]

8.1 Classification

See also: Standard Model

The Standard Model is the theoretical framework describing all the currently known elementary particles. This model contains six flavors of quarks (q), named up (u), down (d), strange (s), charm (c), bottom (b), and top (t).[4] Antiparticles of quarks are called *antiquarks*, and are denoted by a bar over the symbol for the corresponding quark, such as u for an up antiquark. As with antimatter in general, antiquarks have the same mass, mean lifetime, and spin as their respective quarks, but the electric charge and other charges have the opposite sign.[8]

Quarks are spin-$\frac{1}{2}$ particles, implying that they are fermions according to the spin-statistics theorem. They are subject to the Pauli exclusion principle, which states that no two identical fermions can simultaneously occupy the same quantum state. This is in contrast to bosons (particles with integer spin), any number of which can be in the same state.[9] Unlike

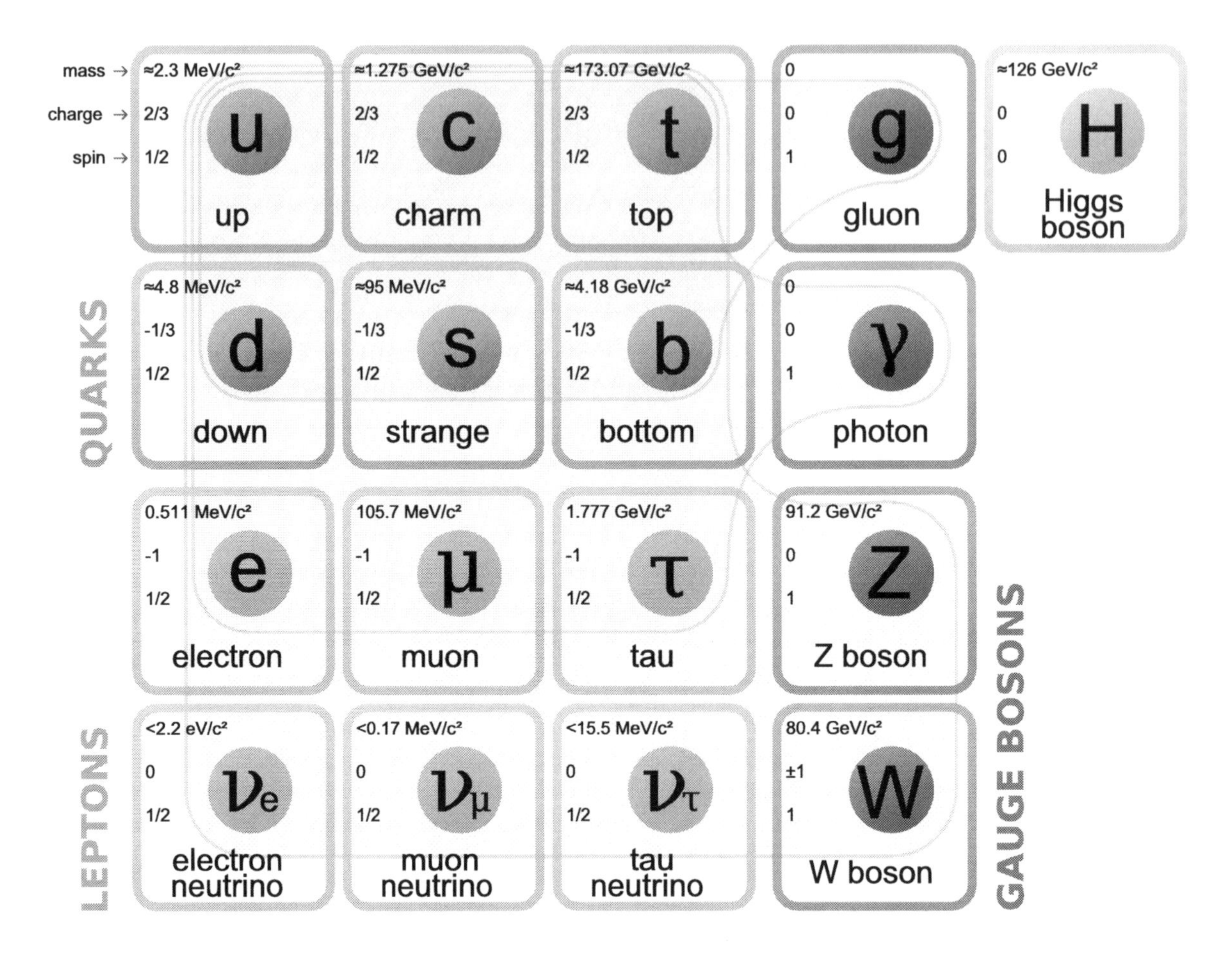

Six of the particles in the Standard Model are quarks (shown in purple). Each of the first three columns forms a generation *of matter.*

leptons, quarks possess color charge, which causes them to engage in the strong interaction. The resulting attraction between different quarks causes the formation of composite particles known as *hadrons* (see "Strong interaction and color charge" below).

The quarks which determine the quantum numbers of hadrons are called *valence quarks*; apart from these, any hadron may contain an indefinite number of virtual (or *sea*) quarks, antiquarks, and gluons which do not influence its quantum numbers.[10] There are two families of hadrons: baryons, with three valence quarks, and mesons, with a valence quark and an antiquark.[11] The most common baryons are the proton and the neutron, the building blocks of the atomic nucleus.[12] A great number of hadrons are known (see list of baryons and list of mesons), most of them differentiated by their quark content and the properties these constituent quarks confer. The existence of “exotic” hadrons with more valence quarks, such as tetraquarks (qqqq) and pentaquarks (qqqqq), has been conjectured[13] but not proven.[nb 1][13][14] However, on 13 July 2015, the LHCb collaboration at CERN reported results consistent with pentaquark states.[15]

Elementary fermions are grouped into three generations, each comprising two leptons and two quarks. The first generation includes up and down quarks, the second strange and charm quarks, and the third bottom and top quarks. All searches for a fourth generation of quarks and other elementary fermions have failed,[16] and there is strong indirect evidence that no more than three generations exist.[nb 2][17] Particles in higher generations generally have greater mass and less stability, causing them to decay into lower-generation particles by means of weak interactions. Only first-generation (up and down) quarks occur commonly in nature. Heavier quarks can only be created in high-energy collisions (such as in those involving cosmic rays), and decay quickly; however, they are thought to have been present during the first fractions of a second after the Big Bang, when the universe was in an extremely hot and dense phase (the quark epoch). Studies of heavier quarks are conducted in artificially created conditions, such as in particle accelerators.[18]

Having electric charge, mass, color charge, and flavor, quarks are the only known elementary particles that engage in all four fundamental interactions of contemporary physics: electromagnetism, gravitation, strong interaction, and weak interaction.[12] Gravitation is too weak to be relevant to individual particle interactions except at extremes of energy (Planck energy) and distance scales (Planck distance). However, since no successful quantum theory of gravity exists, gravitation is not described by the Standard Model.

See the table of properties below for a more complete overview of the six quark flavors' properties.

8.2 History

The quark model was independently proposed by physicists Murray Gell-Mann[19] (pictured) and George Zweig[20][21] in 1964.[5] The proposal came shortly after Gell-Mann's 1961 formulation of a particle classification system known as the *Eightfold Way*—or, in more technical terms, SU(3) flavor symmetry.[22] Physicist Yuval Ne'eman had independently developed a scheme similar to the Eightfold Way in the same year.[23][24]

At the time of the quark theory's inception, the "particle zoo" included, amongst other particles, a multitude of hadrons. Gell-Mann and Zweig posited that they were not elementary particles, but were instead composed of combinations of quarks and antiquarks. Their model involved three flavors of quarks, up, down, and strange, to which they ascribed properties such as spin and electric charge.[19][20][21] The initial reaction of the physics community to the proposal was mixed. There was particular contention about whether the quark was a physical entity or a mere abstraction used to explain concepts that were not fully understood at the time.[25]

In less than a year, extensions to the Gell-Mann–Zweig model were proposed. Sheldon Lee Glashow and James Bjorken predicted the existence of a fourth flavor of quark, which they called *charm*. The addition was proposed because it allowed for a better description of the weak interaction (the mechanism that allows quarks to decay), equalized the number of known quarks with the number of known leptons, and implied a mass formula that correctly reproduced the masses of the known mesons.[26]

In 1968, deep inelastic scattering experiments at the Stanford Linear Accelerator Center (SLAC) showed that the proton contained much smaller, point-like objects and was therefore not an elementary particle.[6][7][27] Physicists were reluctant to firmly identify these objects with quarks at the time, instead calling them "partons"—a term coined by Richard Feynman.[28][29][30] The objects that were observed at SLAC would later be identified as up and down quarks as the other flavors were discovered.[31] Nevertheless, "parton" remains in use as a collective term for the constituents of hadrons (quarks, antiquarks, and gluons).

The strange quark's existence was indirectly validated by SLAC's scattering experiments: not only was it a necessary component of Gell-Mann and Zweig's three-quark model, but it provided an explanation for the kaon (K) and pion (π) hadrons discovered in cosmic rays in 1947.[32]

In a 1970 paper, Glashow, John Iliopoulos and Luciano Maiani presented further reasoning for the existence of the as-yet undiscovered charm quark.[33][34] The number of supposed quark flavors grew to the current six in 1973, when Makoto Kobayashi and Toshihide Maskawa noted that the experimental observation of CP violation[nb 3][35] could be explained if there were another pair of quarks.

Charm quarks were produced almost simultaneously by two teams in November 1974 (see November Revolution)—one at SLAC under Burton Richter, and one at Brookhaven National Laboratory under Samuel Ting. The charm quarks were observed bound with charm antiquarks in mesons. The two parties had assigned the discovered meson two different symbols, J and ψ; thus, it became formally known as the J/ψ meson. The discovery finally convinced the physics community of the quark model's validity.[30]

In the following years a number of suggestions appeared for extending the quark model to six quarks. Of these, the 1975 paper by Haim Harari[36] was the first to coin the terms *top* and *bottom* for the additional quarks.[37]

In 1977, the bottom quark was observed by a team at Fermilab led by Leon Lederman.[38][39] This was a strong indicator of the top quark's existence: without the top quark, the bottom quark would have been without a partner. However, it was not until 1995 that the top quark was finally observed, also by the CDF[40] and DØ[41] teams at Fermilab.[5] It had a mass much larger than had been previously expected,[42] almost as large as that of a gold atom.[43]

Murray Gell-Mann at TED in 2007. Gell-Mann and George Zweig proposed the quark model in 1964.

8.3 Etymology

For some time, Gell-Mann was undecided on an actual spelling for the term he intended to coin, until he found the word *quark* in James Joyce's book *Finnegans Wake*:

> Three quarks for Muster Mark!
> Sure he has not got much of a bark

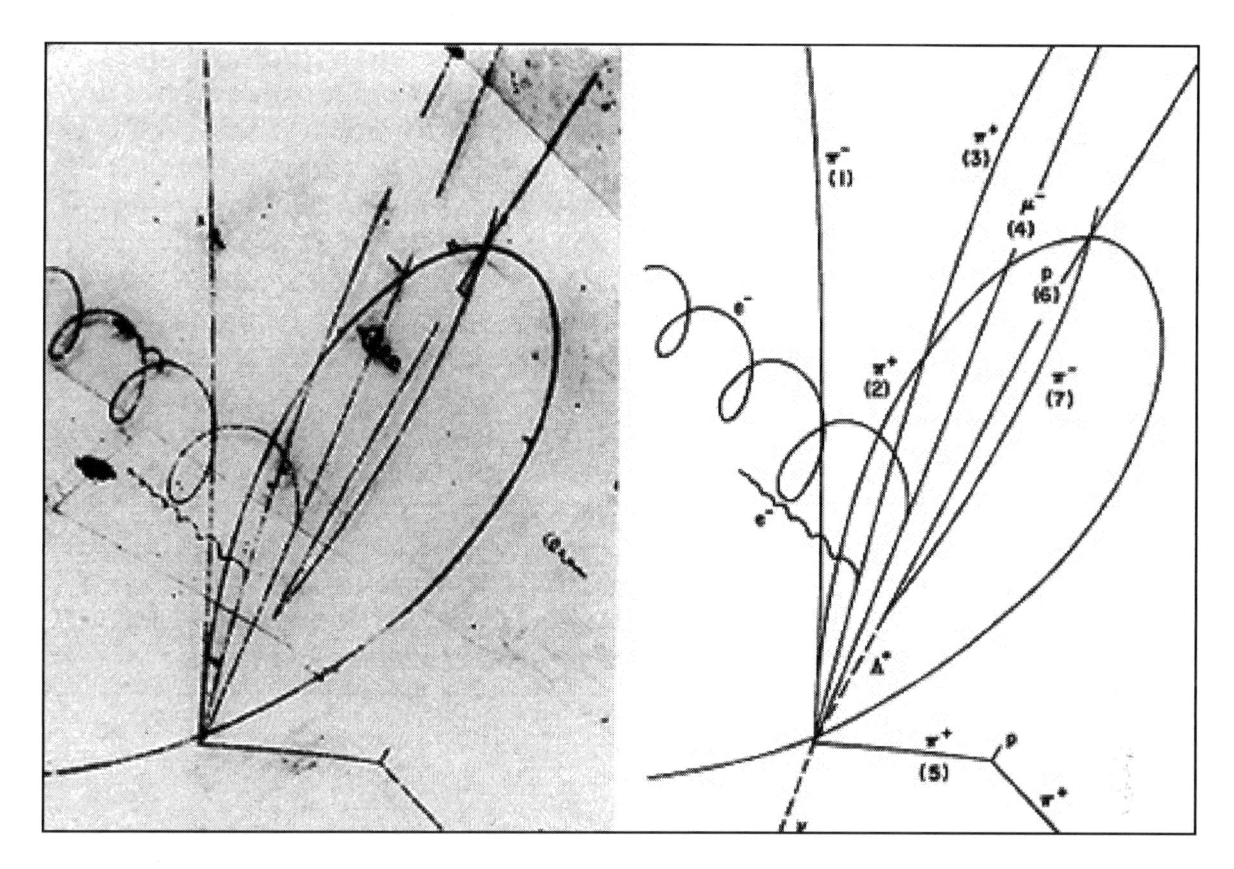

Photograph of the event that led to the discovery of the Σ++ c baryon, at the Brookhaven National Laboratory in 1974

> And sure any he has it's all beside the mark.
> —James Joyce, *Finnegans Wake*[44]

Gell-Mann went into further detail regarding the name of the quark in his book *The Quark and the Jaguar*:[45]

> In 1963, when I assigned the name "quark" to the fundamental constituents of the nucleon, I had the sound first, without the spelling, which could have been "kwork". Then, in one of my occasional perusals of *Finnegans Wake*, by James Joyce, I came across the word "quark" in the phrase "Three quarks for Muster Mark". Since "quark" (meaning, for one thing, the cry of the gull) was clearly intended to rhyme with "Mark", as well as "bark" and other such words, I had to find an excuse to pronounce it as "kwork". But the book represents the dream of a publican named Humphrey Chimpden Earwicker. Words in the text are typically drawn from several sources at once, like the "portmanteau" words in "Through the Looking-Glass". From time to time, phrases occur in the book that are partially determined by calls for drinks at the bar. I argued, therefore, that perhaps one of the multiple sources of the cry "Three quarks for Muster Mark" might be "Three quarts for Mister Mark", in which case the pronunciation "kwork" would not be totally unjustified. In any case, the number three fitted perfectly the way quarks occur in nature.

Zweig preferred the name *ace* for the particle he had theorized, but Gell-Mann's terminology came to prominence once the quark model had been commonly accepted.[46]

The quark flavors were given their names for a number of reasons. The up and down quarks are named after the up and down components of isospin, which they carry.[47] Strange quarks were given their name because they were discovered to be components of the strange particles discovered in cosmic rays years before the quark model was proposed; these particles were deemed "strange" because they had unusually long lifetimes.[48] Glashow, who coproposed charm quark with Bjorken, is quoted as saying, “We called our construct the 'charmed quark', for we were fascinated and pleased by the symmetry it brought to the subnuclear world.”[49] The names “bottom” and “top”, coined by Harari, were chosen because they are “logical partners for up and down quarks”.[36][37][48] In the past, bottom and top quarks were sometimes referred to as “beauty” and “truth” respectively, but these names have somewhat fallen out of use.[50] While “truth” never did catch on, accelerator complexes devoted to massive production of bottom quarks are sometimes called "beauty factories".[51]

8.4 Properties

8.4.1 Electric charge

See also: Electric charge

Quarks have fractional electric charge values – either $\frac{1}{3}$ or $\frac{2}{3}$ times the elementary charge (e), depending on flavor. Up, charm, and top quarks (collectively referred to as *up-type quarks*) have a charge of $+\frac{2}{3}$ e, while down, strange, and bottom quarks (*down-type quarks*) have $-\frac{1}{3}$ e. Antiquarks have the opposite charge to their corresponding quarks; up-type antiquarks have charges of $-\frac{2}{3}$ e and down-type antiquarks have charges of $+\frac{1}{3}$ e. Since the electric charge of a hadron is the sum of the charges of the constituent quarks, all hadrons have integer charges: the combination of three quarks (baryons), three antiquarks (antibaryons), or a quark and an antiquark (mesons) always results in integer charges.[52] For example, the hadron constituents of atomic nuclei, neutrons and protons, have charges of 0 e and +1 e respectively; the neutron is composed of two down quarks and one up quark, and the proton of two up quarks and one down quark.[12]

8.4.2 Spin

See also: Spin (physics)

Spin is an intrinsic property of elementary particles, and its direction is an important degree of freedom. It is sometimes visualized as the rotation of an object around its own axis (hence the name "spin"), though this notion is somewhat misguided at subatomic scales because elementary particles are believed to be point-like.[53]

Spin can be represented by a vector whose length is measured in units of the reduced Planck constant $\hbar$ (pronounced “h bar”). For quarks, a measurement of the spin vector component along any axis can only yield the values $+\hbar/2$ or $-\hbar/2$; for this reason quarks are classified as spin-$\frac{1}{2}$ particles.[54] The component of spin along a given axis – by convention the z axis – is often denoted by an up arrow ↑ for the value $+\frac{1}{2}$ and down arrow ↓ for the value $-\frac{1}{2}$, placed after the symbol for flavor. For example, an up quark with a spin of $+\frac{1}{2}$ along the z axis is denoted by u↑.[55]

8.4.3 Weak interaction

Main article: Weak interaction

A quark of one flavor can transform into a quark of another flavor only through the weak interaction, one of the four fundamental interactions in particle physics. By absorbing or emitting a W boson, any up-type quark (up, charm, and top quarks) can change into any down-type quark (down, strange, and bottom quarks) and vice versa. This flavor transformation mechanism causes the radioactive process of beta decay, in which a neutron (n) “splits” into a proton (p), an electron (e−) and an electron antineutrino (v
e) (see picture). This occurs when one of the down quarks in the neutron (udd) decays into an up quark by emitting a virtual W− boson, transforming the neutron into a proton (uud). The W− boson then decays into an electron and an electron antineutrino.[56]

Both beta decay and the inverse process of *inverse beta decay* are routinely used in medical applications such as positron emission tomography (PET) and in experiments involving neutrino detection.

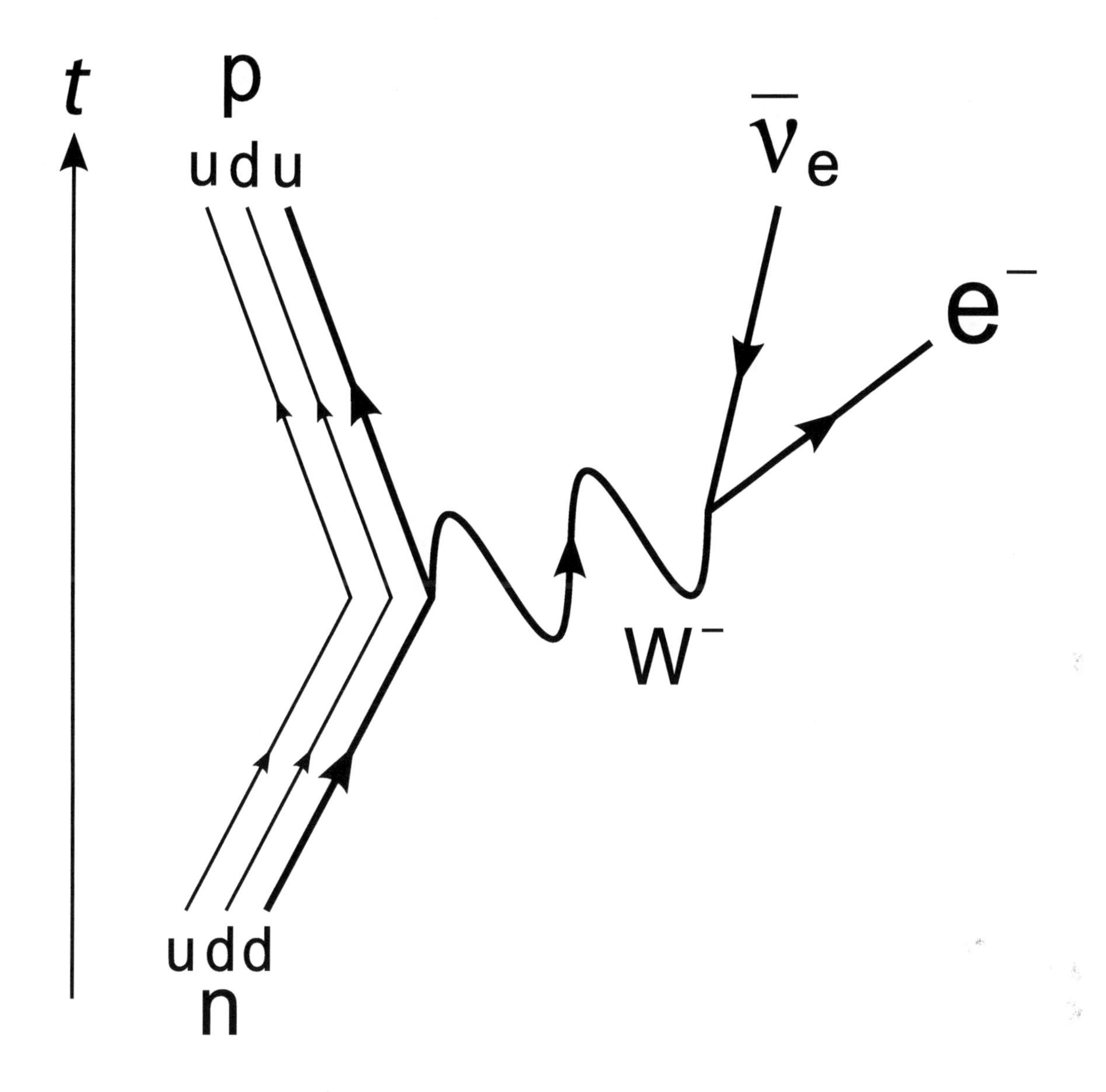

Feynman diagram of beta decay with time flowing upwards. The CKM matrix (discussed below) encodes the probability of this and other quark decays.

While the process of flavor transformation is the same for all quarks, each quark has a preference to transform into the quark of its own generation. The relative tendencies of all flavor transformations are described by a mathematical table, called the Cabibbo–Kobayashi–Maskawa matrix (CKM matrix). Enforcing unitarity, the approximate magnitudes of the entries of the CKM matrix are:[57]

$$\begin{bmatrix} |V_{\mathrm{ud}}| & |V_{\mathrm{us}}| & |V_{\mathrm{ub}}| \\ |V_{\mathrm{cd}}| & |V_{\mathrm{cs}}| & |V_{\mathrm{cb}}| \\ |V_{\mathrm{td}}| & |V_{\mathrm{ts}}| & |V_{\mathrm{tb}}| \end{bmatrix} \approx \begin{bmatrix} 0.974 & 0.225 & 0.003 \\ 0.225 & 0.973 & 0.041 \\ 0.009 & 0.040 & 0.999 \end{bmatrix},$$

where Vij represents the tendency of a quark of flavor i to change into a quark of flavor j (or vice versa).[nb 4]

There exists an equivalent weak interaction matrix for leptons (right side of the W boson on the above beta decay diagram), called the Pontecorvo–Maki–Nakagawa–Sakata matrix (PMNS matrix).[58] Together, the CKM and PMNS matrices

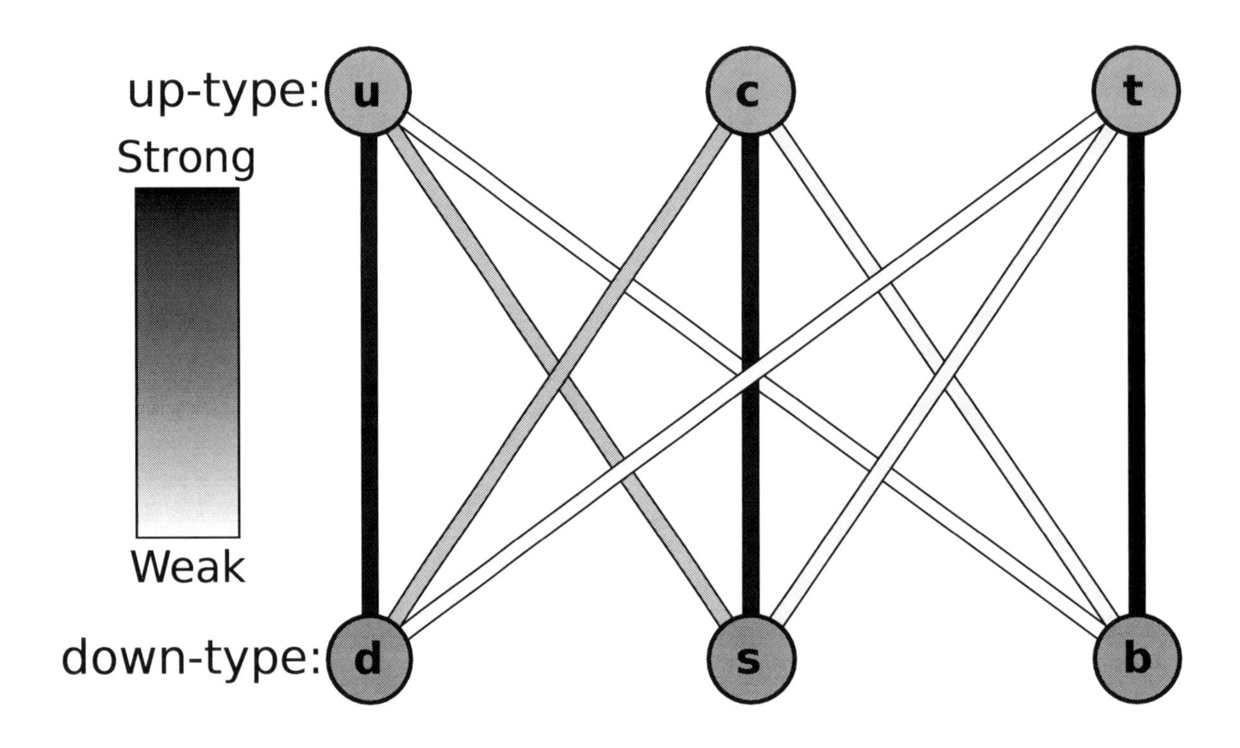

The strengths of the weak interactions between the six quarks. The "intensities" of the lines are determined by the elements of the CKM matrix.

describe all flavor transformations, but the links between the two are not yet clear.[59]

8.4.4 Strong interaction and color charge

See also: Color charge and Strong interaction

According to quantum chromodynamics (QCD), quarks possess a property called *color charge*. There are three types of color charge, arbitrarily labeled *blue*, *green*, and *red*.[nb 5] Each of them is complemented by an anticolor – *antiblue*, *antigreen*, and *antired*. Every quark carries a color, while every antiquark carries an anticolor.[60]

The system of attraction and repulsion between quarks charged with different combinations of the three colors is called strong interaction, which is mediated by force carrying particles known as *gluons*; this is discussed at length below. The theory that describes strong interactions is called quantum chromodynamics (QCD). A quark, which will have a single color value, can form a bound system with an antiquark carrying the corresponding anticolor. The result of two attracting quarks will be color neutrality: a quark with color charge ξ plus an antiquark with color charge $-\xi$ will result in a color charge of 0 (or "white" color) and the formation of a meson. This is analogous to the additive color model in basic optics. Similarly, the combination of three quarks, each with different color charges, or three antiquarks, each with anticolor charges, will result in the same "white" color charge and the formation of a baryon or antibaryon.[61]

In modern particle physics, gauge symmetries – a kind of symmetry group – relate interactions between particles (see gauge theories). Color SU(3) (commonly abbreviated to $SU(3)_c$) is the gauge symmetry that relates the color charge in quarks and is the defining symmetry for quantum chromodynamics.[62] Just as the laws of physics are independent of which directions in space are designated x, y, and z, and remain unchanged if the coordinate axes are rotated to a new orientation, the physics of quantum chromodynamics is independent of which directions in three-dimensional color space are identified as blue, red, and green. $SU(3)_c$ color transformations correspond to "rotations" in color space (which, mathematically speaking, is a complex space). Every quark flavor f, each with subtypes fB, fG, fR corresponding to the quark colors,[63] forms a triplet: a three-component quantum field which transforms under the fundamental representation of $SU(3)_c$.[64] The requirement that $SU(3)_c$ should be local – that is, that its transformations be allowed to vary with space and time – determines the properties of the strong interaction, in particular the existence of eight gluon types to act as its force carriers.[62][65]

8.4.5 Mass

See also: Invariant mass

Two terms are used in referring to a quark's mass: *current quark mass* refers to the mass of a quark by itself, while *constituent quark mass* refers to the current quark mass plus the mass of the gluon particle field surrounding the quark.[66] These masses typically have very different values. Most of a hadron's mass comes from the gluons that bind the constituent quarks together, rather than from the quarks themselves. While gluons are inherently massless, they possess energy – more specifically, quantum chromodynamics binding energy (QCBE) – and it is this that contributes so greatly to the overall mass of the hadron (see mass in special relativity). For example, a proton has a mass of approximately 938 MeV/c^2, of which the rest mass of its three valence quarks only contributes about 11 MeV/c^2; much of the remainder can be attributed to the gluons' QCBE.[67][68]

The Standard Model posits that elementary particles derive their masses from the Higgs mechanism, which is related to the Higgs boson. Physicists hope that further research into the reasons for the top quark's large mass of ~173 GeV/c^2, almost the mass of a gold atom,[67][69] might reveal more about the origin of the mass of quarks and other elementary particles.[70]

8.4.6 Table of properties

See also: Flavor (particle physics)

The following table summarizes the key properties of the six quarks. Flavor quantum numbers (isospin (I_3), charm (C), strangeness (S, not to be confused with spin), topness (T), and bottomness (B')) are assigned to certain quark flavors, and denote qualities of quark-based systems and hadrons. The baryon number (B) is $+\frac{1}{3}$ for all quarks, as baryons are made of three quarks. For antiquarks, the electric charge (Q) and all flavor quantum numbers (B, I_3, C, S, T, and B') are of opposite sign. Mass and total angular momentum (J; equal to spin for point particles) do not change sign for the antiquarks.

J = total angular momentum, B = baryon number, Q = electric charge, I_3 = isospin, C = charm, S = strangeness, T = topness, B' = bottomness.
* Notation such as 4190+180
−60 denotes measurement uncertainty. In the case of the top quark, the first uncertainty is statistical in nature, and the second is systematic.

8.5 Interacting quarks

See also: Color confinement and Gluon

As described by quantum chromodynamics, the strong interaction between quarks is mediated by gluons, massless vector gauge bosons. Each gluon carries one color charge and one anticolor charge. In the standard framework of particle interactions (part of a more general formulation known as perturbation theory), gluons are constantly exchanged between quarks through a virtual emission and absorption process. When a gluon is transferred between quarks, a color change occurs in both; for example, if a red quark emits a red–antigreen gluon, it becomes green, and if a green quark absorbs a red–antigreen gluon, it becomes red. Therefore, while each quark's color constantly changes, their strong interaction is preserved.[71][72][73]

Since gluons carry color charge, they themselves are able to emit and absorb other gluons. This causes *asymptotic freedom*: as quarks come closer to each other, the chromodynamic binding force between them weakens.[74] Conversely, as the distance between quarks increases, the binding force strengthens. The color field becomes stressed, much as an elastic band is stressed when stretched, and more gluons of appropriate color are spontaneously created to strengthen the field. Above a certain energy threshold, pairs of quarks and antiquarks are created. These pairs bind with the quarks being separated, causing new hadrons to form. This phenomenon is known as *color confinement*: quarks never appear in isolation.[72][75] This process of hadronization occurs before quarks, formed in a high energy collision, are able to interact in any other way. The only exception is the top quark, which may decay before it hadronizes.[76]

8.5.1 Sea quarks

Hadrons, along with the *valence quarks* (q
v) that contribute to their quantum numbers, contain virtual quark–antiquark (qq) pairs known as *sea quarks* (q
s). Sea quarks form when a gluon of the hadron's color field splits; this process also works in reverse in that the annihilation of two sea quarks produces a gluon. The result is a constant flux of gluon splits and creations colloquially known as "the sea".[77] Sea quarks are much less stable than their valence counterparts, and they typically annihilate each other within the interior of the hadron. Despite this, sea quarks can hadronize into baryonic or mesonic particles under certain circumstances.[78]

8.5.2 Other phases of quark matter

Main article: QCD matter
Under sufficiently extreme conditions, quarks may become deconfined and exist as free particles. In the course of asymptotic freedom, the strong interaction becomes weaker at higher temperatures. Eventually, color confinement would be lost and an extremely hot plasma of freely moving quarks and gluons would be formed. This theoretical phase of matter is called quark–gluon plasma.[81] The exact conditions needed to give rise to this state are unknown and have been the subject of a great deal of speculation and experimentation. A recent estimate puts the needed temperature at $(1.90\pm0.02)\times10^{12}$ kelvin.[82] While a state of entirely free quarks and gluons has never been achieved (despite numerous attempts by CERN in the 1980s and 1990s),[83] recent experiments at the Relativistic Heavy Ion Collider have yielded evidence for liquid-like quark matter exhibiting "nearly perfect" fluid motion.[84]

The quark–gluon plasma would be characterized by a great increase in the number of heavier quark pairs in relation to the number of up and down quark pairs. It is believed that in the period prior to 10^{-6} seconds after the Big Bang (the quark epoch), the universe was filled with quark–gluon plasma, as the temperature was too high for hadrons to be stable.[85]

Given sufficiently high baryon densities and relatively low temperatures – possibly comparable to those found in neutron stars – quark matter is expected to degenerate into a Fermi liquid of weakly interacting quarks. This liquid would be characterized by a condensation of colored quark Cooper pairs, thereby breaking the local $SU(3)_c$ symmetry. Because quark Cooper pairs harbor color charge, such a phase of quark matter would be color superconductive; that is, color charge would be able to pass through it with no resistance.[86]

8.6 See also

- Color–flavor locking
- Neutron magnetic moment
- Leptons
- Preons – Hypothetical particles which were once postulated to be subcomponents of quarks and leptons

- Quarkonium – Mesons made of a quark and antiquark of the same flavor
- Quark star – A hypothetical degenerate neutron star with extreme density
- Quark–lepton complementarity – Possible fundamental relation between quarks and leptons

8.7 Notes

[1] Several research groups claimed to have proven the existence of tetraquarks and pentaquarks in the early 2000s. While the status of tetraquarks is still under debate, all known pentaquark candidates have previously been established as non-existent.

[2] The main evidence is based on the resonance width of the Z0 boson, which constrains the 4th generation neutrino to have a mass greater than ~45 GeV/c^2. This would be highly contrasting with the other three generations' neutrinos, whose masses cannot exceed 2 MeV/c^2.

[3] CP violation is a phenomenon which causes weak interactions to behave differently when left and right are swapped (P symmetry) and particles are replaced with their corresponding antiparticles (C symmetry).

[4] The actual probability of decay of one quark to another is a complicated function of (amongst other variables) the decaying quark's mass, the masses of the decay products, and the corresponding element of the CKM matrix. This probability is directly proportional (but not equal) to the magnitude squared ($|Vij|^2$) of the corresponding CKM entry.

[5] Despite its name, color charge is not related to the color spectrum of visible light.

8.8 References

[1] "Quark (subatomic particle)". *Encyclopædia Britannica*. Retrieved 2008-06-29.

[2] R. Nave. "Confinement of Quarks". *HyperPhysics*. Georgia State University, Department of Physics and Astronomy. Retrieved 2008-06-29.

[3] R. Nave. "Bag Model of Quark Confinement". *HyperPhysics*. Georgia State University, Department of Physics and Astronomy. Retrieved 2008-06-29.

[4] R. Nave. "Quarks". *HyperPhysics*. Georgia State University, Department of Physics and Astronomy. Retrieved 2008-06-29.

[5] B. Carithers, P. Grannis (1995). "Discovery of the Top Quark" (PDF). *Beam Line* (SLAC) **25** (3): 4–16. Retrieved 2008-09-23.

[6] E.D. Bloom et al. (1969). "High-Energy Inelastic *e–p* Scattering at 6° and 10°". *Physical Review Letters* **23** (16): 930–934. Bibcode:1969PhRvL..23..930B. doi:10.1103/PhysRevLett.23.930.

[7] M. Breidenbach et al. (1969). "Observed Behavior of Highly Inelastic Electron–Proton Scattering". *Physical Review Letters* **23** (16): 935–939. Bibcode:1969PhRvL..23..935B. doi:10.1103/PhysRevLett.23.935.

[8] S.S.M. Wong (1998). *Introductory Nuclear Physics* (2nd ed.). Wiley Interscience. p. 30. ISBN 0-471-23973-9.

[9] K.A. Peacock (2008). *The Quantum Revolution*. Greenwood Publishing Group. p. 125. ISBN 0-313-33448-X.

[10] B. Povh, C. Scholz, K. Rith, F. Zetsche (2008). *Particles and Nuclei*. Springer. p. 98. ISBN 3-540-79367-4.

[11] Section 6.1. in P.C.W. Davies (1979). *The Forces of Nature*. Cambridge University Press. ISBN 0-521-22523-X.

[12] M. Munowitz (2005). *Knowing*. Oxford University Press. p. 35. ISBN 0-19-516737-6.

[13] W.-M. Yao (Particle Data Group) et al. (2006). "Review of Particle Physics: Pentaquark Update" (PDF). *Journal of Physics G* **33** (1): 1–1232. arXiv:astro-ph/0601168. Bibcode:2006JPhG...33....1Y. doi:10.1088/0954-3899/33/1/001.

[14] C. Amsler (Particle Data Group) et al. (2008). "Review of Particle Physics: Pentaquarks" (PDF). *Physics Letters B* **667** (1): 1–1340. Bibcode:2008PhLB..667....1P. doi:10.1016/j.physletb.2008.07.018.
C. Amsler (Particle Data Group) et al. (2008). "Review of Particle Physics: New Charmonium-Like States" (PDF). *Physics Letters B* **667** (1): 1–1340. Bibcode:2008PhLB..667....1P. doi:10.1016/j.physletb.2008.07.018.
E.V. Shuryak (2004). *The QCD Vacuum, Hadrons and Superdense Matter*. World Scientific. p. 59. ISBN 981-238-574-6.

[15] R. Aaij et al. (LHCb collaboration) (2015). "Observation of J/ψp resonances consistent with pentaquark states in Λ0 b→J/ψK−
p decays". *Physical Review Letters* **115** (7). doi:10.1103/PhysRevLett.115.072001.

[16] C. Amsler (Particle Data Group) et al. (2008). "Review of Particle Physics: b′ (4th Generation) Quarks, Searches for" (PDF). *Physics Letters B* **667** (1): 1–1340. Bibcode:2008PhLB..667....1P. doi:10.1016/j.physletb.2008.07.018.
C. Amsler (Particle Data Group) et al. (2008). "Review of Particle Physics: t′ (4th Generation) Quarks, Searches for" (PDF). *Physics Letters B* **667** (1): 1–1340. Bibcode:2008PhLB..667....1P. doi:10.1016/j.physletb.2008.07.018.

[17] D. Decamp; Deschizeaux, B.; Lees, J.-P.; Minard, M.-N.; Crespo, J.M.; Delfino, M.; Fernandez, E.; Martinez, M. et al. (1989). "Determination of the number of light neutrino species". *Physics Letters B* **231** (4): 519. Bibcode:1989PhLB..231..519D. doi:10.1016/0370-2693(89)90704-1.
A. Fisher (1991). "Searching for the Beginning of Time: Cosmic Connection". *Popular Science* **238** (4): 70.
J.D. Barrow (1997) [1994]. "The Singularity and Other Problems". *The Origin of the Universe* (Reprint ed.). Basic Books. ISBN 978-0-465-05314-8.

[18] D.H. Perkins (2003). *Particle Astrophysics*. Oxford University Press. p. 4. ISBN 0-19-850952-9.

[19] M. Gell-Mann (1964). "A Schematic Model of Baryons and Mesons". *Physics Letters* **8** (3): 214–215. Bibcode:1964PhL.....8..214G. doi:10.1016/S0031-9163(64)92001-3.

[20] G. Zweig (1964). "An SU(3) Model for Strong Interaction Symmetry and its Breaking" (PDF). *CERN Report No.8182/TH.401*.

[21] G. Zweig (1964). "An SU(3) Model for Strong Interaction Symmetry and its Breaking: II" (PDF). *CERN Report No.8419/TH.412*.

[22] M. Gell-Mann (2000) [1964]. "The Eightfold Way: A theory of strong interaction symmetry". In M. Gell-Mann, Y. Ne'eman. *The Eightfold Way*. Westview Press. p. 11. ISBN 0-7382-0299-1.
Original: M. Gell-Mann (1961). "The Eightfold Way: A theory of strong interaction symmetry". *Synchrotron Laboratory Report CTSL-20* (California Institute of Technology).

[23] Y. Ne'eman (2000) [1964]. "Derivation of strong interactions from gauge invariance". In M. Gell-Mann, Y. Ne'eman. *The Eightfold Way*. Westview Press. ISBN 0-7382-0299-1.
Original Y. Ne'eman (1961). "Derivation of strong interactions from gauge invariance". *Nuclear Physics* **26** (2): 222. Bibcode:1961NucPh.. doi:10.1016/0029-5582(61)90134-1.

[24] R.C. Olby, G.N. Cantor (1996). *Companion to the History of Modern Science*. Taylor & Francis. p. 673. ISBN 0-415-14578-3.

[25] A. Pickering (1984). *Constructing Quarks*. University of Chicago Press. pp. 114–125. ISBN 0-226-66799-5.

[26] B.J. Bjorken, S.L. Glashow; Glashow (1964). "Elementary Particles and SU(4)". *Physics Letters* **11** (3): 255–257. Bibcode:1964PhL....11.. doi:10.1016/0031-9163(64)90433-0.

[27] J.I. Friedman. "The Road to the Nobel Prize". Hue University. Retrieved 2008-09-29.

[28] R.P. Feynman (1969). "Very High-Energy Collisions of Hadrons". *Physical Review Letters* **23** (24): 1415–1417. Bibcode:1969PhRvL..23. doi:10.1103/PhysRevLett.23.1415.

[29] S. Kretzer et al. (2004). "CTEQ6 Parton Distributions with Heavy Quark Mass Effects". *Physical Review D* **69** (11): 114005. arXiv:hep-ph/0307022. Bibcode:2004PhRvD..69k4005K. doi:10.1103/PhysRevD.69.114005.

[30] D.J. Griffiths (1987). *Introduction to Elementary Particles*. John Wiley & Sons. p. 42. ISBN 0-471-60386-4.

[31] M.E. Peskin, D.V. Schroeder (1995). *An introduction to quantum field theory*. Addison–Wesley. p. 556. ISBN 0-201-50397-2.

[32] V.V. Ezhela (1996). *Particle physics*. Springer. p. 2. ISBN 1-56396-642-5.

[33] S.L. Glashow, J. Iliopoulos, L. Maiani; Iliopoulos; Maiani (1970). "Weak Interactions with Lepton–Hadron Symmetry". *Physical Review D* **2** (7): 1285–1292. Bibcode:1970PhRvD...2.1285G. doi:10.1103/PhysRevD.2.1285.

[34] D.J. Griffiths (1987). *Introduction to Elementary Particles*. John Wiley & Sons. p. 44. ISBN 0-471-60386-4.

[35] M. Kobayashi, T. Maskawa; Maskawa (1973). "CP-Violation in the Renormalizable Theory of Weak Interaction". *Progress of Theoretical Physics* **49** (2): 652–657. Bibcode:1973PThPh..49..652K. doi:10.1143/PTP.49.652.

[36] H. Harari (1975). "A new quark model for hadrons". *Physics Letters B* **57B** (3): 265. Bibcode:1975PhLB...57..265H. doi:10.1016/0370-2693(75)90072-6.

[37] K.W. Staley (2004). *The Evidence for the Top Quark*. Cambridge University Press. pp. 31–33. ISBN 978-0-521-82710-2.

[38] S.W. Herb et al. (1977). "Observation of a Dimuon Resonance at 9.5 GeV in 400-GeV Proton-Nucleus Collisions". *Physical Review Letters* **39** (5): 252. Bibcode:1977PhRvL..39..252H. doi:10.1103/PhysRevLett.39.252.

[39] M. Bartusiak (1994). *A Positron named Priscilla*. National Academies Press. p. 245. ISBN 0-309-04893-1.

[40] F. Abe (CDF Collaboration) et al. (1995). "Observation of Top Quark Production in pp Collisions with the Collider Detector at Fermilab". *Physical Review Letters* **74** (14): 2626–2631. Bibcode:1995PhRvL..74.2626A. doi:10.1103/PhysRevLett.74.2626. PMID 10057978.

[41] S. Abachi (DØ Collaboration) et al. (1995). "Search for High Mass Top Quark Production in pp Collisions at √s = 1.8 TeV". *Physical Review Letters* **74** (13): 2422–2426. Bibcode:1995PhRvL..74.2422A. doi:10.1103/PhysRevLett.74.2422.

[42] K.W. Staley (2004). *The Evidence for the Top Quark*. Cambridge University Press. p. 144. ISBN 0-521-82710-8.

[43] "New Precision Measurement of Top Quark Mass". Brookhaven National Laboratory News. 2004. Retrieved 2013-11-03.

[44] J. Joyce (1982) [1939]. *Finnegans Wake*. Penguin Books. p. 383. ISBN 0-14-006286-6.

[45] M. Gell-Mann (1995). *The Quark and the Jaguar: Adventures in the Simple and the Complex*. Henry Holt and Co. p. 180. ISBN 978-0-8050-7253-2.

[46] J. Gleick (1992). *Genius: Richard Feynman and modern physics*. Little Brown and Company. p. 390. ISBN 0-316-90316-7.

[47] J.J. Sakurai (1994). S.F Tuan, ed. *Modern Quantum Mechanics* (Revised ed.). Addison–Wesley. p. 376. ISBN 0-201-53929-2.

[48] D.H. Perkins (2000). *Introduction to high energy physics*. Cambridge University Press. p. 8. ISBN 0-521-62196-8.

[49] M. Riordan (1987). *The Hunting of the Quark: A True Story of Modern Physics*. Simon & Schuster. p. 210. ISBN 978-0-671-50466-3.

[50] F. Close (2006). *The New Cosmic Onion*. CRC Press. p. 133. ISBN 1-58488-798-2.

[51] J.T. Volk et al. (1987). "Letter of Intent for a Tevatron Beauty Factory" (PDF). Fermilab Proposal #783.

[52] G. Fraser (2006). *The New Physics for the Twenty-First Century*. Cambridge University Press. p. 91. ISBN 0-521-81600-9.

[53] "The Standard Model of Particle Physics". BBC. 2002. Retrieved 2009-04-19.

[54] F. Close (2006). *The New Cosmic Onion*. CRC Press. pp. 80–90. ISBN 1-58488-798-2.

[55] D. Lincoln (2004). *Understanding the Universe*. World Scientific. p. 116. ISBN 981-238-705-6.

[56] "Weak Interactions". *Virtual Visitor Center*. Stanford Linear Accelerator Center. 2008. Retrieved 2008-09-28.

[57] K. Nakamura et al. (2010). "Review of Particles Physics: The CKM Quark-Mixing Matrix" (PDF). *J. Phys. G* **37** (75021): 150.

[58] Z. Maki, M. Nakagawa, S. Sakata (1962). "Remarks on the Unified Model of Elementary Particles". *Progress of Theoretical Physics* **28** (5): 870. Bibcode:1962PThPh..28..870M. doi:10.1143/PTP.28.870.

[59] B.C. Chauhan, M. Picariello, J. Pulido, E. Torrente-Lujan (2007). "Quark–lepton complementarity, neutrino and standard model data predict θPMNS
13 = 9°+1°
−2°". *European Physical Journal* **C50** (3): 573–578. arXiv:hep-ph/0605032. Bibcode:2007EPJC...50..573C. doi:10.1140/epjc/s10052-007-0212-z.

[60] R. Nave. "The Color Force". *HyperPhysics*. Georgia State University, Department of Physics and Astronomy. Retrieved 2009-04-26.

[61] B.A. Schumm (2004). *Deep Down Things*. Johns Hopkins University Press. pp. 131–132. ISBN 0-8018-7971-X. OCLC 55229065.

[62] Part III of M.E. Peskin, D.V. Schroeder (1995). *An Introduction to Quantum Field Theory*. Addison–Wesley. ISBN 0-201-50397-2.

[63] V. Icke (1995). *The force of symmetry*. Cambridge University Press. p. 216. ISBN 0-521-45591-X.

[64] M.Y. Han (2004). *A story of light*. World Scientific. p. 78. ISBN 981-256-034-3.

[65] C. Sutton. "Quantum chromodynamics (physics)". *Encyclopædia Britannica Online*. Retrieved 2009-05-12.

[66] A. Watson (2004). *The Quantum Quark*. Cambridge University Press. pp. 285–286. ISBN 0-521-82907-0.

[67] K.A. Olive *et al.* (Particle Data Group), Chin. Phys. **C38**, 090001 (2014) (URL: http://pdg.lbl.gov)

[68] W. Weise, A.M. Green (1984). *Quarks and Nuclei*. World Scientific. pp. 65–66. ISBN 9971-966-61-1.

[69] D. McMahon (2008). *Quantum Field Theory Demystified*. McGraw–Hill. p. 17. ISBN 0-07-154382-1.

[70] S.G. Roth (2007). *Precision electroweak physics at electron–positron colliders*. Springer. p. VI. ISBN 3-540-35164-7.

[71] R.P. Feynman (1985). *QED: The Strange Theory of Light and Matter* (1st ed.). Princeton University Press. pp. 136–137. ISBN 0-691-08388-6.

[72] M. Veltman (2003). *Facts and Mysteries in Elementary Particle Physics*. World Scientific. pp. 45–47. ISBN 981-238-149-X.

[73] F. Wilczek, B. Devine (2006). *Fantastic Realities*. World Scientific. p. 85. ISBN 981-256-649-X.

[74] F. Wilczek, B. Devine (2006). *Fantastic Realities*. World Scientific. pp. 400ff. ISBN 981-256-649-X.

[75] T. Yulsman (2002). *Origin*. CRC Press. p. 55. ISBN 0-7503-0765-X.

[76] F. Garberson (2008). "Top Quark Mass and Cross Section Results from the Tevatron". arXiv:0808.0273 [hep-ex].

[77] J. Steinberger (2005). *Learning about Particles*. Springer. p. 130. ISBN 3-540-21329-5.

[78] C.-Y. Wong (1994). *Introduction to High-energy Heavy-ion Collisions*. World Scientific. p. 149. ISBN 981-02-0263-6.

[79] S.B. Rüester, V. Werth, M. Buballa, I.A. Shovkovy, D.H. Rischke; Werth; Buballa; Shovkovy; Rischke (2005). "The phase diagram of neutral quark matter: Self-consistent treatment of quark masses". *Physical Review D* **72** (3): 034003. arXiv:hep-ph/0503184. Bibcode:2005PhRvD..72c4004R. doi:10.1103/PhysRevD.72.034004.

[80] M.G. Alford, K. Rajagopal, T. Schaefer, A. Schmitt; Schmitt; Rajagopal; Schäfer (2008). "Color superconductivity in dense quark matter". *Reviews of Modern Physics* **80** (4): 1455–1515. arXiv:0709.4635. Bibcode:2008RvMP...80.1455A. doi:10.1103/RevModF

[81] S. Mrowczynski (1998). "Quark–Gluon Plasma". *Acta Physica Polonica B* **29**: 3711. arXiv:nucl-th/9905005. Bibcode:1998AcPPB..29.37

[82] Z. Fodor, S.D. Katz; Katz (2004). "Critical point of QCD at finite T and μ, lattice results for physical quark masses". *Journal of High Energy Physics* **2004** (4): 50. arXiv:hep-lat/0402006. Bibcode:2004JHEP...04..050F. doi:10.1088/1126-6708/2004/04/050.

[83] U. Heinz, M. Jacob (2000). "Evidence for a New State of Matter: An Assessment of the Results from the CERN Lead Beam Programme". arXiv:nucl-th/0002042.

[84] "RHIC Scientists Serve Up "Perfect" Liquid". Brookhaven National Laboratory News. 2005. Retrieved 2009-05-22.

[85] T. Yulsman (2002). *Origins: The Quest for Our Cosmic Roots*. CRC Press. p. 75. ISBN 0-7503-0765-X.

[86] A. Sedrakian, J.W. Clark, M.G. Alford (2007). *Pairing in fermionic systems*. World Scientific. pp. 2–3. ISBN 981-256-907-3.

8.9 Further reading

- A. Ali, G. Kramer; Kramer (2011). "JETS and QCD: A historical review of the discovery of the quark and gluon jets and its impact on QCD". *European Physical Journal H* **36** (2): 245. arXiv:1012.2288. Bibcode:2011EPJH...36..245A. doi:10.1140/epjh/e2011-10047-1.
- D.J. Griffiths (2008). *Introduction to Elementary Particles* (2nd ed.). Wiley–VCH. ISBN 3-527-40601-8.
- I.S. Hughes (1985). *Elementary particles* (2nd ed.). Cambridge University Press. ISBN 0-521-26092-2.
- R. Oerter (2005). *The Theory of Almost Everything: The Standard Model, the Unsung Triumph of Modern Physics.* Pi Press. ISBN 0-13-236678-9.
- A. Pickering (1984). *Constructing Quarks: A Sociological History of Particle Physics.* The University of Chicago Press. ISBN 0-226-66799-5.
- B. Povh (1995). *Particles and Nuclei: An Introduction to the Physical Concepts.* Springer–Verlag. ISBN 0-387-59439-6.
- M. Riordan (1987). *The Hunting of the Quark: A true story of modern physics.* Simon & Schuster. ISBN 0-671-64884-5.
- B.A. Schumm (2004). *Deep Down Things: The Breathtaking Beauty of Particle Physics.* Johns Hopkins University Press. ISBN 0-8018-7971-X.

8.10 External links

- 1969 Physics Nobel Prize lecture by Murray Gell-Mann
- 1976 Physics Nobel Prize lecture by Burton Richter
- 1976 Physics Nobel Prize lecture by Samuel C.C. Ting
- 2008 Physics Nobel Prize lecture by Makoto Kobayashi
- 2008 Physics Nobel Prize lecture by Toshihide Maskawa
- The Top Quark And The Higgs Particle by T.A. Heppenheimer – A description of CERN's experiment to count the families of quarks.
- Bowley, Roger; Copeland, Ed. "Quarks". *Sixty Symbols.* Brady Haran for the University of Nottingham.

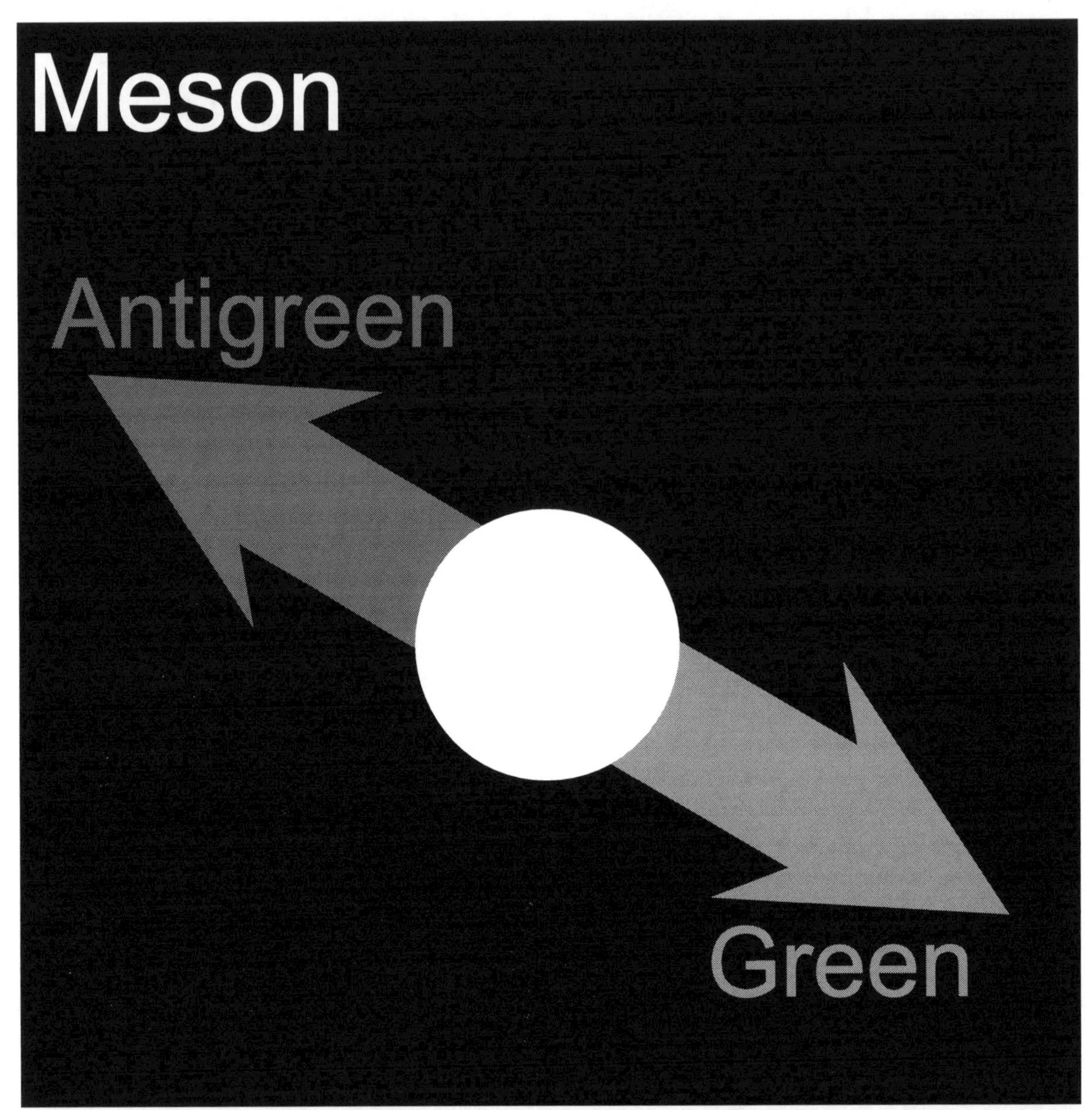
Meson
Antigreen
Green

Baryon
Red

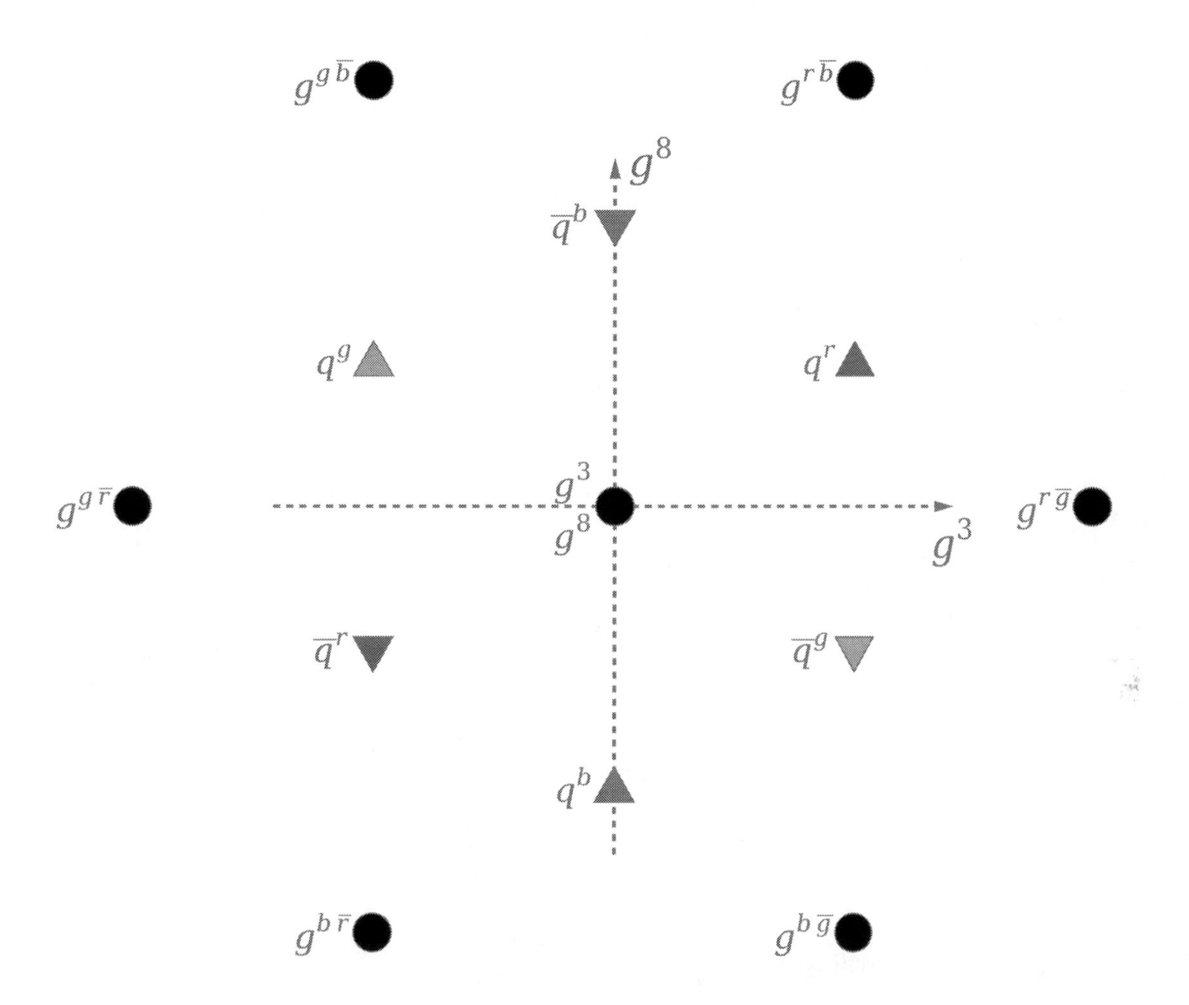

The pattern of strong charges for the three colors of quark, three antiquarks, and eight gluons (with two of zero charge overlapping).

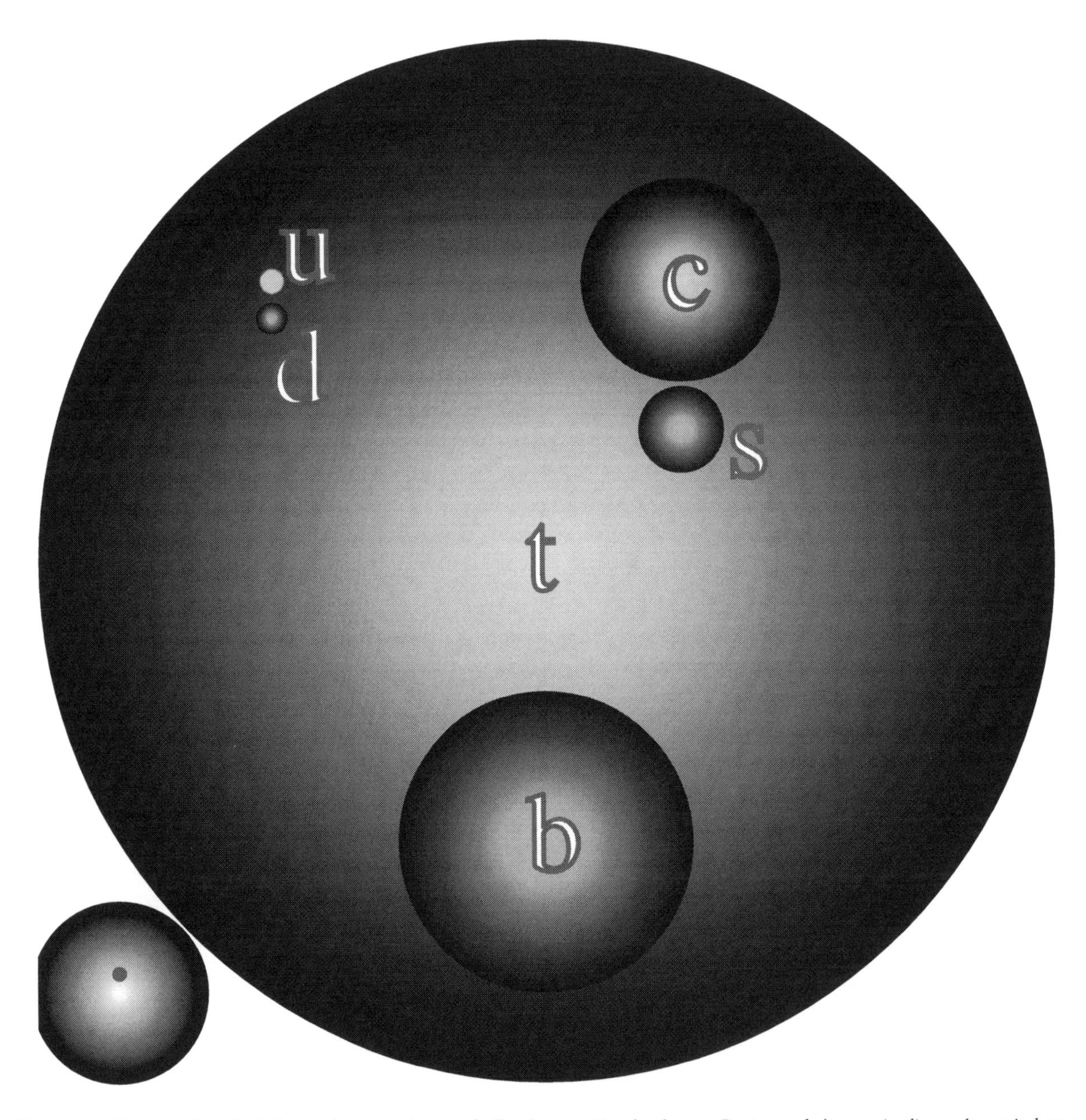

Current quark masses for all six flavors in comparison, as balls of proportional volumes. Proton and electron (red) are shown in bottom left corner for scale

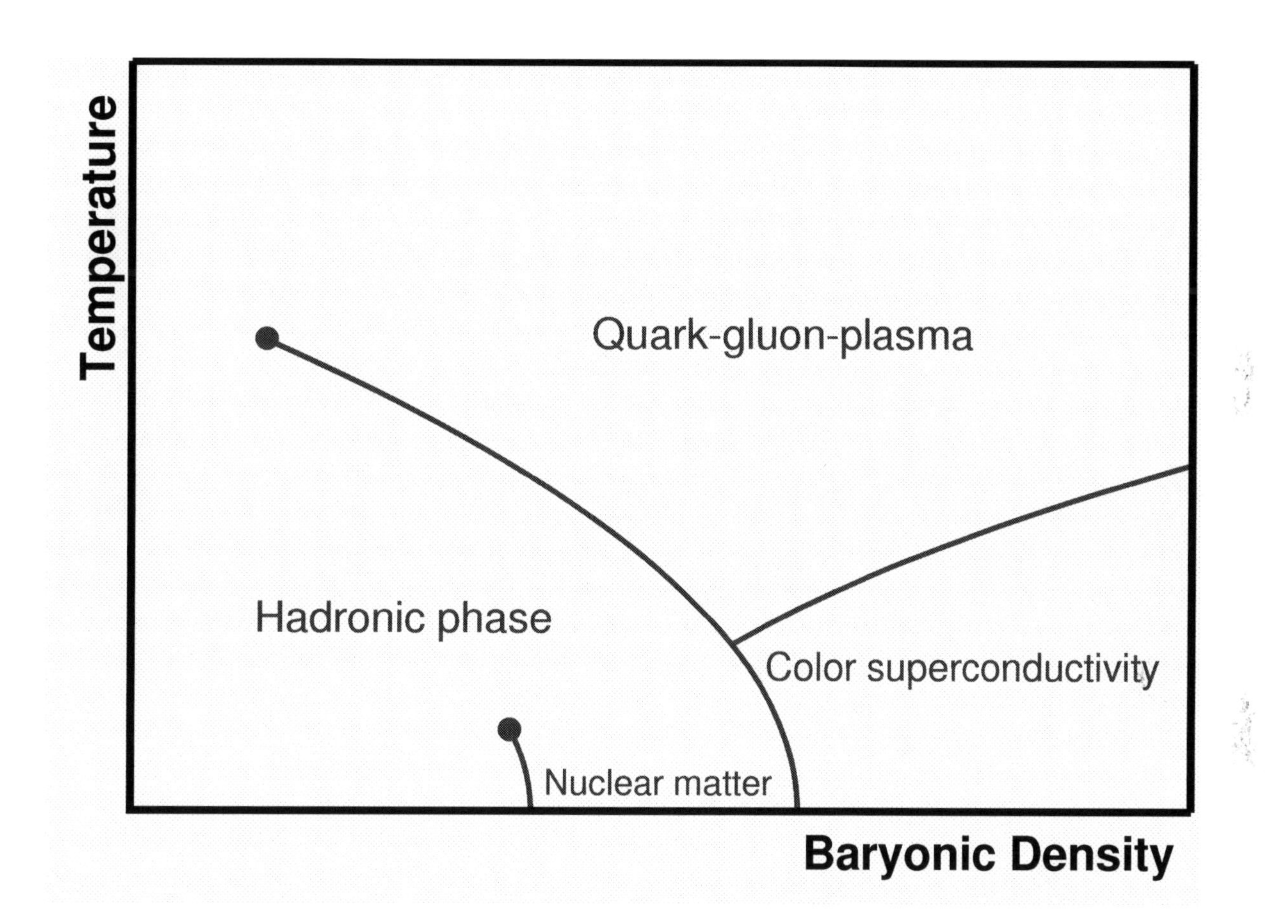

A qualitative rendering of the phase diagram of quark matter. The precise details of the diagram are the subject of ongoing research.[79][80]

Chapter 9

Gauge boson

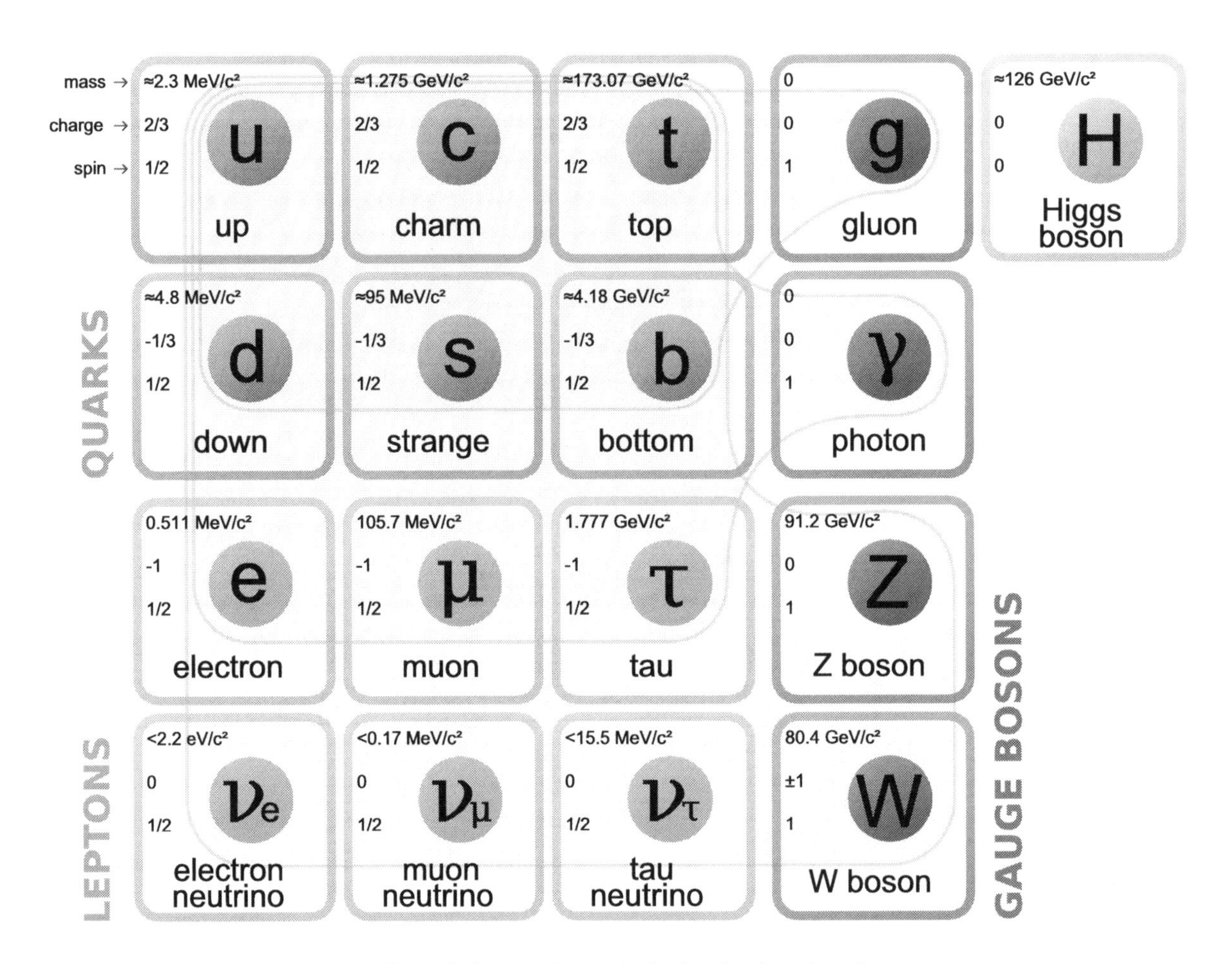

The Standard Model of elementary particles, with the gauge bosons in the fourth column in red

In particle physics, a **gauge boson** is a force carrier, a bosonic particle that carries any of the fundamental interactions of nature.[1][2] Elementary particles, whose interactions are described by a gauge theory, interact with each other by the exchange of gauge bosons—usually as virtual particles.

9.1 Gauge bosons in the Standard Model

The Standard Model of particle physics recognizes four kinds of gauge bosons: photons, which carry the electromagnetic interaction; W and Z bosons, which carry the weak interaction; and gluons, which carry the strong interaction.[3]

Isolated gluons do not occur at low energies because they are color-charged, and subject to color confinement.

9.1.1 Multiplicity of gauge bosons

In a quantized gauge theory, gauge bosons are quanta of the gauge fields. Consequently, there are as many gauge bosons as there are generators of the gauge field. In quantum electrodynamics, the gauge group is $U(1)$; in this simple case, there is only one gauge boson. In quantum chromodynamics, the more complicated group $SU(3)$ has eight generators, corresponding to the eight gluons. The three W and Z bosons correspond (roughly) to the three generators of $SU(2)$ in GWS theory.

9.1.2 Massive gauge bosons

For technical reasons involving gauge invariance, gauge bosons are described mathematically by field equations for massless particles. Therefore, at a naïve theoretical level all gauge bosons are required to be massless, and the forces that they describe are required to be long-ranged. The conflict between this idea and experimental evidence that the weak interaction has a very short range requires further theoretical insight.

According to the Standard Model, the W and Z bosons gain mass via the Higgs mechanism. In the Higgs mechanism, the four gauge bosons (of $SU(2)\times U(1)$ symmetry) of the unified electroweak interaction couple to a Higgs field. This field undergoes spontaneous symmetry breaking due to the shape of its interaction potential. As a result, the universe is permeated by a nonzero Higgs vacuum expectation value (VEV). This VEV couples to three of the electroweak gauge bosons (the Ws and Z), giving them mass; the remaining gauge boson remains massless (the photon). This theory also predicts the existence of a scalar Higgs boson, which has been observed in experiments that were reported on 4 July 2012.[4]

9.2 Beyond the Standard Model

9.2.1 Grand unification theories

A grand unified theory predicts additional gauge bosons named X and Y bosons. The hypothetical X and Y bosons direct interactions between quarks and leptons, hence violating conservation of baryon number and causing proton decay. Such bosons would be even more massive than W and Z bosons due to symmetry breaking. Analysis of data collected from such sources as the Super-Kamiokande neutrino detector has yielded no evidence of X and Y bosons.

9.2.2 Gravitons

The fourth fundamental interaction, gravity, may also be carried by a boson, called the graviton. In the absence of experimental evidence and a mathematically coherent theory of quantum gravity, it is unknown whether this would be a gauge boson or not. The role of gauge invariance in general relativity is played by a similar symmetry: diffeomorphism invariance.

9.2.3 W' and Z' bosons

Main article: W' and Z' bosons

W' and Z' bosons refer to hypothetical new gauge bosons (named in analogy with the Standard Model W and Z bosons).

9.3 See also

- 1964 PRL symmetry breaking papers
- Boson
- Glueball
- Quantum chromodynamics
- Quantum electrodynamics

9.4 References

[1] Gribbin, John (2000). *Q is for Quantum – An Encyclopedia of Particle Physics*. Simon & Schuster. ISBN 0-684-85578-X.

[2] Clark, John, E.O. (2004). *The Essential Dictionary of Science*. Barnes & Noble. ISBN 0-7607-4616-8.

[3] Veltman, Martinus (2003). *Facts and Mysteries in Elementary Particle Physics*. World Scientific. ISBN 981-238-149-X.

[4] "CERN experiments observe particle consistent with long-sought Higgs boson". CERN. Retrieved 4 July 2012.

9.5 External links

- Explanation of gauge boson and gauge fields by Christopher T. Hill

Chapter 10

Higgs boson

The **Higgs boson** or **Higgs particle** is an elementary particle in the Standard Model of particle physics. Observation of the particle allows scientists to explore the **Higgs field**[6][7]—a fundamental field of crucial importance to particle physics theory,[7] first suspected to exist in the 1960s, that unlike other known fields such as the electromagnetic field, takes a non-zero constant value almost everywhere. For several decades the question of the Higgs Field's existence was the last unverified part of the Standard Model of particle physics and "the central problem in particle physics".[8][9] The presence of this field, now believed to be confirmed, explains why some fundamental particles have mass when, based on the symmetries controlling their interactions, they should be massless. It also solves several other long-standing puzzles, such as the reason for the weak force's extremely short range.

Despite being present everywhere, the existence of the Higgs field is very hard to confirm. It can be detected through its excitations (i.e. Higgs particles), but these are extremely hard to produce and detect. The importance of this fundamental question led to a 40 year search, and the construction of one of the world's most expensive and complex experimental facilities to date, CERN's Large Hadron Collider,[10] able to create Higgs bosons and other particles for observation and study. On 4 July 2012, the discovery of a new particle with a mass between 125 and 127 GeV/c^2 was announced; physicists suspected that it was the Higgs boson.[11][12][13] By March 2013, the particle had been proven to behave, interact and decay in many of the ways predicted by the Standard Model, and was also tentatively confirmed to have even parity and zero spin,[1] two fundamental attributes of a Higgs boson. This appears to be the first elementary scalar particle discovered in nature.[14] More data is needed to know if the discovered particle exactly matches the predictions of the Standard Model, or whether, as predicted by some theories, multiple Higgs bosons exist.[3]

The Higgs boson is named after Peter Higgs, one of six physicists who, in 1964, proposed the mechanism that suggested the existence of such a particle. On December 10, 2013, two of them, Peter Higgs and François Englert, were awarded the Nobel Prize in Physics for their work and prediction (Englert's co-researcher Robert Brout had died in 2011 and the Nobel Prize is not ordinarily given posthumously).[15] Although Higgs's name has come to be associated with this theory, several researchers between about 1960 and 1972 each independently developed different parts of it. In mainstream media the Higgs boson has often been called the "God particle", from a 1993 book on the topic; the nickname is strongly disliked by many physicists, including Higgs, who regard it as inappropriate sensationalism.[16][17][18]

In the Standard Model, the Higgs particle is a boson with no spin, electric charge, or colour charge. It is also very unstable, decaying into other particles almost immediately. It is a quantum excitation of one of the four components of the Higgs field. The latter constitutes a scalar field, with two neutral and two electrically charged components, and forms a complex doublet of the weak isospin SU(2) symmetry. The Higgs field is tachyonic (this does not refer to faster-than-light speeds, it means that symmetry-breaking through condensation of a particle must occur under certain conditions), and has a "Mexican hat" shaped potential with nonzero strength everywhere (including otherwise empty space), which in its vacuum state breaks the weak isospin symmetry of the electroweak interaction. When this happens, three components of the Higgs field are "absorbed" by the SU(2) and U(1) gauge bosons (the "Higgs mechanism") to become the longitudinal components of the now-massive W and Z bosons of the weak force. The remaining electrically neutral component separately couples to other particles known as fermions (via Yukawa couplings), causing these to acquire mass as well. Some versions of the theory predict more than one kind of Higgs fields and bosons. Alternative "Higgsless" models would have been considered if the Higgs boson was not discovered.

10.1 A non-technical summary

10.1.1 "Higgs" terminology

10.1.2 Overview

In particle physics, elementary particles and forces give rise to the world around us. Nowadays, physicists explain the behaviour of these particles and how they interact using the Standard Model—a widely accepted and "remarkably" accurate[21] framework based on gauge invariance and symmetries, believed to explain almost everything in the world we see, other than gravity.[22]

But by around 1960 all attempts to create a gauge invariant theory for two of the four fundamental forces had consistently failed at one crucial point: although gauge invariance seemed extremely important, it seemed to make any theory of electromagnetism and the weak force go haywire, by demanding that either many particles with mass were massless or that non-existent forces and massless particles had to exist. Scientists had no idea how to get past this point.

In 1962 physicist Philip Anderson wrote a paper that built upon work by Yoichiro Nambu concerning "broken symmetries" in superconductivity and particle physics. He suggested that "broken symmetries" might also be the missing piece needed to solve the problems of gauge invariance. In 1964 a theory was created almost simultaneously by 3 different groups of researchers, that showed Anderson's suggestion was possible - the gauge theory and "mass problems" could indeed be resolved if an unusual kind of field existed throughout the universe; if this kind of field did exist, it would apparently cause existing particles to acquire mass instead of new massless particles being formed. Although these ideas did not gain much initial support or attention, by 1972 it had been developed into a comprehensive theory and proved capable of giving "sensible" results that were extremely accurate, including very accurate predictions of several other particles discovered during the following years.[Note 7] During the 1970s these theories rapidly became the "standard model" favoured by physicists and used to describe particle physics and particle interactions in nature. There was not yet any direct evidence that this field actually existed, but even without proof of the field, the accuracy of its predictions led scientists to believe the theory might be true. By the 1980s the question whether or not such a field existed and whether this was the correct explanation, was considered to be one of the most important unanswered questions in particle physics, and by the 1990s two of the largest experimental installations ever created were being designed and constructed to find the answer.

If this new kind of field did exist in nature, it would be a monumental discovery for science and human knowledge, and would open doorways to new knowledge in many disciplines. If not, then other more complicated theories would need to be explored. The simplest solution to whether the field existed was by searching for a new kind of particle it would have to give off, known as "Higgs bosons" or the "Higgs particle". These would be extremely difficult to find, so it was only many years later that experimental technology became sophisticated enough to answer the question.

While several symmetries in nature are spontaneously broken through a form of the Higgs mechanism, in the context of the Standard Model the term "Higgs mechanism" almost always means symmetry breaking of the electroweak field. It is considered confirmed, but revealing the exact cause has been difficult.

Various analogies have also been invented to describe the Higgs field and boson, including analogies with well-known symmetry breaking effects such as the rainbow and prism, electric fields, ripples, and resistance of macro objects moving through media, like people moving through crowds or some objects moving through syrup or molasses. However, analogies based on simple resistance to motion are inaccurate as the Higgs field does not work by resisting motion.

10.2 Significance

10.2.1 Scientific impact

Evidence of the Higgs field and its properties has been extremely significant scientifically, for many reasons. The Higgs boson's importance is largely that it is able to be examined using existing knowledge and experimental technology, as a way to confirm and study the entire Higgs field theory.[6][7] Conversely, proof that the Higgs field and boson do not exist would also have been significant. In discussion form, the relevance includes:

10.2.2 "Real world" impact

As yet, there are no known immediate technological benefits of finding the Higgs particle. However, a common pattern for fundamental discoveries is for practical applications to follow later, once the discovery has been explored further, at which point they become the basis for social change and new technologies.[44][45][46]

Other observers have highlighted that the challenges in particle physics have furthered major technological and in turn sociological developments. For example, the World Wide Web began as a project to improve CERN's communication system. Another example is the contribution to the fields of distributed and cloud computing due to CERN's requirement to process massive amounts of data produced by the Large Hadron Collider .

10.3 History

See also: 1964 PRL symmetry breaking papers, Higgs mechanism and History of quantum field theory
Particle physicists study matter made from fundamental particles whose interactions are mediated by exchange particles - gauge bosons - acting as force carriers. At the beginning of the 1960s a number of these particles had been discovered or proposed, along with theories suggesting how they relate to each other, some of which had already been reformulated as field theories in which the objects of study are not particles and forces, but quantum fields and their symmetries.[47]:150 However, attempts to unify known fundamental forces such as the electromagnetic force and the weak nuclear force were known to be incomplete. One known omission was that gauge invariant approaches, including non-abelian models such as Yang–Mills theory (1954), which held great promise for unified theories, also seemed to predict known massive particles as massless.[48] Goldstone's theorem, relating to continuous symmetries within some theories, also appeared to rule out many obvious solutions,[49] since it appeared to show that zero-mass particles would have to also exist that were "simply not seen".[50] According to Guralnik, physicists had "no understanding" how these problems could be overcome.[50]

Particle physicist and mathematician Peter Woit summarised the state of research at the time:

> "Yang and Mills work on non-abelian gauge theory had one huge problem: in perturbation theory it has massless particles which don't correspond to anything we see. One way of getting rid of this problem is now fairly well-understood, the phenomenon of confinement realized in QCD, where the strong interactions get rid of the massless "gluon" states at long distances. By the very early sixties, people had begun to understand another source of massless particles: spontaneous symmetry breaking of a continuous symmetry. What Philip Anderson realized and worked out in the summer of 1962 was that, when you have *both* gauge symmetry *and* spontaneous symmetry breaking, the Nambu–Goldstone massless mode can combine with the massless gauge field modes to produce a physical massive vector field. This is what happens in superconductivity, a subject about which Anderson was (and is) one of the leading experts." *[text condensed]* [48]

The Higgs mechanism is a process by which vector bosons can get rest mass *without* explicitly breaking gauge invariance, as a byproduct of spontaneous symmetry breaking.[51][52] The mathematical theory behind spontaneous symmetry breaking was initially conceived and published within particle physics by Yoichiro Nambu in 1960,[53] the concept that such a mechanism could offer a possible solution for the "mass problem" was originally suggested in 1962 by Philip Anderson (who had previously written papers on broken symmetry and its outcomes in superconductivity[54] and concluded in his 1963 paper on Yang-Mills theory that *"considering the superconducting analog... [t]hese two types of bosons seem capable of canceling each other out... leaving finite mass bosons"*),[55]:4–5[56] and Abraham Klein and Benjamin Lee showed in March 1964 that Goldstone's theorem could be avoided this way in at least some non-relativistic cases and speculated it might be possible in truly relativistic cases.[57]

These approaches were quickly developed into a full relativistic model, independently and almost simultaneously, by three groups of physicists: by François Englert and Robert Brout in August 1964;[58] by Peter Higgs in October 1964;[59] and by Gerald Guralnik, Carl Hagen, and Tom Kibble (GHK) in November 1964.[60] Higgs also wrote a short but important[51]

Nobel Prize Laureate Peter Higgs in Stockholm, December 2013

response published in September 1964 to an objection by Gilbert,[61] which showed that if calculating within the radiation gauge, Goldstone's theorem and Gilbert's objection would become inapplicable.[Note 11] (Higgs later described Gilbert's objection as prompting his own paper.[62]) Properties of the model were further considered by Guralnik in 1965,[63] by Higgs in 1966,[64] by Kibble in 1967,[65] and further by GHK in 1967.[66] The original three 1964 papers showed that when a gauge theory is combined with an additional field that spontaneously breaks the symmetry, the gauge bosons can consistently acquire a finite mass.[51][52][67] In 1967, Steven Weinberg[68] and Abdus Salam[69] independently showed how a Higgs mechanism could be used to break the electroweak symmetry of Sheldon Glashow's unified model for the weak and electromagnetic interactions[70] (itself an extension of work by Schwinger), forming what became the Standard Model of particle physics. Weinberg was the first to observe that this would also provide mass terms for the fermions.[71] [Note 12]

However, the seminal papers on spontaneous breaking of gauge symmetries were at first largely ignored, because it was widely believed that the (non-Abelian gauge) theories in question were a dead-end, and in particular that they could not be renormalised. In 1971–72, Martinus Veltman and Gerard 't Hooft proved renormalisation of Yang–Mills was possible in two papers covering massless, and then massive, fields.[71] Their contribution, and others' work on the renormalization group - including “substantial” theoretical work by Russian physicists Ludvig Faddeev, Andrei Slavnov, Efim Fradkin and Igor Tyutin[72] - was eventually “enormously profound and influential”,[73] but even with all key elements of the eventual theory published there was still almost no wider interest. For example, Coleman found in a study that “essentially no-one paid any attention” to Weinberg’s paper prior to 1971[74] and discussed by David Politzer in his 2004 Nobel speech.[73] – now the most cited in particle physics[75] – and even in 1970 according to Politzer, Glashow’s teaching of the weak interaction contained no mention of Weinberg’s, Salam’s, or Glashow’s own work.[73] In practice, Politzer states, almost everyone learned of the theory due to physicist Benjamin Lee, who combined the work of Veltman and 't Hooft with insights by others, and popularised the completed theory.[73] In this way, from 1971, interest and acceptance “exploded” [73] and the ideas were quickly absorbed in the mainstream.[71][73]

The resulting electroweak theory and Standard Model have correctly predicted (among other discoveries) weak neutral currents, three bosons, the top and charm quarks, and with great precision, the mass and other properties of some of these.[Note 7] Many of those involved eventually won Nobel Prizes or other renowned awards. A 1974 paper and comprehensive review in *Reviews of Modern Physics* commented that “while no one doubted the [mathematical] correctness of these arguments, no one quite believed that nature was diabolically clever enough to take advantage of them”,[76]:9 adding that the theory had so far produced meaningful answers that accorded with experiment, but it was unknown whether the theory was actually correct.[76]:9,36(footnote),43–44,47 By 1986 and again in the 1990s it became possible to write that understanding and proving the Higgs sector of the Standard Model was “the central problem today in particle physics”.[8][9]

10.3.1 Summary and impact of the PRL papers

The three papers written in 1964 were each recognised as milestone papers during *Physical Review Letters* 's 50th anniversary celebration.[67] Their six authors were also awarded the 2010 J. J. Sakurai Prize for Theoretical Particle Physics for this work.[77] (A controversy also arose the same year, because in the event of a Nobel Prize only up to three scientists could be recognised, with six being credited for the papers.[78]) Two of the three PRL papers (by Higgs and by GHK) contained equations for the hypothetical field that eventually would become known as the Higgs field and its hypothetical quantum, the Higgs boson.[59][60] Higgs’ subsequent 1966 paper showed the decay mechanism of the boson; only a massive boson can decay and the decays can prove the mechanism.

In the paper by Higgs the boson is massive, and in a closing sentence Higgs writes that “an essential feature” of the theory “is the prediction of incomplete multiplets of scalar and vector bosons".[59] (Frank Close comments that 1960s gauge theorists were focused on the problem of massless *vector* bosons, and the implied existence of a massive *scalar* boson was not seen as important; only Higgs directly addressed it.[79]:154, 166, 175) In the paper by GHK the boson is massless and decoupled from the massive states.[60] In reviews dated 2009 and 2011, Guralnik states that in the GHK model the boson is massless only in a lowest-order approximation, but it is not subject to any constraint and acquires mass at higher orders, and adds that the GHK paper was the only one to show that there are no massless Goldstone bosons in the model and to give a complete analysis of the general Higgs mechanism.[50][80] All three reached similar conclusions, despite their very different approaches: Higgs’ paper essentially used classical techniques, Englert and Brout’s involved calculating vacuum polarization in perturbation theory around an assumed symmetry-breaking vacuum state, and GHK used operator formalism and conservation laws to explore in depth the ways in which Goldstone’s theorem may be worked around.[51]

10.4 Theoretical properties

Main article: Higgs mechanism

10.4.1 Theoretical need for the Higgs

Gauge invariance is an important property of modern particle theories such as the Standard Model, partly due to its success in other areas of fundamental physics such as electromagnetism and the strong interaction (quantum chromodynamics). However, there were great difficulties in developing gauge theories for the weak nuclear force or a possible

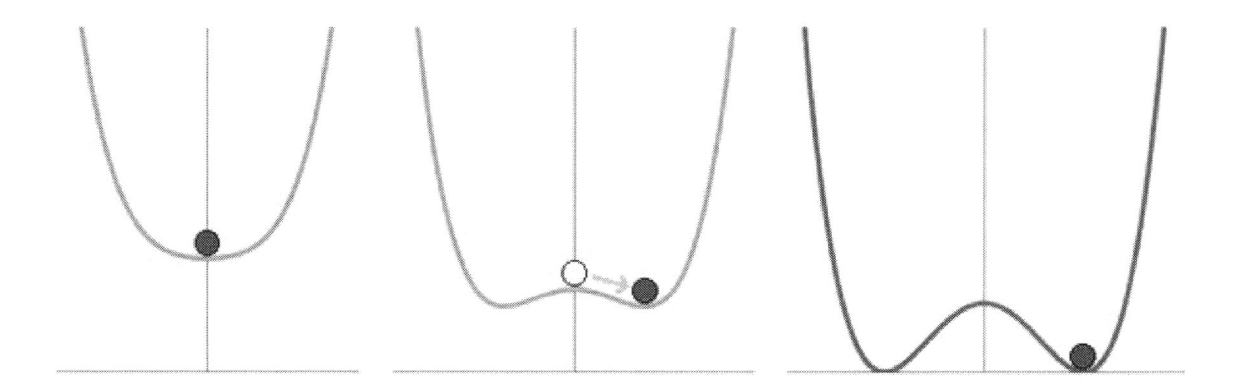

"Symmetry breaking illustrated": – At high energy levels (left) *the ball settles in the center, and the result is symmetrical. At lower energy levels* (right), *the overall "rules" remain symmetrical, but the "Mexican hat" potential comes into effect: "local" symmetry inevitably becomes broken since eventually the ball must at random roll one way or another.*

unified electroweak interaction. Fermions with a mass term would violate gauge symmetry and therefore cannot be gauge invariant. (This can be seen by examining the Dirac Lagrangian for a fermion in terms of left and right handed components; we find none of the spin-half particles could ever flip helicity as required for mass, so they must be massless.[Note 13]) W and Z bosons are observed to have mass, but a boson mass term contains terms, which clearly depend on the choice of gauge and therefore these masses too cannot be gauge invariant. Therefore, it seems that *none* of the standard model fermions *or* bosons could "begin" with mass as an inbuilt property except by abandoning gauge invariance. If gauge invariance were to be retained, then these particles had to be acquiring their mass by some other mechanism or interaction. Additionally, whatever was giving these particles their mass, had to not "break" gauge invariance as the basis for other parts of the theories where it worked well, *and* had to not require or predict unexpected massless particles and long-range forces (seemingly an inevitable consequence of Goldstone's theorem) which did not actually seem to exist in nature.

A solution to all of these overlapping problems came from the discovery of a previously unnoticed borderline case hidden in the mathematics of Goldstone's theorem,[Note 11] that under certain conditions it *might* theoretically be possible for a symmetry to be broken *without* disrupting gauge invariance and *without* any new massless particles or forces, and having "sensible" (renormalisable) results mathematically: this became known as the Higgs mechanism.

The Standard Model hypothesizes a field which is responsible for this effect, called the Higgs field (symbol: ϕ), which has the unusual property of a non-zero amplitude in its ground state; i.e., a non-zero vacuum expectation value. It can have this effect because of its unusual "Mexican hat" shaped potential whose lowest "point" is not at its "centre". Below a certain extremely high energy level the existence of this non-zero vacuum expectation spontaneously breaks electroweak gauge symmetry which in turn gives rise to the Higgs mechanism and triggers the acquisition of mass by those particles interacting with the field. This effect occurs because scalar field components of the Higgs field are "absorbed" by the massive bosons as degrees of freedom, and couple to the fermions via Yukawa coupling, thereby producing the expected mass terms. In effect when symmetry breaks under these conditions, the Goldstone bosons that arise *interact* with the Higgs field (and with other particles capable of interacting with the Higgs field) instead of becoming new massless particles, the intractable problems of both underlying theories "neutralise" each other, and the residual outcome is that elementary particles acquire a consistent mass based on how strongly they interact with the Higgs field. It is the simplest known process capable of giving mass to the gauge bosons while remaining compatible with gauge theories.[81] Its quantum would be a scalar boson, known as the Higgs boson.[82]

10.4.2 Properties of the Higgs field

In the Standard Model, the Higgs field is a scalar tachyonic field – 'scalar' meaning it does not transform under Lorentz transformations, and 'tachyonic' meaning the field (but not the particle) has imaginary mass and in certain configurations must undergo symmetry breaking. It consists of four components, two neutral ones and two charged component fields. Both of the charged components and one of the neutral fields are Goldstone bosons, which act as the longitudinal third-polarization components of the massive W^+, W^-, and Z bosons. The quantum of the remaining neutral component

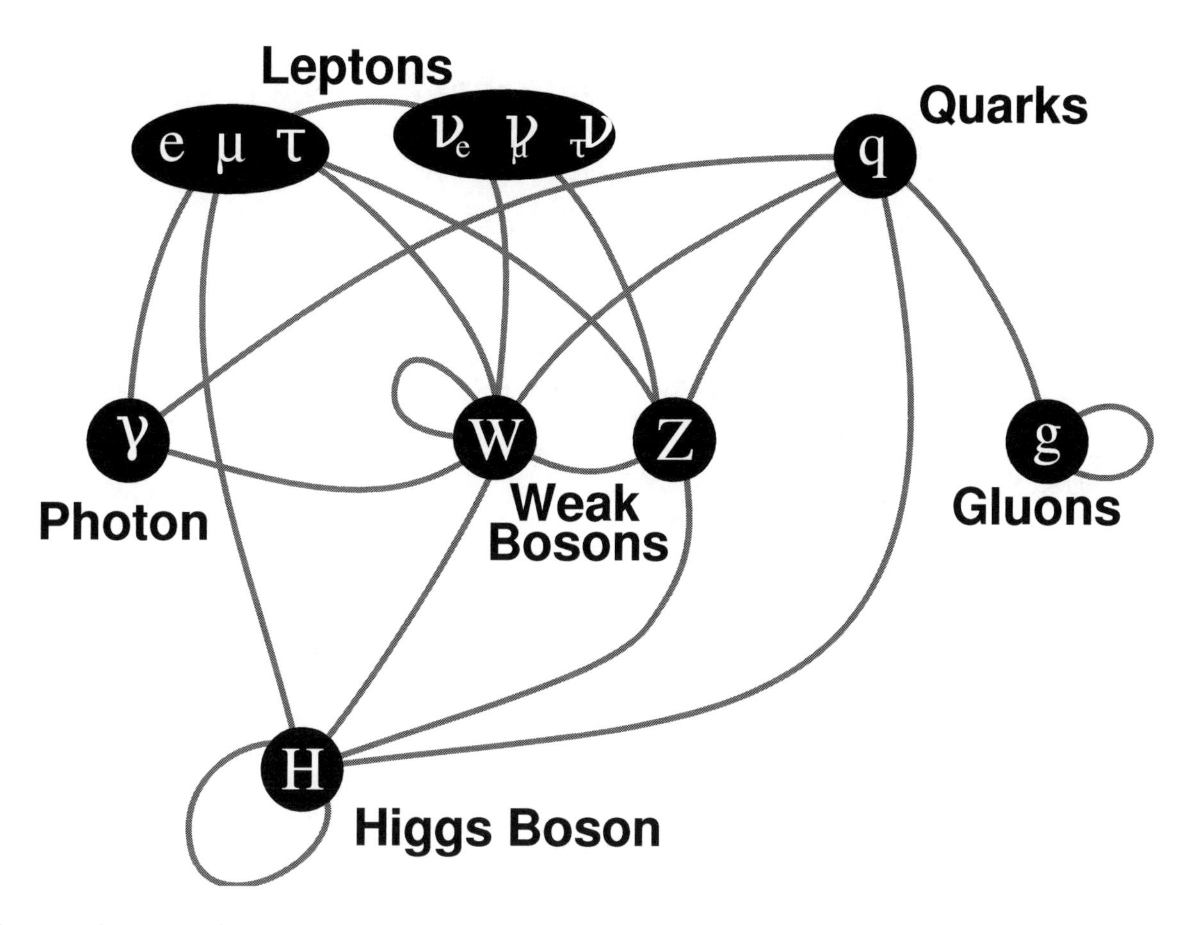

Summary of interactions between certain particles described by the Standard Model.

corresponds to (and is theoretically realised as) the massive Higgs boson,[83] this component can interact with fermions via Yukawa coupling to give them mass, as well.

Mathematically, the Higgs field has imaginary mass and is therefore a tachyonic field.[84] While tachyons (particles that move faster than light) are a purely hypothetical concept, fields with imaginary mass have come to play an important role in modern physics.[85][86] Under no circumstances do any excitations ever propagate faster than light in such theories — the presence or absence of a tachyonic mass has no effect whatsoever on the maximum velocity of signals (there is no violation of causality).[87] Instead of faster-than-light particles, the imaginary mass creates an instability:- any configuration in which one or more field excitations are tachyonic must spontaneously decay, and the resulting configuration contains no physical tachyons. This process is known as tachyon condensation, and is now believed to be the explanation for how the Higgs mechanism itself arises in nature, and therefore the reason behind electroweak symmetry breaking.

Although the notion of imaginary mass might seem troubling, it is only the field, and not the mass itself, that is quantized. Therefore, the field operators at spacelike separated points still commute (or anticommute), and information and particles still do not propagate faster than light.[88] Tachyon condensation drives a physical system that has reached a local limit and might naively be expected to produce physical tachyons, to an alternate stable state where no physical tachyons exist. Once a tachyonic field such as the Higgs field reaches the minimum of the potential, its quanta are not tachyons any more but rather are ordinary particles such as the Higgs boson.[89]

10.4.3 Properties of the Higgs boson

Since the Higgs field is scalar, the Higgs boson has no spin. The Higgs boson is also its own antiparticle and is CP-even, and has zero electric and colour charge.[90]

The Minimal Standard Model does not predict the mass of the Higgs boson.[91] If that mass is between 115 and 180

GeV/c^2, then the Standard Model can be valid at energy scales all the way up to the Planck scale (10^{19} GeV).[92] Many theorists expect new physics beyond the Standard Model to emerge at the TeV-scale, based on unsatisfactory properties of the Standard Model.[93] The highest possible mass scale allowed for the Higgs boson (or some other electroweak symmetry breaking mechanism) is 1.4 TeV; beyond this point, the Standard Model becomes inconsistent without such a mechanism, because unitarity is violated in certain scattering processes.[94]

It is also possible, although experimentally difficult, to estimate the mass of the Higgs boson indirectly. In the Standard Model, the Higgs boson has a number of indirect effects; most notably, Higgs loops result in tiny corrections to masses of W and Z bosons. Precision measurements of electroweak parameters, such as the Fermi constant and masses of W/Z bosons, can be used to calculate constraints on the mass of the Higgs. As of July 2011, the precision electroweak measurements tell us that the mass of the Higgs boson is likely to be less than about 161 GeV/c^2 at 95% confidence level (this upper limit would increase to 185 GeV/c^2 if the lower bound of 114.4 GeV/c^2 from the LEP-2 direct search is allowed for[95]). These indirect constraints rely on the assumption that the Standard Model is correct. It may still be possible to discover a Higgs boson above these masses if it is accompanied by other particles beyond those predicted by the Standard Model.[96]

10.4.4 Production

If Higgs particle theories are correct, then a Higgs particle can be produced much like other particles that are studied, in a particle collider. This involves accelerating a large number of particles to extremely high energies and extremely close to the speed of light, then allowing them to smash together. Protons and lead ions (the bare nuclei of lead atoms) are used at the LHC. In the extreme energies of these collisions, the desired esoteric particles will occasionally be produced and this can be detected and studied; any absence or difference from theoretical expectations can also be used to improve the theory. The relevant particle theory (in this case the Standard Model) will determine the necessary kinds of collisions and detectors. The Standard Model predicts that Higgs bosons could be formed in a number of ways,[97][98][99] although the probability of producing a Higgs boson in any collision is always expected to be very small—for example, only 1 Higgs boson per 10 billion collisions in the Large Hadron Collider.[Note 14] The most common expected processes for Higgs boson production are:

- *Gluon fusion.* If the collided particles are hadrons such as the proton or antiproton—as is the case in the LHC and Tevatron—then it is most likely that two of the gluons binding the hadron together collide. The easiest way to produce a Higgs particle is if the two gluons combine to form a loop of virtual quarks. Since the coupling of particles to the Higgs boson is proportional to their mass, this process is more likely for heavy particles. In practice it is enough to consider the contributions of virtual top and bottom quarks (the heaviest quarks). This process is the dominant contribution at the LHC and Tevatron being about ten times more likely than any of the other processes.[97][98]

- *Higgs Strahlung.* If an elementary fermion collides with an anti-fermion—e.g., a quark with an anti-quark or an electron with a positron—the two can merge to form a virtual W or Z boson which, if it carries sufficient energy, can then emit a Higgs boson. This process was the dominant production mode at the LEP, where an electron and a positron collided to form a virtual Z boson, and it was the second largest contribution for Higgs production at the Tevatron. At the LHC this process is only the third largest, because the LHC collides protons with protons, making a quark-antiquark collision less likely than at the Tevatron. Higgs Strahlung is also known as *associated production*.[97][98][99]

- *Weak boson fusion.* Another possibility when two (anti-)fermions collide is that the two exchange a virtual W or Z boson, which emits a Higgs boson. The colliding fermions do not need to be the same type. So, for example, an up quark may exchange a Z boson with an anti-down quark. This process is the second most important for the production of Higgs particle at the LHC and LEP.[97][99]

- *Top fusion.* The final process that is commonly considered is by far the least likely (by two orders of magnitude). This process involves two colliding gluons, which each decay into a heavy quark–antiquark pair. A quark and antiquark from each pair can then combine to form a Higgs particle.[97][98]

10.4.5 Decay

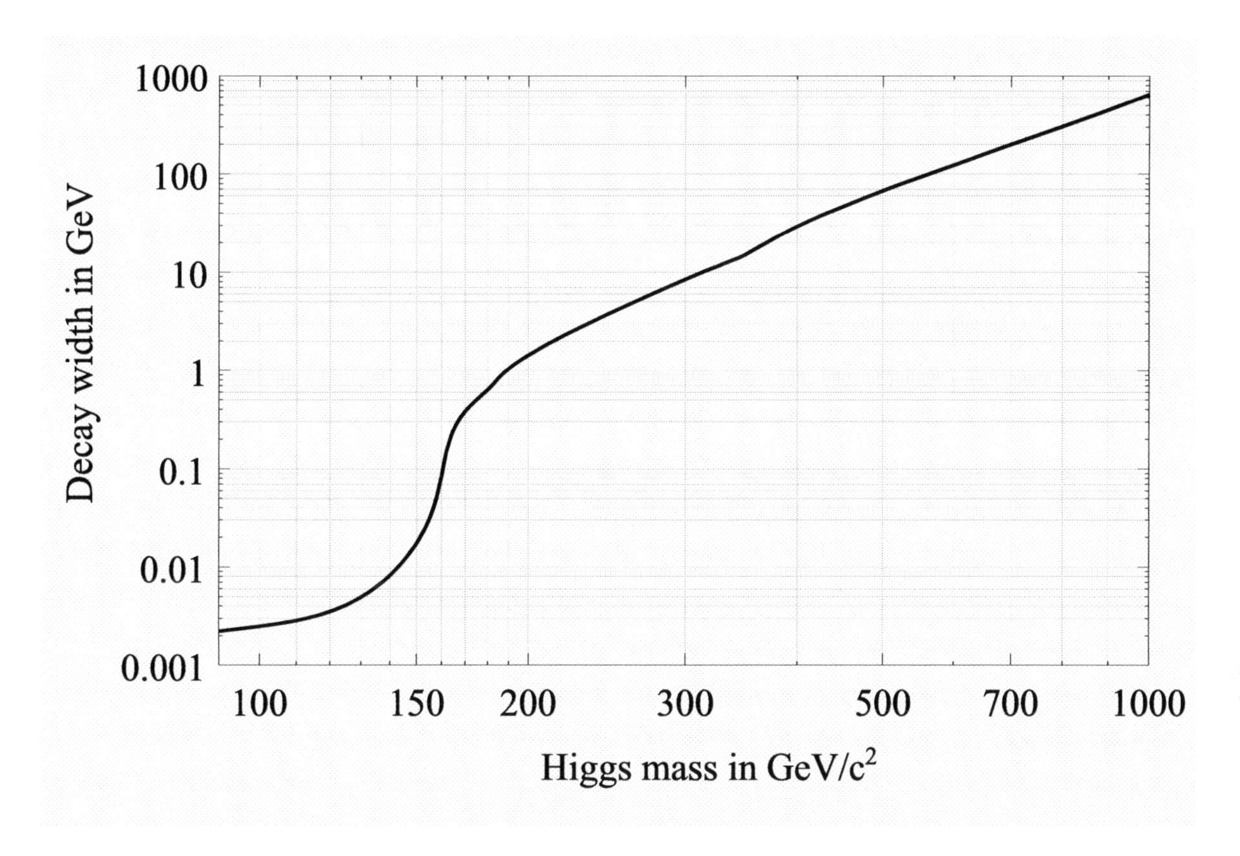

The Standard Model prediction for the decay width of the Higgs particle depends on the value of its mass.

Quantum mechanics predicts that if it is possible for a particle to decay into a set of lighter particles, then it will eventually do so.[101] This is also true for the Higgs boson. The likelihood with which this happens depends on a variety of factors including: the difference in mass, the strength of the interactions, etc. Most of these factors are fixed by the Standard Model, except for the mass of the Higgs boson itself. For a Higgs boson with a mass of 126 GeV/c^2 the SM predicts a mean life time of about 1.6×10^{-22} s.[Note 2]

Since it interacts with all the massive elementary particles of the SM, the Higgs boson has many different processes through which it can decay. Each of these possible processes has its own probability, expressed as the *branching ratio*; the fraction of the total number decays that follows that process. The SM predicts these branching ratios as a function of the Higgs mass (see plot).

One way that the Higgs can decay is by splitting into a fermion–antifermion pair. As general rule, the Higgs is more likely to decay into heavy fermions than light fermions, because the mass of a fermion is proportional to the strength of its interaction with the Higgs.[102] By this logic the most common decay should be into a top–antitop quark pair. However, such a decay is only possible if the Higgs is heavier than ~346 GeV/c^2, twice the mass of the top quark. For a Higgs mass of 126 GeV/c^2 the SM predicts that the most common decay is into a bottom–antibottom quark pair, which happens 56.1% of the time.[5] The second most common fermion decay at that mass is a tau–antitau pair, which happens only about 6% of the time.[5]

Another possibility is for the Higgs to split into a pair of massive gauge bosons. The most likely possibility is for the Higgs to decay into a pair of W bosons (the light blue line in the plot), which happens about 23.1% of the time for a Higgs boson with a mass of 126 GeV/c^2.[5] The W bosons can subsequently decay either into a quark and an antiquark or into a charged lepton and a neutrino. However, the decays of W bosons into quarks are difficult to distinguish from the background, and the decays into leptons cannot be fully reconstructed (because neutrinos are impossible to detect in particle collision experiments). A cleaner signal is given by decay into a pair of Z-bosons (which happens about 2.9% of the time for a Higgs with a mass of 126 GeV/c^2),[5] if each of the bosons subsequently decays into a pair of easy-to-detect

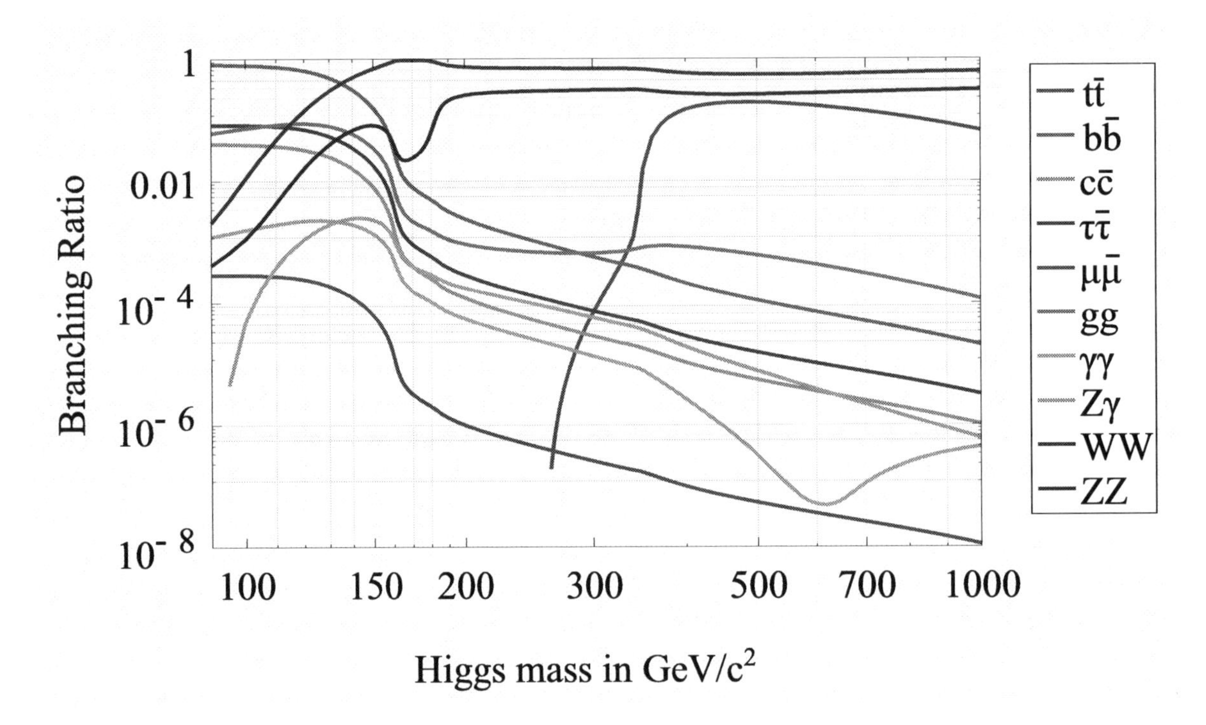

The Standard Model prediction for the branching ratios of the different decay modes of the Higgs particle depends on the value of its mass.

charged leptons (electrons or muons).

Decay into massless gauge bosons (i.e., gluons or photons) is also possible, but requires intermediate loop of virtual heavy quarks (top or bottom) or massive gauge bosons.[102] The most common such process is the decay into a pair of gluons through a loop of virtual heavy quarks. This process, which is the reverse of the gluon fusion process mentioned above, happens approximately 8.5% of the time for a Higgs boson with a mass of 126 GeV/c^2.[5] Much rarer is the decay into a pair of photons mediated by a loop of W bosons or heavy quarks, which happens only twice for every thousand decays.[5] However, this process is very relevant for experimental searches for the Higgs boson, because the energy and momentum of the photons can be measured very precisely, giving an accurate reconstruction of the mass of the decaying particle.[102]

10.4.6 Alternative models

Main article: Alternatives to the Standard Model Higgs

The Minimal Standard Model as described above is the simplest known model for the Higgs mechanism with just one Higgs field. However, an extended Higgs sector with additional Higgs particle doublets or triplets is also possible, and many extensions of the Standard Model have this feature. The non-minimal Higgs sector favoured by theory are the two-Higgs-doublet models (2HDM), which predict the existence of a quintet of scalar particles: two CP-even neutral Higgs bosons h^0 and H^0, a CP-odd neutral Higgs boson A^0, and two charged Higgs particles $H^{\pm}$. Supersymmetry ("SUSY") also predicts relations between the Higgs-boson masses and the masses of the gauge bosons, and could accommodate a 125 GeV/c^2 neutral Higgs boson.

The key method to distinguish between these different models involves study of the particles' interactions ("coupling") and exact decay processes ("branching ratios"), which can be measured and tested experimentally in particle collisions. In the Type-I 2HDM model one Higgs doublet couples to up and down quarks, while the second doublet does not couple to quarks. This model has two interesting limits, in which the lightest Higgs couples to just fermions ("gauge-phobic") or just gauge bosons ("fermiophobic"), but not both. In the Type-II 2HDM model, one Higgs doublet only couples to up-type quarks, the other only couples to down-type quarks.[103] The heavily researched Minimal Supersymmetric Standard

Model (MSSM) includes a Type-II 2HDM Higgs sector, so it could be disproven by evidence of a Type-I 2HDM Higgs.

In other models the Higgs scalar is a composite particle. For example, in technicolor the role of the Higgs field is played by strongly bound pairs of fermions called techniquarks. Other models, feature pairs of top quarks (see top quark condensate). In yet other models, there is no Higgs field at all and the electroweak symmetry is broken using extra dimensions.[104][105]

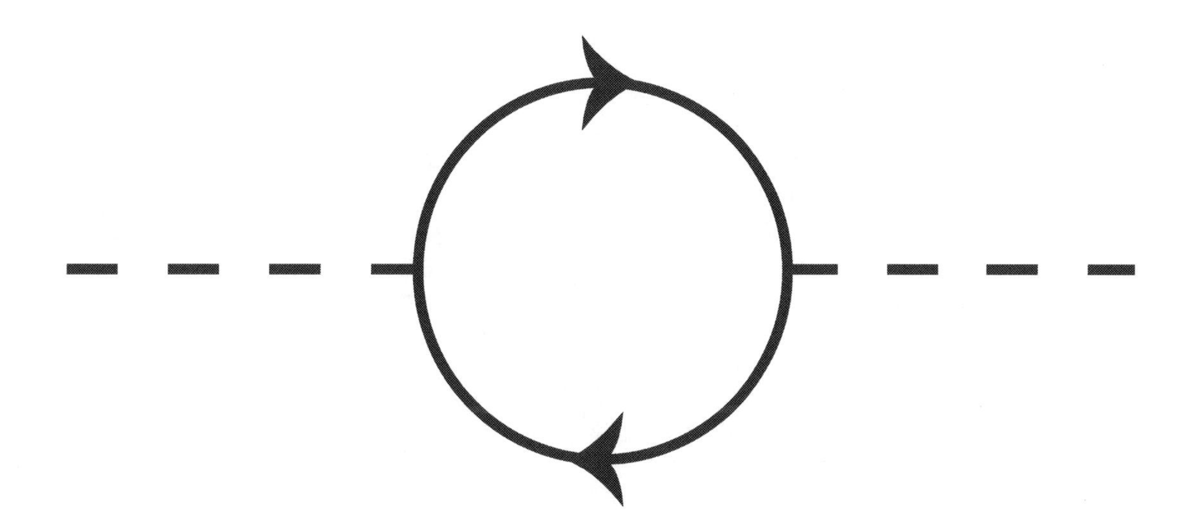

A one-loop Feynman diagram of the first-order correction to the Higgs mass. In the Standard Model the effects of these corrections are potentially enormous, giving rise to the so-called hierarchy problem.

10.4.7 Further theoretical issues and hierarchy problem

Main articles: Hierarchy problem and Hierarchy problem § The Higgs mass

The Standard Model leaves the mass of the Higgs boson as a parameter to be measured, rather than a value to be calculated. This is seen as theoretically unsatisfactory, particularly as quantum corrections (related to interactions with virtual particles) should apparently cause the Higgs particle to have a mass immensely higher than that observed, but at the same time the Standard Model requires a mass of the order of 100 to 1000 GeV to ensure unitarity (in this case, to unitarise longitudinal vector boson scattering).[106] Reconciling these points appears to require explaining why there is an almost-perfect cancellation resulting in the visible mass of ~ 125 GeV, and it is not clear how to do this. Because the weak force is about 10^{32} times stronger than gravity, and (linked to this) the Higgs boson's mass is so much less than the Planck mass or the grand unification energy, it appears that either there is some underlying connection or reason for these observations which is unknown and not described by the Standard Model, or some unexplained and extremely precise fine-tuning of parameters – however at present neither of these explanations is proven. This is known as a hierarchy problem.[107] More broadly, the hierarchy problem amounts to the worry that a future theory of fundamental particles and interactions should not have excessive fine-tunings or unduly delicate cancellations, and should allow masses of particles such as the Higgs boson to be calculable. The problem is in some ways unique to spin-0 particles (such as the Higgs boson), which can give rise to issues related to quantum corrections that do not affect particles with spin.[106] A number of solutions have been proposed, including supersymmetry, conformal solutions and solutions via extra dimensions such as braneworld models.

There are also issues of quantum triviality, which suggests that it may not be possible to create a consistent quantum field theory involving elementary scalar particles.

10.5 Experimental search

Main article: Search for the Higgs boson

To produce Higgs bosons, two beams of particles are accelerated to very high energies and allowed to collide within a particle detector. Occasionally, although rarely, a Higgs boson will be created fleetingly as part of the collision byproducts. Because the Higgs boson decays very quickly, particle detectors cannot detect it directly. Instead the detectors register all the decay products (the *decay signature*) and from the data the decay process is reconstructed. If the observed decay products match a possible decay process (known as a *decay channel*) of a Higgs boson, this indicates that a Higgs boson may have been created. In practice, many processes may produce similar decay signatures. Fortunately, the Standard Model precisely predicts the likelihood of each of these, and each known process, occurring. So, if the detector detects more decay signatures consistently matching a Higgs boson than would otherwise be expected if Higgs bosons did not exist, then this would be strong evidence that the Higgs boson exists.

Because Higgs boson production in a particle collision is likely to be very rare (1 in 10 billion at the LHC),[Note 14] and many other possible collision events can have similar decay signatures, the data of hundreds of trillions of collisions needs to be analysed and must "show the same picture" before a conclusion about the existence of the Higgs boson can be reached. To conclude that a new particle has been found, particle physicists require that the statistical analysis of two independent particle detectors each indicate that there is lesser than a one-in-a-million chance that the observed decay signatures are due to just background random Standard Model events—i.e., that the observed number of events is more than 5 standard deviations (sigma) different from that expected if there was no new particle. More collision data allows better confirmation of the physical properties of any new particle observed, and allows physicists to decide whether it is indeed a Higgs boson as described by the Standard Model or some other hypothetical new particle.

To find the Higgs boson, a powerful particle accelerator was needed, because Higgs bosons might not be seen in lower-energy experiments. The collider needed to have a high luminosity in order to ensure enough collisions were seen for conclusions to be drawn. Finally, advanced computing facilities were needed to process the vast amount of data (25 petabytes per year as at 2012) produced by the collisions.[109] For the announcement of 4 July 2012, a new collider known as the Large Hadron Collider was constructed at CERN with a planned eventual collision energy of 14 TeV—over seven times any previous collider—and over 300 trillion (3×10^{14}) LHC proton–proton collisions were analysed by the LHC Computing Grid, the world's largest computing grid (as of 2012), comprising over 170 computing facilities in a worldwide network across 36 countries.[109][110][111]

10.5.1 Search prior to 4 July 2012

The first extensive search for the Higgs boson was conducted at the Large Electron–Positron Collider (LEP) at CERN in the 1990s. At the end of its service in 2000, LEP had found no conclusive evidence for the Higgs.[Note 15] This implied that if the Higgs boson were to exist it would have to be heavier than 114.4 GeV/c^2.[112]

The search continued at Fermilab in the United States, where the Tevatron—the collider that discovered the top quark in 1995—had been upgraded for this purpose. There was no guarantee that the Tevatron would be able to find the Higgs, but it was the only supercollider that was operational since the Large Hadron Collider (LHC) was still under construction and the planned Superconducting Super Collider had been cancelled in 1993 and never completed. The Tevatron was only able to exclude further ranges for the Higgs mass, and was shut down on 30 September 2011 because it no longer could keep up with the LHC. The final analysis of the data excluded the possibility of a Higgs boson with a mass between 147 GeV/c^2 and 180 GeV/c^2. In addition, there was a small (but not significant) excess of events possibly indicating a Higgs boson with a mass between 115 GeV/c^2 and 140 GeV/c^2.[113]

The Large Hadron Collider at CERN in Switzerland, was designed specifically to be able to either confirm or exclude the existence of the Higgs boson. Built in a 27 km tunnel under the ground near Geneva originally inhabited by LEP, it was designed to collide two beams of protons, initially at energies of 3.5 TeV per beam (7 TeV total), or almost 3.6 times that of the Tevatron, and upgradeable to 2 × 7 TeV (14 TeV total) in future. Theory suggested if the Higgs boson existed, collisions at these energy levels should be able to reveal it. As one of the most complicated scientific instruments ever built, its operational readiness was delayed for 14 months by a magnet quench event nine days after its inaugural tests, caused by a faulty electrical connection that damaged over 50 superconducting magnets and contaminated the vacuum system.[114][115][116]

Data collection at the LHC finally commenced in March 2010.[117] By December 2011 the two main particle detectors at the LHC, ATLAS and CMS, had narrowed down the mass range where the Higgs could exist to around 116-130 GeV (ATLAS) and 115-127 GeV (CMS).[118][119] There had also already been a number of promising event excesses that had "evaporated" and proven to be nothing but random fluctuations. However, from around May 2011,[120] both experiments had seen among their results, the slow emergence of a small yet consistent excess of gamma and 4-lepton decay signatures and several other particle decays, all hinting at a new particle at a mass around 125 GeV.[120] By around November 2011, the anomalous data at 125 GeV was becoming "too large to ignore" (although still far from conclusive), and the team leaders at both ATLAS and CMS each privately suspected they might have found the Higgs.[120] On November 28, 2011, at an internal meeting of the two team leaders and the director general of CERN, the latest analyses were discussed outside their teams for the first time, suggesting both ATLAS and CMS might be converging on a possible shared result at 125 GeV, and initial preparations commenced in case of a successful finding.[120] While this information was not known publicly at the time, the narrowing of the possible Higgs range to around 115–130 GeV and the repeated observation of small but consistent event excesses across multiple channels at both ATLAS and CMS in the 124-126 GeV region (described as "tantalising hints" of around 2-3 sigma) were public knowledge with "a lot of interest".[121] It was therefore widely anticipated around the end of 2011, that the LHC would provide sufficient data to either exclude or confirm the finding of a Higgs boson by the end of 2012, when their 2012 collision data (with slightly higher 8 TeV collision energy) had been examined.[121][122]

10.5.2 Discovery of candidate boson at CERN

On 22 June 2012 CERN announced an upcoming seminar covering tentative findings for 2012,[126][127] and shortly afterwards (from around 1 July 2012 according to an analysis of the spreading rumour in social media[128]) rumours began to spread in the media that this would include a major announcement, but it was unclear whether this would be a stronger signal or a formal discovery.[129][130] Speculation escalated to a "fevered" pitch when reports emerged that Peter Higgs, who proposed the particle, was to be attending the seminar,[131][132] and that "five leading physicists" had been invited – generally believed to signify the five living 1964 authors – with Higgs, Englert, Guralnik, Hagen attending and Kibble confirming his invitation (Brout having died in 2011).[133][134]

On 4 July 2012 both of the CERN experiments announced they had independently made the same discovery:[135] CMS of a previously unknown boson with mass 125.3 ± 0.6 GeV/c^2[136][137] and ATLAS of a boson with mass 126.0 ± 0.6 GeV/c^2.[138][139] Using the combined analysis of two interaction types (known as 'channels'), both experiments independently reached a local significance of 5 sigma — implying that the probability of getting at least as strong a result by chance alone is less than 1 in 3 million. When additional channels were taken into account, the CMS significance was reduced to 4.9 sigma.[137]

The two teams had been working 'blinded' from each other from around late 2011 or early 2012,[120] meaning they did not discuss their results with each other, providing additional certainty that any common finding was genuine validation of a particle.[109] This level of evidence, confirmed independently by two separate teams and experiments, meets the formal level of proof required to announce a confirmed discovery.

On 31 July 2012, the ATLAS collaboration presented additional data analysis on the "observation of a new particle", including data from a third channel, which improved the significance to 5.9 sigma (1 in 588 million chance of obtaining at least as strong evidence by random background effects alone) and mass 126.0 ± 0.4 (stat) ± 0.4 (sys) GeV/c^2, [139] and CMS improved the significance to 5-sigma and mass 125.3 ± 0.4 (stat) ± 0.5 (sys) GeV/c^2.[136]

10.5.3 The new particle tested as a possible Higgs boson

Following the 2012 discovery, it was still unconfirmed whether or not the 125 GeV/c^2 particle was a Higgs boson. On one hand, observations remained consistent with the observed particle being the Standard Model Higgs boson, and the particle decayed into at least some of the predicted channels. Moreover, the production rates and branching ratios for the observed channels broadly matched the predictions by the Standard Model within the experimental uncertainties. However, the experimental uncertainties currently still left room for alternative explanations, meaning an announcement of the discovery of a Higgs boson would have been premature.[102] To allow more opportunity for data collection, the

LHC's proposed 2012 shutdown and 2013–14 upgrade were postponed by 7 weeks into 2013.[140]

In November 2012, in a conference in Kyoto researchers said evidence gathered since July was falling into line with the basic Standard Model more than its alternatives, with a range of results for several interactions matching that theory's predictions.[141] Physicist Matt Strassler highlighted "considerable" evidence that the new particle is not a pseudoscalar negative parity particle (consistent with this required finding for a Higgs boson), "evaporation" or lack of increased significance for previous hints of non-Standard Model findings, expected Standard Model interactions with W and Z bosons, absence of "significant new implications" for or against supersymmetry, and in general no significant deviations to date from the results expected of a Standard Model Higgs boson.[142] However some kinds of extensions to the Standard Model would also show very similar results;[143] so commentators noted that based on other particles that are still being understood long after their discovery, it may take years to be sure, and decades to fully understand the particle that has been found.[141][142]

These findings meant that as of January 2013, scientists were very sure they had found an unknown particle of mass ~ 125 GeV/c^2, and had not been misled by experimental error or a chance result. They were also sure, from initial observations, that the new particle was some kind of boson. The behaviours and properties of the particle, so far as examined since July 2012, also seemed quite close to the behaviours expected of a Higgs boson. Even so, it could still have been a Higgs boson or some other unknown boson, since future tests could show behaviours that do not match a Higgs boson, so as of December 2012 CERN still only stated that the new particle was "consistent with" the Higgs boson,[11][13] and scientists did not yet positively say it was the Higgs boson.[144] Despite this, in late 2012, widespread media reports announced (incorrectly) that a Higgs boson had been confirmed during the year.[Note 16]

In January 2013, CERN director-general Rolf-Dieter Heuer stated that based on data analysis to date, an answer could be possible 'towards' mid-2013,[150] and the deputy chair of physics at Brookhaven National Laboratory stated in February 2013 that a "definitive" answer might require "another few years" after the collider's 2015 restart.[151] In early March 2013, CERN Research Director Sergio Bertolucci stated that confirming spin-0 was the major remaining requirement to determine whether the particle is at least some kind of Higgs boson.[152]

10.5.4 Preliminary confirmation of existence and current status

On 14 March 2013 CERN confirmed that:

> "CMS and ATLAS have compared a number of options for the spin-parity of this particle, and these all prefer no spin and even parity [two fundamental criteria of a Higgs boson consistent with the Standard Model]. This, coupled with the measured interactions of the new particle with other particles, strongly indicates that it is a Higgs boson." [1]

This also makes the particle the first elementary scalar particle to be discovered in nature.[14]

Examples of tests used to validate whether the 125 GeV particle is a Higgs boson:[142][153]

10.6 Public discussion

10.6.1 Naming

Names used by physicists

The name most strongly associated with the particle and field is the Higgs boson[79]:168 and Higgs field. For some time the particle was known by a combination of its PRL author names (including at times Anderson), for example the Brout–Englert–Higgs particle, the Anderson-Higgs particle, or the Englert–Brout–Higgs–Guralnik–Hagen–Kibble mechanism,[Note 17] and these are still used at times.[51][160] Fueled in part by the issue of recognition and a potential shared Nobel Prize,[160][161] the most appropriate name is still occasionally a topic of debate as at 2012.[160] (Higgs himself prefers to call the particle either by an acronym of all those involved, or "the scalar boson", or "the so-called Higgs particle".[161])

A considerable amount has been written on how Higgs' name came to be exclusively used. Two main explanations are offered.

Nickname

The Higgs boson is often referred to as the "God particle" in popular media outside the scientific community.[170][171][172][173][174] The nickname comes from the title of the 1993 book on the Higgs boson and particle physics - The God Particle: If the Universe Is the Answer, What Is the Question? by Nobel Physics prizewinner and Fermilab director Leon Lederman.[21] Lederman wrote it in the context of failing US government support for the Superconducting Super Collider,[175] a part-constructed titanic[176][177] competitor to the Large Hadron Collider with planned collision energies of 2 × 20 TeV that was championed by Lederman since its 1983 inception[175][178][179] and shut down in 1993. The book sought in part to promote awareness of the significance and need for such a project in the face of its possible loss of funding.[180] Lederman, a leading researcher in the field, wanted to title his book "The Goddamn Particle: If the Universe is the Answer, What is the Question?" But his editor decided that the title was too controversial and convinced Lederman to change the title to "The God Particle: If the Universe is the Answer, What is the Question?"[181]

And since the Higgs Boson deals with how matter was formed at the time of the big bang, and since newspapers loved the term, the term "God particle" was used.

While media use of this term may have contributed to wider awareness and interest,[182] many scientists feel the name is inappropriate[16][17][183] since it is sensational hyperbole and misleads readers;[184] the particle also has nothing to do with God, leaves open numerous questions in fundamental physics, and does not explain the ultimate origin of the universe. Higgs, an atheist, was reported to be displeased and stated in a 2008 interview that he found it "embarrassing" because it was "the kind of misuse... which I think might offend some people".[184][185][186] Science writer Ian Sample stated in his 2010 book on the search that the nickname is "universally hate[d]" by physicists and perhaps the "worst derided" in the history of physics, but that (according to Lederman) the publisher rejected all titles mentioning "Higgs" as unimaginative and too unknown.[187]

Lederman begins with a review of the long human search for knowledge, and explains that his tongue-in-cheek title draws an analogy between the impact of the Higgs field on the fundamental symmetries at the Big Bang, and the apparent chaos of structures, particles, forces and interactions that resulted and shaped our present universe, with the biblical story of Babel in which the primordial single language of early Genesis was fragmented into many disparate languages and cultures.[188]

> Today ... we have the standard model, which reduces all of reality to a dozen or so particles and four forces. ... It's a hard-won simplicity [...and...] remarkably accurate. But it is also incomplete and, in fact, internally inconsistent... This boson is so central to the state of physics today, so crucial to our final understanding of the structure of matter, yet so elusive, that I have given it a nickname: the God Particle. Why God Particle? Two reasons. One, the publisher wouldn't let us call it the Goddamn Particle, though that might be a more appropriate title, given its villainous nature and the expense it is causing. And two, there is a connection, of sorts, to another book, a *much* older one...
> —Leon M. Lederman and Dick Teresi, *The God Particle: If the Universe is the Answer, What is the Question*[21] p. 22

Lederman asks whether the Higgs boson was added just to perplex and confound those seeking knowledge of the universe, and whether physicists will be confounded by it as recounted in that story, or ultimately surmount the challenge and understand "how beautiful is the universe [God has] made".[189]

Other proposals

A renaming competition by British newspaper *The Guardian* in 2009 resulted in their science correspondent choosing the name "the champagne bottle boson" as the best submission: "The bottom of a champagne bottle is in the shape of the Higgs potential and is often used as an illustration in physics lectures. So it's not an embarrassingly grandiose name, it is memorable, and [it] has some physics connection too."[190] The name *Higgson* was suggested as well, in an opinion piece in the Institute of Physics' online publication *physicsworld.com*.[191]

10.6.2 Media explanations and analogies

There has been considerable public discussion of analogies and explanations for the Higgs particle and how the field creates mass,[192][193] including coverage of explanatory attempts in their own right and a competition in 1993 for the best popular explanation by then-UK Minister for Science Sir William Waldegrave[194] and articles in newspapers worldwide.

An educational collaboration involving an LHC physicist and a High School Teachers at CERN educator suggests that dispersion of light – responsible for the rainbow and dispersive prism – is a useful analogy for the Higgs field's symmetry breaking and mass-causing effect.[195]

Matt Strassler uses electric fields as an analogy:[196]

> Some particles interact with the Higgs field while others don't. Those particles that feel the Higgs field act as if they have mass. Something similar happens in an electric field – charged objects are pulled around and neutral objects can sail through unaffected. So you can think of the Higgs search as an attempt to make waves in the Higgs field *[create Higgs bosons]* to prove it's really there.

A similar explanation was offered by *The Guardian*:[197]

> The Higgs boson is essentially a ripple in a field said to have emerged at the birth of the universe and to span the cosmos to this day ... The particle is crucial however: it is the smoking gun, the evidence required to show the theory is right.

The Higgs field's effect on particles was famously described by physicist David Miller as akin to a room full of political party workers spread evenly throughout a room: the crowd gravitates to and slows down famous people but does not slow down others.[Note 18] He also drew attention to well-known effects in solid state physics where an electron's effective mass can be much greater than usual in the presence of a crystal lattice.[198]

Analogies based on drag effects, including analogies of "syrup" or "molasses" are also well known, but can be somewhat misleading since they may be understood (incorrectly) as saying that the Higgs field simply resists some particles' motion but not others' – a simple resistive effect could also conflict with Newton's third law.[200]

10.6.3 Recognition and awards

There has been considerable discussion of how to allocate the credit if the Higgs boson is proven, made more pointed as a Nobel prize had been expected, and the very wide basis of people entitled to consideration. These include a range of theoreticians who made the Higgs mechanism theory possible, the theoreticians of the 1964 PRL papers (including Higgs himself), the theoreticians who derived from these, a working electroweak theory and the Standard Model itself, and also the experimentalists at CERN and other institutions who made possible the proof of the Higgs field and boson in reality. The Nobel prize has a limit of 3 persons to share an award, and some possible winners are already prize holders for other work, or are deceased (the prize is only awarded to persons in their lifetime). Existing prizes for works relating to the Higgs field, boson, or mechanism include:

Photograph of light passing through a dispersive prism: the rainbow effect arises because photons are not all affected to the same degree by the dispersive material of the prism.

- Nobel Prize in Physics (1979) – Glashow, Salam, and Weinberg, *for contributions to the theory of the unified weak and electromagnetic interaction between elementary particles* [201]

- Nobel Prize in Physics (1999) – 't Hooft and Veltman, *for elucidating the quantum structure of electroweak inter-*

actions in physics [202]

- Nobel Prize in Physics (2008) – Nambu (shared), *for the discovery of the mechanism of spontaneous broken symmetry in subatomic physics* [53]
- J. J. Sakurai Prize for Theoretical Particle Physics (2010) – Hagen, Englert, Guralnik, Higgs, Brout, and Kibble, *for elucidation of the properties of spontaneous symmetry breaking in four-dimensional relativistic gauge theory and of the mechanism for the consistent generation of vector boson masses* [77] (for the 1964 papers described above)
- Wolf Prize (2004) – Englert, Brout, and Higgs
- Nobel Prize in Physics (2013) - Peter Higgs and François Englert, *for the theoretical discovery of a mechanism that contributes to our understanding of the origin of mass of subatomic particles, and which recently was confirmed through the discovery of the predicted fundamental particle, by the ATLAS and CMS experiments at CERN's Large Hadron Collider* [203]

Additionally Physical Review Letters' 50-year review (2008) recognized the 1964 PRL symmetry breaking papers and Weinberg's 1967 paper *A model of Leptons* (the most cited paper in particle physics, as of 2012) "milestone Letters".[75]

Following reported observation of the Higgs-like particle in July 2012, several Indian media outlets reported on the supposed neglect of credit to Indian physicist Satyendra Nath Bose after whose work in the 1920s the class of particles "bosons" is named[204][205] (although physicists have described Bose's connection to the discovery as tenuous).[206]

10.7 Technical aspects and mathematical formulation

See also: Standard Model (mathematical formulation)

In the Standard Model, the Higgs field is a four-component scalar field that forms a complex doublet of the weak isospin SU(2) symmetry:

while the field has charge +1/2 under the weak hypercharge U(1) symmetry (in the convention where the electric charge, *Q*, the weak isospin, I_3, and the weak hypercharge, *Y*, are related by $Q = I_3 + Y$).[207]

The Higgs part of the Lagrangian is[207]

where W_μ^a and B_μ are the gauge bosons of the SU(2) and U(1) symmetries, g and g' their respective coupling constants, $\tau^a = \sigma^a/2$ (where σ^a are the Pauli matrices) a complete set generators of the SU(2) symmetry, and $\lambda > 0$ and $\mu^2 > 0$, so that the ground state breaks the SU(2) symmetry (see figure). The ground state of the Higgs field (the bottom of the potential) is degenerate with different ground states related to each other by a SU(2) gauge transformation. It is always possible to pick a gauge such that in the ground state $\phi^1 = \phi^2 = \phi^3 = 0$. The expectation value of ϕ^0 in the ground state (the vacuum expectation value or vev) is then $\langle\phi^0\rangle = v$, where $v = \frac{|\mu|}{\sqrt{\lambda}}$. The measured value of this parameter is ~246 GeV/c^2.[102] It has units of mass, and is the only free parameter of the Standard Model that is not a dimensionless number. Quadratic terms in W_μ and B_μ arise, which give masses to the *W* and *Z* bosons:[207]

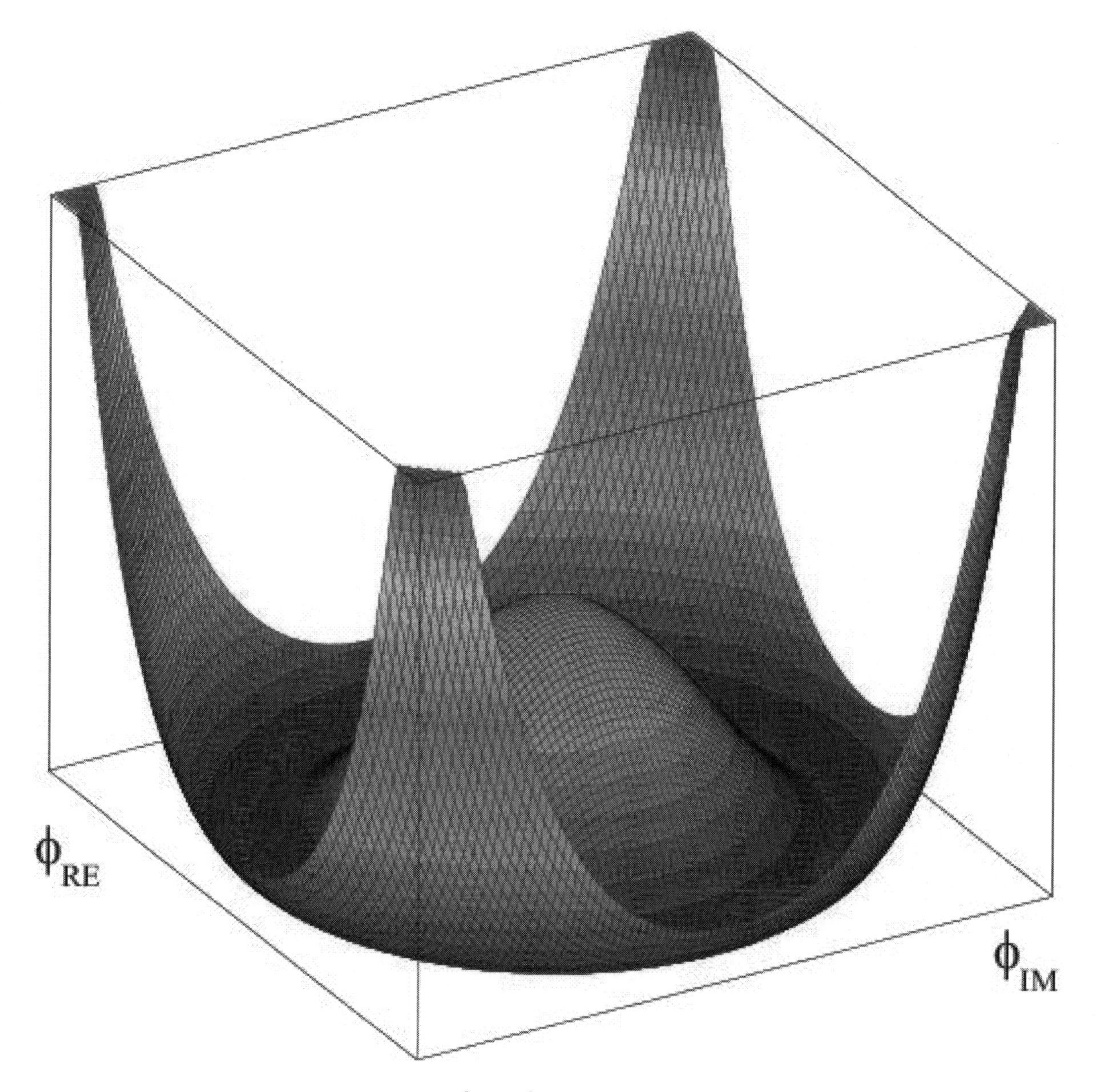

The potential for the Higgs field, plotted as function of ϕ^0 and ϕ^3. It has a Mexican-hat *or* champagne-bottle profile *at the ground.*

with their ratio determining the Weinberg angle, $\cos\theta_W = \frac{M_W}{M_Z} = \frac{|g|}{\sqrt{g^2+g'^2}}$, and leave a massless U(1) photon, γ.

The quarks and the leptons interact with the Higgs field through Yukawa interaction terms:

where $(d, u, e, \nu)^i_{L,R}$ are left-handed and right-handed quarks and leptons of the *i*th generation, $\lambda^{ij}_{u,d,e}$ are matrices of Yukawa couplings where h.c. denotes the hermitian conjugate terms. In the symmetry breaking ground state, only the terms containing ϕ^0 remain, giving rise to mass terms for the fermions. Rotating the quark and lepton fields to the basis where the matrices of Yukawa couplings are diagonal, one gets

where the masses of the fermions are $m^i_{u,d,e} = \lambda^i_{u,d,e} v/\sqrt{2}$, and $\lambda^i_{u,d,e}$ denote the eigenvalues of the Yukawa matrices.[207]

10.8 See also

Standard Model

- Quantum gauge theory
- History of quantum field theory
- Introduction to quantum mechanics
- Noncommutative standard model and noncommutative geometry generally
- Standard Model (mathematical formulation) (and especially Standard Model fields overview and mass terms and the Higgs mechanism)

Other

- Bose–Einstein statistics
- Dalitz plot
- Higgs boson in fiction
- Quantum triviality
- ZZ diboson
- Scalar boson
- Stueckelberg action
- Tachyonic field

10.9 Notes

[1] Note that such events also occur due to other processes. Detection involves a statistically significant excess of such events at specific energies.

[2] In the Standard Model, the total decay width of a Higgs boson with a mass of 126 GeV/c^2 is predicted to be 4.21×10^{-3} GeV.[5] The mean lifetime is given by $\tau = \hbar/\Gamma$.

[3] The range of a force is inversely proportional to the mass of the particles transmitting it.[19] In the Standard Model, forces are carried by virtual particles. These particles' movement and interactions with each other are limited by the energy–time uncertainty principle. As a result, the more massive a single virtual particle is, the greater its energy, and therefore the shorter the distance it can travel. A particle's mass therefore determines the maximum distance at which it can interact with other particles and on any force it mediates. By the same token, the reverse is also true: massless and near-massless particles can carry long distance forces. *(See also: Compton wavelength and Static forces and virtual-particle exchange)* Since experiments have shown that the weak force acts over only a very short range, this implies that there must exist massive gauge bosons. And indeed, their masses have since been confirmed by measurement.

[4] It is quite common for a law of physics to hold true only if certain assumptions held true or only under certain conditions. For example, Newton's laws of motion apply only at speeds where relativistic effects are negligible; and laws related to conductivity, gases, and classical physics (as opposed to quantum mechanics) may apply only within certain ranges of size, temperature, pressure, or other conditions.

[5] Electroweak symmetry is broken by the Higgs field in its lowest energy state, called its "ground state". At high energy levels this does not happen, and the gauge bosons of the weak force would therefore be expected to be massless.

[6] By the 1960s, many had already started to see gauge theories as failing to explain particle physics because theorists had been unable to solve the mass problem or even explain how gauge theory could provide a solution. So the idea that the Standard Model – which relied on a "Higgs field" not yet proved to exist – could be fundamentally incorrect was far from fanciful. Against this, once the entire model was developed around 1972, no better theory existed, and its predictions and solutions were so accurate, that it became the preferred theory anyway. It then became crucial to science, to know whether it was *correct*.

[7] The success of the Higgs-based electroweak theory and Standard Model is illustrated by their predictions of the mass of two particles later detected: the W boson (predicted mass: 80.390 ± 0.018 GeV, experimental measurement: 80.387 ± 0.019 GeV), and the Z boson (predicted mass: 91.1874 ± 0.0021, experimental measurement: 91.1876 ± 0.0021 GeV). The existence of the Z boson was itself another prediction. Other correct predictions included the weak neutral current, the gluon, and the top and charm quarks, all later proven to exist as the theory said.

[8] For example, Huffington Post/Reuters[35] and others[36][37]

[9] The bubble's effects would be expected to propagate across the universe at the speed of light from wherever it occurred. However space is vast – with even the nearest galaxy being over 2 million lightyears from us, and others being many billions of lightyears distant, so the effect of such an event would be unlikely to arise here for billions of years after first occurring.[39][40]

[10] If the Standard Model is correct, then the particles and forces we observe in our universe exist as they do, because of underlying quantum fields. Quantum fields can have states of differing stability, including 'stable', 'unstable' and 'metastable' states (the latter remain stable unless sufficiently perturbed). If a more stable vacuum state were able to arise, then existing particles and forces would no longer arise as they presently do. Different particles or forces would arise from (and be shaped by) whatever new quantum states arose. The world we know depends upon these particles and forces, so if this happened, everything around us, from subatomic particles to galaxies, and all fundamental forces, would be reconstituted into new fundamental particles and forces and structures. The universe would potentially lose all of its present structures and become inhabited by new ones (depending upon the exact states involved) based upon the same quantum fields.

[11] Goldstone's theorem only applies to gauges having manifest Lorentz covariance, a condition that took time to become questioned. But the process of quantisation requires a gauge to be fixed and at this point it becomes possible to choose a gauge such as the 'radiation' gauge which is not invariant over time, so that these problems can be avoided. According to Bernstein (1974, p.8):

> "the "radiation gauge" condition ∇·A(*x*) = 0 is clearly noncovariant, which means that if we wish to maintain transversality of the photon in all Lorentz frames, the photon field Aμ(*x*) cannot transform like a four-vector. This is no catastrophe, since the photon *field* is not an observable, and one can readily show that the S-matrix elements, which *are* observable have covariant structures .... in gauge theories one might arrange things so that one had a symmetry breakdown because of the noninvariance of the vacuum; but, because the Goldstone *et al.* proof breaks down, the zero mass Goldstone mesons need not appear." [Emphasis in original]

Bernstein (1974) contains an accessible and comprehensive background and review of this area, see external links

[12] A field with the "Mexican hat" potential $V(\phi) = \mu^2\phi^2 + \lambda\phi^4$ and $\mu^2 < 0$ has a minimum not at zero but at some non-zero value ϕ_0 . By expressing the action in terms of the field $\tilde{\phi} = \phi - \phi_0$ (where ϕ_0 is a constant independent of position), we find the Yukawa term has a component $g\phi_0\bar{\psi}\psi$. Since both g and ϕ_0 are constants, this looks exactly like the mass term for a fermion of mass $g\phi_0$. The field $\tilde{\phi}$ is then the Higgs field.

[13] In the Standard Model, the mass term arising from the Dirac Lagrangian for any fermion ψ is $-m\bar{\psi}\psi$. This is *not* invariant under the electroweak symmetry, as can be seen by writing ψ in terms of left and right handed components:

$$-m\bar{\psi}\psi = -m(\bar{\psi}_L\psi_R + \bar{\psi}_R\psi_L)$$

i.e., contributions from $\bar{\psi}_L\psi_L$ and $\bar{\psi}_R\psi_R$ terms do not appear. We see that the mass-generating interaction is achieved by constant flipping of particle chirality. Since the spin-half particles have no right/left helicity pair with the same SU(2) and SU(3) representation and the same weak hypercharge, then assuming these gauge charges are conserved in the vacuum, none of the spin-half particles could ever swap helicity. Therefore, in the absence of some other cause, all fermions must be massless.

[14] The example is based on the production rate at the LHC operating at 7 TeV. The total cross-section for producing a Higgs boson at the LHC is about 10 picobarn,[97] while the total cross-section for a proton–proton collision is 110 millibarn.[100]

[15] Just before LEP's shut down, some events that hinted at a Higgs were observed, but it was not judged significant enough to extend its run and delay construction of the LHC.

[16] Announced in articles in *Time*,[145] Forbes,[146] *Slate*,[147] *NPR*,[148] and others.[149]

[17] Other names have included: the "Anderson–Higgs" mechanism,[159] "Higgs–Kibble" mechanism (by Abdus Salam)[79] and "ABEGHHK'tH" mechanism [for Anderson, Brout, Englert, Guralnik, Hagen, Higgs, Kibble and 't Hooft] (by Peter Higgs).[79]

[18] In Miller's analogy, the Higgs field is compared to political party workers spread evenly throughout a room. There will be some people (in Miller's example an anonymous person) who pass through the crowd with ease, paralleling the interaction between the field and particles that do not interact with it, such as massless photons. There will be other people (in Miller's example the British prime minister) who would find their progress being continually slowed by the swarm of admirers crowding around, paralleling the interaction for particles that do interact with the field and by doing so, acquire a finite mass.[198][199]

10.10 References

[1] O'Luanaigh, C. (14 March 2013). "New results indicate that new particle is a Higgs boson". CERN. Retrieved 2013-10-09.

[2] Bryner, J. (14 March 2013). "Particle confirmed as Higgs boson". *NBC News*. Retrieved 2013-03-14.

[3] Heilprin, J. (14 March 2013). "Higgs Boson Discovery Confirmed After Physicists Review Large Hadron Collider Data at CERN". *The Huffington Post*. Retrieved 2013-03-14.

[4] ATLAS; CMS (26 March 2015). "Combined Measurement of the Higgs Boson Mass in pp Collisions at √s=7 and 8 TeV with the ATLAS and CMS Experiments". arXiv:1503.07589.

[5] LHC Higgs Cross Section Working Group; Dittmaier; Mariotti; Passarino; Tanaka; Alekhin; Alwall; Bagnaschi; Banfi (2012). "Handbook of LHC Higgs Cross Sections: 2. Differential Distributions". *CERN Report 2 (Tables A.1 – A.20)* **1201**: 3084. arXiv:1201.3084. Bibcode:2012arXiv1201.3084L.

[6] Onyisi, P. (23 October 2012). "Higgs boson FAQ". University of Texas ATLAS group. Retrieved 2013-01-08.

[7] Strassler, M. (12 October 2012). "The Higgs FAQ 2.0". *ProfMattStrassler.com*. Retrieved 2013-01-08. [Q] Why do particle physicists care so much about the Higgs particle?
[A] Well, actually, they don't. What they really care about is the Higgs *field*, because it is *so* important. [emphasis in original]

[8] José Luis Lucio and Arnulfo Zepeda (1987). *Proceedings of the II Mexican School of Particles and Fields, Cuernavaca-Morelos, 1986*. World Scientific. p. 29. ISBN 9971504340.

[9] Gunion, Dawson, Kane, and Haber (199). *The Higgs Hunter's Guide (1st ed.)*. pp. 11 (?). ISBN 9780786743186. – quoted as being in the first (1990) edition of the book by Peter Higgs in his talk "My Life as a Boson", 2001, ref#25.

[10] Strassler, M. (8 October 2011). "The Known Particles – If The Higgs Field Were Zero". *ProfMattStrassler.com*. Retrieved 13 November 2012. The Higgs field: so important it merited an entire experimental facility, the Large Hadron Collider, dedicated to understanding it.

[11] Biever, C. (6 July 2012). "It's a boson! But we need to know if it's the Higgs". *New Scientist*. Retrieved 2013-01-09. 'As a layman, I would say, I think we have it,' said Rolf-Dieter Heuer, director general of CERN at Wednesday's seminar announcing the results of the search for the Higgs boson. But when pressed by journalists afterwards on what exactly 'it' was, things got more complicated. 'We have discovered a boson – now we have to find out what boson it is'
Q: 'If we don't know the new particle is a Higgs, what do we know about it?' We know it is some kind of boson, says Vivek Sharma of CMS [...]
Q: 'are the CERN scientists just being too cautious? What would be enough evidence to call it a Higgs boson?' As there could be many different kinds of Higgs bosons, there's no straight answer.
[emphasis in original]

[12] Siegfried, T. (20 July 2012). "Higgs Hysteria". *Science News*. Retrieved 2012-12-09. In terms usually reserved for athletic achievements, news reports described the finding as a monumental milestone in the history of science.

[13] Del Rosso, A. (19 November 2012). "Higgs: The beginning of the exploration". *CERN Bulletin* (47–48). Retrieved 2013-01-09. Even in the most specialized circles, the new particle discovered in July is not yet being called the "Higgs boson". Physicists still hesitate to call it that before they have determined that its properties fit with those the Higgs theory predicts the Higgs boson has.

[14] Naik, G. (14 March 2013). "New Data Boosts Case for Higgs Boson Find". *The Wall Street Journal*. Retrieved 2013-03-15. 'We've never seen an elementary particle with spin zero,' said Tony Weidberg, a particle physicist at the University of Oxford who is also involved in the CERN experiments.

[15] Overbye, D. (8 October 2013). "For Nobel, They Can Thank the 'God Particle'". *The New York Times*. Retrieved 2013-11-03.

[16] Sample, I. (29 May 2009). "Anything but the God particle". *The Guardian*. Retrieved 2009-06-24.

[17] Evans, R. (14 December 2011). "The Higgs boson: Why scientists hate that you call it the 'God particle'". *National Post*. Retrieved 2013-11-03.

[18] The nickname occasionally has been satirised in mainstream media as well. Borowitz, Andy (July 13, 2012). "5 questions for the Higgs boson". *The New Yorker*.

[19] Shu, F. H. (1982). *The Physical Universe: An Introduction to Astronomy*. University Science Books. pp. 107–108. ISBN 978-0-935702-05-7.

[20] Shu, F. H. (1982). *The Physical Universe: An Introduction to Astronomy*. University Science Books. pp. 107–108. ISBN 978-0-935702-05-7.

[21] Leon M. Lederman and Dick Teresi (1993). *The God Particle: If the Universe is the Answer, What is the Question*. Houghton Mifflin Company.

[22] Heath, Nick, *The Cern tech that helped track down the God particle*, TechRepublic, 4 July 2012

[23] Rao, Achintya (2 July 2012). "Why would I care about the Higgs boson?". *CMS Public Website*. CERN. Retrieved 18 July 2012.

[24] Max Jammer, *Concepts of Mass in Contemporary Physics and Philosophy* (Princeton, NJ: Princeton University Press, 2000) pp.162–163, who provides many references in support of this statement.

[25] The Large Hadron Collider: Shedding Light on the Early Universe – lecture by R.-D. Heuer, CERN, Chios, Greece, 28 September 2011

[26] Alekhin, Djouadi and Moch, S.; Djouadi, A.; Moch, S. (2012-08-13). "The top quark and Higgs boson masses and the stability of the electroweak vacuum". *Physics Letters B* **716**: 214. arXiv:1207.0980. Bibcode:2012PhLB..716..214A. doi:10.1016/j.physletb.2012.08.024. Retrieved 20 February 2013.

[27] M.S. Turner, F. Wilczek (1982). "Is our vacuum metastable?". *Nature* **298** (5875): 633–634. Bibcode:1982Natur.298..633T. doi:10.1038/298633a0.

[28] S. Coleman and F. De Luccia (1980). "Gravitational effects on and of vacuum decay". *Physical Review* **D21** (12): 3305. Bibcode:1980PhRvD..21.3305C. doi:10.1103/PhysRevD.21.3305.

[29] M. Stone (1976). "Lifetime and decay of excited vacuum states". *Phys. Rev. D* **14** (12): 3568–3573. Bibcode:1976PhRvD..14.3568S. doi:10.1103/PhysRevD.14.3568.

[30] P.H. Frampton (1976). "Vacuum Instability and Higgs Scalar Mass". *Phys. Rev. Lett.* **37** (21): 1378–1380. Bibcode:1976PhRvL..37.1378F. doi:10.1103/PhysRevLett.37.1378.

[31] P.H. Frampton (1977). "Consequences of Vacuum Instability in Quantum Field Theory". *Phys. Rev.* **D15** (10): 2922–28. Bibcode:1977PhRvD..15.2922F. doi:10.1103/PhysRevD.15.2922.

[32] Ellis, Espinosa, Giudice, Hoecker, & Riotto, J.; Espinosa, J.R.; Giudice, G.F.; Hoecker, A.; Riotto, A. (2009). "The Probable Fate of the Standard Model". *Phys. Lett. B* **679** (4): 369–375. arXiv:0906.0954. Bibcode:2009PhLB..679..369E. doi:10.1016/j.physletb.2009.07.054.

[33] Masina, Isabella (2013-02-12). "Higgs boson and top quark masses as tests of electroweak vacuum stability". *Phys. Rev. D* **87** (5): 53001. arXiv:1209.0393. Bibcode:2013PhRvD..87e3001M. doi:10.1103/PhysRevD.87.053001.

[34] Buttazzo, Degrassi, Giardino, Giudice, Sala, Salvio, Strumia (2013-07-12). "Investigating the near-criticality of the Higgs boson". *JHEP 1312 (2013) 089*. arXiv:1307.3536. Bibcode:2013JHEP...12..089B. doi:10.1007/JHEP12(2013)089.

[35] Irene Klotz (editing by David Adams and Todd Eastham) (2013-02-18). "Universe Has Finite Lifespan, Higgs Boson Calculations Suggest". *Huffington Post*. Reuters. Retrieved 21 February 2013. Earth will likely be long gone before any Higgs boson particles set off an apocalyptic assault on the universe

[36] Hoffman, Mark (2013-02-19). "Higgs Boson Will Destroy The Universe Eventually". *ScienceWorldReport*. Retrieved 21 February 2013.

[37] "Higgs boson will aid in creation of the universe – and how it will end". *Catholic Online/NEWS CONSORTIUM*. 2013-02-20. Retrieved 21 February 2013. [T]he Earth will likely be long gone before any Higgs boson particles set off an apocalyptic assault on the universe

[38] Salvio, Alberto (2015-04-09). "A Simple Motivated Completion of the Standard Model below the Planck Scale: Axions and Right-Handed Neutrinos". *Physics Letters B* **743**: 428. arXiv:1501.03781. Bibcode:2015PhLB..743..428S. doi:10.1016/j.physletb.2015.0

[39] Boyle, Alan (2013-02-19). "Will our universe end in a 'big slurp'? Higgs-like particle suggests it might". *NBC News' Cosmic log*. Retrieved 21 February 2013. [T]he bad news is that its mass suggests the universe will end in a fast-spreading bubble of doom. The good news? It'll probably be tens of billions of years. The article quotes Fermilab's Joseph Lykken: "[T]he parameters for our universe, including the Higgs [and top quark's masses] suggest that we're just at the edge of stability, in a "metastable" state. Physicists have been contemplating such a possibility for more than 30 years. Back in 1982, physicists Michael Turner and Frank Wilczek wrote in Nature that "without warning, a bubble of true vacuum could nucleate somewhere in the universe and move outwards..."

[40] Peralta, Eyder (2013-02-19). "If Higgs Boson Calculations Are Right, A Catastrophic 'Bubble' Could End Universe". *npr – two way*. Retrieved 21 February 2013. Article cites Fermilab's Joseph Lykken: "The bubble forms through an unlikely quantum fluctuation, at a random time and place," Lykken tells us. "So in principle it could happen tomorrow, but then most likely in a very distant galaxy, so we are still safe for billions of years before it gets to us."

[41] Bezrukov; Shaposhnikov (2007-10-19). "The Standard Model Higgs boson as the inflaton". *Phys.Lett. B659 (2008) 703-706*. arXiv:0710.3755. Bibcode:2008PhLB..659..703B. doi:10.1016/j.physletb.2007.11.072.

[42] Salvio, Alberto (2013-08-09). "Higgs Inflation at NNLO after the Boson Discovery". *Phys.Lett. B727 (2013) 234-239*. arXiv:1308.2244. Bibcode:2013PhLB..727..234S. doi:10.1016/j.physletb.2013.10.042.

[43] Cole, K. (2000-12-14). "One Thing Is Perfectly Clear: Nothingness Is Perfect". *Los Angeles Times*. p. 'Science File'. Retrieved 17 January 2013. [T]he Higgs' influence (or the influence of something like it) could reach much further. For example, something like the Higgs—if not exactly the Higgs itself—may be behind many other unexplained "broken symmetries" in the universe as well ... In fact, something very much like the Higgs may have been behind the collapse of the symmetry that led to the Big Bang, which created the universe. When the forces first began to separate from their primordial sameness—taking on the distinct characters they have today—they released energy in the same way as water releases energy when it turns to ice. Except in this case, the freezing packed enough energy to blow up the universe. ... However it happened, the moral is clear: Only when the perfection shatters can everything else be born.

[44] Higgs Matters – Kathy Sykes, 30 Nove 2012

[45] Why the public should care about the Higgs Boson – Jodi Lieberman, American Physical Society (APS)

[46] Matt Strassler's blog – Why the Higgs particle matters 2 July 2012

[47] Sean Carroll (13 November 2012). *The Particle at the End of the Universe: How the Hunt for the Higgs Boson Leads Us to the Edge of a New World*. Penguin Group US. ISBN 978-1-101-60970-5.

[48] Woit, Peter (13 November 2010). "The Anderson–Higgs Mechanism". Dr. Peter Woit (Senior Lecturer in Mathematics Columbia University and Ph.D. particle physics). Retrieved 12 November 2012.

[49] Goldstone, J; Salam, Abdus; Weinberg, Steven (1962). "Broken Symmetries". *Physical Review* **127** (3): 965–970. Bibcode:1962PhRv..127 doi:10.1103/PhysRev.127.965.

[50] Guralnik, G. S. (2011). "The Beginnings of Spontaneous Symmetry Breaking in Particle Physics". arXiv:1110.2253 [physics.hist-ph].

[51] Kibble, T. W. B. (2009). "Englert–Brout–Higgs–Guralnik–Hagen–Kibble Mechanism". *Scholarpedia* **4** (1): 6441. Bibcode:2009SchpJ...4 doi:10.4249/scholarpedia.6441. Retrieved 2012-11-23.

[52] Kibble, T. W. B. "History of Englert–Brout–Higgs–Guralnik–Hagen–Kibble Mechanism (history)". *Scholarpedia* **4** (1): 8741. Bibcode:2009SchpJ...4.8741K. doi:10.4249/scholarpedia.8741. Retrieved 2012-11-23.

[53] The Nobel Prize in Physics 2008 – official Nobel Prize website.

[54] List of Anderson 1958–1959 papers referencing 'symmetry', at APS Journals

[55] Higgs, Peter (2010-11-24). "My Life as a Boson" (PDF). Talk given by Peter Higgs at Kings College, London, Nov 24 2010, expanding on a paper originally presented in 2001. Retrieved 17 January 2013. – the original 2001 paper can be found at: Duff and Liu, ed. (2003) [year of publication]. *2001 A Spacetime Odyssey: Proceedings of the Inaugural Conference of the Michigan Center for Theoretical Physics, Michigan, USA, 21–25 May 2001*. World Scientific. pp. 86–88. ISBN 9812382313. Retrieved 17 January 2013.

[56] Anderson, P. (1963). "Plasmons, gauge invariance and mass". *Physical Review* **130**: 439. Bibcode:1963PhRv..130..439A. doi:10.1103/PhysRev.130.439.

[57] Klein, A.; Lee, B. (1964). "Does Spontaneous Breakdown of Symmetry Imply Zero-Mass Particles?". *Physical Review Letters* **12** (10): 266. Bibcode:1964PhRvL..12..266K. doi:10.1103/PhysRevLett.12.266.

[58] Englert, François; Brout, Robert (1964). "Broken Symmetry and the Mass of Gauge Vector Mesons". *Physical Review Letters* **13** (9): 321–23. Bibcode:1964PhRvL..13..321E. doi:10.1103/PhysRevLett.13.321.

[59] Higgs, Peter (1964). "Broken Symmetries and the Masses of Gauge Bosons". *Physical Review Letters* **13** (16): 508–509. Bibcode:1964PhRvL..13..508H. doi:10.1103/PhysRevLett.13.508.

[60] Guralnik, Gerald; Hagen, C. R.; Kibble, T. W. B. (1964). "Global Conservation Laws and Massless Particles". *Physical Review Letters* **13** (20): 585–587. Bibcode:1964PhRvL..13..585G. doi:10.1103/PhysRevLett.13.585.

[61] Higgs, Peter (1964). "Broken symmetries, massless particles and gauge fields". *Physics Letters* **12** (2): 132–133. Bibcode:1964PhL....12..132H. doi:10.1016/0031-9163(64)91136-9.

[62] Higgs, Peter (2010-11-24). "My Life as a Boson" (PDF). Talk given by Peter Higgs at Kings College, London, Nov 24 2010. Retrieved 17 January 2013. Gilbert ... wrote a response to [Klein and Lee's paper] saying 'No, you cannot do that in a relativistic theory. You cannot have a preferred unit time-like vector like that.' This is where I came in, because the next month was when I responded to Gilbert's paper by saying 'Yes, you can have such a thing' but only in a gauge theory with a gauge field coupled to the current.

[63] G.S. Guralnik (2011). "Gauge invariance and the Goldstone theorem – 1965 Feldafing talk". *Modern Physics Letters A* **26** (19): 1381–1392. arXiv:1107.4592. Bibcode:2011MPLA...26.1381G. doi:10.1142/S0217732311036188.

[64] Higgs, Peter (1966). "Spontaneous Symmetry Breakdown without Massless Bosons". *Physical Review* **145** (4): 1156–1163. Bibcode:1966PhRv..145.1156H. doi:10.1103/PhysRev.145.1156.

[65] Kibble, Tom (1967). "Symmetry Breaking in Non-Abelian Gauge Theories". *Physical Review* **155** (5): 1554–1561. Bibcode:1967PhRv..155.1554 doi:10.1103/PhysRev.155.1554.

[66] "Guralnik, G S; Hagen, C R and Kibble, T W B (1967). Broken Symmetries and the Goldstone Theorem. Advances in Physics, vol. 2" (PDF).

[67] "Physical Review Letters – 50th Anniversary Milestone Papers". Physical Review Letters.

[68] S. Weinberg (1967). "A Model of Leptons". *Physical Review Letters* **19** (21): 1264–1266. Bibcode:1967PhRvL..19.1264W. doi:10.1103/PhysRevLett.19.1264.

[69] A. Salam (1968). N. Svartholm, ed. *Elementary Particle Physics: Relativistic Groups and Analyticity*. Eighth Nobel Symposium. Stockholm: Almquvist and Wiksell. p. 367.

[70] S.L. Glashow (1961). "Partial-symmetries of weak interactions". *Nuclear Physics* **22** (4): 579–588. Bibcode:1961NucPh..22..579G. doi:10.1016/0029-5582(61)90469-2.

[71] Ellis, John; Gaillard, Mary K.; Nanopoulos, Dimitri V. (2012). "A Historical Profile of the Higgs Boson". arXiv:1201.6045 [hep-ph].

[72] "Martin Veltman Nobel Lecture, December 12, 1999, p.391" (PDF). Retrieved 2013-10-09.

[73] Politzer, David. "The Dilemma of Attribution". *Nobel Prize lecture, 2004*. Nobel Prize. Retrieved 22 January 2013. Sidney Coleman published in Science magazine in 1979 a citation search he did documenting that essentially no one paid any attention to Weinberg's Nobel Prize winning paper until the work of 't Hooft (as explicated by Ben Lee). In 1971 interest in Weinberg's paper exploded. I had a parallel personal experience: I took a one-year course on weak interactions from Shelly Glashow in 1970, and he never even mentioned the Weinberg–Salam model or his own contributions.

[74] Coleman, Sidney (1979-12-14). "The 1979 Nobel Prize in Physics". *Science* **206** (4424): 1290–1292. Bibcode:1979Sci...206.1290C. doi:10.1126/science.206.4424.1290.

[75] Letters from the Past – A PRL Retrospective (50 year celebration, 2008)

[76] Jeremy Bernstein (January 1974). "Spontaneous symmetry breaking, gauge theories, the Higgs mechanism and all that" (PDF). *Reviews of Modern Physics* **46** (1): 7. Bibcode:1974RvMP...46....7B. doi:10.1103/RevModPhys.46.7. Retrieved 2012-12-10.

[77] American Physical Society – "J. J. Sakurai Prize for Theoretical Particle Physics".

[78] Merali, Zeeya (4 August 2010). "Physicists get political over Higgs". *Nature Magazine*. Retrieved 28 December 2011.

[79] Close, Frank (2011). *The Infinity Puzzle: Quantum Field Theory and the Hunt for an Orderly Universe*. Oxford: Oxford University Press. ISBN 978-0-19-959350-7.

[80] G.S. Guralnik (2009). "The History of the Guralnik, Hagen and Kibble development of the Theory of Spontaneous Symmetry Breaking and Gauge Particles". *International Journal of Modern Physics A* **24** (14): 2601–2627. arXiv:0907.3466. Bibcode:2009IJMPA..24.2601G. doi:10.1142/S0217751X09045431.

[81] Peskin, Michael E.; Schroeder, Daniel V. (1995). *Introduction to Quantum Field Theory*. Reading, MA: Addison-Wesley Publishing Company. pp. 717–719 and 787–791. ISBN 0-201-50397-2.

[82] Peskin & Schroeder 1995, pp. 715–716

[83] Gunion, John (2000). *The Higgs Hunter's Guide* (illustrated, reprint ed.). Westview Press. pp. 1–3. ISBN 9780738203058.

[84] Lisa Randall, *Warped Passages: Unraveling the Mysteries of the Universe's Hidden Dimensions*, p.286: "People initially thought of tachyons as particles travelling faster than the speed of light...But we now know that a tachyon indicates an instability in a theory that contains it. Regrettably for science fiction fans, tachyons are not real physical particles that appear in nature."

[85] Sen, Ashoke (April 2002). "Rolling Tachyon". *J. High Energy Phys.* **2002** (0204): 048. arXiv:hep-th/0203211. Bibcode:2002JHEP...04..0 doi:10.1088/1126-6708/2002/04/048.

[86] Kutasov, David; Marino, Marcos & Moore, Gregory W. (2000). "Some exact results on tachyon condensation in string field theory". *JHEP* **0010**: 045.

[87] Aharonov, Y.; Komar, A.; Susskind, L. (1969). "Superluminal Behavior, Causality, and Instability". *Phys. Rev.* (American Physical Society) **182** (5): 1400–1403. Bibcode:1969PhRv..182.1400A. doi:10.1103/PhysRev.182.1400.

[88] Feinberg, Gerald (1967). "Possibility of Faster-Than-Light Particles". *Physical Review* **159** (5): 1089–1105. Bibcode:1967PhRv..159.108 doi:10.1103/PhysRev.159.1089.

[89] Michael E. Peskin and Daniel V. Schroeder (1995). *An Introduction to Quantum Field Theory*, Perseus books publishing.

[90] Flatow, Ira (6 July 2012). "At Long Last, The Higgs Particle... Maybe". *NPR*. Retrieved 10 July 2012.

[91] "Explanatory Figures for the Higgs Boson Exclusion Plots". *ATLAS News*. CERN. Retrieved 6 July 2012.

[92] Bernardi, G.; Carena, M.; Junk, T. (2012). "Higgs Bosons: Theory and Searches" (PDF). p. 7.

[93] Lykken, Joseph D. (2009). "Beyond the Standard Model". *Proceedings of the 2009 European School of High-Energy Physics, Bautzen, Germany, 14 – 27 June 2009*. arXiv:1005.1676.

[94] Plehn, Tilman (2012). *Lectures on LHC Physics*. Lecture Notes is Physics **844**. Springer. Sec. 1.2.2. arXiv:0910.4122. ISBN 3642240399.

[95] "LEP Electroweak Working Group".

[96] Peskin, Michael E.; Wells, James D. (2001). "How Can a Heavy Higgs Boson be Consistent with the Precision Electroweak Measurements?". *Physical Review D* **64** (9): 093003. arXiv:hep-ph/0101342. Bibcode:2001PhRvD..64i3003P. doi:10.1103/PhysRevD.64.09

[97] Baglio, Julien; Djouadi, Abdelhak (2011). "Higgs production at the lHC". *Journal of High Energy Physics* **1103** (3): 055. arXiv:1012.0530. Bibcode:2011JHEP...03..055B. doi:10.1007/JHEP03(2011)055.

[98] Baglio, Julien; Djouadi, Abdelhak (2010). "Predictions for Higgs production at the Tevatron and the associated uncertainties". *Journal of High Energy Physics* **1010** (10): 063. arXiv:1003.4266. Bibcode:2010JHEP...10..064B. doi:10.1007/JHEP10(2010)064.

[99] Teixeira-Dias (LEP Higgs working group), P. (2008). “Higgs boson searches at LEP”. *Journal of.Physics: Conference Series* **110** (4): 042030. arXiv:0804.4146. Bibcode:2008JPhCS.110d2030T. doi:10.1088/1742-6596/110/4/042030.

[100] “Collisions”. *LHC Machine Outreach*. CERN. Retrieved 26 July 2012.

[101] Asquith, Lily (22 June 2012). “Why does the Higgs decay?". *Life and Physics* (London: The Guardian). Retrieved 14 August 2012.

[102] “Higgs bosons: theory and searches” (PDF). *PDGLive*. Particle Data Group. 12 July 2012. Retrieved 15 August 2012.

[103] Branco, G. C.; Ferreira, P.M.; Lavoura, L.; Rebelo, M.N.; Sher, Marc; Silva, João P. (July 2012). “Theory and phenomenology of two-Higgs-doublet models”. *Physics Reports* (Elsevier) **516** (1): 1–102. arXiv:1106.0034. Bibcode:2012PhR...516....1B. doi:10.1016/j.physrep.2012.02.002.

[104] Csaki, C.; Grojean, C.; Pilo, L.; Terning, J. (2004). “Towards a realistic model of Higgsless electroweak symmetry breaking”. *Physical Review Letters* **92** (10): 101802. arXiv:hep-ph/0308038. Bibcode:2004PhRvL..92j1802C. doi:10.1103/PhysRevLett.92.101802. PMID 15089195.

[105] Csaki, C.; Grojean, C.; Pilo, L.; Terning, J.; Terning, John (2004). “Gauge theories on an interval: Unitarity without a Higgs”. *Physical Review D* **69** (5): 055006. arXiv:hep-ph/0305237. Bibcode:2004PhRvD..69e5006C. doi:10.1103/PhysRevD.69.055006.

[106] “The Hierarchy Problem: why the Higgs has a snowball’s chance in hell”. Quantum Diaries. 2012-07-01. Retrieved 19 March 2013.

[107] “The Hierarchy Problem | Of Particular Significance”. Profmattstrassler.com. Retrieved 2013-10-09.

[108] “Collisions”. *LHC Machine Outreach*. CERN. Retrieved 26 July 2012.

[109] “Hunt for Higgs boson hits key decision point”. MSNBC. 2012-12-06. Retrieved 2013-01-19.

[110] Worldwide LHC Computing Grid main page 14 November 2012: *"[A] global collaboration of more than 170 computing centres in 36 countries ... to store, distribute and analyse the ~25 Petabytes (25 million Gigabytes) of data annually generated by the Large Hadron Collider"*

[111] What is the Worldwide LHC Computing Grid? (Public 'About' page) 14 November 2012: *“Currently WLCG is made up of more than 170 computing centers in 36 countries...The WLCG is now the world’s largest computing grid”*

[112] W.-M. Yao et al. (2006). “Review of Particle Physics” (PDF). *Journal of Physics G* **33**: 1. arXiv:astro-ph/0601168. Bibcode:2006JPhG...33....1Y. doi:10.1088/0954-3899/33/1/001.

[113] The CDF Collaboration, the D0 Collaboration, the Tevatron New Physics, Higgs Working Group (2012). “Updated Combination of CDF and D0 Searches for Standard Model Higgs Boson Production with up to 10.0 fb^{-1} of Data”. arXiv:1207.0449 [hep-ex].

[114] “Interim Summary Report on the Analysis of the 19 September 2008 Incident at the LHC” (PDF). CERN. 15 October 2008. EDMS 973073. Retrieved 28 September 2009.

[115] “CERN releases analysis of LHC incident” (Press release). CERN Press Office. 16 October 2008. Retrieved 28 September 2009.

[116] “LHC to restart in 2009” (Press release). CERN Press Office. 5 December 2008. Retrieved 8 December 2008.

[117] “LHC progress report”. *The Bulletin*. CERN. 3 May 2010. Retrieved 7 December 2011.

[118] “ATLAS experiment presents latest Higgs search status”. *ATLAS homepage*. CERN. 13 December 2011. Retrieved 13 December 2011.

[119] Taylor, Lucas (13 December 2011). “CMS search for the Standard Model Higgs Boson in LHC data from 2010 and 2011”. *CMS public website*. CERN. Retrieved 13 December 2011.

[120] Overbye, D. (5 March 2013). “Chasing The Higgs Boson”. *The New York Times*. Retrieved 2013-03-05.

[121] “ATLAS and CMS experiments present Higgs search status” (Press release). CERN Press Office. 13 December 2011. Retrieved 14 September 2012. the statistical significance is not large enough to say anything conclusive. As of today what we see is consistent either with a background fluctuation or with the presence of the boson. Refined analyses and additional data delivered in 2012 by this magnificent machine will definitely give an answer

[122] "WLCG Public Website". CERN. Retrieved 29 October 2012.

[123] CMS collaboration (2014). "Precise determination of the mass of the Higgs boson and tests of compatibility of its couplings with the standard model predictions using proton collisions at 7 and 8 TeV". arXiv:1412.8662.

[124] ATLAS collaboration (2014). "Measurements of Higgs boson production and couplings in the four-lepton channel in pp collisions at center-of-mass energies of 7 and 8 TeV with the ATLAS detector". arXiv:1408.5191.

[125] ATLAS collaboration (2014). "Measurement of Higgs boson production in the diphoton decay channel in pp collisions at center-of-mass energies of 7 and 8 TeV with the ATLAS detector". arXiv:1408.7084.

[126] "Press Conference: Update on the search for the Higgs boson at CERN on 4 July 2012". Indico.cern.ch. 22 June 2012. Retrieved 4 July 2012.

[127] "CERN to give update on Higgs search". CERN. 22 June 2012. Retrieved 2 July 2011.

[128] "Scientists analyse global Twitter gossip around Higgs boson discovery". *phys.org (from arXiv).* 2013-01-23. Retrieved 6 February 2013. – stated to be *" the first time scientists have been able to analyse the dynamics of social media on a global scale before, during and after the announcement of a major scientific discovery."* For the paper itself see: De Domenico, M.; Lima, A.; Mougel, P.; Musolesi, M. (2013). "The Anatomy of a Scientific Gossip". arXiv:1301.2952. Bibcode:2013NatSR...3E2980D. doi:10.1038/srep02980.

[129] "Higgs boson particle results could be a quantum leap". Times LIVE. 28 June 2012. Retrieved 4 July 2012.

[130] CERN prepares to deliver Higgs particle findings, Australian Broadcasting Corporation. Retrieved 4 July 2012.

[131] "God Particle Finally Discovered? Higgs Boson News At Cern Will Even Feature Scientist It's Named After". Huffington-post.co.uk. Retrieved 2013-01-19.

[132] Our Bureau (2012-07-04). "Higgs on way, theories thicken". Calcutta, India: Telegraphindia.com. Retrieved 2013-01-19.

[133] Thornhill, Ted (2013-07-03). "God Particle Finally Discovered? Higgs Boson News At Cern Will Even Feature Scientist It's Named After". *Huffington Post.* Retrieved 23 July 2013.

[134] Cooper, Rob (2013-07-01) [updated subsequently]. "God particle is 'found': Scientists at Cern expected to announce on Wednesday Higgs boson particle has been discovered". *Daily Mail* (London). Retrieved 23 July 2013. - States that "*"Five leading theoretical physicists have been invited to the event on Wednesday - sparking speculation that the particle has been discovered."*, including Higgs and Englert, and that Kibble - who was invited but unable to attend - "told the Sunday Times: 'My guess is that is must be a pretty positive result for them to be asking us out there'."

[135] Adrian Cho (13 July 2012). "Higgs Boson Makes Its Debut After Decades-Long Search". *Science* **337** (6091): 141–143. doi:10.1126/science.337.6091.141. PMID 22798574.

[136] CMS collaboration (2012). "Observation of a new boson at a mass of 125 GeV with the CMS experiment at the LHC". *Physics Letters B* **716** (1): 30–61. arXiv:1207.7235. Bibcode:2012PhLB..716...30C. doi:10.1016/j.physletb.2012.08.021.

[137] Taylor, Lucas (4 July 2012). "Observation of a New Particle with a Mass of 125 GeV". *CMS Public Website.* CERN. Retrieved 4 July 2012.

[138] "Latest Results from ATLAS Higgs Search". *ATLAS News.* CERN. 4 July 2012. Retrieved 4 July 2012.

[139] ATLAS collaboration (2012). "Observation of a New Particle in the Search for the Standard Model Higgs Boson with the ATLAS Detector at the LHC". *Physics Letters B* **716** (1): 1–29. arXiv:1207.7214. Bibcode:2012PhLB..716....1A. doi:10.1016/j.physletb.201

[140] Gillies, James (23 July 2012). "LHC 2012 proton run extended by seven weeks". *CERN bulletin.* Retrieved 29 August 2012.

[141] "Higgs boson behaving as expected". *3 News NZ.* 15 November 2012.

[142] Strassler, Matt (2012-11-14). "Higgs Results at Kyoto". *Of Particular Significance: Conversations About Science with Theoretical Physicist Matt Strassler.* Prof. Matt Strassler's personal particle physics website. Retrieved 10 January 2013. ATLAS and CMS only just co-discovered this particle in July ... We will not know after today whether it is a Higgs at all, whether it is a Standard Model Higgs or not, or whether any particular speculative idea...is now excluded. [...] Knowledge about nature does not come easy. We discovered the top quark in 1995, and we are still learning about its properties today... we will still be learning important things about the Higgs during the coming few decades. We've no choice but to be patient.

[143] Sample, Ian (14 November 2012). "Higgs particle looks like a bog Standard Model boson, say scientists". *The Guardian* (London). Retrieved 15 November 2012.

[144] "CERN experiments observe particle consistent with long-sought Higgs boson". CERN press release. 4 July 2012. Retrieved 4 July 2012.

[145] "Person Of The Year 2012". *Time*. 19 December 2012.

[146] "Higgs Boson Discovery Has Been Confirmed". Forbes. Retrieved 2013-10-09.

[147] Slate Video Staff (2012-09-11). "Higgs Boson Confirmed; CERN Discovery Passes Test". Slate.com. Retrieved 2013-10-09.

[148] "The Year Of The Higgs, And Other Tiny Advances In Science". NPR. 2013-01-01. Retrieved 2013-10-09.

[149] "Confirmed: the Higgs boson does exist". *The Sydney Morning Herald*. 4 July 2012.

[150] "AP CERN chief: Higgs boson quest could wrap up by midyear". *MSNBC*. Associated Press. 2013-01-27. Retrieved 20 February 2013. Rolf Heuer, director of [CERN], said he is confident that "towards the middle of the year, we will be there." – Interview by AP, at the World Economic Forum, 26 Jan 2013.

[151] Boyle, Alan (2013-02-16). "Will our universe end in a 'big slurp'? Higgs-like particle suggests it might". *NBCNews.com – cosmic log*. Retrieved 20 February 2013. 'it's going to take another few years' after the collider is restarted to confirm definitively that the newfound particle is the Higgs boson.

[152] Gillies, James (2013-03-06). "A question of spin for the new boson". CERN. Retrieved 7 March 2013.

[153] Adam Falkowski (writing as 'Jester') (2013-02-27). "When shall we call it Higgs?". Résonaances particle physics blog. Retrieved 7 March 2013.

[154] CMS Collaboration (February 2013). "Study of the Mass and Spin-Parity of the Higgs Boson Candidate via Its Decays to Z Boson Pairs". *Phys. Rev. Lett.* (American Physical Society) **110** (8): 081803. arXiv:1212.6639. Bibcode:2013PhRvL.110h1803C. doi:10.1103/PhysRevLett.110.081803. Retrieved 15 September 2014.

[155] ATLAS Collaboration (7 October 2013). "Evidence for the spin-0 nature of the Higgs boson using ATLAS data". *Phys. Lett. B* (American Physical Society) **726** (1-3): 120–144. Bibcode:2013PhLB..726..120A. doi:10.1016/j.physletb.2013.08.026. Retrieved 15 September 2014.

[156] "Higgs-like Particle in a Mirror". American Physical Society. Retrieved 26 February 2013.

[157] The CMS Collaboration (2014-06-22). "Evidence for the direct decay of the 125 GeV Higgs boson to fermions". Nature Publishing Group doi= 10.1038/nphys3005.

[158] Adam Falkowski (writing as 'Jester') (2012-12-13). "Twin Peaks in ATLAS". Résonaances particle physics blog. Retrieved 24 February 2013.

[159] Liu, G. Z.; Cheng, G. (2002). "Extension of the Anderson-Higgs mechanism". *Physical Review B* **65** (13): 132513. arXiv:cond-mat/0106070. Bibcode:2002PhRvB..65m2513L. doi:10.1103/PhysRevB.65.132513.

[160] Editorial (2012-03-21). "Mass appeal: As physicists close in on the Higgs boson, they should resist calls to change its name". *Nature*. 483, 374 (7390): 374. Bibcode:2012Natur.483..374.. doi:10.1038/483374a. Retrieved 21 January 2013.

[161] Becker, Kate (2012-03-29). "A Higgs by Any Other Name". "NOVA" (PBS) physics. Retrieved 21 January 2013.

[162] "Frequently Asked Questions: The Higgs!". *The Bulletin*. CERN. Retrieved 18 July 2012.

[163] Woit's physics blog *"Not Even Wrong"*: Anderson on Anderson-Higgs 2013-04-13

[164] Sample, Ian (2012-07-04). "Higgs boson's many great minds cause a Nobel prize headache". *The Guardian* (London). Retrieved 23 July 2013.

[165] "Rochester's Hagen Sakurai Prize Announcement" (Press release). University of Rochester. 2010.

[166] *C.R. Hagen Sakurai Prize Talk* (YouTube). 2010.

[167] Cho, A (2012-09-14). “Particle physics. Why the 'Higgs’?" (PDF). *Science* **337** (6100): 1287. doi:10.1126/science.337.6100.1287. PMID 22984044. Lee ... apparently used the term 'Higgs Boson' as early as 1966... but what may have made the term stick is a seminal paper Steven Weinberg...published in 1967...Weinberg acknowledged the mix-up in an essay in the *New York Review of Books* in May 2012. (See also the original article in *New York Review of Books*[168] and Frank Close’s 2011 book *The Infinity Puzzle*[791:372])

[168] Weinberg, Steven (2012-05-10). “The Crisis of Big Science”. *The New York Review of Books* (footnote 1). Retrieved 12 February 2013.

[169] Examples of early papers using the term “Higgs boson” include 'A phenomenological profile of the Higgs boson' (Ellis, Gaillard and Nanopoulos, 1976), 'Weak interaction theory and neutral currents’ (Bjorken, 1977), and 'Mass of the Higgs boson' (Wienberg, received 1975)

[170] Leon Lederman; Dick Teresi (2006). *The God Particle: If the Universe Is the Answer, What Is the Question?*. Houghton Mifflin Harcourt. ISBN 0-547-52462-5.

[171] Kelly Dickerson (September 8, 2014). “Stephen Hawking Says 'God Particle' Could Wipe Out the Universe”. livescience.com.

[172] Jim Baggott (2012). *Higgs: The invention and discovery of the 'God Particle'*. Oxford University Press. ISBN 978-0-19-165003-1.

[173] Scientific American Editors (2012). *The Higgs Boson: Searching for the God Particle*. Macmillan. ISBN 978-1-4668-2413-3.

[174] Ted Jaeckel (2007). *The God Particle: The Discovery and Modeling of the Ultimate Prime Particle*. Universal-Publishers. ISBN 978-1-58112-959-5.

[175] Aschenbach, Joy (1993-12-05). “No Resurrection in Sight for Moribund Super Collider : Science: Global financial partnerships could be the only way to salvage such a project. Some feel that Congress delivered a fatal blow”. *Los Angeles Times*. Retrieved 16 January 2013. 'We have to keep the momentum and optimism and start thinking about international collaboration,' said Leon M. Lederman, the Nobel Prize-winning physicist who was the architect of the super collider plan

[176] “A Supercompetition For Illinois”. *Chicago Tribune*. 1986-10-31. Retrieved 16 January 2013. The SSC, proposed by the U.S. Department of Energy in 1983, is a mind-bending project ... this gigantic laboratory ... this titanic project

[177] Diaz, Jesus (2012-12-15). “This Is [The] World’s Largest Super Collider That Never Was”. *Gizmodo*. Retrieved 16 January 2013. ...this titanic complex...

[178] Abbott, Charles (June 1987). “Illinois Issues journal, June 1987”. p. 18. Lederman, who considers himself an unofficial propagandist for the super collider, said the SSC could reverse the physics brain drain in which bright young physicists have left America to work in Europe and elsewhere.

[179] Kevles, Dan. “Good-bye to the SSC: On the Life and Death of the Superconducting Super Collider” (PDF). *California Institute of Technology: “Engineering & Science”*. 58 no. 2 (Winter 1995): 16–25. Retrieved 16 January 2013. Lederman, one of the principal spokesmen for the SSC, was an accomplished high-energy experimentalist who had made Nobel Prize-winning contributions to the development of the Standard Model during the 1960s (although the prize itself did not come until 1988). He was a fixture at congressional hearings on the collider, an unbridled advocate of its merits.

[180] Calder, Nigel (2005). *Magic Universe:A Grand Tour of Modern Science*. pp. 369–370. ISBN 9780191622359. The possibility that the next big machine would create the Higgs became a carrot to dangle in front of funding agencies and politicians. A prominent American physicist, Leon lederman *[sic]*, advertised the Higgs as The God Particle in the title of a book published in 1993 ...Lederman was involved in a campaign to persuade the US government to continue funding the Superconducting Super Collider... the ink was not dry on Lederman’s book before the US Congress decided to write off the billions of dollars already spent

[181] Lederman, Leon (1993). *The God Particle If the Universe Is the Answer, What Is the Question?* (PDF). Dell Publishing. p. Chapter 2, Page 2. ISBN 0-385-31211-3. Retrieved 30 July 2015.

[182] Alister McGrath, Higgs boson: the particle of faith, *The Daily Telegraph*, Published 15 December 2011. Retrieved 15 December 2011.

[183] Sample, Ian (3 March 2009). “Father of the God particle: Portrait of Peter Higgs unveiled”. London: The Guardian. Retrieved 24 June 2009.

[184] Chivers, Tom (2011-12-13). “How the 'God particle' got its name”. *The Telegraph* (London). Retrieved 2012-12-03.

[185] Key scientist sure "God particle" will be found soon Reuters news story. 7 April 2008.

[186] "Interview: the man behind the 'God particle'", New Scientist 13 September 2008, pp. 44–5 (original interview in the Guardian: Father of the 'God Particle', June 30, 2008)

[187] Sample, Ian (2010). *Massive: The Hunt for the God Particle*. pp. 148–149 and 278–279. ISBN 9781905264957.

[188] Cole, K. (2000-12-14). "One Thing Is Perfectly Clear: Nothingness Is Perfect". *Los Angeles Times*. p. 'Science File'. Retrieved 17 January 2013. Consider the early universe–a state of pure, perfect nothingness; a formless fog of undifferentiated stuff ... 'perfect symmetry' ... What shattered this primordial perfection? One likely culprit is the so-called Higgs field ... Physicist Leon Lederman compares the way the Higgs operates to the biblical story of Babel [whose citizens] all spoke the same language ... Like God, says Lederman, the Higgs differentiated the perfect sameness, confusing everyone (physicists included) ... [Nobel Prizewinner Richard] Feynman wondered why the universe we live in was so obviously askew ... Perhaps, he speculated, total perfection would have been unacceptable to God. And so, just as God shattered the perfection of Babel, 'God made the laws only nearly symmetrical'

[189] Lederman, p. 22 *et seq*:

> "Something we cannot yet detect and which, one might say, has been put there to test and confuse us ... The issue is whether physicists will be confounded by this puzzle or whether, in contrast to the unhappy Babylonians, we will continue to build the tower and, as Einstein put it, 'know the mind of God'."
> "And the Lord said, Behold the people are un-confounding my confounding. And the Lord sighed and said, Go to, let us go down, and there give them the God Particle so that they may see how beautiful is the universe I have made".

[190] Sample, Ian (12 June 2009). "Higgs competition: Crack open the bubbly, the God particle is dead". *The Guardian* (London). Retrieved 4 May 2010.

[191] Gordon, Fraser (5 July 2012). "Introducing the higgson". *physicsworld.com*. Retrieved 25 August 2012.

[192] Wolchover, Natalie (2012-07-03). "Higgs Boson Explained: How 'God Particle' Gives Things Mass". *Huffington Post*. Retrieved 21 January 2013.

[193] Oliver, Laura (2012-07-04). "Higgs boson: how would you explain it to a seven-year-old?". *The Guardian* (London). Retrieved 21 January 2013.

[194] Zimmer, Ben (2012-07-15). "Higgs boson metaphors as clear as molasses". *The Boston Globe*. Retrieved 21 January 2013.

[195] "The Higgs particle: an analogy for Physics classroom (section)". www.lhc-closer.es (a collaboration website of LHCb physicist Xabier Vidal and High School Teachers at CERN educator Ramon Manzano). Retrieved 2013-01-09.

[196] Flam, Faye (2012-07-12). "Finally – A Higgs Boson Story Anyone Can Understand". *The Philadelphia Inquirer (philly.com)*. Retrieved 21 January 2013.

[197] Sample, Ian (2011-04-28). "How will we know when the Higgs particle has been detected?". *The Guardian* (London). Retrieved 21 January 2013.

[198] Miller, David. "A quasi-political Explanation of the Higgs Boson; for Mr Waldegrave, UK Science Minister 1993". Retrieved 10 July 2012.

[199] Kathryn Grim. "Ten things you may not know about the Higgs boson". Symmetry Magazine. Retrieved 10 July 2012.

[200] David Goldberg, Associate Professor of Physics, Drexel University (2010-10-17). "What's the Matter with the Higgs Boson?". io9.com "Ask a physicist". Retrieved 21 January 2013.

[201] The Nobel Prize in Physics 1979 – official Nobel Prize website.

[202] The Nobel Prize in Physics 1999 – official Nobel Prize website.

[203] – official Nobel Prize website.

[204] Daigle, Katy (10 July 2012). "India: Enough about Higgs, let's discuss the boson". *AP News*. Retrieved 10 July 2012.

[205] Bal, Hartosh Singh (19 September 2012). "The Bose in the Boson". New York Times. Retrieved 21 September 2012.

[206] Alikhan, Anvar (16 July 2012). "The Spark In A Crowded Field". *Outlook India*. Retrieved 10 July 2012.

[207] Peskin & Schroeder 1995, Chapter 20

10.11 Further reading

- Nambu, Yoichiro; Jona-Lasinio, Giovanni (1961). "Dynamical Model of Elementary Particles Based on an Analogy with Superconductivity". *Physical Review* **122**: 345–358. Bibcode:1961PhRv..122..345N. doi:10.1103/PhysRev.122.34
- Klein, Abraham; Lee, Benjamin W. (1964). "Does Spontaneous Breakdown of Symmetry Imply Zero-Mass Particles?". *Physical Review Letters* **12** (10): 266. Bibcode:1964PhRvL..12..266K. doi:10.1103/PhysRevLett.12.266.
- Anderson, Philip W. (1963). "Plasmons, Gauge Invariance, and Mass". *Physical Review* **130**: 439. Bibcode:1963PhRv..130.. doi:10.1103/PhysRev.130.439.
- Gilbert, Walter (1964). "Broken Symmetries and Massless Particles". *Physical Review Letters* **12** (25): 713. Bibcode:1964PhRvL..12..713G. doi:10.1103/PhysRevLett.12.713.
- Higgs, Peter (1964). "Broken Symmetries, Massless Particles and Gauge Fields". *Physics Letters* **12** (2): 132–133. Bibcode:1964PhL....12..132H. doi:10.1016/0031-9163(64)91136-9.
- Guralnik, Gerald S.; Hagen, C.R.; Kibble, Tom W.B. (1968). "Broken Symmetries and the Goldstone Theorem". In R.L. Cool and R.E. Marshak. *Advances in Physics, Vol. 2*. Interscience Publishers. pp. 567–708. ISBN 978-0470170571.

10.12 External links

10.12.1 Popular science, mass media, and general coverage

- Hunting the Higgs Boson at C.M.S. Experiment, at CERN
- The Higgs Boson" by the CERN exploratorium.
- "Particle Fever", documentary film about the search for the Higgs Boson.
- "The Atom Smashers", documentary film about the search for the Higgs Boson at Fermilab.
- Collected Articles at the *Guardian*
- Video (04:38) – CERN Announcement on 4 July 2012, of the discovery of a particle which is suspected will be a Higgs Boson.
- Video1 (07:44) + Video2 (07:44) – Higgs Boson Explained by CERN Physicist, Dr. Daniel Whiteson (16 June 2011).
- HowStuffWorks: What exactly is the Higgs Boson?
- Carroll, Sean. "Higgs Boson with Sean Carroll". *Sixty Symbols*. University of Nottingham.
- Overbye, Dennis (2013-03-05). "Chasing the Higgs Boson: How 2 teams of rivals at CERN searched for physics' most elusive particle". *New York Times Science pages*. Retrieved 22 July 2013. - New York Times "behind the scenes" style article on the Higgs' search at ATLAS and CMS
- The story of the Higgs theory by the authors of the PRL papers and others closely associated:
  - Higgs, Peter (2010). "My Life as a Boson" (PDF). Talk given at Kings College, London, Nov 24 2010. Retrieved 17 January 2013. (also:)
  - Kibble, Tom (2009). "Englert–Brout–Higgs–Guralnik–Hagen–Kibble mechanism (history)". Scholarpedia. Retrieved 17 January 2013. (also:)

- Guralnik, Gerald (2009). "The History of the Guralnik, Hagen and Kibble development of the Theory of Spontaneous Symmetry Breaking and Gauge Particles". *International Journal of Modern Physics A* **24** (14): 2601–2627. arXiv:0907.3466. Bibcode:2009IJMPA..24.2601G. doi:10.1142/S0217751X09045431., Guralnik, Gerald (2011). "The Beginnings of Spontaneous Symmetry Breaking in Particle Physics. Proceedings of the DPF-2011 Conference, Providence, RI, 8–13 August 2011". arXiv:1110.2253v1 [physics.hist-ph]., and Guralnik, Gerald (2013). "Heretical Ideas that Provided the Cornerstone for the Standard Model of Particle Physics". SPG MITTEILUNGEN March 2013, No. 39, (p. 14), and Talk at Brown University about the 1964 PRL papers
- Philip Anderson (not one of the PRL authors) on symmetry breaking in superconductivity and its migration into particle physics and the PRL papers

- Cartoon about the search
- Cham, Jorge (2014-02-19). "True Tales from the Road: The Higgs Boson Re-Explained". *Piled Higher and Deeper*. Retrieved 2014-02-25.

10.12.2 Significant papers and other

- Observation of a new particle in the search for the Standard Model Higgs Boson with the ATLAS detector at the LHC
- Observation of a new Boson at a mass of 125 GeV with the CMS experiment at the LHC
- Particle Data Group: Review of searches for Higgs Bosons.
- 2001, a spacetime odyssey: proceedings of the Inaugural Conference of the Michigan Center for Theoretical Physics : Michigan, USA, 21–25 May 2001, (p.86 – 88), ed. Michael J. Duff, James T. Liu, ISBN 978-981-238-231-3, containing Higgs' story of the Higgs Boson.
- A.A. Migdal & A.M. Polyakov, *Spontaneous Breakdown of Strong Interaction Symmetry and the Absence of Massless Particles*, Sov.J.-JETP 24,91 (1966) - example of a 1966 Russian paper on the subject.

10.12.3 Introductions to the field

- Spontaneous symmetry breaking, gauge theories, the Higgs mechanism and all that (Bernstein, *Reviews of Modern Physics* Jan 1974) - an introduction of 47 pages covering the development, history and mathematics of Higgs theories from around 1950 to 1974.

Chapter 11

Renormalization

In quantum field theory, the statistical mechanics of fields, and the theory of self-similar geometric structures, **renormalization** is any of a collection of techniques used to treat infinities arising in calculated quantities.

Renormalization specifies relationships between parameters in the theory when the parameters describing large distance scales differ from the parameters describing small distances. Physically, the pileup of contributions from an infinity of scales involved in a problem may then result in infinities. When describing space and time as a continuum, certain statistical and quantum mechanical constructions are ill defined. To define them, this continuum limit, the removal of the "construction scaffolding" of lattices at various scales, has to be taken carefully, as detailed below.

Renormalization was first developed in quantum electrodynamics (QED) to make sense of infinite integrals in perturbation theory. Initially viewed as a suspect provisional procedure even by some of its originators, renormalization eventually was embraced as an important and self-consistent actual mechanism of scale physics in several fields of physics and mathematics. Today, the point of view has shifted: on the basis of the breakthrough renormalization group insights of Kenneth Wilson, the focus is on variation of physical quantities across contiguous scales, while distant scales are related to each other through "effective" descriptions. *All scales* are linked in a broadly systematic way, and the actual physics pertinent to each is extracted with the suitable specific computational techniques appropriate for each.

11.1 Self-interactions in classical physics

The problem of infinities first arose in the classical electrodynamics of point particles in the 19th and early 20th century.

The mass of a charged particle should include the mass-energy in its electrostatic field (Electromagnetic mass). Assume that the particle is a charged spherical shell of radius r_e. The mass-energy in the field is

$$m_{\text{em}} = \int \frac{1}{2} E^2 \, dV = \int_{r_e}^{\infty} \frac{1}{2} \left(\frac{q}{4\pi r^2} \right)^2 4\pi r^2 \, dr = \frac{q^2}{8\pi r_e},$$

which becomes infinite as $re \to 0$. This implies that the point particle would have infinite inertia, making it unable to be accelerated. Incidentally, the value of r_e that makes m_{em} equal to the electron mass is called the classical electron radius, which (setting $q = e$ and restoring factors of c and ε_0) turns out to be

$$r_e = \frac{e^2}{4\pi\varepsilon_0 m_e c^2} = \alpha \frac{\hbar}{m_e c} \approx 2.8 \times 10^{-15} \text{ m}.$$

where $\alpha \approx 1/137$ is the fine structure constant, and $\hbar/m_e c$ is the Compton wavelength of the electron.

The total effective mass of a spherical charged particle includes the actual bare mass of the spherical shell (in addition to the aforementioned mass associated with its electric field). If the shell's bare mass is allowed to be negative, it might be

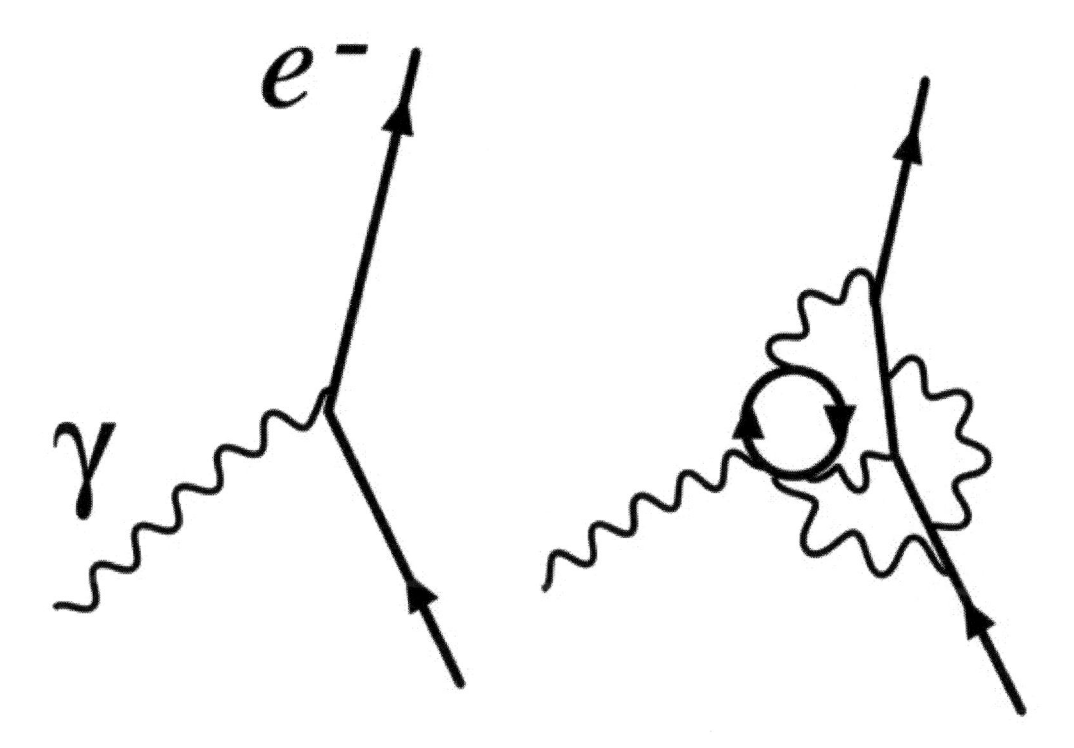

Figure 1. Renormalization in quantum electrodynamics: The simple electron/photon interaction that determines the electron's charge at one renormalization point is revealed to consist of more complicated interactions at another.

possible to take a consistent point limit. This was called *renormalization*, and Lorentz and Abraham attempted to develop a classical theory of the electron this way. This early work was the inspiration for later attempts at regularization and renormalization in quantum field theory.

When calculating the electromagnetic interactions of charged particles, it is tempting to ignore the *back-reaction* of a particle's own field on itself. But this back reaction is necessary to explain the friction on charged particles when they emit radiation. If the electron is assumed to be a point, the value of the back-reaction diverges, for the same reason that the mass diverges, because the field is inverse-square.

The Abraham–Lorentz theory had a noncausal "pre-acceleration". Sometimes an electron would start moving *before* the force is applied. This is a sign that the point limit is inconsistent.

The trouble was worse in classical field theory than in quantum field theory, because in quantum field theory a charged particle experiences Zitterbewegung due to interference with virtual particle-antiparticle pairs, thus effectively smearing out the charge over a region comparable to the Compton wavelength. In quantum electrodynamics at small coupling the electromagnetic mass only diverges as the logarithm of the radius of the particle.

11.2 Divergences in quantum electrodynamics

When developing quantum electrodynamics in the 1930s, Max Born, Werner Heisenberg, Pascual Jordan, and Paul Dirac discovered that in perturbative calculations many integrals were divergent.

One way of describing the divergences was discovered in the 1930s by Ernst Stueckelberg, in the 1940s by Julian Schwinger, Richard Feynman, and Shin'ichiro Tomonaga, and systematized by Freeman Dyson. The divergences ap-

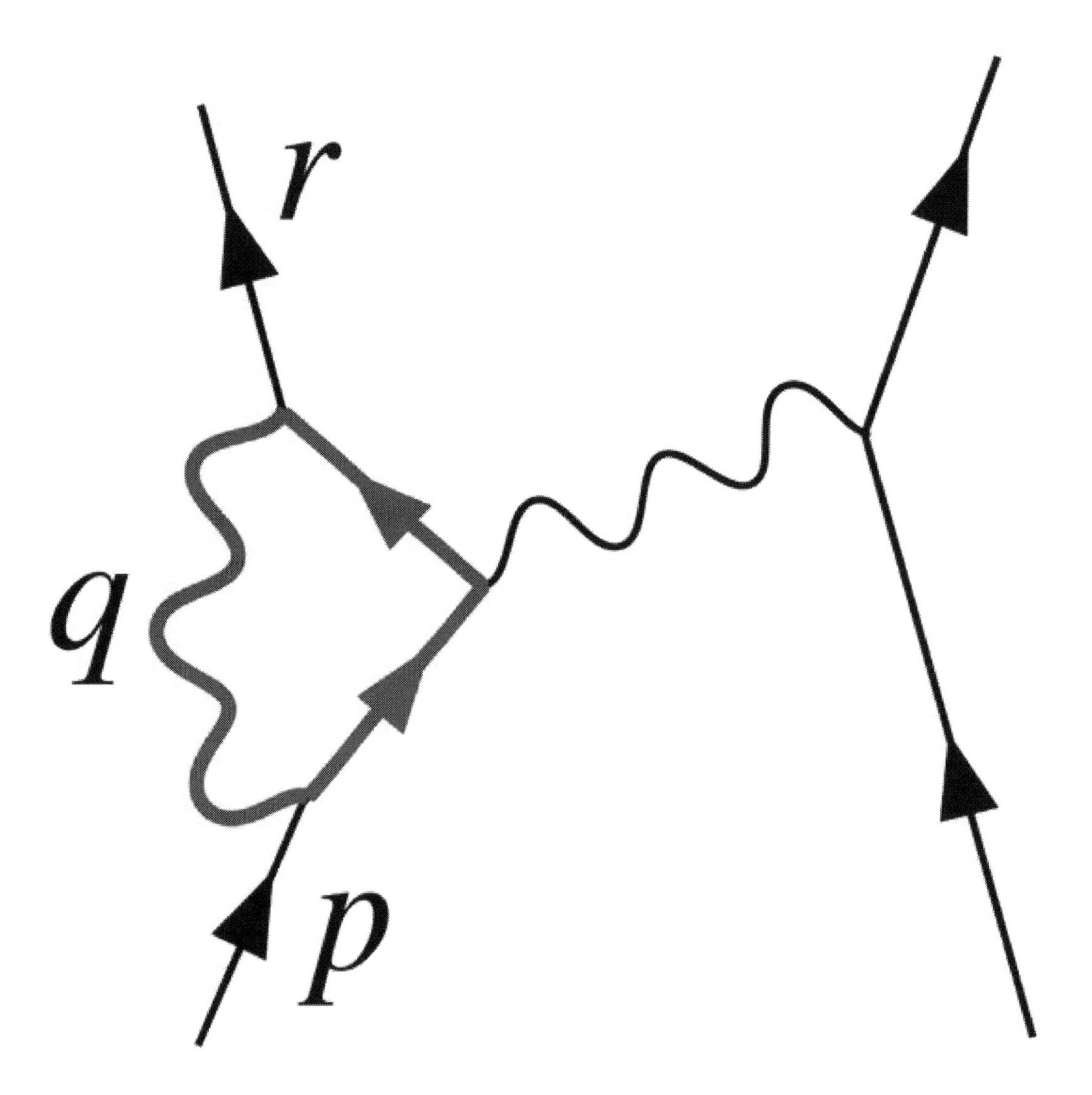

Figure 2. A diagram contributing to electron-electron scattering in QED. The loop has an ultraviolet divergence.

pear in calculations involving Feynman diagrams with closed *loops* of virtual particles in them.

While virtual particles obey conservation of energy and momentum, they can have any energy and momentum, even one that is not allowed by the relativistic energy-momentum relation for the observed mass of that particle. (That is, $E^2 - p^2$ is not necessarily the mass of the particle in that process (e.g. for a photon it could be nonzero).) Such a particle is called off-shell. When there is a loop, the momentum of the particles involved in the loop is not uniquely determined by the energies and momenta of incoming and outgoing particles. A variation in the energy of one particle in the loop must be balanced by an equal and opposite variation in the energy of another particle in the loop. So to find the amplitude for the loop process one must integrate over *all* possible combinations of energy and momentum that could travel around the loop.

These integrals are often *divergent*, that is, they give infinite answers. The divergences which are significant are the "ultraviolet" (UV) ones. An ultraviolet divergence can be described as one which comes from

- the region in the integral where all particles in the loop have large energies and momenta.

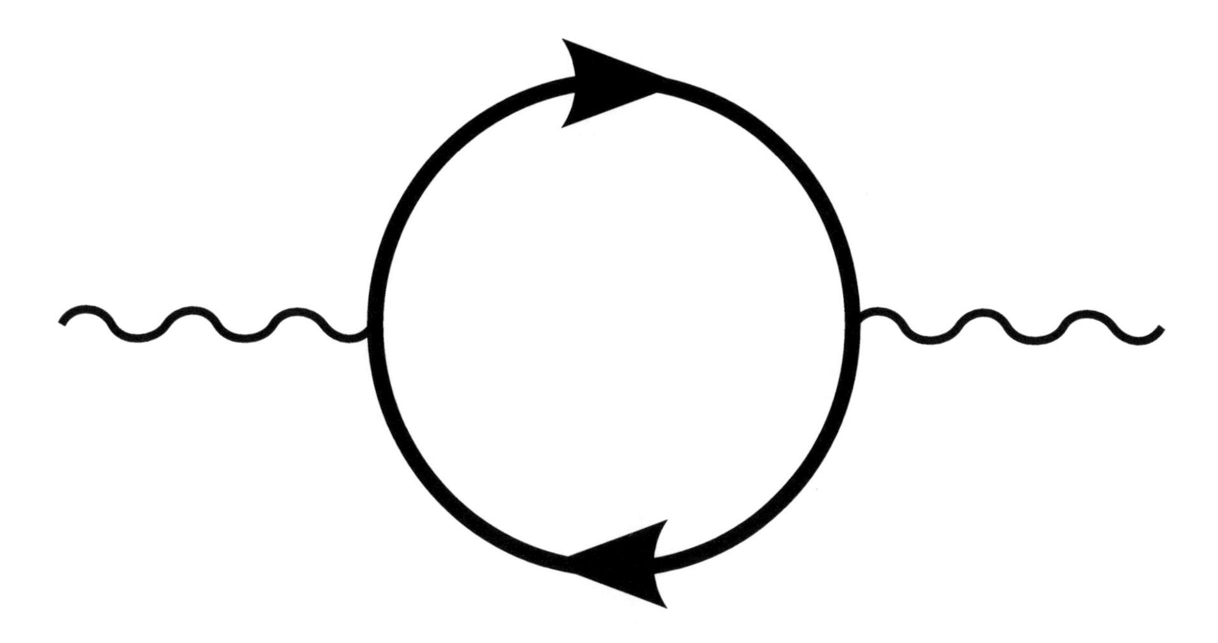

Vacuum polarization, a.k.a. charge screening. This loop has a logarithmic ultraviolet divergence.

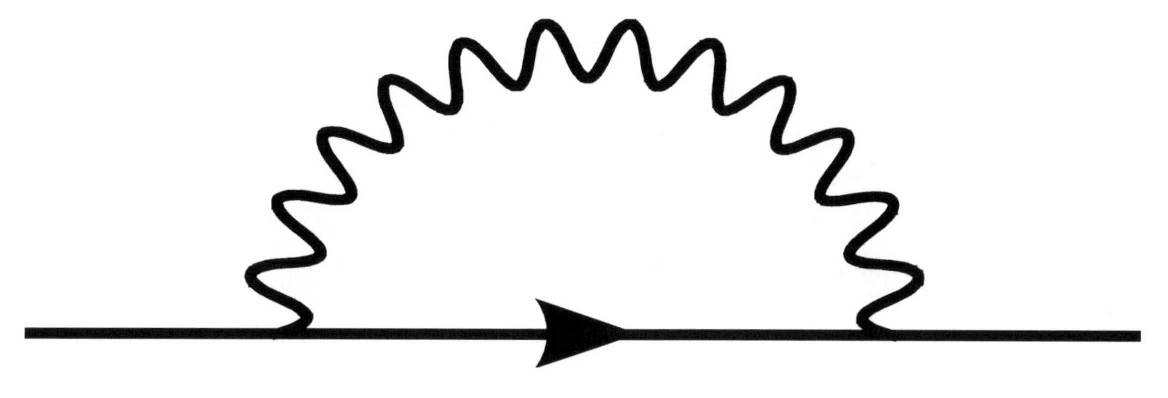

Self energy diagram in QED.

- very short wavelengths and high frequencies fluctuations of the fields, in the path integral for the field.
- Very short proper-time between particle emission and absorption, if the loop is thought of as a sum over particle paths.

So these divergences are short-distance, short-time phenomena.

There are exactly three one-loop divergent loop diagrams in quantum electrodynamics:[1]

1. a photon creates a virtual electron-positron pair which then annihilate, this is a vacuum polarization diagram.
2. an electron which quickly emits and reabsorbs a virtual photon, called a self-energy.
3. An electron emits a photon, emits a second photon, and reabsorbs the first. This process is shown in figure 2, and it is called a vertex renormalization. The Feynman diagram for this is also called a penguin diagram due to its shape remotely resembling a penguin (with the initial and final state electrons as the arms and legs, the second photon as the body and the first looping photon as the head).

The three divergences correspond to the three parameters in the theory:

1. the field normalization Z.
2. the mass of the electron.
3. the charge of the electron.

A second class of divergence, called an infrared divergence, is due to massless particles, like the photon. Every process involving charged particles emits infinitely many coherent photons of infinite wavelength, and the amplitude for emitting any finite number of photons is zero. For photons, these divergences are well understood. For example, at the 1-loop order, the vertex function has both ultraviolet and *infrared* divergences. In contrast to the ultraviolet divergence, the infrared divergence does not require the renormalization of a parameter in the theory. The infrared divergence of the vertex diagram is removed by including a diagram similar to the vertex diagram with the following important difference: the photon connecting the two legs of the electron is cut and replaced by two on shell (i.e. real) photons whose wavelengths tend to infinity; this diagram is equivalent to the bremsstrahlung process. This additional diagram must be included because there is no physical way to distinguish a zero-energy photon flowing through a loop as in the vertex diagram and zero-energy photons emitted through bremsstrahlung. From a mathematical point of view the IR divergences can be regularized by assuming fractional differentiation with respect to a parameter, for example

$$(p^2 - a^2)^{\frac{1}{2}}$$

is well defined at $p = a$ but is UV divergent, if we take the 3/2-th fractional derivative with respect to $-a^2$ we obtain the IR divergence

$$\frac{1}{p^2 - a^2},$$

so we can cure IR divergences by turning them into UV divergences.

11.2.1 A loop divergence

The diagram in Figure 2 shows one of the several one-loop contributions to electron-electron scattering in QED. The electron on the left side of the diagram, represented by the solid line, starts out with four-momentum p^μ and ends up with four-momentum r^μ. It emits a virtual photon carrying $r^\mu - p^\mu$ to transfer energy and momentum to the other electron. But in this diagram, before that happens, it emits another virtual photon carrying four-momentum q^μ, and it reabsorbs this one after emitting the other virtual photon. Energy and momentum conservation do not determine the four-momentum q^μ uniquely, so all possibilities contribute equally and we must integrate.

This diagram's amplitude ends up with, among other things, a factor from the loop of

$$-ie^3 \int \frac{d^4q}{(2\pi)^4} \gamma^\mu \frac{i(\gamma^\alpha(r-q)_\alpha + m)}{(r-q)^2 - m^2 + i\epsilon} \gamma^\rho \frac{i(\gamma^\beta(p-q)_\beta + m)}{(p-q)^2 - m^2 + i\epsilon} \gamma^\nu \frac{-ig_{\mu\nu}}{q^2 + i\epsilon}$$

The various γ^μ factors in this expression are gamma matrices as in the covariant formulation of the Dirac equation; they have to do with the spin of the electron. The factors of e are the electric coupling constant, while the $i\epsilon$ provide a heuristic definition of the contour of integration around the poles in the space of momenta. The important part for our purposes is the dependency on q^μ of the three big factors in the integrand, which are from the propagators of the two electron lines and the photon line in the loop.

This has a piece with two powers of q^μ on top that dominates at large values of q^μ (Pokorski 1987, p. 122):

$$e^3\gamma^\mu\gamma^\alpha\gamma^\rho\gamma^\beta\gamma_\mu \int \frac{d^4q}{(2\pi)^4}\frac{q_\alpha q_\beta}{(r-q)^2(p-q)^2q^2}$$

This integral is divergent, and infinite unless we cut it off at finite energy and momentum in some way.

Similar loop divergences occur in other quantum field theories.

11.3 Renormalized and bare quantities

The solution was to realize that the quantities initially appearing in the theory's formulae (such as the formula for the Lagrangian), representing such things as the electron's electric charge and mass, as well as the normalizations of the quantum fields themselves, did *not* actually correspond to the physical constants measured in the laboratory. As written, they were *bare* quantities that did not take into account the contribution of virtual-particle loop effects to *the physical constants themselves*. Among other things, these effects would include the quantum counterpart of the electromagnetic back-reaction that so vexed classical theorists of electromagnetism. In general, these effects would be just as divergent as the amplitudes under study in the first place; so finite measured quantities would in general imply divergent bare quantities.

In order to make contact with reality, then, the formulae would have to be rewritten in terms of measurable, *renormalized* quantities. The charge of the electron, say, would be defined in terms of a quantity measured at a specific kinematic *renormalization point* or *subtraction point* (which will generally have a characteristic energy, called the *renormalization scale* or simply the energy scale). The parts of the Lagrangian left over, involving the remaining portions of the bare quantities, could then be reinterpreted as *counterterms*, involved in divergent diagrams exactly *canceling out* the troublesome divergences for other diagrams.

11.3.1 Renormalization in QED

For example, in the Lagrangian of QED

$$\mathcal{L} = \bar\psi_B\left[i\gamma_\mu\left(\partial^\mu + ie_BA_B^\mu\right) - m_B\right]\psi_B - \frac{1}{4}F_{B\mu\nu}F_B^{\mu\nu}$$

the fields and coupling constant are really *bare* quantities, hence the subscript B above. Conventionally the bare quantities are written so that the corresponding Lagrangian terms are multiples of the renormalized ones:

$$\left(\bar\psi m\psi\right)_B = Z_0\bar\psi m\psi$$

$$\left(\bar\psi\left(\partial^\mu + ieA^\mu\right)\psi\right)_B = Z_1\bar\psi\left(\partial^\mu + ieA^\mu\right)\psi$$

$$\left(F_{\mu\nu}F^{\mu\nu}\right)_B = Z_3\,F_{\mu\nu}F^{\mu\nu}.$$

Gauge invariance, via a Ward–Takahashi identity, turns out to imply that we can renormalize the two terms of the covariant derivative piece

$$\bar\psi(\partial + ieA)\psi$$

together (Pokorski 1987, p. 115), which is what happened to Z_2; it is the same as Z_1.

A term in this Lagrangian, for example, the electron-photon interaction pictured in Figure 1, can then be written

$$\mathcal{L}_I = -e\bar\psi\gamma_\mu A^\mu\psi - (Z_1 - 1)e\bar\psi\gamma_\mu A^\mu\psi$$

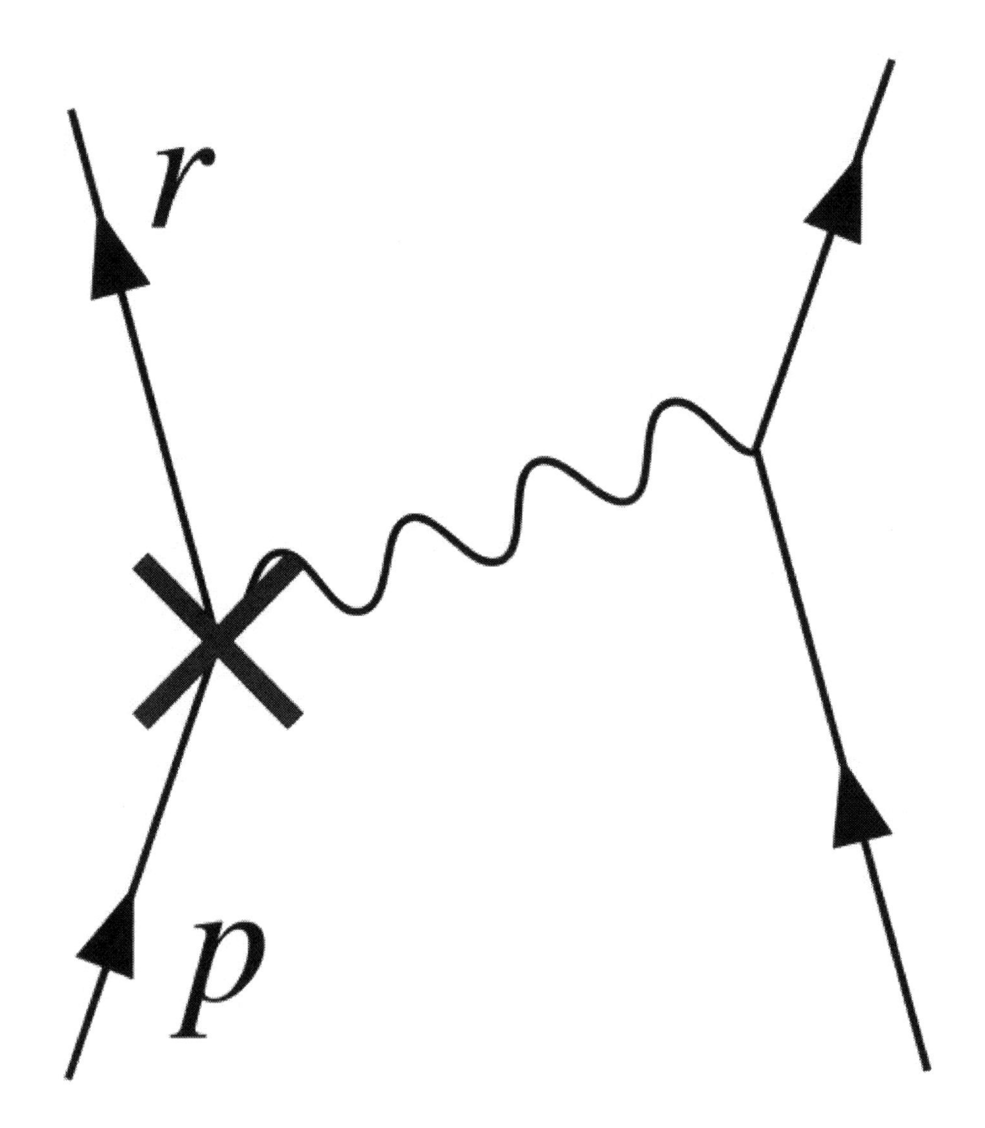

Figure 3. The vertex corresponding to the Z_1 counterterm cancels the divergence in Figure 2.

The physical constant e, the electron's charge, can then be defined in terms of some specific experiment; we set the renormalization scale equal to the energy characteristic of this experiment, and the first term gives the interaction we see in the laboratory (up to small, finite corrections from loop diagrams, providing such exotica as the high-order corrections to the magnetic moment). The rest is the counterterm. If the theory is *renormalizable* (see below for more on this), as it is in QED, the *divergent* parts of loop diagrams can all be decomposed into pieces with three or fewer legs, with an algebraic form that can be canceled out by the second term (or by the similar counterterms that come from Z_0 and Z_3).

The diagram with the Z_1 counterterm's interaction vertex placed as in Figure 3 cancels out the divergence from the loop in Figure 2.

Historically, the splitting of the "bare terms" into the original terms and counterterms came before the renormalization group insights[2] due to Kenneth Wilson. According to such renormalization group insights, detailed in the next section, this splitting is unnatural and actually unphysical, as all scales of the problem enter in systematic continuous ways.

11.3.2 Running couplings

To minimize the contribution of loop diagrams to a given calculation (and therefore make it easier to extract results), one chooses a renormalization point close to the energies and momenta actually exchanged in the interaction. However, the renormalization point is not itself a physical quantity: the physical predictions of the theory, calculated to all orders, should in principle be *independent* of the choice of renormalization point, as long as it is within the domain of application of the theory. Changes in renormalization scale will simply affect how much of a result comes from Feynman diagrams without loops, and how much comes from the leftover finite parts of loop diagrams. One can exploit this fact to calculate the effective variation of physical constants with changes in scale. This variation is encoded by beta-functions, and the general theory of this kind of scale-dependence is known as the renormalization group.

Colloquially, particle physicists often speak of certain physical "constants" as varying with the energy of an interaction, though in fact it is the renormalization scale that is the independent quantity. This *running* does, however, provide a convenient means of describing changes in the behavior of a field theory under changes in the energies involved in an interaction. For example, since the coupling in quantum chromodynamics becomes small at large energy scales, the theory behaves more like a free theory as the energy exchanged in an interaction becomes large, a phenomenon known as asymptotic freedom. Choosing an increasing energy scale and using the renormalization group makes this clear from simple Feynman diagrams; were this not done, the prediction would be the same, but would arise from complicated high-order cancellations.

For example,

$$I = \int_0^a \frac{1}{z}\, dz - \int_0^b \frac{1}{z}\, dz = \ln a - \ln b - \ln 0 + \ln 0$$

is ill defined.

To eliminate the divergence, simply change lower limit of integral into ε_a and ε_b:

$$I = \ln a - \ln b - \ln \varepsilon_a + \ln \varepsilon_b = \ln \tfrac{a}{b} - \ln \tfrac{\varepsilon_b}{\varepsilon_a}$$

Making sure $\varepsilon b/\varepsilon a \to 1$, then $I = \ln a/b$.

11.4 Regularization

Since the quantity $\infty - \infty$ is ill-defined, in order to make this notion of canceling divergences precise, the divergences first have to be tamed mathematically using the theory of limits, in a process known as regularization (Weinberg, 1995).

An essentially arbitrary modification to the loop integrands, or *regulator*, can make them drop off faster at high energies and momenta, in such a manner that the integrals converge. A regulator has a characteristic energy scale known as the cutoff; taking this cutoff to infinity (or, equivalently, the corresponding length/time scale to zero) recovers the original integrals.

With the regulator in place, and a finite value for the cutoff, divergent terms in the integrals then turn into finite but cutoff-dependent terms. After canceling out these terms with the contributions from cutoff-dependent counterterms, the cutoff is taken to infinity and finite physical results recovered. If physics on scales we can measure is independent of what happens at the very shortest distance and time scales, then it should be possible to get cutoff-independent results for calculations.

Many different types of regulator are used in quantum field theory calculations, each with its advantages and disadvantages. One of the most popular in modern use is *dimensional regularization*, invented by Gerardus 't Hooft and Martinus J. G. Veltman,[3] which tames the integrals by carrying them into a space with a fictitious fractional number of dimensions. Another is *Pauli–Villars regularization*, which adds fictitious particles to the theory with very large masses, such that loop integrands involving the massive particles cancel out the existing loops at large momenta.

Yet another regularization scheme is the *Lattice regularization*, introduced by Kenneth Wilson, which pretends that our space-time is constructed by hyper-cubical lattice with fixed grid size. This size is a natural cutoff for the maximal momentum that a particle could possess when propagating on the lattice. And after doing calculation on several lattices with different grid size, the physical result is extrapolated to grid size 0, or our natural universe. This presupposes the existence of a scaling limit.

A rigorous mathematical approach to renormalization theory is the so-called causal perturbation theory, where ultraviolet divergences are avoided from the start in calculations by performing well-defined mathematical operations only within the framework of distribution theory. The disadvantage of the method is the fact that the approach is quite technical and requires a high level of mathematical knowledge.

11.4.1 Zeta function regularization

Julian Schwinger discovered a relationship between zeta function regularization and renormalization, using the asymptotic relation:

$$I(n,\Lambda)=\int_0^\Lambda dp\,p^n\sim 1+2^n+3^n+\cdots+\Lambda^n\to\zeta(-n)$$

as the regulator $\Lambda\to\infty$. Based on this, he considered using the values of $\zeta(-n)$ to get finite results. Although he reached inconsistent results, an improved formula studied by Hartle, J. Garcia, and based on the works by E. Elizalde includes the technique of the zeta regularization algorithm

$$I(n,\Lambda)=\frac{n}{2}I(n-1,\Lambda)+\zeta(-n)-\sum_{r=1}^{\infty}\frac{B_{2r}}{(2r)!}a_{n,r}(n-2r+1)I(n-2r,\Lambda),$$

where the B's are the Bernoulli numbers and

$$a_{n,r}=\frac{\Gamma(n+1)}{\Gamma(n-2r+2)}.$$

So every $I(m,\Lambda)$ can be written as a linear combination of $\zeta(-1)$, $\zeta(-3)$, $\zeta(-5)$, ..., $\zeta(-m)$.

Or simply using Abel–Plana formula we have for every divergent integral:

$$\zeta(-m,\beta)-\frac{\beta^m}{2}-i\int_0^\infty dt\frac{(it+\beta)^m-(-it+\beta)^m}{e^{2\pi t}-1}=\int_0^\infty dp\,(p+\beta)^m$$

valid when $m>0$, Here the zeta function is Hurwitz zeta function and Beta is a positive real number.

The "geometric" analogy is given by, (if we use rectangle method) to evaluate the integral so:

$$\int_0^\infty dx\,(\beta+x)^m\approx\sum_{n=0}^{\infty}h^{m+1}\zeta\left(\beta h^{-1},-m\right)$$

Using Hurwitz zeta regularization plus the rectangle method with step h (not to be confused with Planck's constant).

The logarithmic divergent integral has the regularization

$$\sum_{n=0}^{\infty}\frac{1}{n+a}=-\psi(a)$$

For multi-loop integrals that will depend on several variables $k_1, \cdots, k_n$ we can make a change of variables to polar coordinates and then replace the integral over the angles $\int d\Omega$ by a sum so we have only a divergent integral, that will depend on the modulus $r^2 = k_1^2 + \cdots + k_n^2$ and then we can apply the zeta regularization algorithm, the main idea for multi-loop integrals is to replace the factor $F(q_1, \cdots, q_n)$ after a change to hyperspherical coordinates $F(r, \Omega)$ so the UV overlapping divergences are encoded in variable r. In order to regularize these integrals one needs a regulator, for the case of multi-loop integrals, these regulator can be taken as

$$\left(1 + \sqrt{q_i q^i}\right)^{-s}$$

so the multi-loop integral will converge for big enough s using the Zeta regularization we can analytic continue the variable s to the physical limit where $s = 0$ and then regularize any UV integral, by replacing a divergent integral by a linear combination of divergent series, which can be regularized in terms of the negative values of the Riemann zeta function $\zeta(-m)$.

11.5 Attitudes and interpretation

The early formulators of QED and other quantum field theories were, as a rule, dissatisfied with this state of affairs. It seemed illegitimate to do something tantamount to subtracting infinities from infinities to get finite answers.

Freeman Dyson argued that these infinities are of a basic nature and cannot be eliminated by any formal mathematical procedures, such as the renormalization method.[4][5]

Dirac's criticism was the most persistent.[6] As late as 1975, he was saying:[7]

> Most physicists are very satisfied with the situation. They say: 'Quantum electrodynamics is a good theory and we do not have to worry about it any more.' I must say that I am very dissatisfied with the situation, because this so-called 'good theory' does involve neglecting infinities which appear in its equations, neglecting them in an arbitrary way. This is just not sensible mathematics. Sensible mathematics involves neglecting a quantity when it is small – not neglecting it just because it is infinitely great and you do not want it!

Another important critic was Feynman. Despite his crucial role in the development of quantum electrodynamics, he wrote the following in 1985:[8]

> The shell game that we play ... is technically called 'renormalization'. But no matter how clever the word, it is still what I would call a dippy process! Having to resort to such hocus-pocus has prevented us from proving that the theory of quantum electrodynamics is mathematically self-consistent. It's surprising that the theory still hasn't been proved self-consistent one way or the other by now; I suspect that renormalization is not mathematically legitimate.

While Dirac's criticism was based on the procedure of renormalization itself, Feynman's criticism was very different. Feynman was concerned that all field theories known in the 1960s had the property that the interactions become infinitely strong at short enough distance scales. This property, called a Landau pole, made it plausible that quantum field theories were all inconsistent. In 1974, Gross, Politzer and Wilczek showed that another quantum field theory, quantum chromodynamics, does not have a Landau pole. Feynman, along with most others, accepted that QCD was a fully consistent theory.

The general unease was almost universal in texts up to the 1970s and 1980s. Beginning in the 1970s, however, inspired by work on the renormalization group and effective field theory, and despite the fact that Dirac and various others—all of whom belonged to the older generation—never withdrew their criticisms, attitudes began to change, especially among younger theorists. Kenneth G. Wilson and others demonstrated that the renormalization group is useful in statistical field theory applied to condensed matter physics, where it provides important insights into the behavior of phase transitions. In condensed matter physics, a *physical* short-distance regulator exists: matter ceases to be continuous on the scale of atoms. Short-distance divergences in condensed matter physics do not present a philosophical problem, since the field theory is only an effective, smoothed-out representation of the behavior of matter anyway; there are no infinities since the cutoff is actually always finite, and it makes perfect sense that the bare quantities are cutoff-dependent.

If QFT holds all the way down past the Planck length (where it might yield to string theory, causal set theory or something different), then there may be no real problem with short-distance divergences in particle physics either; *all* field theories could simply be effective field theories. In a sense, this approach echoes the older attitude that the divergences in QFT speak of human ignorance about the workings of nature, but also acknowledges that this ignorance can be quantified and that the resulting effective theories remain useful.

Be that as it may, Salam's remark [9] in 1972 seems still relevant

> Field-theoretic infinities first encountered in Lorentz's computation of electron have persisted in classical electrodynamics for seventy and in quantum electrodynamics for some thirty-five years. These long years of frustration have left in the subject a curious affection for the infinities and a passionate belief that they are an inevitable part of nature; so much so that even the suggestion of a hope that they may after all be circumvented — and finite values for the renormalization constants computed — is considered irrational. Compare Russell's postscript to the third volume of his autobiography *The Final Years, 1944–1969* (George Allen and Unwin, Ltd., London 1969),[10] p.221:
>
>> In the modern world, if communities are unhappy, it is often because they have ignorances, habits, beliefs, and passions, which are dearer to them than happiness or even life. I find many men in our dangerous age who seem to be in love with misery and death, and who grow angry when hopes are suggested to them. They think hope is irrational and that, in sitting down to lazy despair, they are merely facing facts.

In QFT, the value of a physical constant, in general, depends on the scale that one chooses as the renormalization point, and it becomes very interesting to examine the renormalization group running of physical constants under changes in the energy scale. The coupling constants in the Standard Model of particle physics vary in different ways with increasing energy scale: the coupling of quantum chromodynamics and the weak isospin coupling of the electroweak force tend to decrease, and the weak hypercharge coupling of the electroweak force tends to increase. At the colossal energy scale of 10^{15} GeV (far beyond the reach of our current particle accelerators), they all become approximately the same size (Grotz and Klapdor 1990, p. 254), a major motivation for speculations about grand unified theory. Instead of being only a worrisome problem, renormalization has become an important theoretical tool for studying the behavior of field theories in different regimes.

If a theory featuring renormalization (e.g. QED) can only be sensibly interpreted as an effective field theory, i.e. as an approximation reflecting human ignorance about the workings of nature, then the problem remains of discovering a more accurate theory that does not have these renormalization problems. As Lewis Ryder has put it, "In the Quantum Theory, these [classical] divergences do not disappear; on the contrary, they appear to get worse. And despite the comparative success of renormalisation theory the feeling remains that there ought to be a more satisfactory way of doing things."[11]

11.6 Renormalizability

From this philosophical reassessment a new concept follows naturally: the notion of renormalizability. Not all theories lend themselves to renormalization in the manner described above, with a finite supply of counterterms and all quantities becoming cutoff-independent at the end of the calculation. If the Lagrangian contains combinations of field operators of high enough dimension in energy units, the counterterms required to cancel all divergences proliferate to infinite number, and, at first glance, the theory would seem to gain an infinite number of free parameters and therefore lose all predictive power, becoming scientifically worthless. Such theories are called *nonrenormalizable*.

The Standard Model of particle physics contains only renormalizable operators, but the interactions of general relativity become nonrenormalizable operators if one attempts to construct a field theory of quantum gravity in the most straightforward manner (treating the metric in the Einstein-Hilbert Lagrangian as a perturbation about the Minkowski metric), suggesting that perturbation theory is useless in application to quantum gravity.

However, in an effective field theory, "renormalizability" is, strictly speaking, a misnomer. In a nonrenormalizable effective field theory, terms in the Lagrangian do multiply to infinity, but have coefficients suppressed by ever-more-extreme inverse powers of the energy cutoff. If the cutoff is a real, physical quantity—if, that is, the theory is only an effective description of physics up to some maximum energy or minimum distance scale—then these extra terms could represent real physical interactions. Assuming that the dimensionless constants in the theory do not get too large, one can group calculations by inverse powers of the cutoff, and extract approximate predictions to finite order in the cutoff that still have a finite number of free parameters. It can even be useful to renormalize these "nonrenormalizable" interactions.

Nonrenormalizable interactions in effective field theories rapidly become weaker as the energy scale becomes much smaller than the cutoff. The classic example is the Fermi theory of the weak nuclear force, a nonrenormalizable effective theory whose cutoff is comparable to the mass of the W particle. This fact may also provide a possible explanation for *why* almost all of the particle interactions we see are describable by renormalizable theories. It may be that any others that may exist at the GUT or Planck scale simply become too weak to detect in the realm we can observe, with one exception: gravity, whose exceedingly weak interaction is magnified by the presence of the enormous masses of stars and planets.

11.7 Renormalization schemes

In actual calculations, the counterterms introduced to cancel the divergences in Feynman diagram calculations beyond tree level must be *fixed* using a set of *renormalization conditions*. The common renormalization schemes in use include:

- Minimal subtraction (MS) scheme and the related modified minimal subtraction (MS-bar) scheme
- On-shell scheme

11.8 Application in statistical physics

As mentioned in the introduction, the methods of renormalization have been applied to Statistical Physics, namely to the problems of the critical behaviour near second-order phase transitions, in particular at fictitious spatial dimensions just below the number of 4, where the above-mentioned methods could even be sharpened (i.e., instead of "renormalizability" one gets "super-renormalizability"), which allowed extrapolation to the real spatial dimensionality for phase transitions, 3. Details can be found in the book of Zinn-Justin, mentioned below.

For the discovery of these unexpected applications, and working out the details, in 1982 the physics Nobel prize was awarded to Kenneth G. Wilson.

11.9 See also

- Effective field theory
- Landau pole
- Quantum field theory
- Quantum triviality
- Regularization
- Renormalization group
- Ward–Takahashi identity
- Zeta function regularization
- Zeno's paradoxes

11.10 References

[1] See ch. 10 of "An Introduction To Quantum Field Theory", Michael E. Peskin And Daniel V. Schroeder, Sarat Book House, 2005

[2] K.G. Wilson (1975), "The renormalization group: critical phenomena and the Kondo problem", *Rev. Mod. Phys.* **47**, 4, 773.

[3] 't Hooft, G.; Veltman, M. (1972). "Regularization and renormalization of gauge fields". *Nuclear Physics B* **44**: 189. Bibcode:1972NuPhB..4 doi:10.1016/0550-3213(72)90279-9.

[4] F. J. Dyson, *Phys. Rev.* **85** (1952) 631.

[5] A. W. Stern, *Science* **116** (1952) 493.

[6] P.A.M. Dirac, "The Evolution of the Physicist's Picture of Nature," in Scientific American, May 1963, p. 53.

[7] Kragh, Helge; *Dirac: A scientific biography*, CUP 1990, p. 184

[8] Feynman, Richard P. ; *QED, The Strange Theory of Light and Matter*, Penguin 1990, p. 128

[9] C.J.Isham, A.Salam and J.Strathdee, `Infinity Suppression Gravity Modified Quantum Electrodynamics,' Phys. Rev. D5, 2548 (1972)

[10] Russell, Bertrand. *The Autobiography of Bertrand Russell: The Final Years, 1944-1969* (Bantam Books, 1970)

[11] Ryder, Lewis. *Quantum Field Theory*, page 390 (Cambridge University Press 1996).

11.11 Further reading

11.11.1 General introduction

- Delamotte, Bertrand ; *A hint of renormalization*, American Journal of Physics 72 (2004) pp. 170–184. Beautiful elementary introduction to the ideas, no prior knowledge of field theory being necessary. Full text available at: *hep-th/0212049*
- Baez, John ; *Renormalization Made Easy*, (2005). A qualitative introduction to the subject.
- Blechman, Andrew E. ; *Renormalization: Our Greatly Misunderstood Friend*, (2002). Summary of a lecture; has more information about specific regularization and divergence-subtraction schemes.
- Cao, Tian Yu & Schweber, Silvan S. ; *The Conceptual Foundations and the Philosophical Aspects of Renormalization Theory*, Synthese, 97(1) (1993), 33–108.
- Shirkov, Dmitry ; *Fifty Years of the Renormalization Group*, C.E.R.N. Courrier 41(7) (2001). Full text available at : *I.O.P Magazines*.
- E. Elizalde ; *Zeta regularization techniques with Applications*.

11.11.2 Mainly: quantum field theory

- N. N. Bogoliubov, D. V. Shirkov (1959): *The Theory of Quantized Fields*. New York, Interscience. The first text-book on the renormalization group theory.
- Ryder, Lewis H. ; *Quantum Field Theory* (Cambridge University Press, 1985), ISBN 0-521-33859-X Highly readable textbook, certainly the best introduction to relativistic Q.F.T. for particle physics.
- Zee, Anthony ; *Quantum Field Theory in a Nutshell*, Princeton University Press (2003) ISBN 0-691-01019-6. Another excellent textbook on Q.F.T.

- Weinberg, Steven ; *The Quantum Theory of Fields* (3 volumes) Cambridge University Press (1995). A monumental treatise on Q.F.T. written by a leading expert, *Nobel laureate 1979*.
- Pokorski, Stefan ; *Gauge Field Theories*, Cambridge University Press (1987) ISBN 0-521-47816-2.
- 't Hooft, Gerard ; *The Glorious Days of Physics – Renormalization of Gauge theories*, lecture given at Erice (August/September 1998) by the *Nobel laureate 1999* . Full text available at: *hep-th/9812203*.
- Rivasseau, Vincent ; *An introduction to renormalization*, Poincaré Seminar (Paris, Oct. 12, 2002), published in : Duplantier, Bertrand; Rivasseau, Vincent (Eds.) ; *Poincaré Seminar 2002*, Progress in Mathematical Physics 30, Birkhäuser (2003) ISBN 3-7643-0579-7. Full text available in *PostScript*.
- Rivasseau, Vincent ; *From perturbative to constructive renormalization*, Princeton University Press (1991) ISBN 0-691-08530-7. Full text available in *PostScript*.
- Iagolnitzer, Daniel & Magnen, J. ; *Renormalization group analysis*, Encyclopaedia of Mathematics, Kluwer Academic Publisher (1996). Full text available in PostScript and pdf *here*.
- Scharf, Günter; *Finite quantum electrodynamics: The causal approach*, Springer Verlag Berlin Heidelberg New York (1995) ISBN 3-540-60142-2.
- A. S. Švarc (Albert Schwarz), Математические основы квантовой теории поля, (Mathematical aspects of quantum field theory), Atomizdat, Moscow, 1975. 368 pp.

11.11.3 Mainly: statistical physics

- A. N. Vasil'ev *The Field Theoretic Renormalization Group in Critical Behavior Theory and Stochastic Dynamics* (Routledge Chapman & Hall 2004); ISBN 978-0-415-31002-4
- Nigel Goldenfeld ; *Lectures on Phase Transitions and the Renormalization Group*, Frontiers in Physics 85, Westview Press (June, 1992) ISBN 0-201-55409-7. Covering the elementary aspects of the physics of phases transitions and the renormalization group, this popular book emphasizes understanding and clarity rather than technical manipulations.
- Zinn-Justin, Jean ; *Quantum Field Theory and Critical Phenomena*, Oxford University Press (4th edition – 2002) ISBN 0-19-850923-5. A masterpiece on applications of renormalization methods to the calculation of critical exponents in statistical mechanics, following Wilson's ideas (Kenneth Wilson was *Nobel laureate 1982*).
- Zinn-Justin, Jean ; *Phase Transitions & Renormalization Group: from Theory to Numbers*, Poincaré Seminar (Paris, Oct. 12, 2002), published in : Duplantier, Bertrand; Rivasseau, Vincent (Eds.) ; *Poincaré Seminar 2002*, Progress in Mathematical Physics 30, Birkhäuser (2003) ISBN 3-7643-0579-7. Full text available in *PostScript*.
- Domb, Cyril ; *The Critical Point: A Historical Introduction to the Modern Theory of Critical Phenomena*, CRC Press (March, 1996) ISBN 0-7484-0435-X.
- Brown, Laurie M. (Ed.) ; *Renormalization: From Lorentz to Landau (and Beyond)*, Springer-Verlag (New York-1993) ISBN 0-387-97933-6.
- Cardy, John ; *Scaling and Renormalization in Statistical Physics*, Cambridge University Press (1996) ISBN 0-521-49959-3.

11.11.4 Miscellaneous

- Shirkov, Dmitry ; *The Bogoliubov Renormalization Group*, JINR Communication E2-96-15 (1996). Full text available at: *hep-th/9602024*

- García Moreta, José Javier http://prespacetime.com/index.php/pst/article/view/498 The Application of Zeta Regularization Method to the Calculation of Certain Divergent Series and Integrals Refined Higgs, CMB from Planck, Departures in Logic, and GR Issues & Solutions vol 4 Nº 3 prespacetime journal http://prespacetime.com/index.php/pst/issue/view/41/showToc

- Zinn Justin, Jean ; *Renormalization and renormalization group: From the discovery of UV divergences to the concept of effective field theories*, in: de Witt-Morette C., Zuber J.-B. (eds), Proceedings of the NATO ASI on *Quantum Field Theory: Perspective and Prospective*, June 15–26, 1998, Les Houches, France, Kluwer Academic Publishers, NATO ASI Series C 530, 375–388 (1999). Full text available in *PostScript*.

- Connes, Alain ; *Symétries Galoisiennes & Renormalisation*, Poincaré Seminar (Paris, Oct. 12, 2002), published in : Duplantier, Bertrand; Rivasseau, Vincent (Eds.) ; *Poincaré Seminar 2002*, Progress in Mathematical Physics 30, Birkhäuser (2003) ISBN 3-7643-0579-7. French mathematician *Alain Connes* (Fields medallist 1982) describe the mathematical underlying structure (the Hopf algebra) of renormalization, and its link to the Riemann-Hilbert problem. Full text (in French) available at *math/0211199v1*.

Chapter 12

Physics beyond the Standard Model

Physics beyond the Standard Model (BSM) refers to the theoretical developments needed to explain the deficiencies of the Standard Model, such as the origin of mass, the strong CP problem, neutrino oscillations, matter–antimatter asymmetry, and the nature of dark matter and dark energy.[1] Another problem lies within the mathematical framework of the Standard Model itself—the Standard Model is inconsistent with that of general relativity, to the point that one or both theories break down under certain conditions (for example within known space-time singularities like the Big Bang and black hole event horizons).

Theories that lie beyond the Standard Model include various extensions of the standard model through supersymmetry, such as the Minimal Supersymmetric Standard Model (MSSM) and Next-to-Minimal Supersymmetric Standard Model (NMSSM), or entirely novel explanations, such as string theory, M-theory, and extra dimensions. As these theories tend to reproduce the entirety of current phenomena, the question of which theory is the right one, or at least the "best step" towards a Theory of Everything, can only be settled via experiments, and is one of the most active areas of research in both theoretical and experimental physics.

12.1 Problems with the Standard Model

Despite being the most successful theory of particle physics to date, the Standard Model is not perfect.[2] A large share of the published output of theoretical physicists consists of proposals for various forms of "Beyond the Standard Model" new physics proposals that would modify the Standard Model in ways subtle enough to be consistent with existing data, yet address its imperfections materially enough to predict non-Standard Model outcomes of new experiments that can be proposed.

12.1.1 Phenomena not explained

The Standard Model is inherently an incomplete theory. There are fundamental physical phenomena in nature that the Standard Model does not adequately explain:

- *Gravity*. The standard model does not explain gravity. The approach of simply adding a "graviton" (whose properties are the subject of considerable consensus among physicists if it exists) to the Standard Model does not recreate what is observed experimentally without other modifications, as yet undiscovered, to the Standard Model. Moreover, instead, the Standard Model is widely considered to be incompatible with the most successful theory of gravity to date, general relativity.[3]

- *Dark matter and dark energy*. Cosmological observations tell us the standard model explains about 5% of the energy present in the universe. About 26% should be dark matter, which would behave just like other matter, but which only interacts weakly (if at all) with the Standard Model fields. Yet, the Standard Model does not supply

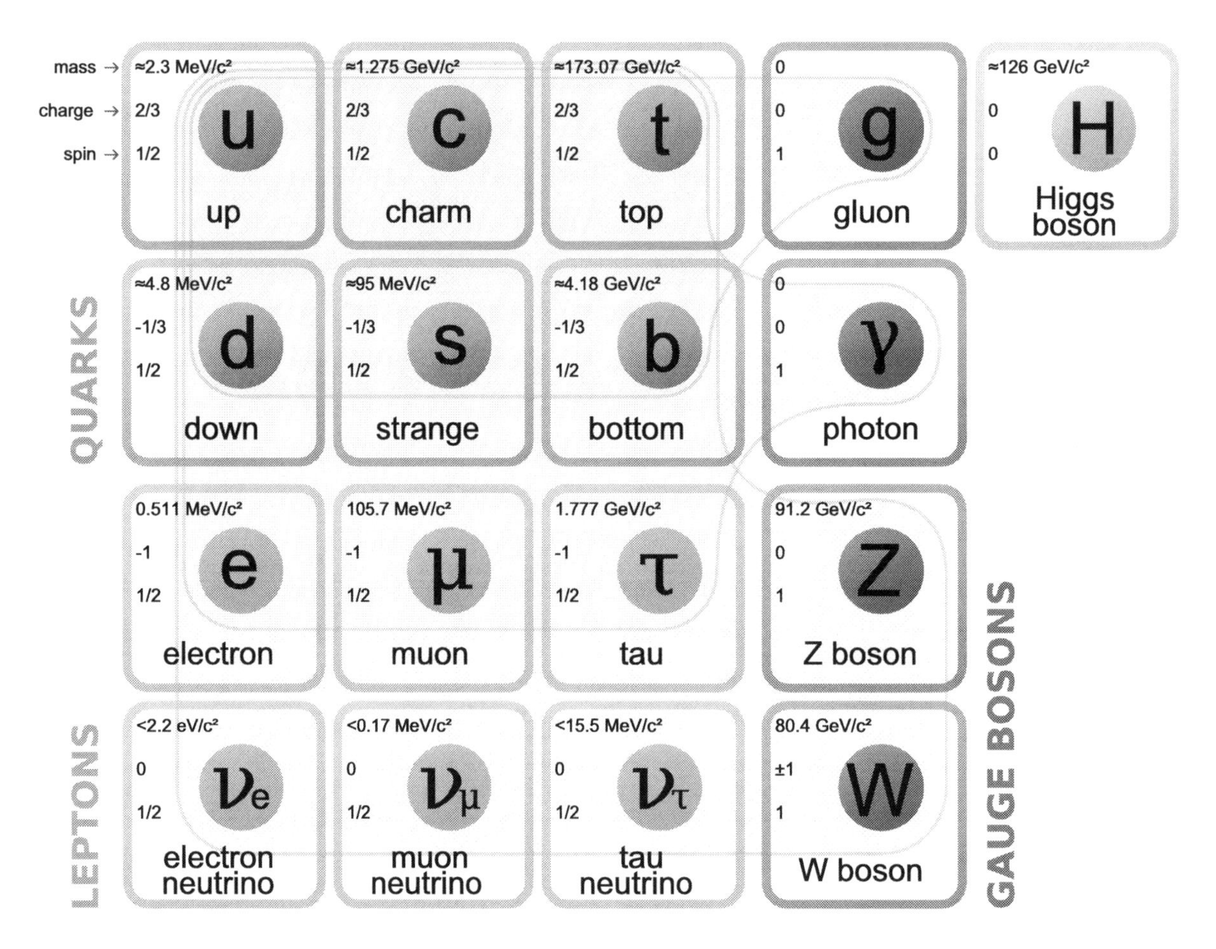

The Standard Model of elementary particles

any fundamental particles that are good dark matter candidates. The rest (69%) should be dark energy, a constant energy density for the vacuum. Attempts to explain dark energy in terms of vacuum energy of the standard model lead to a mismatch of 120 orders of magnitude.[4]

- *Neutrino masses.* According to the standard model, neutrinos are massless particles. However, neutrino oscillation experiments have shown that neutrinos do have mass. Mass terms for the neutrinos can be added to the standard model by hand, but these lead to new theoretical problems. For example, the mass terms need to be extraordinarily small and it is not clear if the neutrino masses would arise in the same way that the masses of other fundamental particles do in the Standard Model.

- *Matter-antimatter asymmetry.* The universe is made out of mostly matter. However, the standard model predicts that matter and antimatter should have been created in (almost) equal amounts if the initial conditions of the universe did not involve disproportionate matter relative to antimatter. Yet, no mechanism sufficient to explain this asymmetry exists in the Standard Model.

12.1.2 Experimental results not explained

No experimental result is widely accepted as contradicting the Standard Model at a level that definitively contradicts it at the "five sigma" (i.e. five standard deviation) level widely considered to be the threshold of a "discovery" in particle physics. But, because every experiment contains some degree of statistical and systemic uncertainty, and the theoretical predictions themselves are also almost never calculated exactly and are subject to uncertainties in measurements of the fundamental constants of the Standard Model (some of which are tiny and others of which are substantial,) it is mathematically expected that some of the hundreds of experimental tests of the Standard Model will deviate to some extent from the Standard Model even if there were no "new physics" beyond the Standard Model to be discovered.

At any given time there are a number of experimental results that are significantly different from the Standard Model expectation, although many of these have been found to be statistical flukes or experimental errors as more data has been collected. On the other hand, any "beyond the Standard Model" physics would necessarily first manifest experimentally as a statistically significant difference between an experiment and a Standard Model theoretical prediction.

In each case, physicists seek to determine if a result is a mere statistical fluke or experimental error on the one hand, or a sign of new physics on the other. More statistically significant results cannot be mere statistical flukes but can still result from experimental error or inaccurate estimates of experimental precision. Frequently, experiments are tailored to be more sensitive to experimental results that would distinguish the Standard Model from theoretical alternatives.

Some of the most notable examples include the following:

- *Muonic hydrogen* — the Standard Model makes precise theoretical predictions regarding the atomic radius size of ordinary hydrogen (a proton-electron system) and that of muonic hydrogen (a proton-muon system in which a muon is a "heavy" variant of an electron). However, the measured atomic radius of muonic hydrogen differs significantly from that of the radius predicted by the Standard Model using existing physical constant measurements by what appears to be as many as seven standard deviations.[5] Doubts about the accuracy of the error estimates in earlier experiments, which are still within 4% of each other in measuring a truly tiny distance, and a lack of a well motivated theory that could explain the discrepancy, have caused physicists to be hesitant to describe these results as contradicting the Standard Model despite the apparent statistical significance of the result and a lack of any clearly identified possible source of experimental error in the results.
- *BaBar Data Suggests Possible Flaws in the Standard Model* — results from a BaBar experiment may suggest a surplus over Standard Model predictions of a type of particle decay called "B to D-star-tau-nu." In this, an electron and positron collide, resulting in a B meson and an antimatter B-bar meson, which then decays into a D meson and a tau lepton as well as a smaller antineutrino. While the level of certainty of the excess (3.4 sigma in statistical language) is not enough to claim a break from the Standard Model, the results are a potential sign of something amiss and are likely to affect existing theories, including those attempting to deduce the properties of Higgs bosons.[6] However, results at LHCb have demonstrated no significant deviation from the Standard Model prediction of very nearly zero asymmetry.[7][8]
- Proton radius — radius measured using electrons is different from radius measured using muons[9]

12.1.3 Theoretical predictions not observed

Observation at particle colliders of all of the fundamental particles predicted by the Standard Model has been confirmed. The Higgs boson is predicted by the Standard Model's explanation of the Higgs mechanism, which describes how the weak SU(2) gauge symmetry is broken and how fundamental particles obtain mass; it was the last particle predicted by the Standard Model to be observed. On July 4, 2012, CERN scientists using the Large Hadron Collider announced the discovery of a particle consistent with the Higgs boson, with a mass of about 126 GeV/c^2. A Higgs boson was confirmed to exist on March 14, 2013, although efforts to confirm that it has all of the properties predicted by the Standard Model are ongoing.[10]

A few hadrons (i.e. composite particles made of quarks) whose existence is predicted by the Standard Model, which can be produced only at very high energies in very low frequencies have not yet been definitively observed, and "glueballs"[11] (i.e. composite particles made of gluons) have also not yet been definitively observed. Some very low frequency particle decays predicted by the Standard Model have also not yet been definitively observed because insufficient data is available to make a statistically significant observation.

12.1.4 Theoretical problems

Some features of the standard model are added in an ad hoc way. These are not problems per se (i.e. the theory works fine with these ad hoc features), but they imply a lack of understanding. These ad hoc features have motivated theorists

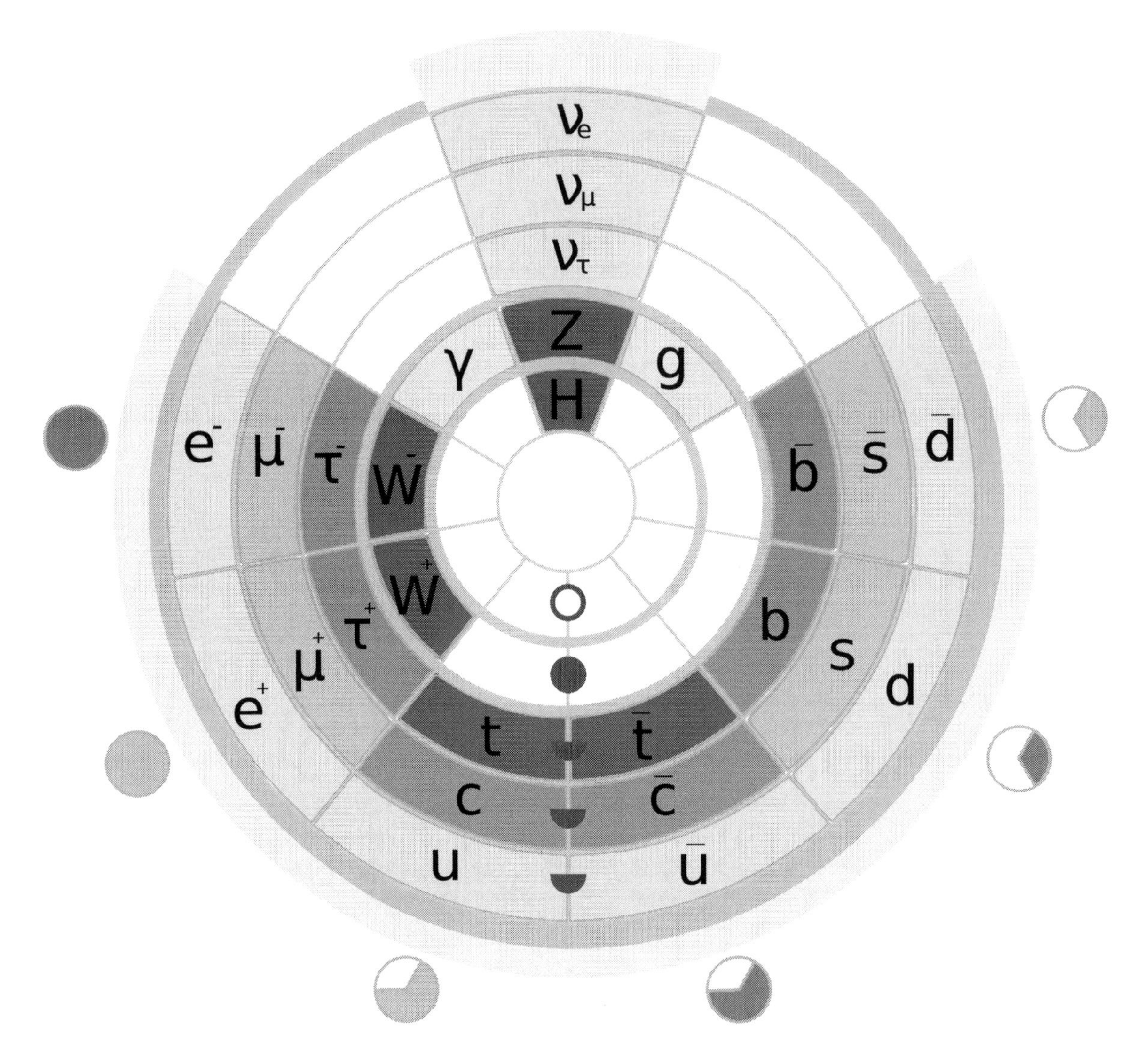

Masses of fundamental particles ----
more than 80 GeV/c²
1-5 GeV/c²
90-110 MeV/c²
less than 16 MeV/c²
Massless

to look for more fundamental theories with fewer parameters. Some of the ad hoc features are:

- *Hierarchy problem* — the standard model introduces particle masses through a process known as spontaneous symmetry breaking caused by the Higgs field. Within the standard model, the mass of the Higgs gets some very large quantum corrections due to the presence of virtual particles (mostly virtual top quarks). These corrections are much larger than the actual mass of the Higgs. This means that the bare mass parameter of the Higgs in the standard model must be fine tuned in such a way that almost completely cancels the quantum corrections. This level of fine-tuning is deemed unnatural by many theorists.

- *Number of parameters* — the standard model depends on 19 numerical parameters. Their values are known from experiment, but the origin of the values is unknown. Some theorists have tried to find relations between different parameters, for example, between the masses of particles in different generations.

- *Quantum triviality* — suggests that it may not be possible to create a consistent quantum field theory involving elementary scalar Higgs particles.
- *Strong CP problem* — theoretically it can be argued that the standard model should contain a term that breaks CP symmetry—relating matter to antimatter—in the strong interaction sector. Experimentally, however, no such violation has been found, implying that the coefficient of this term is very close to zero. This fine tuning is also considered unnatural.

12.2 Grand unified theories

Main article: Grand Unified Theory

The standard model has three gauge symmetries; the colour SU(3), the weak isospin SU(2), and the hypercharge U(1) symmetry, corresponding to the three fundamental forces. Due to renormalization the coupling constants of each of these symmetries vary with the energy at which they are measured. Around 10^{16} GeV these couplings become approximately equal. This has led to speculation that above this energy the three gauge symmetries of the standard model are unified in one single gauge symmetry with a simple group gauge group, and just one coupling constant. Below this energy the symmetry is spontaneously broken to the standard model symmetries.[12] Popular choices for the unifying group are the special unitary group in five dimensions SU(5) and the special orthogonal group in ten dimensions SO(10).[13]

Theories that unify the standard model symmetries in this way are called Grand Unified Theories (or GUTs), and the energy scale at which the unified symmetry is broken is called the GUT scale. Generically, grand unified theories predict the creation of magnetic monopoles in the early universe,[14] and instability of the proton.[15] Neither of which have been observed, and this absence of observation puts limits on the possible GUTs.

12.3 Supersymmetry

Main article: Supersymmetry

Supersymmetry extends the Standard Model by adding another class of symmetries to the Lagrangian. These symmetries exchange fermionic particles with bosonic ones. Such a symmetry predicts the existence of *supersymmetric particles*, abbreviated as *sparticles*, which include the sleptons, squarks, neutralinos and charginos. Each particle in the Standard Model would have a superpartner whose spin differs by 1/2 from the ordinary particle. Due to the breaking of supersymmetry, the sparticles are much heavier than their ordinary counterparts; they are so heavy that existing particle colliders may not be powerful enough to produce them.

12.4 Neutrinos

In the standard model, neutrinos have exactly zero mass. This is a consequence of the standard model containing only left-handed neutrinos. With no suitable right-handed partner, it is impossible to add a renormalizable mass term to the standard model.[16] Measurements however indicated that neutrinos spontaneously change flavour, which implies that neutrinos have a mass. These measurements only give the relative masses of the different flavours. The best constraint on the absolute mass of the neutrinos comes from precision measurements of tritium decay, providing an upper limit 2 eV, which makes them at least five orders of magnitude lighter than the other particles in the standard model.[17] This necessitates an extension of the standard model, which not only needs to explain how neutrinos get their mass, but also why the mass is so small.[18]

One approach to add masses to the neutrinos, the so-called seesaw mechanism, is to add right-handed neutrinos and have these couple to left-handed neutrinos with a Dirac mass term. The right-handed neutrinos have to be sterile, meaning that they do not participate in any of the standard model interactions. Because they have no charges, the right-handed neutrinos can act as their own anti-particles, and have a Majorana mass term. Like the other Dirac masses in the standard model, the neutrino Dirac mass is expected to be generated through the Higgs mechanism, and is therefore unpredictable. The standard model fermion masses differ by many orders of magnitude; the Dirac neutrino mass has at least the same uncertainty. On the other hand, the Majorana mass for the right-handed neutrinos does not arise from the Higgs mechanism, and is therefore expected to be tied to some energy scale of new physics beyond the standard model, for example the Planck scale.[19] Therefore, any process involving right-handed neutrinos will be suppressed at low energies. The correction due to these suppressed processes effectively gives the left-handed neutrinos a mass that is inversely proportional to the right-handed Majorana mass, a mechanism known as the see-saw.[20] The presence of heavy right-handed neutrinos thereby explains both the small mass of the left-handed neutrinos and the absence of the right-handed neutrinos in observations. However, due to the uncertainty in the Dirac neutrino masses, the right-handed neutrino masses can lie anywhere. For example, they could be as light as keV and be dark matter,[21] they can have a mass in the LHC energy range[22][23] and lead to observable lepton number violation,[24] or they can be near the GUT scale, linking the right-handed neutrinos to the possibility of a grand unified theory.[25][26]

The mass terms mix neutrinos of different generations. This mixing is parameterized by the PMNS matrix, which is the neutrino analogue of the CKM quark mixing matrix. Unlike the quark mixing, which is almost minimal, the mixing of the neutrinos appears to be almost maximal. This has led to various speculations of symmetries between the various generations that could explain the mixing patterns.[27] The mixing matrix could also contain several complex phases that break CP invariance, although there has been no experimental probe of these. These phases could potentially create a surplus of leptons over anti-leptons in the early universe, a process known as leptogenesis. This asymmetry could then at a later stage be converted in an excess of baryons over anti-baryons, and explain the matter-antimatter asymmetry in the universe.[13]

The light neutrinos are disfavored as an explanation for the observation of dark matter, due to considerations of large-scale structure formation in the early universe. Simulations of structure formation show that they are too hot—i.e. their kinetic energy is large compared to their mass—while formation of structures similar to the galaxies in our universe requires cold dark matter. The simulations show that neutrinos can at best explain a few percent of the missing dark matter. The heavy sterile right-handed neutrinos are however a possible candidate for a dark matter WIMP.[28]

12.5 Preon Models

Several preon models have been proposed to address the unsolved problem concerning the fact that there are three generations of quarks and leptons. Preon models generally postulate some additional new particles which are further postulated to be able to combine to form the quarks and leptons of the standard model. One of the earliest preon models was the Rishon model.[29][30][31]

To date, no preon model is widely accepted or fully verified.

12.6 Theories of everything

12.6.1 Theory of everything

Main article: Theory of everything

Theoretical physics continues to strive toward a theory of everything, a theory that fully explains and links together all known physical phenomena, and predicts the outcome of any experiment that could be carried out in principle. In practical terms the immediate goal in this regard is to develop a theory which would unify the Standard Model with General Relativity in a theory of quantum gravity. Additional features, such as overcoming conceptual flaws in either theory or accurate prediction of particle masses, would be desired. The challenges in putting together such a theory are not just conceptual - they include the experimental aspects of the very high energies needed to probe exotic realms.

Several notable attempts in this direction are supersymmetry, string theory, and loop quantum gravity.

12.6.2 String theory

Main article: String theory

Extensions, revisions, replacements, and reorganizations of the Standard Model exist in attempt to correct for these and other issues. String theory is one such reinvention, and many theoretical physicists think that such theories are the next theoretical step toward a true Theory of Everything. Theories of quantum gravity such as loop quantum gravity and others are thought by some to be promising candidates to the mathematical unification of quantum field theory and general relativity, requiring less drastic changes to existing theories.[32] However recent work places stringent limits on the putative effects of quantum gravity on the speed of light, and disfavours some current models of quantum gravity.[33]

Among the numerous variants of string theory, M-theory, whose mathematical existence was first proposed at a String Conference in 1995, is believed by many to be a proper "ToE" candidate, notably by physicists Brian Greene and Stephen Hawking. Though a full mathematical description is not yet known, solutions to the theory exist for specific cases.[34] Recent works have also proposed alternate string models, some of which lack the various harder-to-test features of M-theory (e.g. the existence of Calabi–Yau manifolds, many extra dimensions, etc.) including works by well-published physicists such as Lisa Randall.[35][36]

12.7 See also

- *A New Kind of Science*
- Antimatter tests of Lorentz violation
- Fundamental physical constants in the standard model
- Higgsless model
- Holographic principle
- Little Higgs
- Lorentz-violating neutrino oscillations
- Minimal Supersymmetric Standard Model
- Peccei–Quinn theory
- Preon
- Standard-Model Extension
- Supergravity
- Seesaw mechanism
- Supersymmetry
- Superfluid vacuum theory
- String theory
- Technicolor (physics)
- Theory of everything
- Unsolved problems in physics
- Unparticle physics

12.8 References

[1] Womersley, J. (February 2005). "Beyond the Standard Model" (PDF). *Symmetry Magazine*. Retrieved 2010-11-23.

[2] Lykken, J. D. (2010). "Beyond the Standard Model". *CERN Yellow Report*. CERN. pp. 101–109. arXiv:1005.1676. CERN-2010-002.

[3] Sushkov, A. O.; Kim, W. J.; Dalvit, D. A. R.; Lamoreaux, S. K. (2011). "New Experimental Limits on Non-Newtonian Forces in the Micrometer Range". *Physical Review Letters* **107** (17): 171101. arXiv:1108.2547. Bibcode:2011PhRvL.107q1101S. doi:10.1103/PhysRevLett.107.171101. It is remarkable that two of the greatest successes of 20th century physics, general relativity and the standard model, appear to be fundamentally incompatible. But see also Donoghue, John F. (2012). "The effective field theory treatment of quantum gravity". *AIP Conference Proceedings* **1473**: 73. arXiv:1209.3511. doi:10.1063/1.4756964. One can find thousands of statements in the literature to the effect that "general relativity and quantum mechanics are incompatible". These are completely outdated and no longer relevant. Effective field theory shows that general relativity and quantum mechanics work together perfectly normally over a range of scales and curvatures, including those relevant for the world that we see around us. However, effective field theories are only valid over some range of scales. General relativity certainly does have problematic issues at extreme scales. There are important problems which the effective field theory does not solve because they are beyond its range of validity. However, this means that the issue of quantum gravity is not what we thought it to be. Rather than a fundamental incompatibility of quantum mechanics and gravity, we are in the more familiar situation of needing a more complete theory beyond the range of their combined applicability. The usual marriage of general relativity and quantum mechanics is fine at ordinary energies, but we now seek to uncover the modifications that must be present in more extreme conditions. This is the modern view of the problem of quantum gravity, and it represents progress over the outdated view of the past."

[4] Krauss, L. (2009). *A Universe from Nothing*. AAI Conference.

[5] Randolf Pohl, Ronald Gilman, Gerald A. Miller, Krzysztof Pachucki, "Muonic hydrogen and the proton radius puzzle" (May 30, 2013) http://arxiv.org/abs/1301.0905 in print Annu. Rev. Nucl. Part. Sci. Vol 63 (2013) 10.1146/annurev-nucl-102212-170627 ("The recent determination of the proton radius using the measurement of the Lamb shift in the muonic hydrogen atom startled the physics world. The obtained value of 0.84087(39) fm differs by about 4% or 7 standard deviations from the CODATA value of 0.8775(51) fm. The latter is composed from the electronic hydrogenate atom value of 0.8758(77) fm and from a similar value with larger uncertainties determined by electron scattering.")

[6] Lees, J. P.; et al. (BaBar Collaboration) (1970). "Evidence for an excess of $B \to D^{(*)}\tau^{-}\tau\nu$ decays". *Physical Review Letters* **109** (10). arXiv:1205.5442. Bibcode:2012PhRvL.109j1802L. doi:10.1103/PhysRevLett.109.101802.

[7] Article on LHCb results

[8] 2012 LHCb paper

[9] http://arxiv.org/pdf/1502.05314.pdf

[10] O'Luanaigh, C. (14 March 2013). "New results indicate that new particle is a Higgs boson". CERN.

[11] Marco Frasca, "What is a Glueball?" (March 31, 2009) http://marcofrasca.wordpress.com/2009/03/31/what-is-a-glueball-2/

[12] Peskin, M. E.; Schroeder, D. V. (1995). *An introduction to quantum field theory*. Addison-Wesley. pp. 786–791. ISBN 978-0-201-50397-5.

[13] Buchmüller, W. (2002). "Neutrinos, Grand Unification and Leptogenesis". arXiv:hep-ph/0204288 [hep-ph].

[14] Milstead, D.; Weinberg, E.J. (2009). "Magnetic Monopoles" (PDF). Particle Data Group. Retrieved 2010-12-20.

[15] P., Nath; P. F., Perez (2006). "Proton stability in grand unified theories, in strings, and in branes". *Physics Reports* **441** (5–6): 191–317. arXiv:hep-ph/0601023. Bibcode:2007PhR...441..191N. doi:10.1016/j.physrep.2007.02.010.

[16] Peskin, M. E.; Schroeder, D. V. (1995). *An introduction to quantum field theory*. Addison-Wesley. pp. 713–715. ISBN 978-0-201-50397-5.

[17] Nakamura, K.; et al. (Particle Data Group) (2010). "Neutrino Properties". Particle Data Group. Retrieved 2010-12-20.

[18] Mohapatra, R. N.; Pal, P. B. (2007). *Massive neutrinos in physics and astrophysics*. Lecture Notes in Physics **72** (3rd ed.). World Scientific. ISBN 978-981-238-071-5.

[19] Senjanovic, G. (2011). "Probing the Origin of Neutrino Mass: from GUT to LHC". arXiv:1107.5322 [hep-ph].

[20] Grossman, Y. (2003). "TASI 2002 lectures on neutrinos". arXiv:hep-ph/0305245v1 [hep-ph].

[21] Dodelson, S.; Widrow, L. M. (1993). "Sterile neutrinos as dark matter". *Physical Review Letters* **72**: 17. arXiv:hep-ph/9303287. Bibcode:1994PhRvL..72...17D. doi:10.1103/PhysRevLett.72.17.

[22] Minkowski, P. (1977). "$\mu \rightarrow e\gamma$ at a Rate of One Out of 10^9 Muon Decays?". *Physics Letters B* **67** (4): 421. Bibcode:1977PhLB...67..421M. doi:10.1016/0370-2693(77)90435-X.

[23] Mohapatra, R. N.; Senjanovic, G. (1980). "Neutrino mass and spontaneous parity nonconservation". *Physical Review Letters* **44** (14): 912. Bibcode:1980PhRvL..44..912M. doi:10.1103/PhysRevLett.44.912.

[24] Keung, W.-Y.; Senjanovic, G. (1983). "Majorana Neutrinos And The Production Of The Right-handed Charged Gauge Boson". *Physical Review Letters* **50** (19): 1427. Bibcode:1983PhRvL..50.1427K. doi:10.1103/PhysRevLett.50.1427.

[25] Gell-Mann, M.; Ramond, P.; Slansky, R. (1979). P. van Nieuwenhuizen; D. Freedman, eds. *Supergravity*. North Holland.

[26] Glashow, S. L. (1979). M. Levy, ed. *Proceedings of the 1979 Cargèse Summer Institute on Quarks and Leptons*. Plenum Press.

[27] Altarelli, G. (2007). "Lectures on Models of Neutrino Masses and Mixings". arXiv:0711.0161 [hep-ph].

[28] Murayama, H. (2007). "Physics Beyond the Standard Model and Dark Matter". arXiv:0704.2276 [hep-ph].

[29] Harari, H. (1979). "A Schematic Model of Quarks and Leptons". Physics Letters B 86 (1): 83-86.

[30] Shupe, M. A. (1979). "A Composite Model of Leptons and Quarks". Physics Letters B 86 (1): 87-92.

[31] Zenczykowski, P. (2008). "The Harari-Shupe preon model and nonrelativistic quantum phase space". Physics Letters B 660 (5): 567-572.

[32] Smolin, L. (2001). *Three Roads to Quantum Gravity*. Basic Books. ISBN 0-465-07835-4.

[33] Abdo, A. A.; et al. (Fermi GBM/LAT Collaborations) (2009). "A limit on the variation of the speed of light arising from quantum gravity effects". *Nature* **462** (7271): 331–4. arXiv:0908.1832. Bibcode:2009Natur.462..331A. doi:10.1038/nature08574. PMID 19865083.

[34] Maldacena, J.; Strominger, A.; Witten, E. (1997). "Black hole entropy in M-Theory". *Journal of High Energy Physics* **1997** (12): 2. arXiv:hep-th/9711053. Bibcode:1997JHEP...12..002M. doi:10.1088/1126-6708/1997/12/002.

[35] Randall, L.; Sundrum, R. (1999). "Large Mass Hierarchy from a Small Extra Dimension". *Physical Review Letters* **83** (17): 3370. arXiv:hep-ph/9905221. Bibcode:1999PhRvL..83.3370R. doi:10.1103/PhysRevLett.83.3370.

[36] Randall, L.; Sundrum, R. (1999). "An Alternative to Compactification". *Physical Review Letters* **83** (23): 4690. arXiv:hep-th/9906064. Bibcode:1999PhRvL..83.4690R. doi:10.1103/PhysRevLett.83.4690.

12.9 Further reading

- Lisa Randall (2005). *Warped Passages: Unraveling the Mysteries of the Universe's Hidden Dimensions*. HarperCollins. ISBN 0-06-053108-8.

12.10 External resources

- Standard Model Theory @ SLAC
- Scientific American Apr 2006
- LHC. Nature July 2007
- Open Questions
- Working group - schedule
- Les Houches Conference, Summer 2005

Chapter 13

Graviton

This article is about the hypothetical particle. For other uses, see Graviton (disambiguation).

In physics, the **graviton** is a hypothetical elementary particle that mediates the force of gravitation in the framework of quantum field theory. If it exists, the graviton is expected to be massless (because the gravitational force appears to have unlimited range) and must be a spin−2 boson. The spin follows from the fact that the source of gravitation is the stress–energy tensor, a second-rank tensor (compared to electromagnetism's spin-1 photon, the source of which is the four-current, a first-rank tensor). Additionally, it can be shown that any massless spin-2 field would give rise to a force indistinguishable from gravitation, because a massless spin-2 field must couple to (interact with) the stress–energy tensor in the same way that the gravitational field does. Seeing as the graviton is hypothetical, its discovery would unite quantum theory with gravity.[4] This result suggests that, if a massless spin-2 particle is discovered, it must be the graviton, so that the only experimental verification needed for the graviton may simply be the discovery of a massless spin-2 particle.[5]

13.1 Theory

The four other known forces of nature are mediated by elementary particles: electromagnetism by the photon, the strong interaction by the gluons, the Higgs field by the Higgs Boson, and the weak interaction by the W and Z bosons. The hypothesis is that the gravitational interaction is likewise mediated by an – as yet undiscovered – elementary particle, dubbed as *the graviton*. In the classical limit, the theory would reduce to general relativity and conform to Newton's law of gravitation in the weak-field limit.[6][7][8]

13.1.1 Gravitons and renormalization

When describing graviton interactions, the classical theory (i.e., the tree diagrams) and semiclassical corrections (one-loop diagrams) behave normally, but Feynman diagrams with two (or more) loops lead to ultraviolet divergences; that is, infinite results that cannot be removed because the quantized general relativity is not renormalizable, unlike quantum electrodynamics. That is, the usual ways physicists calculate the probability that a particle will emit or absorb a graviton give nonsensical answers and the theory loses its predictive power. These problems, together with some conceptual puzzles, led many physicists to believe that a theory more complete than quantized general relativity must describe the behavior near the Planck scale.

13.1.2 Comparison with other forces

Unlike the force carriers of the other forces, gravitation plays a special role in general relativity in defining the spacetime in which events take place. In some descriptions, matter modifies the 'shape' of spacetime itself, and gravity is a result of this shape, an idea which at first glance may appear hard to match with the idea of a force acting between particles.[9]

Because the diffeomorphism invariance of the theory does not allow any particular space-time background to be singled out as the "true" space-time background, general relativity is said to be background independent. In contrast, the Standard Model is *not* background independent, with Minkowski space enjoying a special status as the fixed background space-time.[10] A theory of quantum gravity is needed in order to reconcile these differences.[11] Whether this theory should be background independent is an open question. The answer to this question will determine our understanding of what specific role gravitation plays in the fate of the universe.[12]

13.1.3 Gravitons in speculative theories

String theory predicts the existence of gravitons and their well-defined interactions. A graviton in perturbative string theory is a closed string in a very particular low-energy vibrational state. The scattering of gravitons in string theory can also be computed from the correlation functions in conformal field theory, as dictated by the AdS/CFT correspondence, or from matrix theory.

A feature of gravitons in string theory is that, as closed strings without endpoints, they would not be bound to branes and could move freely between them. If we live on a brane (as hypothesized by brane theories) this "leakage" of gravitons from the brane into higher-dimensional space could explain why gravitation is such a weak force, and gravitons from other branes adjacent to our own could provide a potential explanation for dark matter. However, if gravitons were to move completely freely between branes this would dilute gravity too much, causing a violation of Newton's inverse square law. To combat this, Lisa Randall found that a three-brane (such as ours) would have a gravitational pull of its own, preventing gravitons from drifting freely, possibly resulting in the diluted gravity we observe while roughly maintaining Newton's inverse square law.[13] See brane cosmology.

A theory by Ahmed Farag Ali and Saurya Das adds quantum mechanical corrections (using Bohm trajectories) to general relativistic geodesics. If gravitons are given a small but non-zero mass, it could explain the cosmological constant without need for dark energy and solve the smallness problem.[14]

13.2 Experimental observation

Unambiguous detection of individual gravitons, though not prohibited by any fundamental law, is impossible with any physically reasonable detector.[15] The reason is the extremely low cross section for the interaction of gravitons with matter. For example, a detector with the mass of Jupiter and 100% efficiency, placed in close orbit around a neutron star, would only be expected to observe one graviton every 10 years, even under the most favorable conditions. It would be impossible to discriminate these events from the background of neutrinos, since the dimensions of the required neutrino shield would ensure collapse into a black hole.[15]

However, experiments to detect gravitational waves, which may be viewed as coherent states of many gravitons, are underway (e.g., LIGO and VIRGO). Although these experiments cannot detect individual gravitons, they might provide information about certain properties of the graviton.[16] For example, if gravitational waves were observed to propagate slower than c (the speed of light in a vacuum), that would imply that the graviton has mass (however, gravitational waves must propagate slower than "c" in a region with non-zero mass density if they are to be detectable).[17] Astronomical observations of the kinematics of galaxies, especially the galaxy rotation problem and modified Newtonian dynamics, might point toward gravitons having non-zero mass.[18]

13.3 Difficulties and outstanding issues

Most theories containing gravitons suffer from severe problems. Attempts to extend the Standard Model or other quantum field theories by adding gravitons run into serious theoretical difficulties at high energies (processes involving energies close to or above the Planck scale) because of infinities arising due to quantum effects (in technical terms, gravitation is nonrenormalizable). Since classical general relativity and quantum mechanics seem to be incompatible at such energies, from a theoretical point of view, this situation is not tenable. One possible solution is to replace particles with strings.

String theories are quantum theories of gravity in the sense that they reduce to classical general relativity plus field theory at low energies, but are fully quantum mechanical, contain a graviton, and are believed to be mathematically consistent.[19]

13.4 See also

- Gravitomagnetism
- Gravitational wave
- Planck mass
- Gravitation
- Static forces and virtual-particle exchange
- Multiverse
- Gravitino

13.5 References

[1] G is used to avoid confusion with gluons (symbol g)

[2] Rovelli, C. (2001). "Notes for a brief history of quantum gravity". arXiv:gr-qc/0006061 [gr-qc].

[3] Blokhintsev, D. I.; Gal'perin, F. M. (1934). "Gipoteza neitrino i zakon sokhraneniya energii" [Neutrino hypothesis and conservation of energy]. *Pod Znamenem Marxisma* (in Russian) **6**: 147–157.

[4] Lightman, A. P.; Press, W. H.; Price, R. H.; Teukolsky, S. A. (1975). "Problem 12.16". *Problem book in Relativity and Gravitation*. Princeton University Press. ISBN 0-691-08162-X.

[5] For a comparison of the geometric derivation and the (non-geometric) spin-2 field derivation of general relativity, refer to box 18.1 (and also 17.2.5) of Misner, C. W.; Thorne, K. S.; Wheeler, J. A. (1973). *Gravitation*. W. H. Freeman. ISBN 0-7167-0344-0.

[6] Feynman, R. P.; Morinigo, F. B.; Wagner, W. G.; Hatfield, B. (1995). *Feynman Lectures on Gravitation*. Addison-Wesley. ISBN 0-201-62734-5.

[7] Zee, A. (2003). *Quantum Field Theory in a Nutshell*. Princeton University Press. ISBN 0-691-01019-6.

[8] Randall, L. (2005). *Warped Passages: Unraveling the Universe's Hidden Dimensions*. Ecco Press. ISBN 0-06-053108-8.

[9] See the other articles on General relativity, Gravitational field, Gravitational wave, etc

[10] Colosi, D. et al. (2005). "Background independence in a nutshell: The dynamics of a tetrahedron". *Classical and Quantum Gravity* **22** (14): 2971. arXiv:gr-qc/0408079. Bibcode:2005CQGra..22.2971C. doi:10.1088/0264-9381/22/14/008.

[11] Witten, E. (1993). "Quantum Background Independence In String Theory". arXiv:hep-th/9306122 [hep-th].

[12] Smolin, L. (2005). "The case for background independence". arXiv:hep-th/0507235 [hep-th].

[13] Kaku, Michio (2006). *Parallel Worlds - The science of alternative universes and our future in the Cosmos*. pp. 218–221.

[14] Ali, Ahmed Farang (2014). "Cosmology from quantum potential". *Physical Letters B* **741**: 276–279. arXiv:1404.3093v3. doi:10.1016/j.physletb.2014.12.057.

[15] Rothman, T.; Boughn, S. (2006). "Can Gravitons be Detected?". *Foundations of Physics* **36** (12): 1801–1825. arXiv:gr-qc/0601043. Bibcode:2006FoPh...36.1801R. doi:10.1007/s10701-006-9081-9.

[16] Freeman Dyson (8 October 2013). "Is a graviton detectable?". *International Journal of Modern Physics A* **28** (25): 1330041–1–1330035–14. Bibcode:2013IJMPA..2830041D. doi:10.1142/S0217751X1330041X.

[17] Will, C. M. (1998). "Bounding the mass of the graviton using gravitational-wave observations of inspiralling compact binaries". *Physical Review D* **57** (4): 2061–2068. arXiv:gr-qc/9709011. Bibcode:1998PhRvD..57.2061W. doi:10.1103/PhysRevD.57.2061.

[18] Trippe, S. (2013), "A Simplified Treatment of Gravitational Interaction on Galactic Scales", J. Kor. Astron. Soc. **46**, 41. arXiv:1211.4692

[19] Sokal, A. (July 22, 1996). "Don't Pull the String Yet on Superstring Theory". *The New York Times*. Retrieved March 26, 2010.

13.6 External links

-
- Graviton on *In Our Time* at the BBC. (listen now)

Chapter 14

Effective field theory

In physics, an **effective field theory** is a type of approximation to (or effective theory for) an underlying physical theory, such as a quantum field theory or a statistical mechanics model. An effective field theory includes the appropriate degrees of freedom to describe physical phenomena occurring at a chosen length scale or energy scale, while ignoring substructure and degrees of freedom at shorter distances (or, equivalently, at higher energies). Intuitively, one averages over the behavior of the underlying theory at shorter length scales to derive a hopefully simplified model at longer length scales. Effective field theories typically work best when there is a large separation between length scale of interest and the length scale of the underlying dynamics. Effective field theories have found use in particle physics, statistical mechanics, condensed matter physics, general relativity, and hydrodynamics. They simplify calculations, and allow treatment of Dissipation and Radiation effects .[1][2]

14.1 The renormalization group

Presently, effective field theories are discussed in the context of the renormalization group (RG) where the process of *integrating out* short distance degrees of freedom is made systematic. Although this method is not sufficiently concrete to allow the actual construction of effective field theories, the gross understanding of their usefulness becomes clear through a RG analysis. This method also lends credence to the main technique of constructing effective field theories, through the analysis of symmetries. If there is a single mass scale **M** in the *microscopic* theory, then the effective field theory can be seen as an expansion in **1/M**. The construction of an effective field theory accurate to some power of **1/M** requires a new set of free parameters at each order of the expansion in **1/M**. This technique is useful for scattering or other processes where the maximum momentum scale **k** satisfies the condition **k/M≪1**. Since effective field theories are not valid at small length scales, they need not be renormalizable. Indeed, the ever expanding number of parameters at each order in **1/M** required for an effective field theory means that they are generally not renormalizable in the same sense as quantum electrodynamics which requires only the renormalization of two parameters.

14.2 Examples of effective field theories

14.2.1 Fermi theory of beta decay

The best-known example of an effective field theory is the Fermi theory of beta decay. This theory was developed during the early study of weak decays of nuclei when only the hadrons and leptons undergoing weak decay were known. The typical reactions studied were:

$$n \to p + e^- + \overline{\nu}_e$$
$$\mu^- \to e^- + \overline{\nu}_e + \nu_\mu.$$

This theory posited a pointlike interaction between the four fermions involved in these reactions. The theory had great phenomenological success and was eventually understood to arise from the gauge theory of electroweak interactions, which forms a part of the standard model of particle physics. In this more fundamental theory, the interactions are mediated by a flavour-changing gauge boson, the $W^{\pm}$. The immense success of the Fermi theory was because the W particle has mass of about 80 GeV, whereas the early experiments were all done at an energy scale of less than 10 MeV. Such a separation of scales, by over 3 orders of magnitude, has not been met in any other situation as yet.

14.2.2 BCS theory of superconductivity

Another famous example is the BCS theory of superconductivity. Here the underlying theory is of electrons in a metal interacting with lattice vibrations called phonons. The phonons cause attractive interactions between some electrons, causing them to form Cooper pairs. The length scale of these pairs is much larger than the wavelength of phonons, making it possible to neglect the dynamics of phonons and construct a theory in which two electrons effectively interact at a point. This theory has had remarkable success in describing and predicting the results of experiments on superconductivity.

14.2.3 Effective Field Theories in Gravity

General relativity itself is expected to be the low energy effective field theory of a full theory of quantum gravity, such as string theory. The expansion scale is the Planck mass. Effective field theories have also been used to simplify problems in General Relativity, in particular in calculating the gravitational wave signature of inspiralling finite-sized objects.[3] The most common EFT in GR is "Non-Relativistic General Relativity" (NRGR),[4][5][6] which is similar to the post-Newtonian expansion.[7] Another common GR EFT is the Extreme Mass Ratio (EMR), which in the context of the inspiralling problem is called EMRI.

14.2.4 Other examples

Presently, effective field theories are written for many situations.

- One major branch of nuclear physics is quantum hadrodynamics, where the interactions of hadrons are treated as a field theory, which should be derivable from the underlying theory of quantum chromodynamics. Due to the smaller separation of length scales here, this effective theory has some classificatory power, but not the spectacular success of the Fermi theory.
- In particle physics the effective field theory of QCD called chiral perturbation theory has had better success.[8] This theory deals with the interactions of hadrons with pions or kaons, which are the Goldstone bosons of spontaneous chiral symmetry breaking. The expansion parameter is the pion energy/momentum.
- For hadrons containing one heavy quark (such as the bottom or charm), an effective field theory which expands in powers of the quark mass, called the heavy-quark effective theory (HQET), has been found useful.
- For hadrons containing two heavy quarks, an effective field theory which expands in powers of the relative velocity of the heavy quarks, called non-relativistic QCD (NRQCD), has been found useful, especially when used in conjunctions with lattice QCD.
- For hadron reactions with light energetic (collinear) particles, the interactions with low-energetic (soft) degrees of freedom are described by the soft-collinear effective theory (SCET).
- Much of condensed matter physics consists of writing effective field theories for the particular property of matter being studied.
- Hydrodynamics can also be treated using Effective Field Theories[9]

14.3 See also

- Form factor (quantum field theory)
- Renormalization group
- Quantum field theory
- Quantum triviality
- Ginzburg–Landau theory

14.4 References

[1] "Classical Mechanics of Nonconservative Systems" by Chad Galley

[2] "Radiation reaction at the level of the action" by Ofek Birnholtz, Shahar Hadar, and Barak Kol

[3] "An Effective Field Theory of Gravity for Extended Objects" by Walter D. Goldberger, Ira Z. Rothstein

[4]

[5] "Non-Relativistic Gravitation: From Newton to Einstein and Back" by Barak Kol & Michael Smolkin

[6]

[7] "Theory of post-Newtonian radiation and reaction" by Ofek Birnholtz, Shahar Hadar, and Barak Kol

[8] On the foundations of chiral perturbation theory, H. Leutwyler (Annals of Physics, v 235, 1994, p 165-203)

[9] "Dissipation in the effective field theory for hydrodynamics: First order effects" by Solomon Endlich, Alberto Nicolis, Rafael A. Porto, Junpu Wang

14.5 External links

- Effective Field Theory, A. Pich, Lectures at the 1997 Les Houches Summer School "Probing the Standard Model of Particle Interactions."
- Effective field theories, reduction and scientific explanation, by S. Hartmann, *Studies in History and Philosophy of Modern Physics* **32B**, 267-304 (2001).
- Aspects of heavy quark theory, by I. Bigi, M. Shifman and N. Uraltsev (Annual Reviews of Nuclear and Particle Science, v 47, 1997, p 591-661)
- Effective field theory (Interactions, Symmetry Breaking and Effective Fields - from Quarks to Nuclei. an Internet Lecture by Jacek Dobaczewski)

Chapter 15

Elementary particle

This article is about the physics concept. For the novel, see The Elementary Particles.

In particle physics, an **elementary particle** or **fundamental particle** is a particle whose substructure is unknown, thus

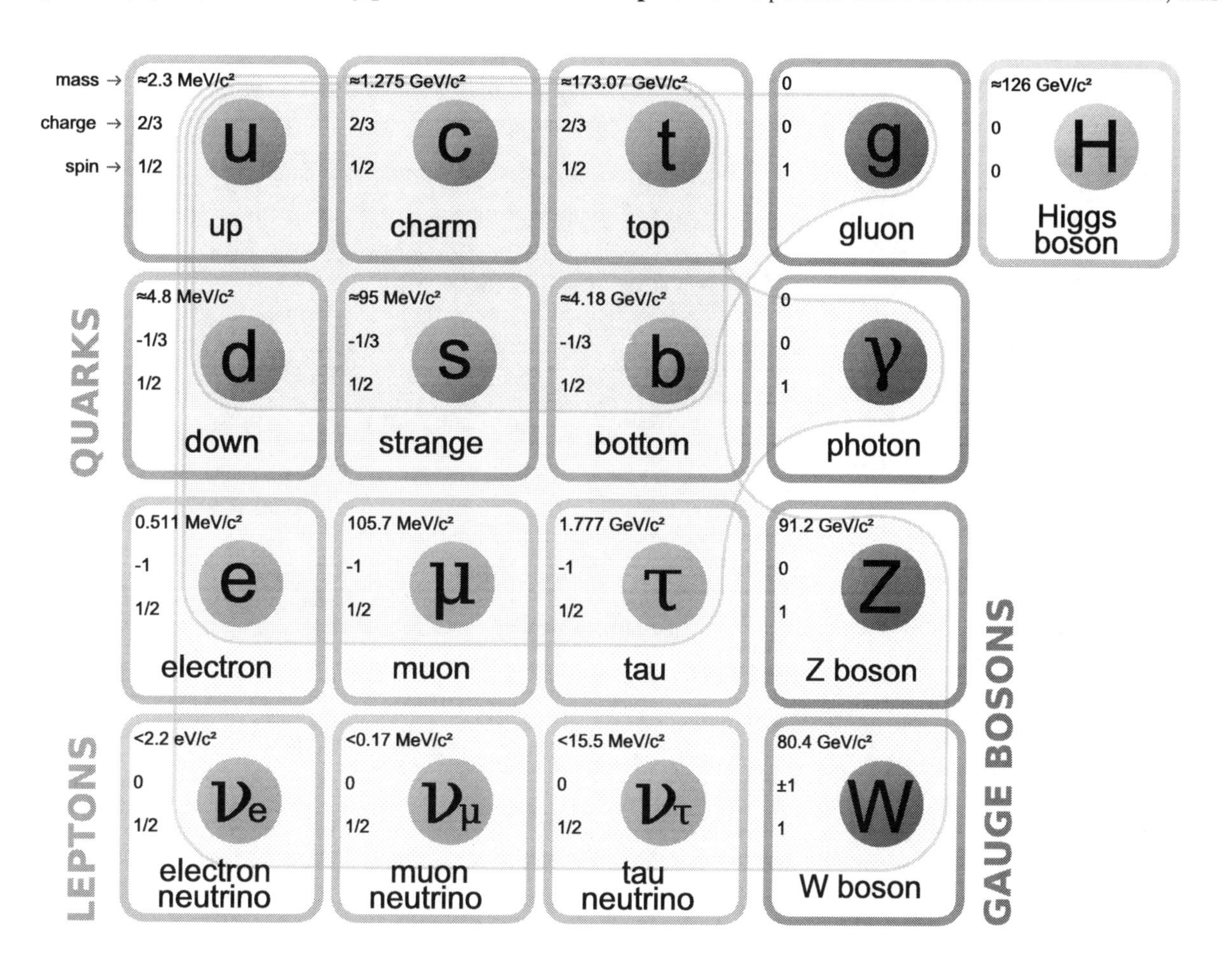

Elementary particles included in the Standard Model

it is unknown whether it is composed of other particles.[1] Known elementary particles include the fundamental fermions (quarks, leptons, antiquarks, and antileptons), which generally are "matter particles" and "antimatter particles", as well as the fundamental bosons (gauge bosons and Higgs boson), which generally are "force particles" that mediate interactions among fermions.[1] A particle containing two or more elementary particles is a *composite particle*.

Everyday matter is composed of atoms, once presumed to be matter's elementary particles—*atom* meaning "indivisible" in Greek—although the atom's existence remained controversial until about 1910, as some leading physicists regarded molecules as mathematical illusions, and matter as ultimately composed of energy.[1][2] Soon, subatomic constituents of the atom were identified. As the 1930s opened, the electron and the proton had been observed, along with the photon, the particle of electromagnetic radiation.[1] At that time, the recent advent of quantum mechanics was radically altering the conception of particles, as a single particle could seemingly span a field as would a wave, a paradox still eluding satisfactory explanation.[3][4][5]

Via quantum theory, protons and neutrons were found to contain quarks—up quarks and down quarks—now considered elementary particles.[1] And within a molecule, the electron's three degrees of freedom (charge, spin, orbital) can separate via wavefunction into three quasiparticles (holon, spinon, orbiton).[6] Yet a free electron—which, not orbiting an atomic nucleus, lacks orbital motion—appears unsplittable and remains regarded as an elementary particle.[6]

Around 1980, an elementary particle's status as indeed elementary—an *ultimate constituent* of substance—was mostly discarded for a more practical outlook,[1] embodied in particle physics' Standard Model, science's most experimentally successful theory.[5][7] Many elaborations upon and theories beyond the Standard Model, including the extremely popular supersymmetry, double the number of elementary particles by hypothesizing that each known particle associates with a "shadow" partner far more massive,[8][9] although all such superpartners remain undiscovered.[7][10] Meanwhile, an elementary boson mediating gravitation—the graviton—remains hypothetical.[1]

15.1 Overview

Main article: Standard Model
See also: Physics beyond the Standard Model
All elementary particles are—depending on their *spin*—either bosons or fermions. These are differentiated via the spin–

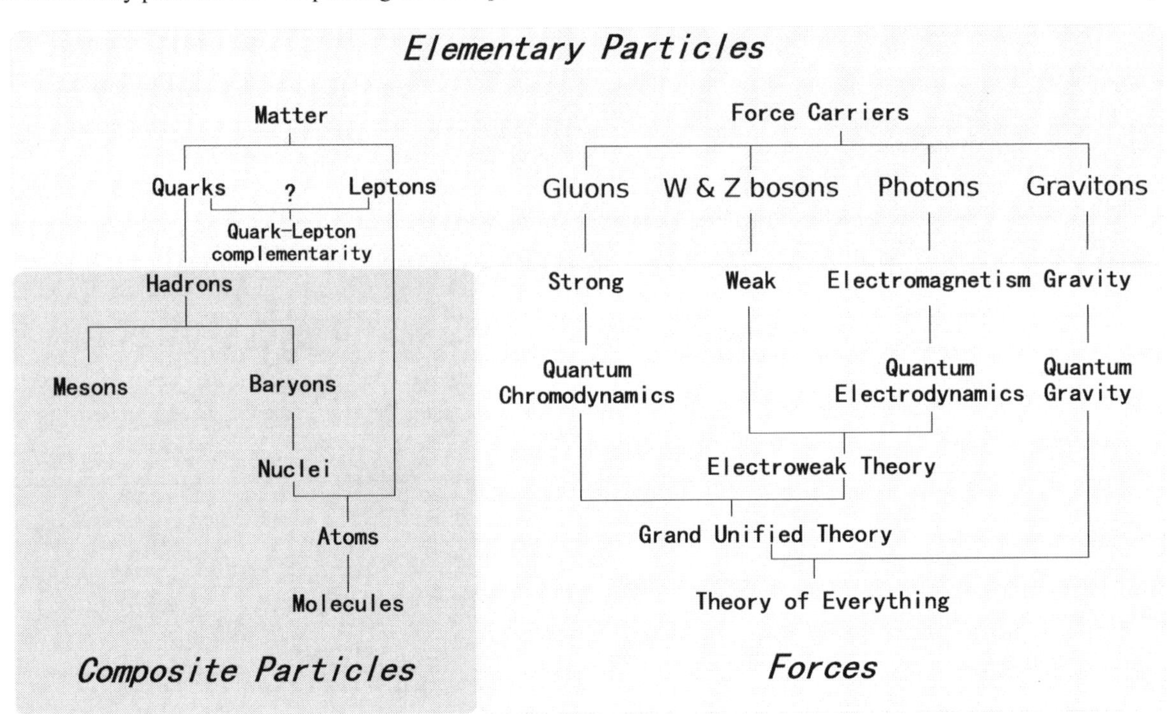

An overview of the various families of elementary and composite particles, and the theories describing their interactions

statistics theorem of quantum statistics. Particles of *half-integer* spin exhibit Fermi–Dirac statistics and are fermions.[1] Particles of *integer* spin, in other words full-integer, exhibit Bose–Einstein statistics and are bosons.[1]

Elementary fermions:

- Matter particles
  - Quarks:
    - up, down
    - charm, strange
    - top, bottom
  - Leptons:
    - electron, electron neutrino (a.k.a., "neutrino")
    - muon, muon neutrino
    - tau, tau neutrino
- Antimatter particles
  - Antiquarks
  - Antileptons

Elementary bosons:

- Force particles (gauge bosons):
  - photon
  - gluon (numbering eight)[1]
  - W^+, W^-, and Z^0 bosons
  - graviton (hypothetical)[1]
- Scalar boson
  - Higgs boson

A particle's mass is quantified in units of energy versus the electron's (electronvolts). Through conversion of energy into mass, any particle can be produced through collision of other particles at high energy,[1][11] although the output particle might not contain the input particles, for instance matter creation from colliding photons. Likewise, the composite fermions protons were collided at nearly light speed to produce a Higgs boson, which elementary boson is far more massive.[11] The most massive elementary particle, the top quark, rapidly decays, but apparently does not contain, lighter particles.

When probed at energies available in experiments, particles exhibit spherical sizes. In operating particle physics' Standard Model, elementary particles are usually represented for predictive utility as point particles, which, as zero-dimensional, lack spatial extension. Though extremely successful, the Standard Model is limited to the microcosm by its omission of gravitation, and has some parameters arbitrarily added but unexplained.[12] Seeking to resolve those shortcomings, string theory posits that elementary particles are ultimately composed of one-dimensional energy strings whose absolute minimum size is the Planck length.

15.2 Common elementary particles

Main article: cosmic abundance of elements

According to the current models of big bang nucleosynthesis, the primordial composition of visible matter of the universe should be about 75% hydrogen and 25% helium-4 (in mass). Neutrons are made up of one up and two down quark, while protons are made of two up and one down quark. Since the other common elementary particles (such as electrons, neutrinos, or weak bosons) are so light or so rare when compared to atomic nuclei, we can neglect their mass contribution to the observable universe's total mass. Therefore, one can conclude that most of the visible mass of the universe consists of protons and neutrons, which, like all baryons, in turn consist of up quarks and down quarks.

Some estimates imply that there are roughly 10^{80} baryons (almost entirely protons and neutrons) in the observable universe.[13][14][15]

The number of protons in the observable universe is called the Eddington number.

In terms of number of particles, some estimates imply that nearly all the matter, excluding dark matter, occurs in neutrinos, and that roughly 10^{86} elementary particles of matter exist in the visible universe, mostly neutrinos.[15] Other estimates imply that roughly 10^{97} elementary particles exist in the visible universe (not including dark matter), mostly photons, gravitons, and other massless force carriers.[15]

15.3 Standard Model

Main article: Standard Model

The Standard Model of particle physics contains 12 flavors of elementary fermions, plus their corresponding antiparticles, as well as elementary bosons that mediate the forces and the Higgs boson, which was reported on July 4, 2012, as having been likely detected by the two main experiments at the LHC (ATLAS and CMS). However, the Standard Model is widely considered to be a provisional theory rather than a truly fundamental one, since it is not known if it is compatible with Einstein's general relativity. There may be hypothetical elementary particles not described by the Standard Model, such as the graviton, the particle that would carry the gravitational force, and sparticles, supersymmetric partners of the ordinary particles.

15.3.1 Fundamental fermions

Main article: Fermion

The 12 fundamental fermionic flavours are divided into three generations of four particles each. Six of the particles are quarks. The remaining six are leptons, three of which are neutrinos, and the remaining three of which have an electric charge of -1: the electron and its two cousins, the muon and the tau.

Antiparticles

Main article: Antimatter

There are also 12 fundamental fermionic antiparticles that correspond to these 12 particles. For example, the antielectron (positron) $e+$ is the electron's antiparticle and has an electric charge of +1.

Quarks

Main article: Quark

Isolated quarks and antiquarks have never been detected, a fact explained by confinement. Every quark carries one of three color charges of the strong interaction; antiquarks similarly carry anticolor. Color-charged particles interact via gluon exchange in the same way that charged particles interact via photon exchange. However, gluons are themselves color-charged, resulting in an amplification of the strong force as color-charged particles are separated. Unlike the electromagnetic force, which diminishes as charged particles separate, color-charged particles feel increasing force.

However, color-charged particles may combine to form color neutral composite particles called hadrons. A quark may pair up with an antiquark: the quark has a color and the antiquark has the corresponding anticolor. The color and anticolor cancel out, forming a color neutral meson. Alternatively, three quarks can exist together, one quark being "red",

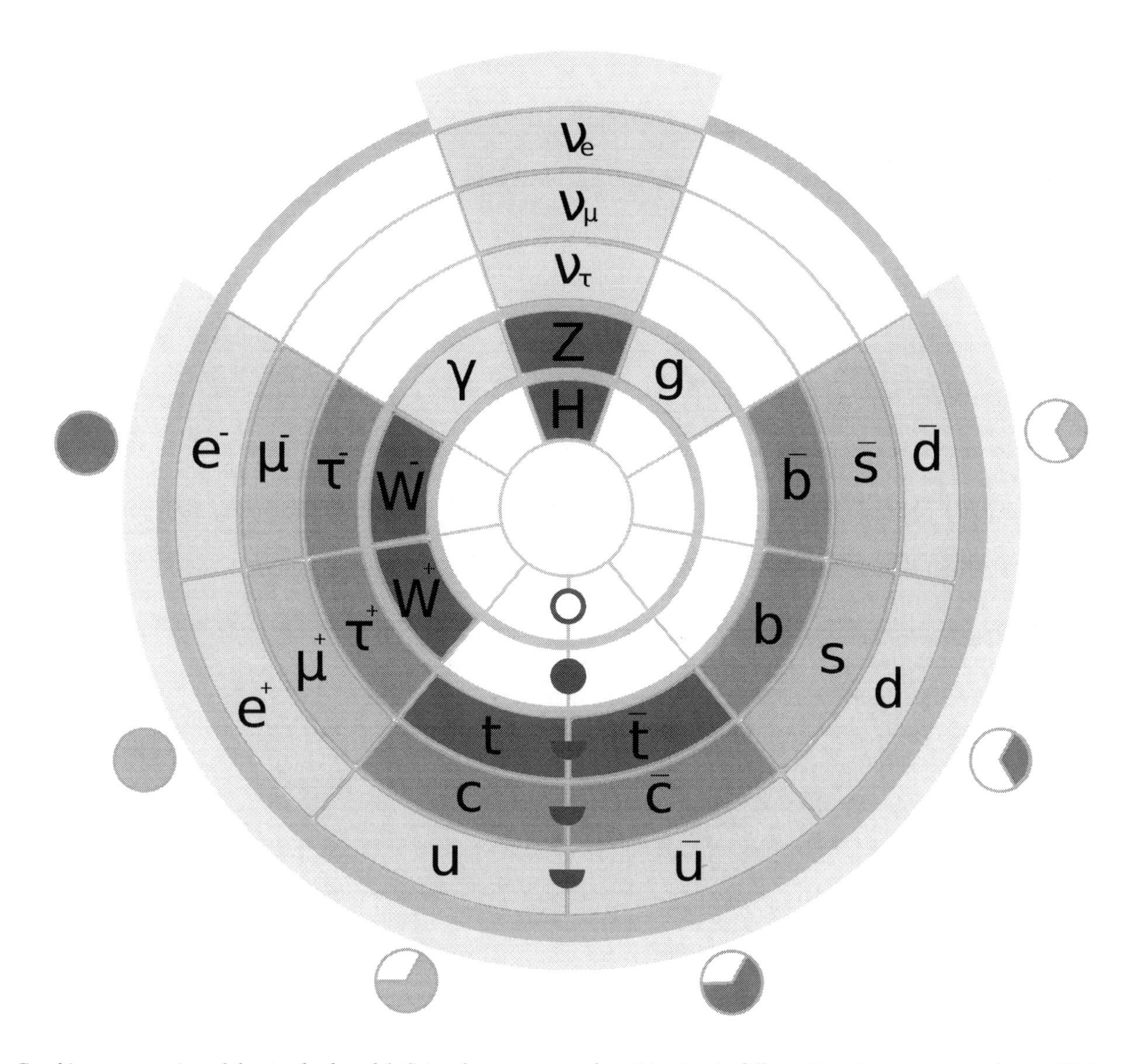

Graphic representation of the standard model. Spin, charge, mass and participation in different force interactions are shown. Click on the image to see the full description

another "blue", another "green". These three colored quarks together form a color-neutral baryon. Symmetrically, three antiquarks with the colors "antired", "antiblue" and "antigreen" can form a color-neutral antibaryon.

Quarks also carry fractional electric charges, but, since they are confined within hadrons whose charges are all integral, fractional charges have never been isolated. Note that quarks have electric charges of either +2/3 or −1/3, whereas antiquarks have corresponding electric charges of either −2/3 or +1/3.

Evidence for the existence of quarks comes from deep inelastic scattering: firing electrons at nuclei to determine the distribution of charge within nucleons (which are baryons). If the charge is uniform, the electric field around the proton should be uniform and the electron should scatter elastically. Low-energy electrons do scatter in this way, but, above a particular energy, the protons deflect some electrons through large angles. The recoiling electron has much less energy and a jet of particles is emitted. This inelastic scattering suggests that the charge in the proton is not uniform but split among smaller charged particles: quarks.

15.3.2 Fundamental bosons

Main article: Boson

In the Standard Model, vector (spin−1) bosons (gluons, photons, and the W and Z bosons) mediate forces, whereas the Higgs boson (spin-0) is responsible for the intrinsic mass of particles. Bosons differ from fermions in the fact that multiple bosons can occupy the same quantum state (Pauli exclusion principle). Also, bosons can be either elementary, like photons, or a combination, like mesons. The spin of bosons are integers instead of half integers.

Gluons

Main article: Gluon

Gluons mediate the strong interaction, which join quarks and thereby form hadrons, which are either baryons (three quarks) or mesons (one quark and one antiquark). Protons and neutrons are baryons, joined by gluons to form the atomic nucleus. Like quarks, gluons exhibit colour and anticolour—unrelated to the concept of visual color—sometimes in combinations, altogether eight variations of gluons.

Electroweak bosons

Main articles: W and Z bosons and Photon

There are three weak gauge bosons: W^{+}, W^{-}, and Z^{0}; these mediate the weak interaction. The W bosons are known for their mediation in nuclear decay. The W^{-} converts a neutron into a proton then decay into an electron and electron antineutrino pair. The Z^{0} does not convert charge but rather changes momentum and is the only mechanism for elastically scattering neutrinos. The weak gauge bosons were discovered due to momentum change in electrons from neutrino-Z exchange. The massless photon mediates the electromagnetic interaction. These four gauge bosons form the electroweak interaction among elementary particles.

Higgs boson

Main article: Higgs boson

Although the weak and electromagnetic forces appear quite different to us at everyday energies, the two forces are theorized to unify as a single electroweak force at high energies. This prediction was clearly confirmed by measurements of cross-sections for high-energy electron-proton scattering at the HERA collider at DESY. The differences at low energies is a consequence of the high masses of the W and Z bosons, which in turn are a consequence of the Higgs mechanism. Through the process of spontaneous symmetry breaking, the Higgs selects a special direction in electroweak space that causes three electroweak particles to become very heavy (the weak bosons) and one to remain massless (the photon). On 4 July 2012, after many years of experimentally searching for evidence of its existence, the Higgs boson was announced to have been observed at CERN's Large Hadron Collider. Peter Higgs who first posited the existence of the Higgs boson was present at the announcement.[16] The Higgs boson is believed to have a mass of approximately 125 GeV.[17] The statistical significance of this discovery was reported as 5-sigma, which implies a certainty of roughly 99.99994%. In particle physics, this is the level of significance required to officially label experimental observations as a discovery. Research into the properties of the newly discovered particle continues.

Graviton

Main article: Graviton

The graviton is hypothesized to mediate gravitation, but remains undiscovered and yet is sometimes included in tables of elementary particles.[1] Its spin would be two—thus a boson—and it would lack charge or mass. Besides mediating an extremely feeble force, the graviton would have its own antiparticle and rapidly annihilate, rendering its detection extremely difficult even if it exists.

15.4 Beyond the Standard Model

Although experimental evidence overwhelmingly confirms the predictions derived from the Standard Model, some of its parameters were added arbitrarily, not determined by a particular explanation, which remain mysteries, for instance the hierarchy problem. Theories beyond the Standard Model attempt to resolve these shortcomings.

15.4.1 Grand unification

Main article: Grand Unified Theory

One extension of the Standard Model attempts to combine the electroweak interaction with the strong interaction into a single 'grand unified theory' (GUT). Such a force would be spontaneously broken into the three forces by a Higgs-like mechanism. The most dramatic prediction of grand unification is the existence of X and Y bosons, which cause proton decay. However, the non-observation of proton decay at the Super-Kamiokande neutrino observatory rules out the simplest GUTs, including SU(5) and SO(10).

15.4.2 Supersymmetry

Main article: Supersymmetry

Supersymmetry extends the Standard Model by adding another class of symmetries to the Lagrangian. These symmetries exchange fermionic particles with bosonic ones. Such a symmetry predicts the existence of supersymmetric particles, abbreviated as *sparticles*, which include the sleptons, squarks, neutralinos, and charginos. Each particle in the Standard Model would have a superpartner whose spin differs by 1/2 from the ordinary particle. Due to the breaking of supersymmetry, the sparticles are much heavier than their ordinary counterparts; they are so heavy that existing particle colliders would not be powerful enough to produce them. However, some physicists believe that sparticles will be detected by the Large Hadron Collider at CERN.

15.4.3 String theory

Main article: String theory

String Theory is a model of physics where all “particles” that make up matter are composed of strings (measuring at the Planck length) that exist in an 11-dimensional (according to M-theory, the leading version) universe. These strings vibrate at different frequencies that determine mass, electric charge, color charge, and spin. A string can be open (a line) or closed in a loop (a one-dimensional sphere, like a circle). As a string moves through space it sweeps out something called a *world sheet*. String theory predicts 1- to 10-branes (a 1-brane being a string and a 10-brane being a 10-dimensional object) that prevent tears in the “fabric” of space using the uncertainty principle (E.g., the electron orbiting a hydrogen atom has the probability, albeit small, that it could be anywhere else in the universe at any given moment).

String theory proposes that our universe is merely a 4-brane, inside which exist the 3 space dimensions and the 1 time dimension that we observe. The remaining 6 theoretical dimensions either are very tiny and curled up (and too small to be macroscopically accessible) or simply do not/cannot exist in our universe (because they exist in a grander scheme called the "multiverse" outside our known universe).

Some predictions of the string theory include existence of extremely massive counterparts of ordinary particles due to vibrational excitations of the fundamental string and existence of a massless spin-2 particle behaving like the graviton.

15.4.4 Technicolor

Main article: Technicolor (physics)

Technicolor theories try to modify the Standard Model in a minimal way by introducing a new QCD-like interaction. This means one adds a new theory of so-called Techniquarks, interacting via so called Technigluons. The main idea is that the Higgs-Boson is not an elementary particle but a bound state of these objects.

15.4.5 Preon theory

Main article: Preon

According to preon theory there are one or more orders of particles more fundamental than those (or most of those) found in the Standard Model. The most fundamental of these are normally called preons, which is derived from “pre-quarks”. In essence, preon theory tries to do for the Standard Model what the Standard Model did for the particle zoo that came before it. Most models assume that almost everything in the Standard Model can be explained in terms of three to half a dozen more fundamental particles and the rules that govern their interactions. Interest in preons has waned since the simplest models were experimentally ruled out in the 1980s.

15.4.6 Acceleron theory

Accelerons are the hypothetical subatomic particles that integrally link the newfound mass of the neutrino and to the dark energy conjectured to be accelerating the expansion of the universe.[18]

In theory, neutrinos are influenced by a new force resulting from their interactions with accelerons. Dark energy results as the universe tries to pull neutrinos apart.[18]

15.5 See also

- Asymptotic freedom
- List of particles
- Physical ontology
- Quantum field theory
- Quantum gravity
- Quantum triviality
- UV fixed point

15.6 Notes

[1] Sylvie Braibant; Giorgio Giacomelli; Maurizio Spurio (2012). *Particles and Fundamental Interactions: An Introduction to Particle Physics* (2nd ed.). Springer. pp. 1–3. ISBN 978-94-007-2463-1.

[2] Ronald Newburgh; Joseph Peidle; Wolfgang Rueckner (2006). "Einstein, Perrin, and the reality of atoms: 1905 revisited" (PDF). *American Journal of Physics*. **74** (6): 478–481. Bibcode:2006AmJPh..74..478N. doi:10.1119/1.2188962.

[3] Friedel Weinert (2004). *The Scientist as Philosopher: Philosophical Consequences of Great Scientific Discoveries*. Springer. p. 43. ISBN 978-3-540-20580-7.

[4] Friedel Weinert (2004). *The Scientist as Philosopher: Philosophical Consequences of Great Scientific Discoveries*. Springer. pp. 57–59. ISBN 978-3-540-20580-7.

[5] Meinard Kuhlmann (24 Jul 2013). "Physicists debate whether the world is made of particles or fields—or something else entirely". *Scientific American*.

[6] Zeeya Merali (18 Apr 2012). "Not-quite-so elementary, my dear electron: Fundamental particle 'splits' into quasiparticles, including the new 'orbiton'". *Nature*. doi:10.1038/nature.2012.10471.

[7] Ian O'Neill (24 Jul 2013). "LHC discovery maims supersymmetry, again". *Discovery News*. Retrieved 2013-08-28.

[8] Particle Data Group. "Unsolved mysteries—supersymmetry". *The Particle Adventure*. Berkeley Lab. Retrieved 2013-08-28.

[9] National Research Council (2006). *Revealing the Hidden Nature of Space and Time: Charting the Course for Elementary Particle Physics*. National Academies Press. p. 68. ISBN 978-0-309-66039-6.

[10] "CERN latest data shows no sign of supersymmetry—yet". *Phys.Org*. 25 Jul 2013. Retrieved 2013-08-28.

[11] Ryan Avent (19 Jul 2012). "The Q&A: Brian Greene—Life after the Higgs". *The Economist*. Retrieved 2013-08-28.

[12] Sylvie Braibant; Giorgio Giacomelli; Maurizio Spurio (2012). *Particles and Fundamental Interactions: An Introduction to Particle Physics* (2nd ed.). Springer. p. 384. ISBN 978-94-007-2463-1.

[13] Frank Heile. "Is the Total Number of Particles in the Universe Stable Over Long Periods of Time?". 2014.

[14] Jared Brooks. "Galaxies and Cosmology". 2014. p. 4, equation 16.

[15] Robert Munafo (24 Jul 2013). "Notable Properties of Specific Numbers". Retrieved 2013-08-28.

[16] Lizzy Davies (4 July 2014). "Higgs boson announcement live: CERN scientists discover subatomic particle". *The Guardian*. Retrieved 2012-07-06.

[17] Lucas Taylor (4 Jul 2014). "Observation of a new particle with a mass of 125 GeV". CMS. Retrieved 2012-07-06.

[18] "New theory links neutrino's slight mass to accelerating Universe expansion". *ScienceDaily*. 28 Jul 2004. Retrieved 2008-06-05.

15.7 Further reading

15.7.1 General readers

- Feynman, R.P. & Weinberg, S. (1987) *Elementary Particles and the Laws of Physics: The 1986 Dirac Memorial Lectures*. Cambridge Univ. Press.
- Ford, Kenneth W. (2005) *The Quantum World*. Harvard Univ. Press.
- Brian Greene (1999). *The Elegant Universe*. W.W.Norton & Company. ISBN 0-393-05858-1.
- John Gribbin (2000) *Q is for Quantum – An Encyclopedia of Particle Physics*. Simon & Schuster. ISBN 0-684-85578-X.
- Oerter, Robert (2006) *The Theory of Almost Everything: The Standard Model, the Unsung Triumph of Modern Physics*. Plume.
- Schumm, Bruce A. (2004) *Deep Down Things: The Breathtaking Beauty of Particle Physics*. Johns Hopkins University Press. ISBN 0-8018-7971-X.

- Martinus Veltman (2003). *Facts and Mysteries in Elementary Particle Physics.* World Scientific. ISBN 981-238-149-X.
- Frank Close (2004). *Particle Physics: A Very Short Introduction.* Oxford: Oxford University Press. ISBN 0-19-280434-0.
- Seiden, Abraham (2005). *Particle Physics – A Comprehensive Introduction.* Addison Wesley. ISBN 0-8053-8736-6.

15.7.2 Textbooks

- Bettini, Alessandro (2008) *Introduction to Elementary Particle Physics.* Cambridge Univ. Press. ISBN 978-0-521-88021-3
- Coughlan, G. D., J. E. Dodd, and B. M. Gripaios (2006) *The Ideas of Particle Physics: An Introduction for Scientists*, 3rd ed. Cambridge Univ. Press. An undergraduate text for those not majoring in physics.
- Griffiths, David J. (1987) *Introduction to Elementary Particles.* John Wiley & Sons. ISBN 0-471-60386-4.
- Kane, Gordon L. (1987). *Modern Elementary Particle Physics.* Perseus Books. ISBN 0-201-11749-5.
- Perkins, Donald H. (2000) *Introduction to High Energy Physics*, 4th ed. Cambridge Univ. Press.

15.8 External links

The most important address about the current experimental and theoretical knowledge about elementary particle physics is the Particle Data Group, where different international institutions collect all experimental data and give short reviews over the contemporary theoretical understanding.

- Particle Data Group

other pages are:

- Greene, Brian, "*Elementary particles*", The Elegant Universe, NOVA (PBS)
- particleadventure.org, a well-made introduction also for non physicists
- CERNCourier: Season of Higgs and melodrama
- Pentaquark information page
- Interactions.org, particle physics news
- Symmetry Magazine, a joint Fermilab/SLAC publication
- “Sized Matter: perception of the extreme unseen”, Michigan University project for artistic visualisation of subatomic particles
- Elementary Particles made thinkable, an interactive visualisation allowing physical properties to be compared

Chapter 16

Generation (particle physics)

In particle physics, a **generation** (or **family**) is a division of the elementary particles. Between generations, particles differ by their (flavour) quantum number and mass, but their interactions are identical.

There are three generations according to the Standard Model of particle physics. Each generation is divided into two types of leptons and two types of quarks. The two leptons may be classified into one with electric charge −1 (electron-like) and one neutral (neutrino); the two quarks may be classified into one with charge $-\frac{1}{3}$ (down-type) and one with charge $+\frac{2}{3}$ (up-type).

16.1 Overview

Each member of a higher generation has greater mass than the corresponding particle of the previous generation, with the possible exception of the neutrinos (whose small but non-zero masses have not been accurately determined). For example, the first-generation electron has a mass of only 0.511 MeV/c^2, the second-generation muon has a mass of 106 MeV/c^2, and the third-generation tau has a mass of 1777 MeV/c^2 (almost twice as heavy as a proton). This mass hierarchy causes particles of higher generations to decay to the first generation, which explains why everyday matter (atoms) is made of particles from the first generation. Electrons surround a nucleus made of protons and neutrons, which contain up and down quarks. The second and third generations of charged particles do not occur in normal matter and are only seen in extremely high-energy environments such as cosmic rays or particle accelerators. The term *generation* was first introduced by Haim Harari in Les Houches Summer School, 1976.[1] [2]

Neutrinos of all generations stream throughout the universe but rarely interact with normal matter.[3] It is hoped that a comprehensive understanding of the relationship between the generations of the leptons may eventually explain the ratio of masses of the fundamental particles, and shed further light on the nature of mass generally, from a quantum perspective.[4]

16.2 Fourth generation

Fourth and further generations are considered to be unlikely. Some of the arguments against the possibility of a fourth generation are based on the subtle modifications of precision electroweak observables that extra generations would induce; such modifications are strongly disfavored by measurements. Furthermore, a fourth generation with a "light" neutrino (one with a mass less than about 45 GeV/c^2) has been ruled out by measurements of the widths of the Z boson at CERN's Large Electron–Positron Collider (LEP).[5] Nonetheless, searches at high-energy colliders for particles from a fourth generation continue, but as yet no evidence has been observed.[6] In such searches, fourth-generation particles are denoted by the same symbols as third-generation ones with an added prime (e.g. b' and t').

According to the results of the statistical analysis by researchers from CERN, and Humboldt University of Berlin, the existence of further fermions can be excluded with a probability of 99.99999% (5.3 sigma). The researchers combined latest data collected by the particle accelerators LHC and Tevatron with many known measurements results relating to particles, such as the Z-boson or the top-quark. The most important data used for this analysis come from the discovery of the Higgs particle. In the Standard Model, the Higgs particle gives all other particles their mass. As additional fermions were not detected directly in accelerator experiments, they have to be heavier than the fermions known so far. Hence, these fermions would also interact with the Higgs particle more strongly. This interaction would have modified the properties of the Higgs particle such that this particle would not have been detected.[7]

16.3 See also

- Metric expansion of space
- Spacetime
- Supersymmetry
- World line

16.4 References

[1] Harari, H. (1977). "Beyond charm". In Balian, R.; Llewellyn-Smith, C.H. *Weak and Electromagnetic Interactions at High Energy, Les Houches, France, Jul 5- Aug 14, 1976*. Les Houches Summer School Proceedings **29**. North-Holland. p. 613.

[2] Harari H. (1977). "Three generations of quarks and leptons" (PDF). In E. van Goeler, Weinstein R. (eds.). *Proceedings of the XII Rencontre de Moriond*. p. 170. SLAC-PUB-1974.

[3] "Experiment confirms famous physics model" (Press release). MIT News Office. 18 April 2007.

[4] M.H. Mac Gregor (2006). "A 'Muon Mass Tree' with α-quantized Lepton, Quark, and Hadron Masses". arXiv:hep-ph/0607233 [hep-ph].

[5] D. Decamp *et al.* (ALEPH collaboration) (1989). "Determination of the number of light neutrino species". *Physics Letters B* **231** (4): 519. Bibcode:1989PhLB..231..519D. doi:10.1016/0370-2693(89)90704-1.

[6] C. Amsler *et al.* (Particle Data Group) (2008). "Review of Particle Physics: b′ (4th Generation) Quarks, Searches for" (PDF). *Physics Letters B* **667** (1): 1–1340. Bibcode:2008PhLB..667....1P. doi:10.1016/j.physletb.2008.07.018.

[7] *12 matter particles suffice in nature* Dec 13, 2012 Phys.Org

Chapter 17

Strong interaction

In particle physics, the **strong interaction** is the mechanism responsible for the strong nuclear force (also called the **strong force**, **nuclear strong force** or **colour force**), one of the four fundamental interactions of nature, the others being electromagnetism, the weak interaction and gravitation. Effective only at a distance of a femtometre, it is approximately 100 times stronger than electromagnetism, a million times stronger than the weak force interaction and 10^{38} times stronger than gravitation at that range.[1] It ensures the stability of ordinary matter, as it confines the quark elementary particles into hadron particles, such as the proton and neutron, the largest components of the mass of ordinary matter. Furthermore, most of the mass-energy of a common proton or neutron is in the form of the strong force field energy; the individual quarks provide only about 1% of the mass-energy of a proton.

The strong interaction is observable in two areas: on a larger scale (about 1 to 3 femtometers (fm)), it is the force that binds protons and neutrons (nucleons) together to form the nucleus of an atom. On the smaller scale (less than about 0.8 fm, the radius of a nucleon), it is the force (carried by gluons) that holds quarks together to form protons, neutrons, and other hadron particles. The strong force inherently has so high a strength that the energy of an object bound by the strong force (a hadron) is high enough to produce new massive particles. Thus, if hadrons are struck by high-energy particles, they give rise to new hadrons instead of emitting freely moving radiation (gluons). This property of the strong force is called colour confinement, and it prevents the free "emission" of the strong force: instead, in practice, jets of massive particles are observed.

In the context of binding protons and neutrons together to form atoms, the strong interaction is called the nuclear force (or *residual strong force*). In this case, it is the residuum of the strong interaction between the quarks that make up the protons and neutrons. As such, the residual strong interaction obeys a quite different distance-dependent behavior between nucleons, from when it is acting to bind quarks within nucleons. The binding energy that is partly released on the breakup of a nucleus is related to the residual strong force and is harnessed in nuclear power and fission-type nuclear weapons.[2][3]

The strong interaction is thought to be mediated by massless particles called gluons, that are exchanged between quarks, antiquarks, and other gluons. Gluons, in turn, are thought to interact with quarks and gluons as all carry a type of charge called colour charge. Colour charge is analogous to electromagnetic charge, but it comes in three types rather than one (+/- red, +/- green, +/- blue) that results in a different type of force, with different rules of behavior. These rules are detailed in the theory of quantum chromodynamics (QCD), which is the theory of quark-gluon interactions.

Just after the Big Bang, and during the electroweak epoch, the electroweak force separated from the strong force. Although it is expected that a Grand Unified Theory exists to describe this, no such theory has been successfully formulated, and the unification remains an unsolved problem in physics.

17.1 History

Before the 1970s, physicists were uncertain about the binding mechanism of the atomic nucleus. It was known that the nucleus was composed of protons and neutrons and that protons possessed positive electric charge, while neutrons were electrically neutral. However, these facts seemed to contradict one another. By physical understanding at that time,

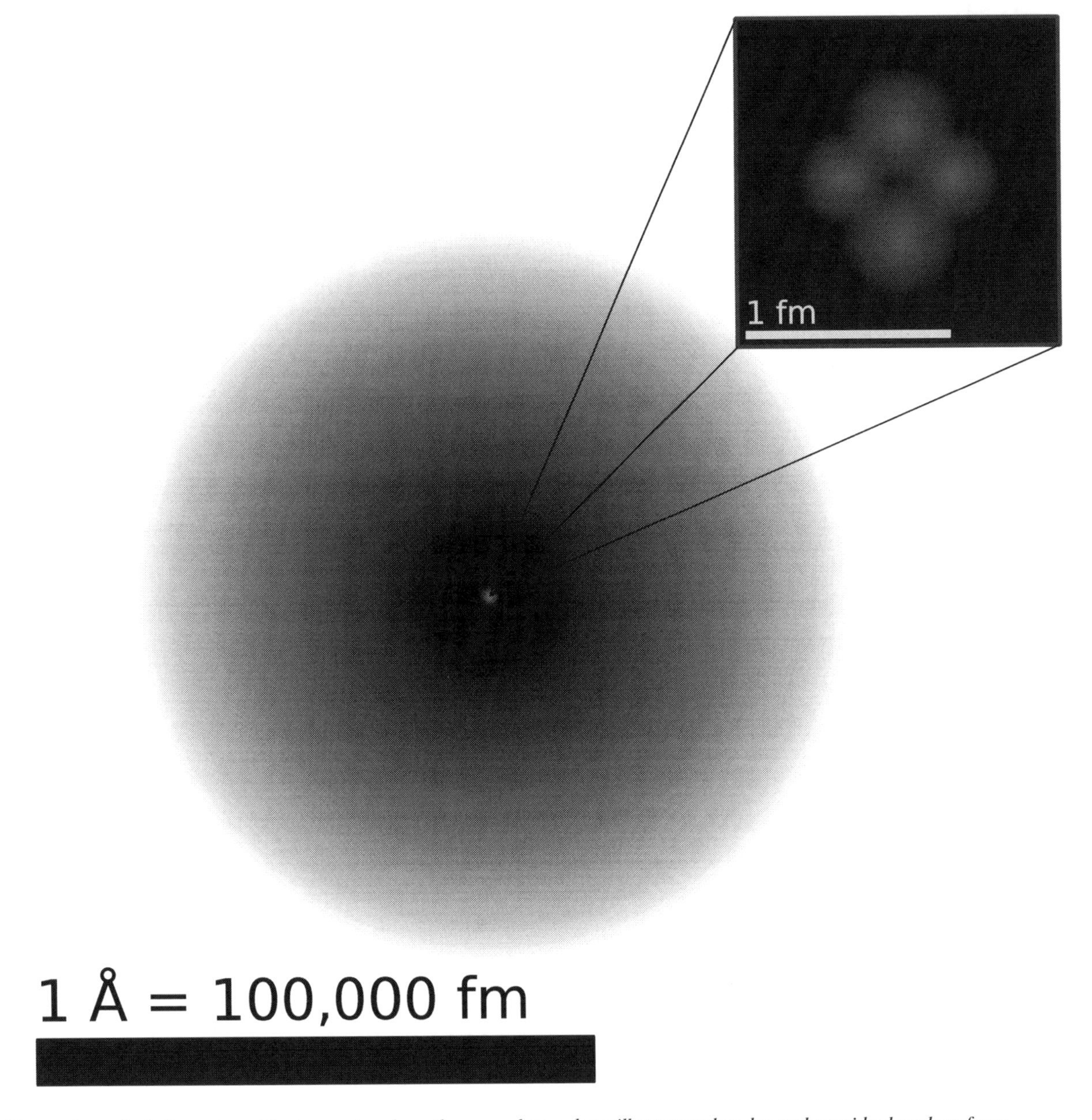

The nucleus of a helium atom. The two protons have the same charge, but still stay together due to the residual nuclear force

positive charges would repel one another and the nucleus should therefore fly apart. However, this was never observed. New physics was needed to explain this phenomenon.

A stronger attractive force was postulated to explain how the atomic nucleus was bound together despite the protons' mutual electromagnetic repulsion. This hypothesized force was called the *strong force*, which was believed to be a fundamental force that acted on the protons and neutrons that make up the nucleus.

It was later discovered that protons and neutrons were not fundamental particles, but were made up of constituent particles called quarks. The strong attraction between nucleons was the side-effect of a more fundamental force that bound the quarks together in the protons and neutrons. The theory of quantum chromodynamics explains that quarks carry what is called a colour charge, although it has no relation to visible colour.[4] Quarks with unlike colour charge attract one another as a result of the **strong interaction**, which is mediated by particles called gluons.

17.2 Details

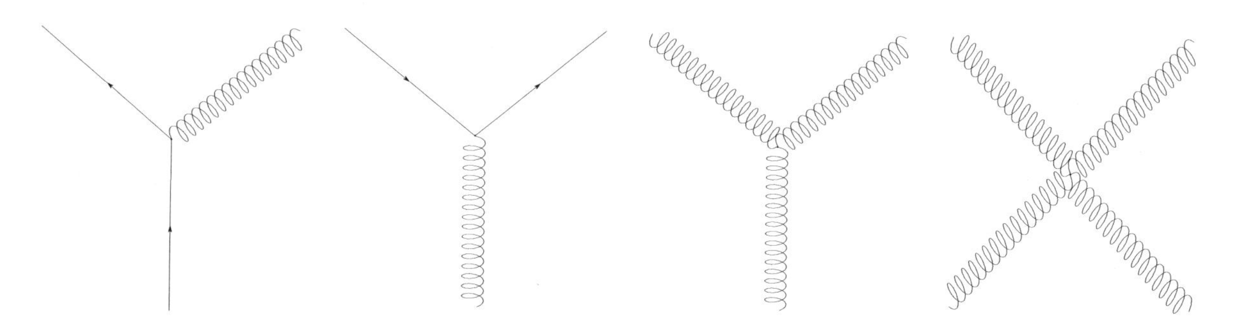

The fundamental couplings of the strong interaction, from left to right: gluon radiation, gluon splitting and gluon self-coupling.

The word *strong* is used since the strong interaction is the "strongest" of the four fundamental forces; its strength is around 10^2 times that of the electromagnetic force, some 10^6 times as great as that of the weak force, and about 10^{39} times that of gravitation, at a distance of a femtometer or less.

17.2.1 Behaviour of the strong force

The contemporary understanding of strong force is described by quantum chromodynamics (QCD), a part of the standard model of particle physics. Mathematically, QCD is a non-Abelian gauge theory based on a local (gauge) symmetry group called SU(3).

Quarks and gluons are the only fundamental particles that carry non-vanishing colour charge, and hence participate in strong interactions. The strong force itself acts directly only on elementary quark and gluon particles.

All quarks and gluons in QCD interact with each other through the strong force. The strength of interaction is parametrized by the strong coupling constant. This strength is modified by the gauge colour charge of the particle, a group theoretical property.

The strong force acts between quarks. Unlike all other forces (electromagnetic, weak, and gravitational), the strong force does not diminish in strength with increasing distance. After a limiting distance (about the size of a hadron) has been reached, it remains at a strength of about 10,000 newtons, no matter how much farther the distance between the quarks.[5] In QCD, this phenomenon is called colour confinement; it implies that only hadrons, not individual free quarks, can be observed. The explanation is that the amount of work done against a force of 10,000 newtons (about the weight of a one-metric ton mass on the surface of the Earth) is enough to create particle-antiparticle pairs within a very short distance of an interaction. In simple terms, the very energy applied to pull two quarks apart will create a pair of new quarks that will pair up with the original ones. The failure of all experiments that have searched for free quarks is considered to be evidence for this phenomenon.

The elementary quark and gluon particles affected are unobservable directly, but they instead emerge as jets of newly created hadrons, whenever energy is deposited into a quark-quark bond, as when a quark in a proton is struck by a very fast quark (in an impacting proton) during a particle accelerator experiment. However, quark–gluon plasmas have been observed.

Every quark in the universe does not attract every other quark in the above distance independent manner, since colour-confinement implies that the strong force acts without distance-diminishment only between pairs of single quarks, and that in collections of bound quarks (i.e., hadrons), the net colour-charge of the quarks cancels out, as seen from far away. Collections of quarks (hadrons) therefore appear (nearly) without colour-charge, and the strong force is therefore nearly absent between these hadrons (i.e., between baryons or mesons). However, the cancellation is not quite perfect. A small residual force remains (described below) known as the **residual strong force**. This residual force *does* diminish rapidly with distance, and is thus very short-range (effectively a few femtometers). It manifests as a force between the "colourless" hadrons, and is therefore sometimes known as the **strong nuclear force** or simply nuclear force.

17.2.2 Residual strong force

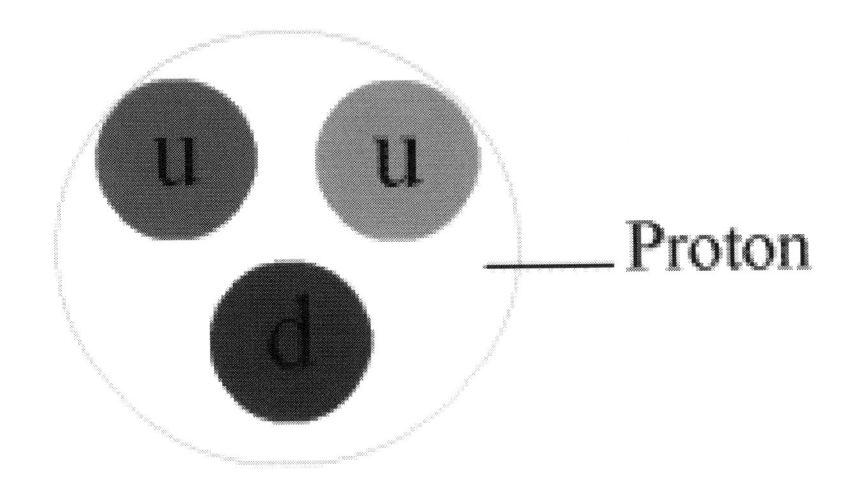

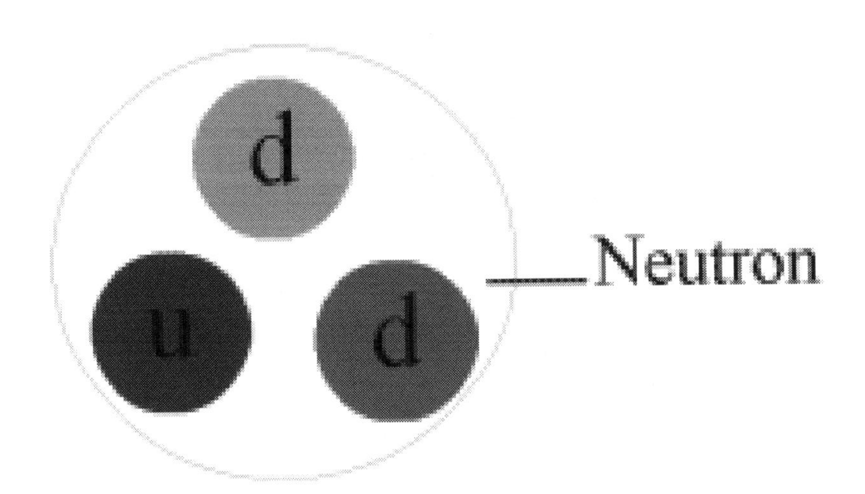

An animation of the nuclear force (or residual strong force) interaction between a proton and a neutron. The small coloured double circles are gluons, which can be seen binding the proton and neutron together. These gluons also hold the quark-antiquark combination called the pion together, and thus help transmit a residual part of the strong force even between colourless hadrons. Anticolours are shown as per this diagram. For a larger version, click here

The residual effect of the strong force is called the nuclear force. The nuclear force acts between hadrons, such as mesons or the nucleons in atomic nuclei. This “residual strong force”, acting indirectly, transmits gluons that form part of the virtual pi and rho mesons, which, in turn, transmit the nuclear force between nucleons.

The residual strong force is thus a minor residuum of the strong force that binds quarks together into protons and neutrons.

This same force is much weaker *between* neutrons and protons, because it is mostly neutralized *within* them, in the same way that electromagnetic forces between neutral atoms (van der Waals forces) are much weaker than the electromagnetic forces that hold the atoms internally together.[6]

Unlike the strong force itself, the nuclear force, or residual strong force, *does* diminish in strength, and in fact diminishes rapidly with distance. The decrease is approximately as a negative exponential power of distance, though there is no simple expression known for this; see Yukawa potential. This fact, together with the less-rapid decrease of the disruptive electromagnetic force between protons with distance, causes the instability of larger atomic nuclei, such as all those with atomic numbers larger than 82 (the element lead).

17.3 See also

- Nuclear binding energy
- Colour charge
- Coupling constant
- Nuclear physics
- QCD matter
- Quantum field theory and Gauge theory
- Standard model of particle physics and Standard Model (mathematical formulation)
- Weak interaction, electromagnetism and gravity
- Intermolecular force
- Vortex
- Yukawa interaction

17.4 References

[1] Relative strength of interaction varies with distance. See for instance Matt Strassler's essay, "The strength of the known forces".

[2] on Binding energy: see Binding Energy, Mass Defect, Furry Elephant physics educational site, retr 2012 7 1

[3] on Binding energy: see Chapter 4 NUCLEAR PROCESSES, THE STRONG FORCE, M. Ragheb 1/27/2012, University of Illinois

[4] Feynman, R. P. (1985). *QED: The Strange Theory of Light and Matter*. Princeton University Press. p. 136. ISBN 0-691-08388-6. The idiot physicists, unable to come up with any wonderful Greek words anymore, call this type of polarization by the unfortunate name of 'colour,' which has nothing to do with colour in the normal sense.

[5] Fritzsch, op. cite, p. 164. The author states that the force between differently coloured quarks remains constant at any distance after they travel only a tiny distance from each other, and is equal to that need to raise one ton, which is 1000 kg x 9.8 m/s^2 = ~10,000 N.

[6] Fritzsch, H. (1983). *Quarks: The Stuff of Matter*. Basic Books. pp. 167–168. ISBN 978-0-465-06781-7.

17.5 Further reading

- Christman, J. R. (2001). "MISN-0-280: *The Strong Interaction*" (PDF). *Project PHYSNET*.
- Griffiths, David (1987). *Introduction to Elementary Particles*. John Wiley & Sons. ISBN 0-471-60386-4.
- Halzen, F.; Martin, A. D. (1984). *Quarks and Leptons: An Introductory Course in Modern Particle Physics*. John Wiley & Sons. ISBN 0-471-88741-2.
- Kane, G. L. (1987). *Modern Elementary Particle Physics*. Perseus Books. ISBN 0-201-11749-5.
- Morris, R. (2003). *The Last Sorcerers: The Path from Alchemy to the Periodic Table*. Joseph Henry Press. ISBN 0-309-50593-3.

17.6 External links

- Strong force at *Encyclopædia Britannica*

Chapter 18

Weak interaction

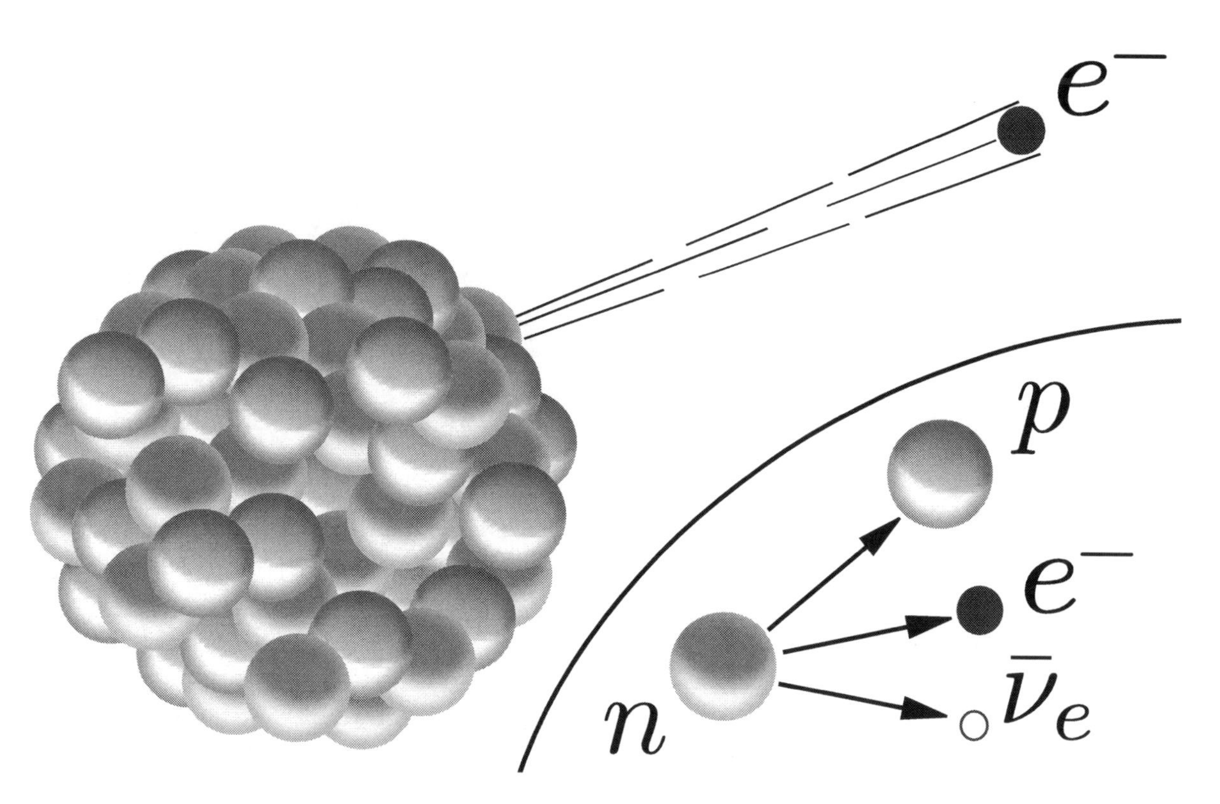

The radioactive beta decay is possible due to the weak interaction, which transforms a neutron into a proton, electron, and an electron antineutrino.

In particle physics, the **weak interaction** is the mechanism responsible for the **weak force** or **weak nuclear force**, one of the four known fundamental interactions of nature, alongside the strong interaction, electromagnetism, and gravitation. The weak interaction is responsible for the radioactive decay of subatomic particles, and it plays an essential role in nuclear fission. The theory of the weak interaction is sometimes called **quantum flavordynamics** (**QFD**), in analogy with the terms QCD and QED, but the term is rarely used because the weak force is best understood in terms of electro-weak theory (EWT).[1]

In the Standard Model of particle physics, the weak interaction is caused by the emission or absorption of W and Z bosons. All known fermions interact through the weak interaction. Fermions are particles that have half-integer spin (one of the fundamental properties of particles). A fermion can be an elementary particle, such as the electron, or it can be a composite particle, such as the proton. The masses of W^+, W^-, and Z bosons are each far greater than that of protons or neutrons, consistent with the short range of the weak force. The force is termed *weak* because its field strength over a given distance is typically several orders of magnitude less than that of the strong nuclear force and electromagnetic force.

During the quark epoch, the electroweak force split into the electromagnetic and weak forces. Important examples of weak interaction include beta decay, and the production, from hydrogen, of deuterium needed to power the sun's thermonuclear process. Most fermions will decay by a weak interaction over time. Such decay also makes radiocarbon dating possible, as carbon-14 decays through the weak interaction to nitrogen-14. It can also create radioluminescence, commonly used in tritium illumination, and in the related field of betavoltaics.[2]

Quarks, which make up composite particles like neutrons and protons, come in six "flavours" – up, down, strange, charm, top and bottom – which give those composite particles their properties. The weak interaction is unique in that it allows for quarks to swap their flavour for another. For example, during beta minus decay, a down quark decays into an up quark, converting a neutron to a proton. Also the weak interaction is the only fundamental interaction that breaks parity-symmetry, and similarly, the only one to break CP-symmetry.

18.1 History

In 1933, Enrico Fermi proposed the first theory of the weak interaction, known as Fermi's interaction. He suggested that beta decay could be explained by a four-fermion interaction, involving a contact force with no range.[3][4]

However, it is better described as a non-contact force field having a finite range, albeit very short. In 1968, Sheldon Glashow, Abdus Salam and Steven Weinberg unified the electromagnetic force and the weak interaction by showing them to be two aspects of a single force, now termed the electro-weak force.

The existence of the W and Z bosons was not directly confirmed until 1983.

18.2 Properties

The weak interaction is unique in a number of respects:

1. It is the only interaction capable of changing the flavor of quarks (i.e., of changing one type of quark into another).
2. It is the only interaction that violates **P** or parity-symmetry. It is also the only one that violates **CP** symmetry.
3. It is propagated by carrier particles (known as gauge bosons) that have significant masses, an unusual feature which is explained in the Standard Model by the Higgs mechanism.

Due to their large mass (approximately 90 GeV/c^2[5]) these carrier particles, termed the W and Z bosons, are short-lived: they have a lifetime of under 1×10^{-24} seconds.[6] The weak interaction has a coupling constant (an indicator of interaction strength) of between 10^{-7} and 10^{-6}, compared to the strong interaction's coupling constant of about 1 and the electromagnetic coupling constant of about 10^{-2};[7] consequently the weak interaction is weak in terms of strength.[8] The weak interaction has a very short range (around 10^{-17}–10^{-16} m[8]).[7] At distances around 10^{-18} meters, the weak interaction has a strength of a similar magnitude to the electromagnetic force, but this starts to decrease exponentially with increasing distance. At distances of around 3×10^{-17} m, the weak interaction is 10,000 times weaker than the electromagnetic.[9]

The weak interaction affects all the fermions of the Standard Model, as well as the Higgs boson; neutrinos interact through gravity and the weak interaction only, and neutrinos were the original reason for the name *weak force*.[8] The weak interaction does not produce bound states (nor does it involve binding energy) – something that gravity does on an astronomical scale, that the electromagnetic force does at the atomic level, and that the strong nuclear force does inside nuclei.[10]

Its most noticeable effect is due to its first unique feature: flavor changing. A neutron, for example, is heavier than a proton (its sister nucleon), but it cannot decay into a proton without changing the flavor (type) of one of its two *down*

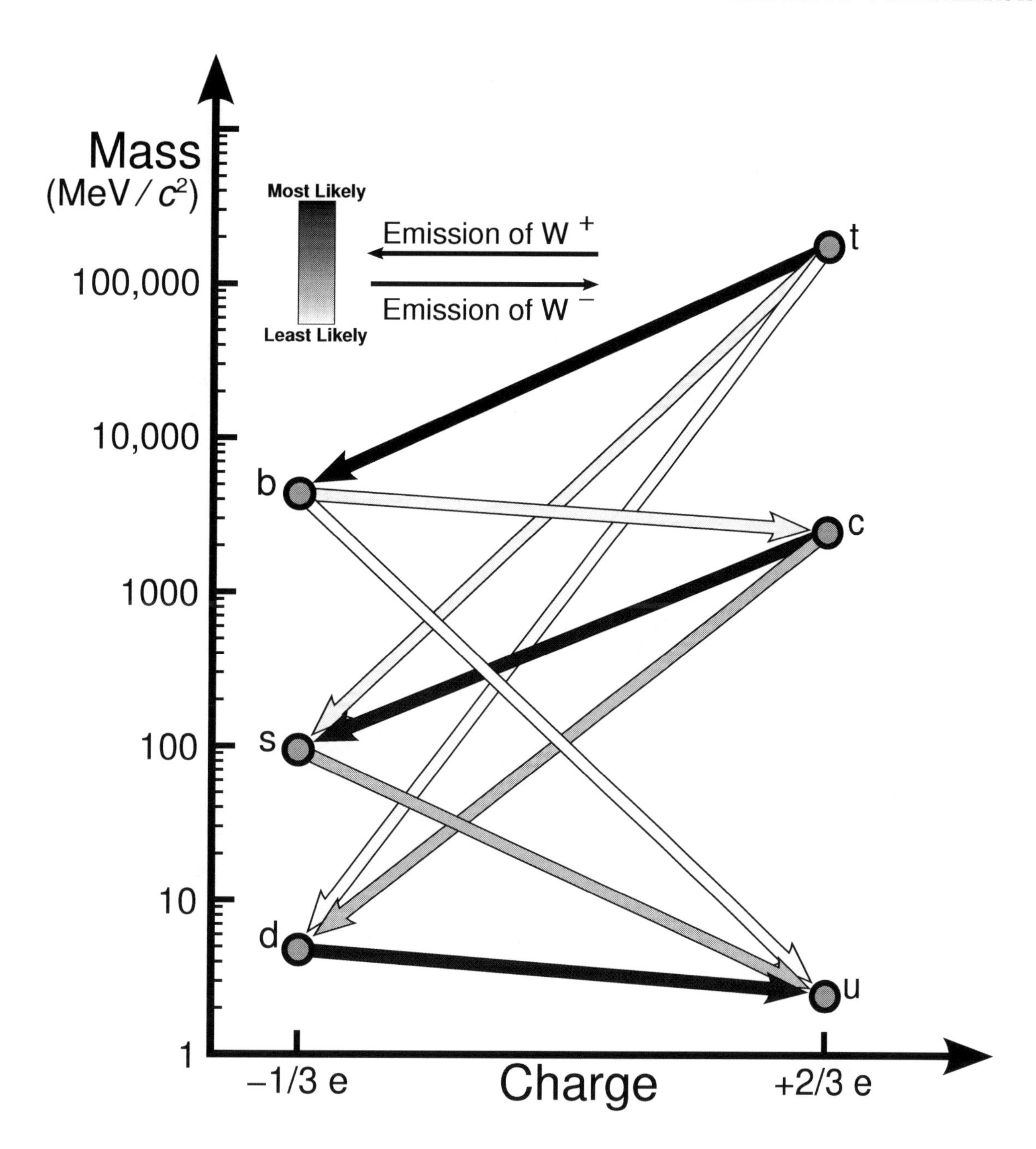

A diagram depicting the various decay routes due to the weak interaction and some indication of their likelihood. The intensity of the lines are given by the CKM parameters.

quarks to *up*. Neither the strong interaction nor electromagnetism permit flavour changing, so this must proceed by **weak decay**; without weak decay, quark properties such as strangeness and charm (associated with the quarks of the same name) would also be conserved across all interactions. All mesons are unstable because of weak decay.[11] In the process known as beta decay, a *down* quark in the neutron can change into an *up* quark by emitting a virtual W− boson which is then converted into an electron and an electron antineutrino.[12] Another example is the electron capture, a common variant of radioactive decay, where a proton (up quark) and an electron within an atom interact, and are changed to a neutron (down quark) and an electron antineutrino.

Due to the large mass of a boson, weak decay is much more unlikely than strong or electromagnetic decay, and hence occurs less rapidly. For example, a neutral pion (which decays electromagnetically) has a life of about 10^{-16} seconds, while a charged pion (which decays through the weak interaction) lives about 10^{-8} seconds, a hundred million times longer.[13] In contrast, a free neutron (which also decays through the weak interaction) lives about 15 minutes.[12]

18.2.1 Weak isospin and weak hypercharge

Main article: Weak isospin

All particles have a property called weak isospin (T_3), which serves as a quantum number and governs how that particle interacts in the weak interaction. Weak isospin therefore plays the same role in the weak interaction as electric charge does in electromagnetism, and color charge in the strong interaction. All fermions have a weak isospin value of either $+^1\!/_2$ or $-^1\!/_2$. For example, the up quark has a T_3 of $+^1\!/_2$ and the down quark $-^1\!/_2$. A quark never decays through the weak interaction into a quark of the same T_3: quarks with a T_3 of $+^1\!/_2$ decay into quarks with a T_3 of $-^1\!/_2$ and vice versa.

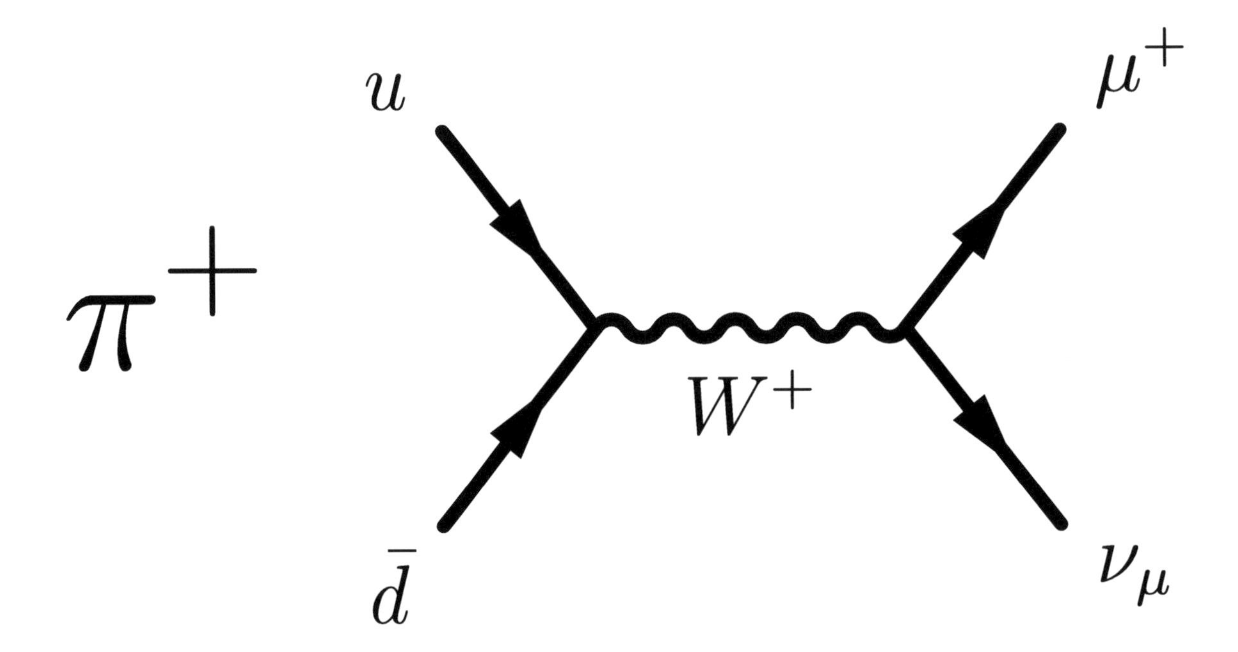

π+ decay through the weak interaction

In any given interaction, weak isospin is conserved: the sum of the weak isospin numbers of the particles entering the interaction equals the sum of the weak isospin numbers of the particles exiting that interaction. For example, a (left-handed) π+, with a weak isospin of 1 normally decays into a ν
μ (+1/2) and a μ+ (as a right-handed antiparticle, +1/2).[13]

Following the development of the electroweak theory, another property, weak hypercharge, was developed. It is dependent on a particle's electrical charge and weak isospin, and is defined as:

$$Y_W = 2(Q - T_3)$$

where *YW* is the weak hypercharge of a given type of particle, *Q* is its electrical charge (in elementary charge units) and T_3 is its weak isospin. Whereas some particles have a weak isospin of zero, all particles, except gluons, have non-zero weak hypercharge. Weak hypercharge is the generator of the U(1) component of the electroweak gauge group.

18.3 Interaction types

There are two types of weak interaction (called *vertices*). The first type is called the "charged-current interaction" because it is mediated by particles that carry an electric charge (the W+ or W− bosons), and is responsible for the beta decay phenomenon. The second type is called the "neutral-current interaction" because it is mediated by a neutral particle, the Z boson.

18.3.1 Charged-current interaction

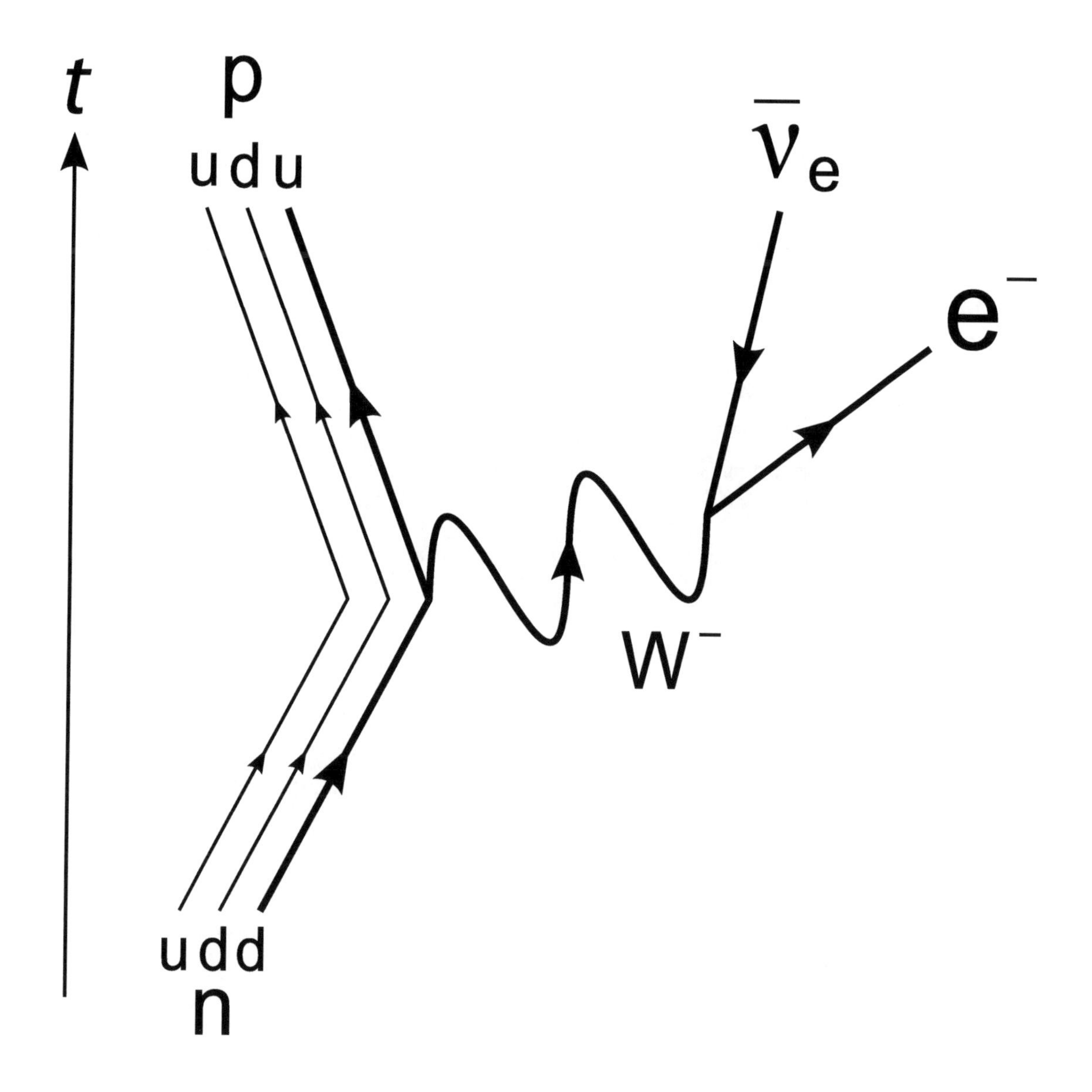

The Feynman diagram for beta-minus decay of a neutron into a proton, electron and electron anti-neutrino, via an intermediate heavy W− boson

In one type of charged current interaction, a charged lepton (such as an electron or a muon, having a charge of −1) can absorb a W+ boson (a particle with a charge of +1) and be thereby converted into a corresponding neutrino (with a charge of 0), where the type ("family") of neutrino (electron, muon or tau) is the same as the type of lepton in the interaction, for example:

$$\mu^- + W^+ \to \nu_\mu$$

Similarly, a down-type quark (d with a charge of −⅓) can be converted into an up-type quark (u, with a charge of +⅔), by emitting a W− boson or by absorbing a W+ boson. More precisely, the down-type quark becomes a quantum superposition of up-type quarks: that is to say, it has a possibility of becoming any one of the three up-type quarks, with the probabilities given in the CKM matrix tables. Conversely, an up-type quark can emit a W+ boson – or absorb a W− boson – and thereby be converted into a down-type quark, for example:

$$d \to u + W^-$$
$$d + W^+ \to u$$
$$c \to s + W^+$$
$$c + W^- \to s$$

The W boson is unstable so will rapidly decay, with a very short lifetime. For example:

$$W^- \to e^- + \bar{\nu}_e$$
$$W^+ \to e^+ + \nu_e$$

Decay of the W boson to other products can happen, with varying probabilities.[15]

In the so-called beta decay of a neutron (see picture, above), a down quark within the neutron emits a virtual W− boson and is thereby converted into an up quark, converting the neutron into a proton. Because of the energy involved in the process (i.e., the mass difference between the down quark and the up quark), the W− boson can only be converted into an electron and an electron-antineutrino.[16] At the quark level, the process can be represented as:

$$d \to u + e^- + \bar{\nu}_e$$

18.3.2 Neutral-current interaction

In neutral current interactions, a quark or a lepton (e.g., an electron or a muon) emits or absorbs a neutral Z boson. For example:

$$e^- \to e^- + Z^0$$

Like the W boson, the Z boson also decays rapidly,[15] for example:

$$Z^0 \to b + \bar{b}$$

18.4 Electroweak theory

Main article: Electroweak interaction

The Standard Model of particle physics describes the electromagnetic interaction and the weak interaction as two different aspects of a single electroweak interaction, the theory of which was developed around 1968 by Sheldon Glashow, Abdus Salam and Steven Weinberg. They were awarded the 1979 Nobel Prize in Physics for their work.[17] The Higgs mechanism provides an explanation for the presence of three massive gauge bosons (the three carriers of the weak interaction) and the massless photon of the electromagnetic interaction.[18]

According to the electroweak theory, at very high energies, the universe has four massless gauge boson fields similar to the photon and a complex scalar Higgs field doublet. However, at low energies, gauge symmetry is spontaneously broken down to the **U**(1) symmetry of electromagnetism (one of the Higgs fields acquires a vacuum expectation value). This symmetry breaking would produce three massless bosons, but they become integrated by three photon-like fields (through the Higgs mechanism) giving them mass. These three fields become the W+, W− and Z bosons of the weak interaction, while the fourth gauge field, which remains massless, is the photon of electromagnetism.[18]

This theory has made a number of predictions, including a prediction of the masses of the Z and W bosons before their discovery. On 4 July 2012, the CMS and the ATLAS experimental teams at the Large Hadron Collider independently announced that they had confirmed the formal discovery of a previously unknown boson of mass between 125–127 GeV/c^2, whose behaviour so far was "consistent with" a Higgs boson, while adding a cautious note that further data and analysis were needed before positively identifying the new boson as being a Higgs boson of some type. By 14 March 2013, the Higgs boson was tentatively confirmed to exist .[19]

18.5 Violation of symmetry

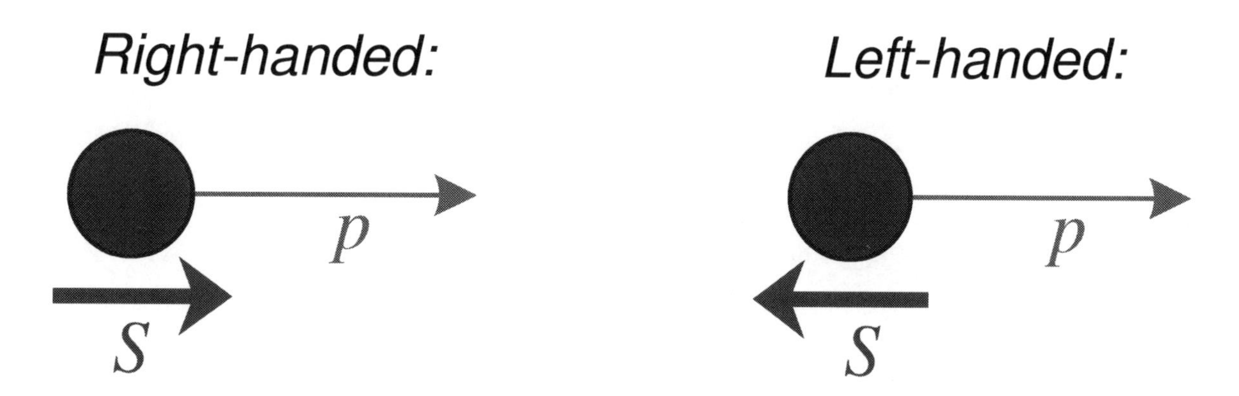

Left- and right-handed particles: p is the particle's momentum and S is its spin. Note the lack of reflective symmetry between the states.

The laws of nature were long thought to remain the same under mirror reflection, the reversal of one spatial axis. The results of an experiment viewed via a mirror were expected to be identical to the results of a mirror-reflected copy of the experimental apparatus. This so-called law of parity conservation was known to be respected by classical gravitation, electromagnetism and the strong interaction; it was assumed to be a universal law.[20] However, in the mid-1950s Chen Ning Yang and Tsung-Dao Lee suggested that the weak interaction might violate this law. Chien Shiung Wu and collaborators in 1957 discovered that the weak interaction violates parity, earning Yang and Lee the 1957 Nobel Prize in Physics.[21]

Although the weak interaction used to be described by Fermi's theory, the discovery of parity violation and renormalization theory suggested that a new approach was needed. In 1957, Robert Marshak and George Sudarshan and, somewhat later, Richard Feynman and Murray Gell-Mann proposed a **V−A** (vector minus axial vector or left-handed) Lagrangian for weak interactions. In this theory, the weak interaction acts only on left-handed particles (and right-handed antiparticles). Since the mirror reflection of a left-handed particle is right-handed, this explains the maximal violation of parity. Interestingly, the **V−A** theory was developed before the discovery of the Z boson, so it did not include the right-handed fields that enter in the neutral current interaction.

However, this theory allowed a compound symmetry **CP** to be conserved. **CP** combines parity **P** (switching left to right) with charge conjugation **C** (switching particles with antiparticles). Physicists were again surprised when in 1964, James

Cronin and Val Fitch provided clear evidence in kaon decays that CP symmetry could be broken too, winning them the 1980 Nobel Prize in Physics.[22] In 1973, Makoto Kobayashi and Toshihide Maskawa showed that CP violation in the weak interaction required more than two generations of particles,[23] effectively predicting the existence of a then unknown third generation. This discovery earned them half of the 2008 Nobel Prize in Physics.[24] Unlike parity violation, CP violation occurs in only a small number of instances, but remains widely held as an answer to the difference between the amount of matter and antimatter in the universe; it thus forms one of Andrei Sakharov's three conditions for baryogenesis.[25]

18.6 See also

- Weakless Universe – the postulate that weak interactions are not anthropically necessary
- Gravity
- Nuclear force
- Electromagnetism

18.7 References

18.7.1 Citations

[1] Griffiths, David (2009). *Introduction to Elementary Particles*. pp. 59–60. ISBN 978-3-527-40601-2.

[2] "The Nobel Prize in Physics 1979: Press Release". *NobelPrize.org*. Nobel Media. Retrieved 22 March 2011.

[3] Fermi, Enrico (1934). "Versuch einer Theorie der β-Strahlen. I". *Zeitschrift für Physik A* **88** (3–4): 161–177. Bibcode:1934ZPhy...88..16 doi:10.1007/BF01351864.

[4] Wilson, Fred L. (December 1968). "Fermi's Theory of Beta Decay". *American Journal of Physics* **36** (12): 1150–1160. Bibcode:1968AmJPh..36.1150W. doi:10.1119/1.1974382.

[5] W.-M. Yao *et al.* (Particle Data Group) (2006). "Review of Particle Physics: Quarks" (PDF). *Journal of Physics G* **33**: 1–1232. arXiv:astro-ph/0601168. Bibcode:2006JPhG...33....1Y. doi:10.1088/0954-3899/33/1/001.

[6] Peter Watkins (1986). *Story of the W and Z*. Cambridge: Cambridge University Press. p. 70. ISBN 978-0-521-31875-4.

[7] "Coupling Constants for the Fundamental Forces". *HyperPhysics*. Georgia State University. Retrieved 2 March 2011.

[8] J. Christman (2001). "The Weak Interaction" (PDF). *Physnet*. Michigan State University.

[9] "Electroweak". *The Particle Adventure*. Particle Data Group. Retrieved 3 March 2011.

[10] Walter Greiner; Berndt Müller (2009). *Gauge Theory of Weak Interactions*. Springer. p. 2. ISBN 978-3-540-87842-1.

[11] Cottingham & Greenwood (1986, 2001), p.29

[12] Cottingham & Greenwood (1986, 2001), p.28

[13] Cottingham & Greenwood (1986, 2001), p.30

[14] Baez, John C.; Huerta, John (2009). "The Algebra of Grand Unified Theories". *Bull.Am.Math.Soc.* **0904**: 483–552. arXiv:0904.1556. Bibcode:2009arXiv0904.1556B. doi:10.1090/s0273-0979-10-01294-2. Retrieved 15 October 2013.

[15] K. Nakamura *et al.* (Particle Data Group) (2010). "Gauge and Higgs Bosons" (PDF). *Journal of Physics G* **37**. doi:10.1088/0954-3899/37/7a/075021.

[16] K. Nakamura *et al.* (Particle Data Group) (2010). "n" (PDF). *Journal of Physics G* **37**: 7. doi:10.1088/0954-3899/37/7a/075021.

[17] "The Nobel Prize in Physics 1979". *NobelPrize.org*. Nobel Media. Retrieved 26 February 2011.

[18] C. Amsler *et al.* (Particle Data Group) (2008). "Review of Particle Physics – Higgs Bosons: Theory and Searches" (PDF). *Physics Letters B* **667**: 1–6. Bibcode:2008PhLB..667....1P. doi:10.1016/j.physletb.2008.07.018.

[19] "New results indicate that new particle is a Higgs boson | CERN". Home.web.cern.ch. Retrieved 20 September 2013.

[20] Charles W. Carey (2006). "Lee, Tsung-Dao". *American scientists*. Facts on File Inc. p. 225. ISBN 9781438108070.

[21] "The Nobel Prize in Physics 1957". *NobelPrize.org*. Nobel Media. Retrieved 26 February 2011.

[22] "The Nobel Prize in Physics 1980". *NobelPrize.org*. Nobel Media. Retrieved 26 February 2011.

[23] M. Kobayashi, T. Maskawa (1973). "CP-Violation in the Renormalizable Theory of Weak Interaction". *Progress of Theoretical Physics* **49** (2): 652–657. Bibcode:1973PThPh..49..652K. doi:10.1143/PTP.49.652.

[24] "The Nobel Prize in Physics 1980". *NobelPrize.org*. Nobel Media. Retrieved 17 March 2011.

[25] Paul Langacker (2001) [1989]. "Cp Violation and Cosmology". In Cecilia Jarlskog. *CP violation*. London, River Edge: World Scientific Publishing Co. p. 552. ISBN 9789971505615.

18.7.2 General readers

- R. Oerter (2006). *The Theory of Almost Everything: The Standard Model, the Unsung Triumph of Modern Physics*. Plume. ISBN 978-0-13-236678-6.
- B.A. Schumm (2004). *Deep Down Things: The Breathtaking Beauty of Particle Physics*. Johns Hopkins University Press. ISBN 0-8018-7971-X.

18.7.3 Texts

- D.A. Bromley (2000). *Gauge Theory of Weak Interactions*. Springer. ISBN 3-540-67672-4.
- G.D. Coughlan, J.E. Dodd, B.M. Gripaios (2006). *The Ideas of Particle Physics: An Introduction for Scientists* (3rd ed.). Cambridge University Press. ISBN 978-0-521-67775-2.
- W. N. Cottingham; D. A. Greenwood (2001) [1986]. *An introduction to nuclear physics* (2nd ed.). Cambridge University Press. p. 30. ISBN 978-0-521-65733-4.
- D.J. Griffiths (1987). *Introduction to Elementary Particles*. John Wiley & Sons. ISBN 0-471-60386-4.
- G.L. Kane (1987). *Modern Elementary Particle Physics*. Perseus Books. ISBN 0-201-11749-5.
- D.H. Perkins (2000). *Introduction to High Energy Physics*. Cambridge University Press. ISBN 0-521-62196-8.

Chapter 19

Electromagnetism

Electromagnetism is the study of the **electromagnetic force** which is a type of physical interaction that occurs between electrically charged particles. The electromagnetic force usually shows electromagnetic fields, such as electric fields, magnetic fields, and light. The electromagnetic force is one of the four fundamental interactions in nature. The other three fundamental interactions are the strong interaction, the weak interaction, and gravitation.[1]

Lightning is an electrostatic discharge that travels between two charged regions.

The word *electromagnetism* is a compound form of two Greek terms, ἤλεκτρον, *ēlektron*, "amber", and μαγνήτης, *magnetic*, from "magnítis líthos" (μαγνήτης λίθος), which means "magnesian stone", a type of iron ore. The science of electromagnetic phenomena is defined in terms of the electromagnetic force, sometimes called the Lorentz force, which includes both electricity and magnetism as elements of one phenomenon.

The electromagnetic force plays a major role in determining the internal properties of most objects encountered in daily life. Ordinary matter takes its form as a result of intermolecular forces between individual molecules in matter. Electrons are bound by electromagnetic wave mechanics into orbitals around atomic nuclei to form atoms, which are the building blocks of molecules. This governs the processes involved in chemistry, which arise from interactions between the electrons of neighboring atoms, which are in turn determined by the interaction between electromagnetic force and the momentum of the electrons.

There are numerous mathematical descriptions of the electromagnetic field. In classical electrodynamics, electric fields are described as electric potential and electric current in Ohm's law, magnetic fields are associated with electromagnetic induction and magnetism, and Maxwell's equations describe how electric and magnetic fields are generated and altered by each other and by charges and currents.

The theoretical implications of electromagnetism, in particular the establishment of the speed of light based on properties of the "medium" of propagation (permeability and permittivity), led to the development of special relativity by Albert Einstein in 1905.

Although electromagnetism is considered one of the four fundamental forces, at high energy the weak force and electromagnetism are unified. In the history of the universe, during the quark epoch, the electroweak force split into the electromagnetic and weak forces.

19.1 History of the theory

See also: History of electromagnetic theory

Originally electricity and magnetism were thought of as two separate forces. This view changed, however, with the publication of James Clerk Maxwell's 1873 *A Treatise on Electricity and Magnetism* in which the interactions of positive and negative charges were shown to be regulated by one force. There are four main effects resulting from these interactions, all of which have been clearly demonstrated by experiments:

1. Electric charges attract or repel one another with a force inversely proportional to the square of the distance between them: unlike charges attract, like ones repel.

2. Magnetic poles (or states of polarization at individual points) attract or repel one another in a similar way and always come in pairs: every north pole is yoked to a south pole.

3. An electric current in a wire creates a circular magnetic field around the wire, its direction (clockwise or counter-clockwise) depending on that of the current.

4. A current is induced in a loop of wire when it is moved towards or away from a magnetic field, or a magnet is moved towards or away from it, the direction of current depending on that of the movement.

While preparing for an evening lecture on 21 April 1820, Hans Christian Ørsted made a surprising observation. As he was setting up his materials, he noticed a compass needle deflected from magnetic north when the electric current from the battery he was using was switched on and off. This deflection convinced him that magnetic fields radiate from all sides of a wire carrying an electric current, just as light and heat do, and that it confirmed a direct relationship between electricity and magnetism.

At the time of discovery, Ørsted did not suggest any satisfactory explanation of the phenomenon, nor did he try to represent the phenomenon in a mathematical framework. However, three months later he began more intensive investigations. Soon thereafter he published his findings, proving that an electric current produces a magnetic field as it flows through a wire. The CGS unit of magnetic induction (oersted) is named in honor of his contributions to the field of electromagnetism.

Hans Christian Ørsted.

His findings resulted in intensive research throughout the scientific community in electrodynamics. They influenced French physicist André-Marie Ampère's developments of a single mathematical form to represent the magnetic forces between current-carrying conductors. Ørsted's discovery also represented a major step toward a unified concept of energy.

This unification, which was observed by Michael Faraday, extended by James Clerk Maxwell, and partially reformulated by Oliver Heaviside and Heinrich Hertz, is one of the key accomplishments of 19th century mathematical physics. It had far-reaching consequences, one of which was the understanding of the nature of light. Unlike what was proposed in Electromagnetism, light and other electromagnetic waves are at the present seen as taking the form of quantized, self-propagating oscillatory electromagnetic field disturbances which have been called photons. Different frequencies of oscillation give rise to the different forms of electromagnetic radiation, from radio waves at the lowest frequencies, to visible light at intermediate frequencies, to gamma rays at the highest frequencies.

Ørsted was not the only person to examine the relation between electricity and magnetism. In 1802 Gian Domenico Romagnosi, an Italian legal scholar, deflected a magnetic needle by electrostatic charges. Actually, no galvanic current existed in the setup and hence no electromagnetism was present. An account of the discovery was published in 1802 in an Italian newspaper, but it was largely overlooked by the contemporary scientific community.[2]

19.2 Fundamental forces

The electromagnetic force is one of the four known fundamental forces. The other fundamental forces are:

- the weak nuclear force, which binds to all known particles in the Standard Model, and causes certain forms of radioactive decay. (In particle physics though, the electroweak interaction is the unified description of two of the four known fundamental interactions of nature: electromagnetism and the weak interaction);
- the strong nuclear force, which binds quarks to form nucleons, and binds nucleons to form nuclei
- the gravitational force.

All other forces (e.g., friction) are ultimately derived from these fundamental forces and momentum carried by the movement of particles.

The electromagnetic force is the one responsible for practically all the phenomena one encounters in daily life above the nuclear scale, with the exception of gravity. Roughly speaking, all the forces involved in interactions between atoms can be explained by the electromagnetic force acting on the electrically charged atomic nuclei and electrons inside and around the atoms, together with how these particles carry momentum by their movement. This includes the forces we experience in "pushing" or "pulling" ordinary material objects, which come from the intermolecular forces between the individual molecules in our bodies and those in the objects. It also includes all forms of chemical phenomena.

A necessary part of understanding the intra-atomic to intermolecular forces is the effective force generated by the momentum of the electrons' movement, and that electrons move between interacting atoms, carrying momentum with them. As a collection of electrons becomes more confined, their minimum momentum necessarily increases due to the Pauli exclusion principle. The behaviour of matter at the molecular scale including its density is determined by the balance between the electromagnetic force and the force generated by the exchange of momentum carried by the electrons themselves.

19.3 Classical electrodynamics

Main article: Classical electrodynamics

The scientist William Gilbert proposed, in his *De Magnete* (1600), that electricity and magnetism, while both capable of causing attraction and repulsion of objects, were distinct effects. Mariners had noticed that lightning strikes had the ability to disturb a compass needle, but the link between lightning and electricity was not confirmed until Benjamin Franklin's proposed experiments in 1752. One of the first to discover and publish a link between man-made electric current and magnetism was Romagnosi, who in 1802 noticed that connecting a wire across a voltaic pile deflected a nearby compass needle. However, the effect did not become widely known until 1820, when Ørsted performed a similar experiment.[3] Ørsted's work influenced Ampère to produce a theory of electromagnetism that set the subject on a mathematical foundation.

A theory of electromagnetism, known as classical electromagnetism, was developed by various physicists over the course of the 19th century, culminating in the work of James Clerk Maxwell, who unified the preceding developments into a single theory and discovered the electromagnetic nature of light. In classical electromagnetism, the electromagnetic field obeys a set of equations known as Maxwell's equations, and the electromagnetic force is given by the Lorentz force law.

One of the peculiarities of classical electromagnetism is that it is difficult to reconcile with classical mechanics, but it is compatible with special relativity. According to Maxwell's equations, the speed of light in a vacuum is a universal constant, dependent only on the electrical permittivity and magnetic permeability of free space. This violates Galilean invariance, a long-standing cornerstone of classical mechanics. One way to reconcile the two theories (electromagnetism and classical mechanics) is to assume the existence of a luminiferous aether through which the light propagates. However, subsequent experimental efforts failed to detect the presence of the aether. After important contributions of Hendrik Lorentz and Henri Poincaré, in 1905, Albert Einstein solved the problem with the introduction of special relativity, which replaces classical kinematics with a new theory of kinematics that is compatible with classical electromagnetism. (For more information, see History of special relativity.)

In addition, relativity theory shows that in moving frames of reference a magnetic field transforms to a field with a nonzero electric component and vice versa; thus firmly showing that they are two sides of the same coin, and thus the term "electromagnetism". (For more information, see Classical electromagnetism and special relativity and Covariant formulation of classical electromagnetism.

19.4 Quantum mechanics

19.4.1 Photoelectric effect

Main article: Photoelectric effect

In another paper published in 1905, Albert Einstein undermined the very foundations of classical electromagnetism. In his theory of the photoelectric effect (for which he won the Nobel prize in physics) and inspired by the idea of Max Planck's "quanta", he posited that light could exist in discrete particle-like quantities as well, which later came to be known as photons. Einstein's theory of the photoelectric effect extended the insights that appeared in the solution of the ultraviolet catastrophe presented by Max Planck in 1900. In his work, Planck showed that hot objects emit electromagnetic radiation in discrete packets ("quanta"), which leads to a finite total energy emitted as black body radiation. Both of these results were in direct contradiction with the classical view of light as a continuous wave. Planck's and Einstein's theories were progenitors of quantum mechanics, which, when formulated in 1925, necessitated the invention of a quantum theory of electromagnetism. This theory, completed in the 1940s-1950s, is known as quantum electrodynamics (or "QED"), and, in situations where perturbation theory is applicable, is one of the most accurate theories known to physics.

19.4.2 Quantum electrodynamics

Main article: Quantum electrodynamics

All electromagnetic phenomena are underpinned by quantum mechanics, specifically by quantum electrodynamics (which includes classical electrodynamics as a limiting case) and this accounts for almost all physical phenomena observable to the unaided human senses, including light and other electromagnetic radiation, all of chemistry, most of mechanics (excepting gravitation), and, of course, magnetism and electricity.

19.4.3 Electroweak interaction

Main article: Electroweak interaction

The **electroweak interaction** is the unified description of two of the four known fundamental interactions of nature: electromagnetism and the weak interaction. Although these two forces appear very different at everyday low energies, the theory models them as two different aspects of the same force. Above the unification energy, on the order of 100 GeV, they would merge into a single **electroweak force**. Thus if the universe is hot enough (approximately 10^{15} K, a temperature exceeded until shortly after the Big Bang) then the electromagnetic force and weak force merge into a combined electroweak force. During the electroweak epoch, the electroweak force separated from the strong force. During the quark epoch, the electroweak force split into the electromagnetic and weak force.

19.5 Quantities and units

See also: List of physical quantities and List of electromagnetism equations

Electromagnetic units are part of a system of electrical units based primarily upon the magnetic properties of electric currents, the fundamental SI unit being the ampere. The units are:

- ampere (electric current)
- coulomb (electric charge)
- farad (capacitance)
- henry (inductance)
- ohm (resistance)
- tesla (magnetic flux density)
- volt (electric potential)
- watt (power)
- weber (magnetic flux)

In the electromagnetic cgs system, electric current is a fundamental quantity defined via Ampère's law and takes the permeability as a dimensionless quantity (relative permeability) whose value in a vacuum is unity. As a consequence, the square of the speed of light appears explicitly in some of the equations interrelating quantities in this system.

Formulas for physical laws of electromagnetism (such as Maxwell's equations) need to be adjusted depending on what system of units one uses. This is because there is no one-to-one correspondence between electromagnetic units in SI and those in CGS, as is the case for mechanical units. Furthermore, within CGS, there are several plausible choices of electromagnetic units, leading to different unit "sub-systems", including Gaussian, "ESU", "EMU", and Heaviside–Lorentz. Among these choices, Gaussian units are the most common today, and in fact the phrase "CGS units" is often used to refer specifically to CGS-Gaussian units.

19.6 See also

- Abraham–Lorentz force
- Aeromagnetic surveys
- Computational electromagnetics
- Double-slit experiment
- Electromagnet

- Electromagnetic wave equation
- Electromechanics
- Magnetostatics
- Magnetoquasistatic field
- Optics
- Relativistic electromagnetism
- Wheeler–Feynman absorber theory

19.7 References

[1] Ravaioli, Fawwaz T. Ulaby, Eric Michielssen, Umberto (2010). *Fundamentals of applied electromagnetics* (6th ed.). Boston: Prentice Hall. p. 13. ISBN 978-0-13-213931-1.

[2] Martins, Roberto de Andrade. "Romagnosi and Volta's Pile: Early Difficulties in the Interpretation of Voltaic Electricity". In Fabio Bevilacqua and Lucio Fregonese (eds). *Nuova Voltiana: Studies on Volta and his Times* (PDF). vol. 3. Università degli Studi di Pavia. pp. 81–102. Retrieved 2010-12-02.

[3] Stern, Dr. David P.; Peredo, Mauricio (2001-11-25). "Magnetic Fields -- History". NASA Goddard Space Flight Center. Retrieved 2009-11-27.

[4] International Union of Pure and Applied Chemistry (1993). *Quantities, Units and Symbols in Physical Chemistry*, 2nd edition, Oxford: Blackwell Science. ISBN 0-632-03583-8. pp. 14–15. Electronic version.

19.8 Further reading

19.8.1 Web sources

- Nave, R. "Electricity and magnetism". *HyperPhysics*. Georgia State University. Retrieved 2013-11-12.

19.8.2 Lecture notes

- Littlejohn, Robert (Spring 2011). "Emission and absorption of radiation" (PDF). *Physics 221B: Quantum mechanics*. University of California Berkeley. Retrieved 2013-11-12.
- Littlejohn, Robert (Spring 2011). "The Classical Electromagnetic Field Hamiltonian" (PDF). *Physics 221B: Quantum mechanics*. University of California Berkeley. Retrieved 2013-11-12.

19.8.3 Textbooks

- G.A.G. Bennet (1974). *Electricity and Modern Physics* (2nd ed.). Edward Arnold (UK). ISBN 0-7131-2459-8.
- Dibner, Bern (2012). *Oersted and the discovery of electromagnetism*. Literary Licensing, LLC. ISBN 9781258335557.
- Durney, Carl H. and Johnson, Curtis C. (1969). *Introduction to modern electromagnetics*. McGraw-Hill. ISBN 0-07-018388-0.
- Feynman, Richard P. (1970). *The Feynman Lectures on Physics Vol II*. Addison Wesley Longman. ISBN 978-0-201-02115-8.

- Fleisch, Daniel (2008). *A Student's Guide to Maxwell's Equations.* Cambridge, UK: Cambridge University Press. ISBN 978-0-521-70147-1.
- I.S. Grant, W.R. Phillips, Manchester Physics (2008). *Electromagnetism* (2nd ed.). John Wiley & Sons. ISBN 978-0-471-92712-9.
- Griffiths, David J. (1998). *Introduction to Electrodynamics* (3rd ed.). Prentice Hall. ISBN 0-13-805326-X.
- Jackson, John D. (1998). *Classical Electrodynamics* (3rd ed.). Wiley. ISBN 0-471-30932-X.
- Moliton, André (2007). *Basic electromagnetism and materials. 430 pages* (New York City: Springer-Verlag New York, LLC). ISBN 978-0-387-30284-3.
- Purcell, Edward M. (1985). *Electricity and Magnetism Berkeley Physics Course Volume 2 (2nd ed.).* McGraw-Hill. ISBN 0-07-004908-4.
- Rao, Nannapaneni N. (1994). *Elements of engineering electromagnetics (4th ed.).* Prentice Hall. ISBN 0-13-948746-8.
- Rothwell, Edward J.; Cloud, Michael J. (2001). *Electromagnetics.* CRC Press. ISBN 0-8493-1397-X.
- Tipler, Paul (1998). *Physics for Scientists and Engineers: Vol. 2: Light, Electricity and Magnetism* (4th ed.). W. H. Freeman. ISBN 1-57259-492-6.
- Wangsness, Roald K.; Cloud, Michael J. (1986). *Electromagnetic Fields (2nd Edition).* Wiley. ISBN 0-471-81186-6.

19.8.4 General references

- A. Beiser (1987). *Concepts of Modern Physics* (4th ed.). McGraw-Hill (International). ISBN 0-07-100144-1.
- L.H. Greenberg (1978). *Physics with Modern Applications.* Holt-Saunders International W.B. Saunders and Co. ISBN 0-7216-4247-0.
- R.G. Lerner, G.L. Trigg (2005). *Encyclopaedia of Physics* (2nd ed.). VHC Publishers, Hans Warlimont, Springer. pp. 12–13. ISBN 978-0-07-025734-4.
- J.B. Marion, W.F. Hornyak (1984). *Principles of Physics.* Holt-Saunders International Saunders College. ISBN 4-8337-0195-2.
- H.J. Pain (1983). *The Physics of Vibrations and Waves* (3rd ed.). John Wiley & Sons,. ISBN 0-471-90182-2.
- C.B. Parker (1994). *McGraw Hill Encyclopaedia of Physics* (2nd ed.). McGraw Hill. ISBN 0-07-051400-3.
- R. Penrose (2007). *The Road to Reality.* Vintage books. ISBN 0-679-77631-1.
- P.A. Tipler, G. Mosca (2008). *Physics for Scientists and Engineers: With Modern Physics* (6th ed.). W.H. Freeman and Co. ISBN 9-781429-202657.
- P.M. Whelan, M.J. Hodgeson (1978). *Essential Principles of Physics* (2nd ed.). John Murray. ISBN 0-7195-3382-1.

19.9 External links

- Oppelt, Arnulf (2006-11-02). “magnetic field strength”. Retrieved 2007-06-04.
- “magnetic field strength converter”. Retrieved 2007-06-04.
- Electromagnetic Force - from Eric Weisstein’s World of Physics
- Goudarzi, Sara (2006-08-15). “Ties That Bind Atoms Weaker Than Thought”. *LiveScience.com*. Retrieved 2013-11-12.
- Quarked Electromagnetic force - A good introduction for kids
- The Deflection of a Magnetic Compass Needle by a Current in a Wire (video) on YouTube
- Electromagnetism abridged

André-Marie Ampère

Michael Faraday

James Clerk Maxwell

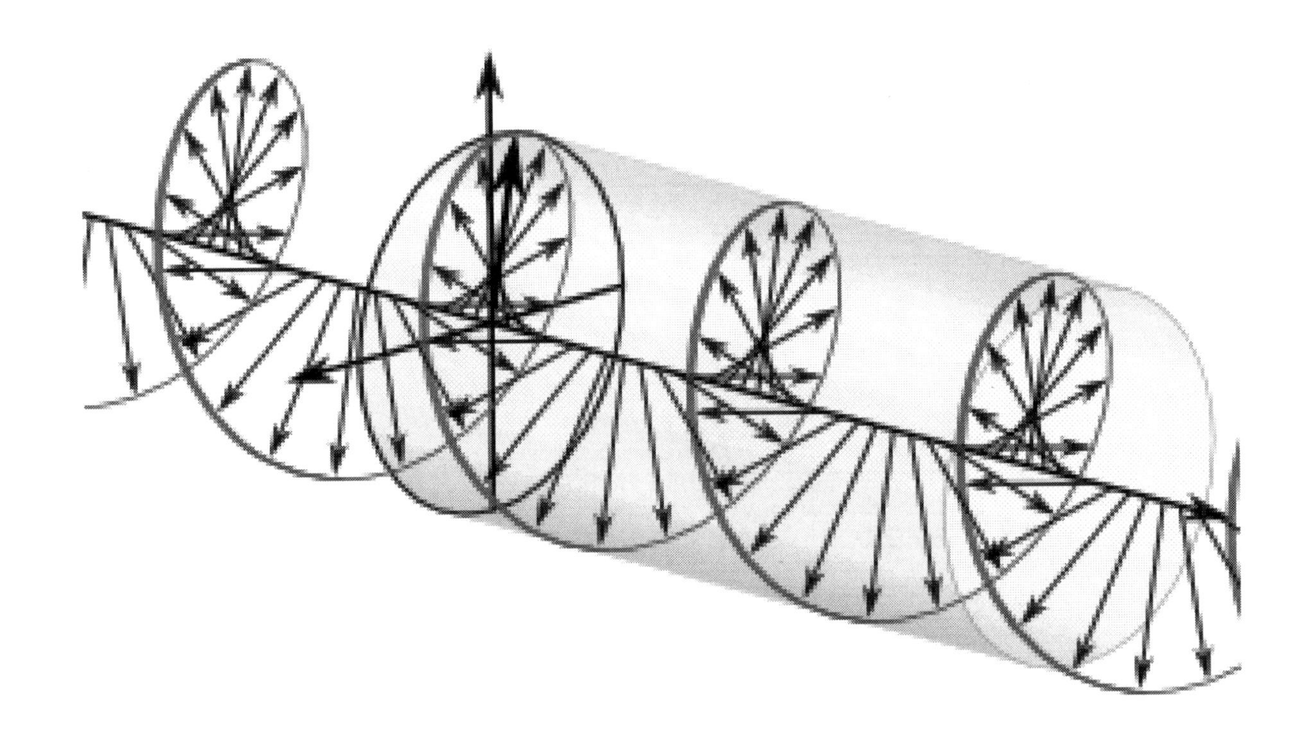

Representation of the electric field vector of a wave of circularly polarized electromagnetic radiation.

Chapter 20

Symmetry (physics)

For other uses, see Symmetry (disambiguation).

In physics, a **symmetry** of a physical system is a physical or mathematical feature of the system (observed or intrinsic) that is preserved or remains unchanged under some transformation.

A family of particular transformations may be *continuous* (such as rotation of a circle) or *discrete* (e.g., reflection of a bilaterally symmetric figure, or rotation of a regular polygon). Continuous and discrete transformations give rise to corresponding types of symmetries. Continuous symmetries can be described by Lie groups while discrete symmetries are described by finite groups (see Symmetry group).

These two concepts, Lie and finite groups, are the foundation for the fundamental theories of modern physics. Symmetries are frequently amenable to mathematical formulations such as group representations and can, in addition, be exploited to simplify many problems.

Arguably the most important example of a symmetry in physics is that the speed of light has the same value in all frames of reference, which is known in mathematical terms as Poincare group, the symmetry group of special relativity. Another important example is the invariance of the form of physical laws under arbitrary differentiable coordinate transformations, which is an important idea in general relativity.

20.1 Symmetry as invariance

Invariance is specified mathematically by transformations that leave some quantity unchanged. This idea can apply to basic real-world observations. For example, temperature may be constant throughout a room. Since the temperature is independent of position within the room, the temperature is *invariant* under a shift in the measurer's position.

Similarly, a uniform sphere rotated about its center will appear exactly as it did before the rotation. The sphere is said to exhibit spherical symmetry. A rotation about any axis of the sphere will preserve how the sphere "looks".

20.1.1 Invariance in force

The above ideas lead to the useful idea of *invariance* when discussing observed physical symmetry; this can be applied to symmetries in forces as well.

For example, an electric field due to a wire is said to exhibit cylindrical symmetry, because the electric field strength at a given distance r from the electrically charged wire of infinite length will have the same magnitude at each point on the surface of a cylinder (whose axis is the wire) with radius r. Rotating the wire about its own axis does not change its position or charge density, hence it will preserve the field. The field strength at a rotated position is the same. Suppose some configuration of charges (may be non-stationary) produce an electric field in some direction, then rotating the configuration of the charges (without disturbing the internal dynamics that produces the particular field) will lead to a net

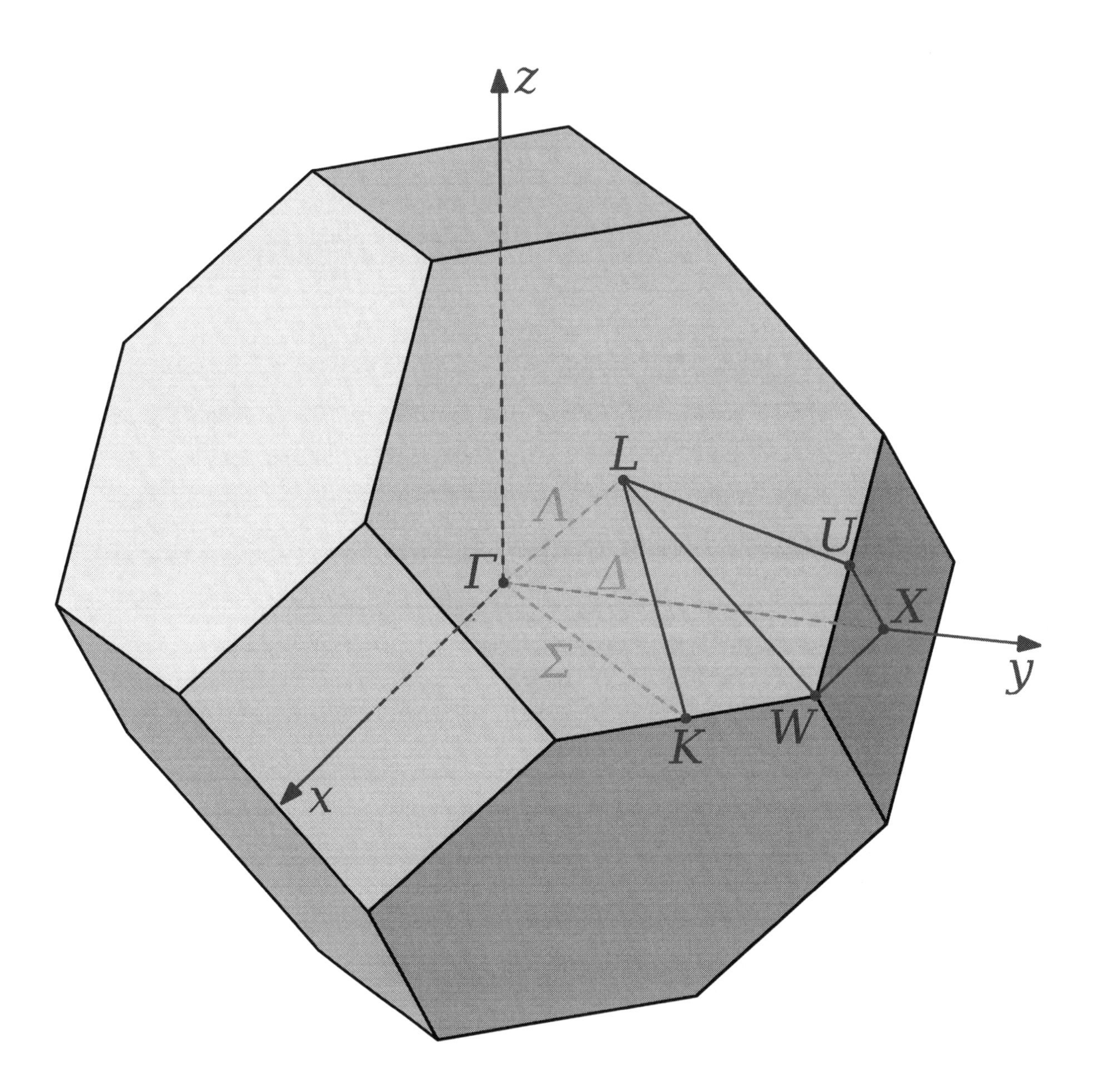

First Brillouin zone of FCC lattice showing symmetry labels

rotation of the direction of the electric field. These two properties are interconnected through the more general property that rotating *any* system of charges causes a corresponding rotation of the electric field.

In Newton's theory of mechanics, given two bodies, each with mass m, starting from rest at the origin and moving along the x-axis in opposite directions, one with speed v_1 and the other with speed v_2 the total kinetic energy of the system (as calculated from an observer at the origin) is $\frac{1}{2}m(v_1{}^2 + v_2{}^2)$ and remains the same if the velocities are interchanged. The total kinetic energy is preserved under a reflection in the y-axis.

The last example above illustrates another way of expressing symmetries, namely through the equations that describe some aspect of the physical system. The above example shows that the total kinetic energy will be the same if v_1 and v_2 are interchanged.

20.2 Local and global symmetries

Main articles: Global symmetry and Local symmetry

Symmetries may be broadly classified as *global* or *local*. A *global symmetry* is one that holds at all points of spacetime, whereas a *local symmetry* is one that has a different symmetry transformation at different points of spacetime; specifically a local symmetry transformation is parameterised by the spacetime co-ordinates. Local symmetries play an important role in physics as they form the basis for gauge theories.

20.3 Continuous symmetries

The two examples of rotational symmetry described above - spherical and cylindrical - are each instances of continuous symmetry. These are characterised by invariance following a continuous change in the geometry of the system. For example, the wire may be rotated through any angle about its axis and the field strength will be the same on a given cylinder. Mathematically, continuous symmetries are described by continuous or smooth functions. An important subclass of continuous symmetries in physics are spacetime symmetries.

20.3.1 Spacetime symmetries

Main article: Spacetime symmetries

Continuous *spacetime symmetries* are symmetries involving transformations of space and time. These may be further classified as *spatial symmetries*, involving only the spatial geometry associated with a physical system; *temporal symmetries*, involving only changes in time; or *spatio-temporal symmetries*, involving changes in both space and time.

- ***Time translation***: A physical system may have the same features over a certain interval of time δt ; this is expressed mathematically as invariance under the transformation $t \to t + a$ for any real numbers *t* and *a* in the interval. For example, in classical mechanics, a particle solely acted upon by gravity will have gravitational potential energy mgh when suspended from a height h above the Earth's surface. Assuming no change in the height of the particle, this will be the total gravitational potential energy of the particle at all times. In other words, by considering the state of the particle at some time (in seconds) t_0 and also at $t_0 + 3$, say, the particle's total gravitational potential energy will be preserved.

- ***Spatial translation***: These spatial symmetries are represented by transformations of the form $\vec{r} \to \vec{r} + \vec{a}$ and describe those situations where a property of the system does not change with a continuous change in location. For example, the temperature in a room may be independent of where the thermometer is located in the room.

- ***Spatial rotation***: These spatial symmetries are classified as proper rotations and improper rotations. The former are just the 'ordinary' rotations; mathematically, they are represented by square matrices with unit determinant. The latter are represented by square matrices with determinant -1 and consist of a proper rotation combined with a spatial reflection (inversion). For example, a sphere has proper rotational symmetry. Other types of spatial rotations are described in the article *Rotation symmetry*.

- ***Poincaré transformations***: These are spatio-temporal symmetries which preserve distances in Minkowski spacetime, i.e. they are isometries of Minkowski space. They are studied primarily in special relativity. Those isometries that leave the origin fixed are called Lorentz transformations and give rise to the symmetry known as Lorentz covariance.

- ***Projective symmetries***: These are spatio-temporal symmetries which preserve the geodesic structure of spacetime. They may be defined on any smooth manifold, but find many applications in the study of exact solutions in general relativity.

- ***Inversion transformations***: These are spatio-temporal symmetries which generalise Poincaré transformations to include other conformal one-to-one transformations on the space-time coordinates. Lengths are not invariant under inversion transformations but there is a cross-ratio on four points that is invariant.

Mathematically, spacetime symmetries are usually described by smooth vector fields on a smooth manifold. The underlying local diffeomorphisms associated with the vector fields correspond more directly to the physical symmetries, but the vector fields themselves are more often used when classifying the symmetries of the physical system.

Some of the most important vector fields are Killing vector fields which are those spacetime symmetries that preserve the underlying metric structure of a manifold. In rough terms, Killing vector fields preserve the distance between any two points of the manifold and often go by the name of isometries.

20.4 Discrete symmetries

Main article: Discrete symmetry

A **discrete symmetry** is a symmetry that describes non-continuous changes in a system. For example, a square possesses discrete rotational symmetry, as only rotations by multiples of right angles will preserve the square's original appearance. Discrete symmetries sometimes involve some type of 'swapping', these swaps usually being called *reflections* or *interchanges*.

- ***Time reversal***: Many laws of physics describe real phenomena when the direction of time is reversed. Mathematically, this is represented by the transformation, $t \rightarrow -t$. For example, Newton's second law of motion still holds if, in the equation $F = m\ddot{r}$, t is replaced by $-t$. This may be illustrated by recording the motion of an object thrown up vertically (neglecting air resistance) and then playing it back. The object will follow the same parabolic trajectory through the air, whether the recording is played normally or in reverse. Thus, position is symmetric with respect to the instant that the object is at its maximum height.

- ***Spatial inversion***: These are represented by transformations of the form $\vec{r} \rightarrow -\vec{r}$ and indicate an invariance property of a system when the coordinates are 'inverted'. Said another way, these are symmetries between a certain object and its mirror image.

- ***Glide reflection***: These are represented by a composition of a translation and a reflection. These symmetries occur in some crystals and in some planar symmetries, known as wallpaper symmetries.

20.4.1 C, P, and T symmetries

The Standard model of particle physics has three related natural near-symmetries. These state that the actual universe about us is indistinguishable from one where:

- Every particle is replaced with its antiparticle. This is C-symmetry (charge symmetry);
- Everything appears as if reflected in a mirror. This is P-symmetry (parity symmetry);
- The direction of time is reversed. This is T-symmetry (time symmetry).

T-symmetry is counterintuitive (surely the future and the past are not symmetrical) but explained by the fact that the Standard model describes local properties, not global ones like entropy. To properly reverse the direction of time, one would have to put the big bang and the resulting low-entropy state in the "future." Since we perceive the "past" ("future") as having lower (higher) entropy than the present (see perception of time), the inhabitants of this hypothetical time-reversed universe would perceive the future in the same way as we perceive the past.

These symmetries are near-symmetries because each is broken in the present-day universe. However, the Standard Model predicts that the combination of the three (that is, the simultaneous application of all three transformations) must be a symmetry, called CPT symmetry. In <ref name=*qm*>G. Kalmbach H.E.: *Quantum Mathematics: WIGRIS.* RGN Publications, Delhi, 2014.</ref> the 4 dimensional matrix description of P,T is through a diagonal matrix, the negative identity, as well as C. Hence CPT is the identity operator. CP violation, the violation of the combination of C- and P-symmetry, is necessary for the presence of significant amounts of baryonic matter in the universe. CP violation is a fruitful area of current research in particle physics.

20.4.2 Supersymmetry

Main article: Supersymmetry

A type of symmetry known as supersymmetry has been used to try to make theoretical advances in the standard model. Supersymmetry is based on the idea that there is another physical symmetry beyond those already developed in the standard model, specifically a symmetry between bosons and fermions. Supersymmetry asserts that each type of boson has, as a supersymmetric partner, a fermion, called a superpartner, and vice versa. Supersymmetry has not yet been experimentally verified: no known particle has the correct properties to be a superpartner of any other known particle. If superpartners exist they must have masses greater than current particle accelerators can generate.

20.5 Mathematics of physical symmetry

Main article: Symmetry group
See also: Symmetry in quantum mechanics and Symmetries in general relativity

The transformations describing physical symmetries typically form a mathematical group. Group theory is an important area of mathematics for physicists.

Continuous symmetries are specified mathematically by *continuous groups* (called Lie groups). Many physical symmetries are isometries and are specified by symmetry groups. Sometimes this term is used for more general types of symmetries. The set of all proper rotations (about any angle) through any axis of a sphere form a Lie group called the special orthogonal group $SO(3)$. (The *3* refers to the three-dimensional space of an ordinary sphere.) Thus, the symmetry group of the sphere with proper rotations is $SO(3)$. Any rotation preserves distances on the surface of the ball. The set of all Lorentz transformations form a group called the Lorentz group (this may be generalised to the Poincaré group).

Discrete symmetries are described by discrete groups. For example, the symmetries of an equilateral triangle are described by the symmetric group S_3 .

An important type of physical theory based on *local* symmetries is called a *gauge* theory and the symmetries natural to such a theory are called gauge symmetries. Gauge symmetries in the Standard model, used to describe three of the fundamental interactions, are based on the SU(3) × SU(2) × U(1) group. (Roughly speaking, the symmetries of the SU(3) group describe the strong force, the SU(2) group describes the weak interaction and the U(1) group describes the electromagnetic force.)

Also, the reduction by symmetry of the energy functional under the action by a group and spontaneous symmetry breaking of transformations of symmetric groups appear to elucidate topics in particle physics (for example, the unification of electromagnetism and the weak force in physical cosmology).

20.5.1 Conservation laws and symmetry

Main article: Noether's theorem

The symmetry properties of a physical system are intimately related to the conservation laws characterizing that system.

Noether's theorem gives a precise description of this relation. The theorem states that each continuous symmetry of a physical system implies that some physical property of that system is conserved. Conversely, each conserved quantity has a corresponding symmetry. For example, the isometry of space gives rise to conservation of (linear) momentum, and isometry of time gives rise to conservation of energy.

The following table summarizes some fundamental symmetries and the associated conserved quantity.

20.6 Mathematics

Continuous symmetries in physics preserve transformations. One can specify a symmetry by showing how a very small transformation affects various particle fields. The commutator of two of these infinitessimal transformations are equivalent to a third infinitessimal transformation of the same kind hence they form a Lie algebra.

A general coordinate transformation (also known as a diffeomorphism) has the infinitessimal effect on a scalar, spinor and vector field for example:

$$\delta\phi(x) = h^{\mu}(x)\partial_{\mu}\phi(x)$$

$$\delta\psi^{\alpha}(x) = h^{\mu}(x)\partial_{\mu}\psi^{\alpha}(x) + \partial_{\mu}h_{\nu}(x)\sigma_{\mu\nu}^{\alpha\beta}\psi^{\beta}(x)$$

$$\delta A_{\mu}(x) = h^{\nu}(x)\partial_{\nu}A_{\mu}(x) + A_{\nu}(x)\partial_{\nu}h_{\mu}(x)$$

for a general field, $h(x)$. Without gravity only the Poincaré symmetries are preserved which restricts $h(x)$ to be of the form:

$$h^{\mu}(x) = M^{\mu\nu}x_{\nu} + P^{\mu}$$

where **M** is an antisymmetric matrix (giving the Lorentz and rotational symmetries) and **P** is a general vector (giving the translational symmetries). Other symmetries affect multiple fields simultaneously. For example local gauge transformations apply to both a vector and spinor field:

$$\delta\psi^{\alpha}(x) = \lambda(x).\tau^{\alpha\beta}\psi^{\beta}(x)$$

$$\delta A_{\mu}(x) = \partial_{\mu}\lambda(x)$$

where τ are generators of a particular Lie group. So far the transformations on the right have only included fields of the same type. Supersymmetries are defined according to how the mix fields of *different* types.

Another symmetry which is part of some theories of physics and not in others is scale invariance which involve Weyl transformations of the following kind:

$$\delta\phi(x) = \Omega(x)\phi(x)$$

If the fields have this symmetry then it can be shown that the field theory is almost certainly conformally invariant also. This means that in the absence of gravity h(x) would restricted to the form:

$$h^{\mu}(x) = M^{\mu\nu}x_{\nu} + P^{\mu} + Dx_{\mu} + K^{\mu}|x|^2 - 2K^{\nu}x_{\nu}x_{\mu}$$

with **D** generating scale transformations and **K** generating special conformal transformations. For example N=4 super-Yang-Mills theory has this symmetry while General Relativity doesn't although other theories of gravity such as conformal gravity do. The 'action' of a field theory is an invariant under all the symmetries of the theory. Much of modern theoretical physics is to do with speculating on the various symmetries the Universe may have and finding the invariants to construct field theories as models.

In string theories, since a string can be decomposed into an infinite number of particle fields, the symmetries on the string world sheet is equivalent to special transformations which mix an infinite number of fields.

20.7 See also

- Conservation law
- Conserved current

- Coordinate-free
- Covariance and contravariance
- Diffeomorphism
- Fictitious force
- Galilean invariance
- Gauge theory
- General covariance
- Harmonic coordinate condition
- Inertial frame of reference
- Lie group
- List of mathematical topics in relativity
- Lorentz covariance
- Noether's theorem
- Poincaré group
- Special relativity
- Spontaneous symmetry breaking
- Standard model
- Standard model (mathematical formulation)
- Symmetry breaking
- Wheeler–Feynman Time-Symmetric Theory

20.8 References

20.8.1 General readers

- Leon Lederman and Christopher T. Hill (2005) *Symmetry and the Beautiful Universe*. Amherst NY: Prometheus Books.
- Schumm, Bruce (2004) *Deep Down Things*. Johns Hopkins Univ. Press.
- Victor J. Stenger (2000) *Timeless Reality: Symmetry, Simplicity, and Multiple Universes*. Buffalo NY: Prometheus Books. Chpt. 12 is a gentle introduction to symmetry, invariance, and conservation laws.
- Anthony Zee (2007) *Fearful Symmetry: The search for beauty in modern physics*, 2nd ed. Princeton University Press. ISBN 978-0-691-00946-9. 1986 1st ed. published by Macmillan.

20.8.2 Technical

- Brading, K., and Castellani, E., eds. (2003) *Symmetries in Physics: Philosophical Reflections*. Cambridge Univ. Press.
- -------- (2007) “Symmetries and Invariances in Classical Physics” in Butterfield, J., and John Earman, eds., *Philosophy of Physic Part B*. North Holland: 1331-68.
- Debs, T. and Redhead, M. (2007) *Objectivity, Invariance, and Convention: Symmetry in Physical Science*. Harvard Univ. Press.
- John Earman (2002) "Laws, Symmetry, and Symmetry Breaking: Invariance, Conservations Principles, and Objectivity." Address to the 2002 meeting of the Philosophy of Science Association.
- G. Kalmbach H.E.: *Quantum Mathematics: WIGRIS*. RGN Publications, Delhi, 2014
- Mainzer, K. (1996) *Symmetries of nature*. Berlin: De Gruyter.
- Mouchet, A. “Reflections on the four facets of symmetry: how physics exemplifies rational thinking”. European Physical Journal H 38 (2013) 661 hal.archives-ouvertes.fr:hal-00637572
- Thompson, William J. (1994) *Angular Momentum: An Illustrated Guide to Rotational Symmetries for Physical Systems*. Wiley. ISBN 0-471-55264-X.
- Bas Van Fraassen (1989) *Laws and symmetry*. Oxford Univ. Press.
- Eugene Wigner (1967) *Symmetries and Reflections*. Indiana Univ. Press.

20.9 External links

- Stanford Encyclopedia of Philosophy: "Symmetry"—by K. Brading and E. Castellani.
- Pedagogic Aids to Quantum Field Theory Click on link to Chapter 6: Symmetry, Invariance, and Conservation for a simplified, step-by-step introduction to symmetry in physics.

Chapter 21

History of quantum field theory

In particle physics, the **history of quantum field theory** starts with its creation by Paul Dirac, when he attempted to quantize the electromagnetic field in the late 1920s. Major advances in the theory were made in the 1950s, and led to the introduction of quantum electrodynamics (QED). QED was so successful and "natural" that efforts were made to use the same basic concepts for the other forces of nature. These efforts were successful in the application of gauge theory to the strong nuclear force and weak nuclear force, producing the modern standard model of particle physics. Efforts to describe gravity using the same techniques have, to date, failed. The study of quantum field theory is alive and flourishing, as are applications of this method to many physical problems. It remains one of the most vital areas of theoretical physics today, providing a common language to many branches of physics.

21.1 History

Quantum field theory originated in the 1920s from the problem of creating a quantum mechanical theory of the electromagnetic field. In particular de Broglie in 1924 introduced the idea of a wave description of elementary systems in the following way: "we proceed in this work from the assumption of the existence of a certain periodic phenomenon of a yet to be determined character, which is to be attributed to each and every isolated energy parcel".[1]

In 1925, Werner Heisenberg, Max Born, and Pascual Jordan constructed such a theory by expressing the field's internal degrees of freedom as an infinite set of harmonic oscillators and by employing the canonical quantization procedure to those oscillators.[2] [3] This theory assumed that no electric charges or currents were present and today would be called a free field theory.

The first reasonably complete theory of quantum electrodynamics, which included both the electromagnetic field and electrically charged matter (specifically, electrons) as quantum mechanical objects, was created by Paul Dirac in 1927.[4] This quantum field theory could be used to model important processes such as the emission of a photon by an electron dropping into a quantum state of lower energy, a process in which the *number of particles changes*—one atom in the initial state becomes an atom plus a photon in the final state. It is now understood that the ability to describe such processes is one of the most important features of quantum field theory.

The final crucial step, where creation and annihilation of fermions came to the fore and quantum field theory was seen to describe particle decay processes so "it was taken seriously" was Enrico Fermi's theory of β-decay (1934). [5]

21.2 Incorporating Special Relativity

It was evident from the beginning that a proper quantum treatment of the electromagnetic field had to somehow incorporate Einstein's relativity theory, which had grown out of the study of classical electromagnetism. This need to put together relativity and quantum mechanics was the second major motivation in the development of quantum field theory. Pascual Jordan and Wolfgang Pauli showed in 1928 that quantum fields could be made to behave in the way predicted by special relativity during coordinate transformations (specifically, they showed that the field commutators were Lorentz invariant). A further boost for quantum field theory came with the discovery of the Dirac equation, which was originally formulated and interpreted as a single-particle equation analogous to the Schrödinger equation, but unlike the Schrödinger equation, the Dirac equation satisfies both the Lorentz invariance, that is, the requirements of special relativity, and the rules of quantum mechanics. The Dirac equation accommodated the spin-1/2 value of the electron and accounted for its magnetic moment as well as giving accurate predictions for the spectra of hydrogen. The attempted interpretation of the Dirac equation as a single-particle equation could not be maintained long, however, and finally it was shown that several of its undesirable properties (such as negative-energy states) could be made sense of by reformulating and reinterpreting the Dirac equation as a true field equation, in this case for the quantized "Dirac field" or the "electron field", with the "negative-energy solutions" pointing to the existence of anti-particles. This work was performed first by Dirac himself with the invention of hole theory in 1930 and by Wendell Furry, Robert Oppenheimer, Vladimir Fock, and others. Schrödinger, during the same period that he discovered his famous equation in 1926, also independently found the relativistic generalization of it known as the Klein–Gordon equation but dismissed it since, without spin, it predicted impossible properties for the hydrogen spectrum. (See Oskar Klein and Walter Gordon.) All relativistic wave equations that describe spin-zero particles are said to be of the Klein–Gordon type.

21.3 Role of Soviet scientists

Of great importance are the studies of Soviet physicists, Viktor Ambartsumian and Dmitri Ivanenko, in particular the Ambarzumian-Ivanenko hypothesis of creation of massive particles (published in 1930) which is the cornerstone of the contemporary quantum field theory.[6] The idea is that not only the quanta of the electromagnetic field, photons, but also other particles (including particles having nonzero rest mass) may be born and disappear as a result of their interaction with other particles. This idea of Ambartsumian and Ivanenko formed the basis of modern quantum field theory and theory of elementary particles.[7][8]

21.4 Uncertainty, again

A subtle and careful analysis in 1933 and later in 1950 by Niels Bohr and Leon Rosenfeld showed that there is a fundamental limitation on the ability to simultaneously measure the electric and magnetic field strengths that enter into the description of charges in interaction with radiation, imposed by the uncertainty principle, which must apply to all canonically conjugate quantities. This limitation is crucial for the successful formulation and interpretation of a quantum field theory of photons and electrons (quantum electrodynamics), and indeed, any perturbative quantum field theory. The analysis of Bohr and Rosenfeld explains fluctuations in the values of the electromagnetic field that differ from the classically "allowed" values distant from the sources of the field. Their analysis was crucial to showing that the limitations and physical implications of the uncertainty principle apply to all dynamical systems, whether fields or material particles. Their analysis also convinced most physicists that any notion of returning to a fundamental description of nature based on classical field theory, such as what Einstein aimed at with his numerous and failed attempts at a classical unified field theory, was simply out of the question.

21.5 Second quantization

The third thread in the development of quantum field theory was the need to handle the statistics of many-particle systems consistently and with ease. In 1927, Pascual Jordan tried to extend the canonical quantization of fields to the many-body wave functions of identical particles, a procedure that is sometimes called second quantization. In 1928, Jordan and Eugene Wigner found that the quantum field describing electrons, or other fermions, had to be expanded using anti-commuting creation and annihilation operators due to the Pauli exclusion principle. This thread of development was incorporated into many-body theory and strongly influenced condensed matter physics and nuclear physics.

21.6 Problem of infinities

Despite its early successes quantum field theory was plagued by several serious theoretical difficulties. Basic physical quantities, such as the self-energy of the electron, the energy shift of electron states due to the presence of the electromagnetic field, gave infinite, divergent contributions—a nonsensical result—when computed using the perturbative techniques available in the 1930s and most of the 1940s. The electron self-energy problem was already a serious issue in the classical electromagnetic field theory, where the attempt to attribute to the electron a finite size or extent (the classical electron-radius) led immediately to the question of what non-electromagnetic stresses would need to be invoked, which would presumably hold the electron together against the Coulomb repulsion of its finite-sized "parts". The situation was dire, and had certain features that reminded many of the "Rayleigh-Jeans difficulty". What made the situation in the 1940s so desperate and gloomy, however, was the fact that the correct ingredients (the second-quantized Maxwell-Dirac field equations) for the theoretical description of interacting photons and electrons were well in place, and no major conceptual change was needed analogous to that which was necessitated by a finite and physically sensible account of the radiative behavior of hot objects, as provided by the Planck radiation law.

21.7 Normalizing renormalization

This "divergence problem" was solved in the case of quantum electrodynamics during the late 1940s and early 1950s by Hans Bethe, Tomonaga, Schwinger, Feynman, and Dyson, through the procedure known as renormalization. Great progress was made after realizing that all infinities in quantum electrodynamics are related to two effects: the self-energy of the electron/positron, and vacuum polarization. Renormalization concerns the business of paying very careful attention to just what is meant by, for example, the very concepts "charge" and "mass" as they occur in the pure, non-interacting field-equations. The "vacuum" is itself polarizable and, hence, populated by virtual particle (on shell and off shell) pairs, and, hence, is a seething and busy dynamical system in its own right. This was a critical step in identifying the source of "infinities" and "divergences". The "bare mass" and the "bare charge" of a particle, the values that appear in the free-field equations (non-interacting case), are abstractions that are simply not realized in experiment (in interaction). What we measure, and hence, what we must take account of with our equations, and what the solutions must account for, are the "renormalized mass" and the "renormalized charge" of a particle. That is to say, the "shifted" or "dressed" values these quantities must have when due care is taken to include all deviations from their "bare values" is dictated by the very nature of quantum fields themselves.

21.8 Gauge invariance

The first approach that bore fruit is known as the "interaction representation", (see the article Interaction picture) a Lorentz covariant and gauge-invariant generalization of time-dependent perturbation theory used in ordinary quantum mechanics, and developed by Tomonaga and Schwinger, generalizing earlier efforts of Dirac, Fock and Podolsky. Tomonaga and Schwinger invented a relativistically covariant scheme for representing field commutators and field operators intermediate between the two main representations of a quantum system, the Schrödinger and the Heisenberg representations (see the article on quantum mechanics). Within this scheme, field commutators at separated points can be evaluated in terms of "bare" field creation and annihilation operators. This allows for keeping track of the time-evolution of both the "bare" and "renormalized", or perturbed, values of the Hamiltonian and expresses everything in terms of the coupled, gauge invariant "bare" field-equations. Schwinger gave the most elegant formulation of this approach. The next and most famous development is due to Feynman, who, with his brilliant rules for assigning a "graph"/"diagram" to the terms in the scattering matrix (See S-Matrix Feynman diagrams). These directly corresponded (through the Schwinger-Dyson equation) to the measurable physical processes (cross sections, probability amplitudes, decay widths and lifetimes of excited states) one needs to be able to calculate. This revolutionized how quantum field theory calculations are carried-out in practice.

Two classic text-books from the 1960s, J.D. Bjorken and S.D. Drell, *Relativistic Quantum Mechanics* (1964) and J.J. Sakurai, *Advanced Quantum Mechanics* (1967), thoroughly developed the Feynman graph expansion techniques using physically intuitive and practical methods following from the correspondence principle, without worrying about the technicalities involved in deriving the Feynman rules from the superstructure of quantum field theory itself. Although both Feynman's heuristic and pictorial style of dealing with the infinities, as well as the formal methods of Tomonaga and Schwinger, worked extremely well, and gave spectacularly accurate answers, the true analytical nature of the question of "renormalizability", that is, whether ANY theory formulated as a "quantum field theory" would give finite answers, was not worked-out until much later, when the urgency of trying to formulate finite theories for the strong and electro-weak (and gravitational interactions) demanded its solution.

Renormalization in the case of QED was largely fortuitous due to the smallness of the coupling constant, the fact that the coupling has no dimensions involving mass, the so-called fine structure constant, and also the zero-mass of the gauge boson involved, the photon, rendered the small-distance/high-energy behavior of QED manageable. Also, electromagnetic processes are very "clean" in the sense that they are not badly suppressed/damped and/or hidden by the other gauge interactions. By 1958 Sidney Drell observed: "Quantum electrodynamics (QED) has achieved a status of peaceful coexistence with its divergences...".

The unification of the electromagnetic force with the weak force encountered initial difficulties due to the lack of accelerator energies high enough to reveal processes beyond the Fermi interaction range. Additionally, a satisfactory theoretical understanding of hadron substructure had to be developed, culminating in the quark model.

21.9 The strong force

In the case of the strong interactions, progress concerning their short-distance/high-energy behavior was much slower and more frustrating. For strong interactions with the electro-weak fields, there were difficult issues regarding the strength of coupling, the mass generation of the force carriers as well as their non-linear, self interactions. Although there has been theoretical progress toward a grand unified quantum field theory incorporating the electro-magnetic force, the weak force and the strong force, empirical verification is still pending. Superunification, incorporating the gravitational force, is still very speculative, and is under intensive investigation by many of the best minds in contemporary theoretical physics. Gravitation is a tensor field description of a spin-2 gauge-boson, the "graviton", and is further discussed in the articles on general relativity and quantum gravity.

21.10 Quantum gravity?

From the point of view of the techniques of (four-dimensional) quantum field theory, and as the numerous and heroic efforts to formulate a consistent quantum gravity theory by some very able minds attests, gravitational quantization was, and is still, the reigning champion for bad behavior. There are problems and frustrations stemming from the fact that the gravitational coupling constant has dimensions involving inverse powers of mass, and as a simple consequence, it is plagued by badly behaved (in the sense of perturbation theory) non-linear and violent self-interactions. Gravity, basically, gravitates, which in turn...gravitates...and so on, (i.e., gravity is itself a source of gravity,...,) thus creating a nightmare at all orders of perturbation theory. Also, gravity couples to all energy equally strongly, as per the equivalence principle, so this makes the notion of ever really "switching-off", "cutting-off" or separating, the gravitational interaction from other interactions ambiguous and impossible since, with gravitation, we are dealing with the very structure of space-time itself. (See general covariance and, for a modest, yet highly non-trivial and significant account of the interplay between (QFT) and gravitation (spacetime), see the article Hawking radiation and references cited therein. Also see Quantum field theory in curved spacetime).

21.11 An Abelian gauge theory

Thanks to the somewhat brute-force, clanky and heuristic methods of Feynman, and the elegant and abstract methods of Tomonaga/Schwinger, from the period of early renormalization, we do have the modern theory of quantum electrodynamics (QED). It is still the most accurate physical theory known, the prototype of a successful quantum field theory. Beginning in the 1950s with the work of Yang and Mills, as well as Ryoyu Utiyama, following the previous lead of Weyl and Pauli, deep explorations illuminated the types of symmetries and invariances any field theory must satisfy. QED, and indeed, all field theories, were generalized to a class of quantum field theories known as gauge theories. Quantum electrodynamics is the most famous example of what is known as an Abelian gauge theory. It relies on the symmetry group **U**(1) and has one massless gauge field, the **U**(1) gauge symmetry, dictating the form of the interactions involving the electromagnetic field, with the photon being the gauge boson. That symmetries dictate, limit and necessitate the form of interaction between particles is the essence of the "gauge theory revolution". Yang and Mills formulated the first explicit example of a non-Abelian gauge theory, Yang–Mills theory, with an attempted explanation of the strong interactions in mind. The strong interactions were then (incorrectly) understood in the mid-1950s, to be mediated by the pi-mesons, the particles predicted by Hideki Yukawa in 1935, based on his profound reflections concerning the reciprocal connection between the mass of any force-mediating particle and the range of the force it mediates. This was allowed by the uncertainty principle. The 1960s and 1970s saw the formulation of a gauge theory now known as the Standard Model of particle physics, which systematically describes the elementary particles and the interactions between them.

21.12 Electroweak unification

The electroweak interaction part of the standard model was formulated by Sheldon Glashow in the years 1958-60 with his discovery of the SU(2)xU(1) group structure of the theory. Steven Weinberg and Abdus Salam brilliantly invoked the Anderson-Higgs mechanism for the generation of the W's and Z masses (the intermediate vector boson(s) responsible for the weak interactions and neutral-currents) and keeping the mass of the photon zero. The Goldstone/Higgs idea for generating mass in gauge theories was sparked in the late 1950s and early 1960s when a number of theoreticians (including Yoichiro Nambu, Steven Weinberg, Jeffrey Goldstone, François Englert, Robert Brout, G. S. Guralnik, C. R. Hagen, Tom Kibble and Philip Warren Anderson) noticed a possibly useful analogy to the (spontaneous) breaking of the U(1) symmetry of electromagnetism in the formation of the BCS ground-state of a superconductor. The gauge boson involved in this situation, the photon, behaves as though it has acquired a finite mass. There is a further possibility that the physical vacuum (ground-state) does not respect the symmetries implied by the "unbroken" electroweak Lagrangian (see the article Electroweak interaction for more details) from which one arrives at the field equations. The electroweak theory of Weinberg and Salam was shown to be renormalizable (finite) and hence consistent by Gerardus 't Hooft and Martinus Veltman. The Glashow–Weinberg–Salam theory (GWS-Theory) is a triumph and, in certain applications, gives an accuracy on a par with quantum electrodynamics.

21.13 A grand synthesis

Also during the 1970s, parallel developments in the study of phase transitions in condensed matter physics led Leo Kadanoff, Michael Fisher and Kenneth Wilson (extending work of Ernst Stueckelberg, Andre Peterman, Murray Gell-Mann, and Francis Low) to a set of ideas and methods monitor changes of the behavior of the theory with scale, known as the renormalization group. By providing a deep physical understanding of the formal renormalization procedure invented in the 1940s, the renormalization group sparked what has been called the "grand synthesis" of theoretical physics, uniting the quantum field theoretical techniques used in particle physics and condensed matter physics into a single theoretical framework. The gauge field theory of the strong interactions, quantum chromodynamics, QCD, relies crucially on this renormalization group for its distinguishing characteristic features, asymptotic freedom and color confinement.

21.14 Modern developments

- Topological quantum field theory (TQFT)
- Axiomatic quantum field theory
- Local quantum field theory
- Algebraic quantum field theory

21.15 See also

- History of physics
- Electrodynamics
- History of quantum mechanics
- Quantum electrodynamics-QED
- Quantum field theory
- QED vacuum

21.16 Notes

[1] Recherches sur la theorrie des quanta (ann. de Phys., 10, III, 1925; translation by A. F. Kracklauer)

[2] Todorov, Ivan (2012). "Quantization is a mystery", *Bulg. J. Phys.* **39** (2012) 107-149 ; arXiv: 1206.3116 [math-ph]

[3] Born, M.; Heisenberg, W.; Jordan, P. (1926). "Zur Quantenmechanik II". *Zeitschrift für Physik* **35** (8–9): 557–615. Bibcode:1926ZPhy... doi:10.1007/BF01379806. The paper was received on 16 November 1925. [English translation in: van der Waerden 1968, 15 "On Quantum Mechanics II"]

[4] Dirac, P.A.M. (1927). *The Quantum Theory of the Emission and Absorption of Radiation*, Proceedings of the Royal Society of London, Series A, Vol. 114, p. 243.

[5] Chen Ning Yang (2012). "Fermi's β-decay Theory", *Asia Pac. Phys. Newslett.* **01**, p 27 . doi: 10.1142/S2251158X12000045 online

[6] G-sardanashvily.ru

[7] Vaprize.sci.am

[8] Sciteclibrary.ru

21.17 Further reading

- Pais, Abraham ; *Inward Bound - Of Matter & Forces in the Physical World*, Oxford University Press (1986) [ISBN 0-19-851997-4] Written by a former Einstein assistant at Princeton, this is a beautiful detailed history of modern fundamental physics, from 1895 (discovery of X-rays) to 1983 (discovery of vectors bosons at C.E.R.N.).
- Richard Feynman; *Lecture Notes in Physics*. Princeton University Press: Princeton, (1986).
- Richard Feynman; *QED*. Princeton University Press: Princeton, (1982).
- Weinberg, Steven ; *The Quantum Theory of Fields - Foundations (vol. I)*, Cambridge University Press (1995) [ISBN 0-521-55001-7] The first chapter (pp. 1–40) of Weinberg's monumental treatise gives a brief history of Q.F.T., pp. 608.
- Weinberg, Steven; *The Quantum Theory of Fields - Modern Applications* (vol. II), Cambridge University Press: Cambridge, U.K. (1996) [ISBN 0-521-55001-7], pp. 489.
- Weinberg, Steven; *The Quantum Theory of Fields - Supersymmetry* (vol. III), Cambridge University Press:Cambridge, U.K. (2000) [ISBN 0-521-55002-5], pp. 419.
- Schweber, Silvan S. ; *Q.E.D. and the men who made it: Dyson, Feynman, Schwinger, and Tomonaga*, Princeton University Press (1994) [ISBN 0-691-03327-7]

- Yndurain, Francisco Jose ; *Quantum Chromodynamics: An Introduction to the Theory of Quarks and Gluons*, Springer Verlag, New York, 1983. [ISBN 0-387-11752-0]
- Miller, Arthur I. ; *Early Quantum Electrodynamics : A Sourcebook*, Cambridge University Press (1995) [ISBN 0-521-56891-9]
- Schwinger, Julian ; *Selected Papers on Quantum Electrodynamics*, Dover Publications, Inc. (1958) [ISBN 0-486-60444-6]
- O'Raifeartaigh, Lochlainn ; *The Dawning of Gauge Theory*, Princeton University Press (May 5, 1997) [ISBN 0-691-02977-6]
- Cao, Tian Yu ; *Conceptual Developments of 20th Century Field Theories*, Cambridge University Press (1997) [ISBN 0-521-63420-2]
- Darrigol, Olivier ; *La genèse du concept de champ quantique*, Annales de Physique (France) 9 (1984) pp. 433–501. Text in French, adapted from the author's Ph.D. thesis.

Chapter 22

Bra–ket notation

In quantum mechanics, **bra–ket notation** is a standard notation for describing quantum states, composed of angle brackets and vertical bars. It can also be used to denote abstract vectors and linear functionals in mathematics. It is so called because the inner product (or dot product on a complex vector space) of two states is denoted by

$$\langle \phi \mid \psi \rangle$$

consisting of a left part, $\langle \phi|$ called the **bra** /brɑː/, and a right part, $|\psi\rangle$, called the **ket** /kɛt/. The notation was introduced in 1939 by Paul Dirac[1] and is also known as **Dirac notation**, though the notation has precursors in Grassmann's use of the notation $[\phi \mid \psi]$ for his inner products nearly 100 years earlier.[2][3]

Bra–ket notation is widespread in quantum mechanics: almost every phenomenon that is explained using quantum mechanics—including a large portion of modern physics — is usually explained with the help of bra-ket notation. Part of the appeal of the notation is the abstract representation-independence it encodes, together with its versatility in producing a specific representation (e.g. x, or p, or eigenfunction base) without much ado, or excessive reliance on the nature of the linear spaces involved. The overlap expression $\langle \phi \mid \psi \rangle$ is typically interpreted as the probability amplitude for the state ψ to collapse into the state φ.

22.1 Vector spaces

22.1.1 Background: Vector spaces

Main article: Vector space

In physics, basis vectors allow any Euclidean vector to be represented geometrically using angles and lengths, in different directions, i.e. in terms of the spatial orientations. It is simpler to see the notational equivalences between ordinary notation and bra-ket notation; so, for now, consider a vector **A** starting at the origin and ending at an element of 3-d Euclidean space; the vector then is specified by this end-point, a triplet of elements in the field of real numbers, symbolically dubbed as $\mathbf{A} \in \mathbb{R}^3$.

The vector **A** can be written using any set of basis vectors and corresponding coordinate system. Informally basis vectors are like “building blocks of a vector": they are added together to compose a vector, and the coordinates are the numerical coefficients of basis vectors in each direction. Two useful representations of a vector are simply a linear combination of basis vectors, and column matrices. Using the familiar Cartesian basis, a vector **A** may be written as

$$\mathbf{A} \doteq A_x\mathbf{e}_x + A_y\mathbf{e}_y + A_z\mathbf{e}_z = A_x\begin{pmatrix}1\\0\\0\end{pmatrix} + A_y\begin{pmatrix}0\\1\\0\end{pmatrix} + A_z\begin{pmatrix}0\\0\\1\end{pmatrix}$$

$$= \begin{pmatrix}A_x\\0\\0\end{pmatrix} + \begin{pmatrix}0\\A_y\\0\end{pmatrix} + \begin{pmatrix}0\\0\\A_z\end{pmatrix} = \begin{pmatrix}A_x\\A_y\\A_z\end{pmatrix}$$

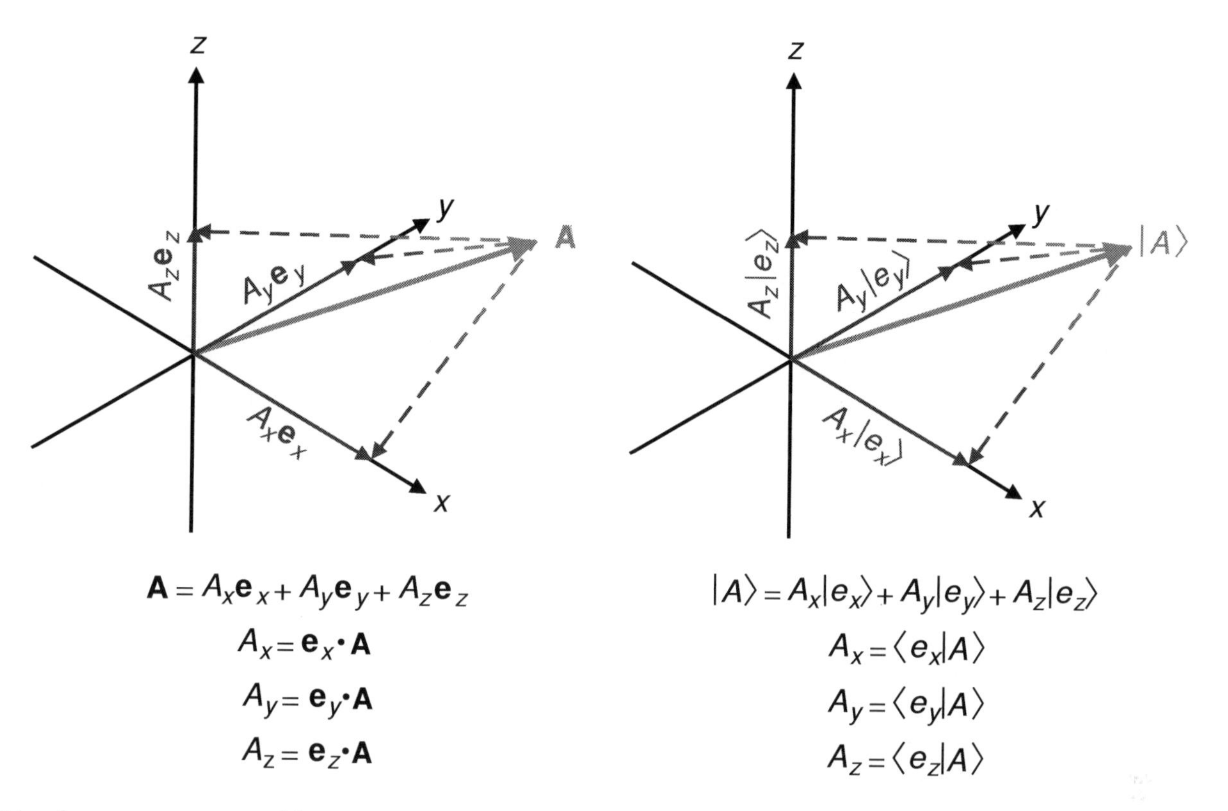

3d real vector components and bases projection; similarities between vector calculus notation and ***Dirac notation****. Projection is an important feature of the Dirac notation.*

respectively, where **e***x*, **e***y*, **e***z* denote the Cartesian basis vectors (all are orthogonal unit vectors) and *Ax*, *Ay*, *Az* are the corresponding coordinates, in the *x*, *y*, *z* directions. In a more general notation, for any basis in 3-d space one writes

$$\mathbf{A} \doteq A_1\mathbf{e}_1 + A_2\mathbf{e}_2 + A_3\mathbf{e}_3 = \begin{pmatrix}A_1\\A_2\\A_3\end{pmatrix}$$

Generalizing further, consider a vector **A** in an N-dimensional vector space over the field of complex numbers $\mathbb{C}$, symbolically stated as $\mathbf{A} \in \mathbb{C}^N$. The vector **A** is still conventionally represented by a linear combination of basis vectors or a column matrix:

$$\mathbf{A} \doteq \sum_{n=1}^{N} A_n\mathbf{e}_n = \begin{pmatrix}A_1\\A_2\\\vdots\\A_N\end{pmatrix}$$

though the coordinates are now all complex-valued.

Even more generally, **A** can be a vector in a complex Hilbert space. Some Hilbert spaces, like $\mathbb{C}^N$, have finite dimension, while others have infinite dimension. In an infinite-dimensional space, the column-vector representation of **A** would be a list of infinitely many complex numbers.

22.1.2 Ket notation for vectors

Rather than boldtype, over arrows, underscores etc. conventionally used elsewhere, **A**, $\vec{A}$, $\underline{A}$, Dirac's notation for a vector uses vertical bars and angular brackets: $|A\rangle$. When this notation is used, these vectors are called "ket", read as "ket-A".[4] This applies to all vectors, the resultant vector and the basis. The previous vectors are now written

$$|A\rangle = A_x|e_x\rangle + A_y|e_y\rangle + A_z|e_z\rangle \doteq \begin{pmatrix} A_x \\ A_y \\ A_z \end{pmatrix},$$

or in a more easily generalized notation,

$$|A\rangle = A_1|e_1\rangle + A_2|e_2\rangle + A_3|e_3\rangle \doteq \begin{pmatrix} A_1 \\ A_2 \\ A_3 \end{pmatrix},$$

The last one may be written in short as

$$|A\rangle = A_1|1\rangle + A_2|2\rangle + A_3|3\rangle \ .$$

Note how any symbols, letters, numbers, or even words—whatever serves as a convenient label—can be used as the label inside a ket. In other words, the symbol " $|A\rangle$ " has a specific and universal mathematical meaning, while just the "A" by itself does not. Nevertheless, for convenience, there is usually some logical scheme behind the labels inside kets, such as the common practice of labeling energy eigenkets in quantum mechanics through a listing of their quantum numbers. Further note that a ket and its representation by a coordinate vector are not the same mathematical object: a ket does not require specification of a basis, whereas the coordinate vector needs a basis in order to be well defined (the same holds for an operator and its representation by a matrix).[5] In this context, one should best use a symbol different than the equal sign, for example the symbol $\doteq$, read as "is represented by".

22.1.3 Inner products and bras

Main article: Inner product

An inner product is a generalization of the dot product. The inner product of two vectors is a scalar. bra-ket notation uses a specific notation for inner products:

$$\langle A|B\rangle = \text{ ket of product inner the}|A\rangle \text{ ket with } |B\rangle$$

For example, in three-dimensional complex Euclidean space,

$$\langle A|B\rangle \doteq A_x^* B_x + A_y^* B_y + A_z^* B_z$$

where A_i^* denotes the complex conjugate of A_i. A special case is the inner product of a vector with itself, which is the square of its norm (magnitude):

$$\langle A|A\rangle \doteq |A_x|^2 + |A_y|^2 + |A_z|^2$$

bra-ket notation splits this inner product (also called a "bracket") into two pieces, the "bra" and the "ket":

$$\langle A|B\rangle = (\,\langle A|\,)\;(\,|B\rangle\,)$$

where $\langle A|$ is called a bra, read as "bra-A", and $|B\rangle$ is a ket as above.

The purpose of "splitting" the inner product into a bra and a ket is that *both* the bra $\langle A|$ and the ket $|B\rangle$ are meaningful *on their own*, and can be used in other contexts besides within an inner product. There are two main ways to think about the meanings of separate bras and kets:

Bras and kets as row and column vectors

For a finite-dimensional vector space, using a fixed orthonormal basis, the inner product can be written as a matrix multiplication of a row vector with a column vector:

$$\langle A|B\rangle \doteq A_1^*B_1 + A_2^*B_2 + \cdots + A_N^*B_N = \begin{pmatrix} A_1^* & A_2^* & \cdots & A_N^* \end{pmatrix} \begin{pmatrix} B_1 \\ B_2 \\ \vdots \\ B_N \end{pmatrix}$$

Based on this, the bras and kets can be defined as:

$$\langle A| \doteq \begin{pmatrix} A_1^* & A_2^* & \cdots & A_N^* \end{pmatrix}$$

$$|B\rangle \doteq \begin{pmatrix} B_1 \\ B_2 \\ \vdots \\ B_N \end{pmatrix}$$

and then it is understood that a bra next to a ket implies matrix multiplication.

The conjugate transpose (also called *Hermitian conjugate*) of a bra is the corresponding ket and vice versa:

$$\langle A|^\dagger = |A\rangle, \quad |A\rangle^\dagger = \langle A|$$

because if one starts with the bra

$$\begin{pmatrix} A_1^* & A_2^* & \cdots & A_N^* \end{pmatrix},$$

then performs a complex conjugation, and then a matrix transpose, one ends up with the ket

$$\begin{pmatrix} A_1 \\ A_2 \\ \vdots \\ A_N \end{pmatrix}$$

Bras as linear operators on kets

Main articles: Dual space and Riesz representation theorem

A more abstract definition, which is equivalent but more easily generalized to infinite-dimensional spaces, is to say that bras are linear functionals on kets, i.e. operators that input a ket and output a complex number. The bra operators are defined to be consistent with the inner product.

In mathematics terminology, the vector space of bras is the dual space to the vector space of kets, and corresponding bras and kets are related by the Riesz representation theorem.

22.1.4 Non-normalizable states and non-Hilbert spaces

bra-ket notation can be used even if the vector space is not a Hilbert space.

In quantum mechanics, it is common practice to write down kets which have infinite norm, i.e. non-normalisable wavefunctions. Examples include states whose wavefunctions are Dirac delta functions or infinite plane waves. These do not, technically, belong to the Hilbert space itself. However, the definition of "Hilbert space" can be broadened to accommodate these states (see the Gelfand–Naimark–Segal construction or rigged Hilbert spaces). The bra-ket notation continues to work in an analogous way in this broader context.

For a rigorous treatment of the Dirac inner product of non-normalizable states, see the definition given by D. Carfì.[6][7] For a rigorous definition of basis with a continuous set of indices and consequently for a rigorous definition of position and momentum basis, see.[8] For a rigorous statement of the expansion of an S-diagonalizable operator, or observable, in its eigenbasis or in another basis, see.[9]

Banach spaces are a different generalization of Hilbert spaces. In a Banach space B, the vectors may be notated by kets and the continuous linear functionals by bras. Over any vector space without topology, we may also notate the vectors by kets and the linear functionals by bras. In these more general contexts, the bracket does not have the meaning of an inner product, because the Riesz representation theorem does not apply.

22.2 Usage in quantum mechanics

The mathematical structure of quantum mechanics is based in large part on linear algebra:

- Wave functions and other quantum states can be represented as vectors in a complex Hilbert space. (The exact structure of this Hilbert space depends on the situation.) In bra-ket notation, for example, an electron might be in the "state" $|\psi\rangle$. (Technically, the quantum states are *rays* of vectors in the Hilbert space, as $c|\psi\rangle$ corresponds to the same state for any nonzero complex number c.)

- Quantum superpositions can be described as vector sums of the constituent states. For example, an electron in the state $|1\rangle + i\,|2\rangle$ is in a quantum superposition of the states $|1\rangle$ and $|2\rangle$.

- Measurements are associated with linear operators (called observables) on the Hilbert space of quantum states.

- Dynamics are also described by linear operators on the Hilbert space. For example, in the Schrödinger picture, there is a linear time evolution operator U with the property that if an electron is in state $|\psi\rangle$ right now, then in one second it will be in the state $U|\psi\rangle$, the same U for every possible $|\psi\rangle$.

- Wave function normalization is scaling a wave function so that its norm is 1.

Since virtually every calculation in quantum mechanics involves vectors and linear operators, it can involve, and often *does* involve, bra-ket notation. A few examples follow:

22.2.1 Spinless position–space wave function

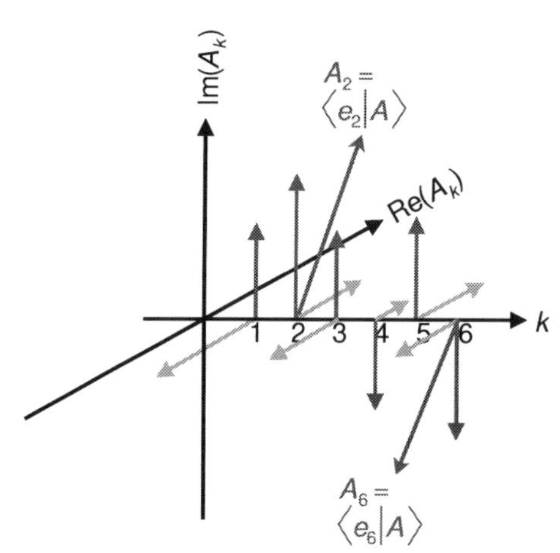

Discrete components Ak of a complex vector $|A\rangle = \sum k\ Ak|ek\rangle$, which belongs to a *countably infinite*-dimensional Hilbert space; there are countably infinitely many k values and basis vectors $|ek\rangle$.

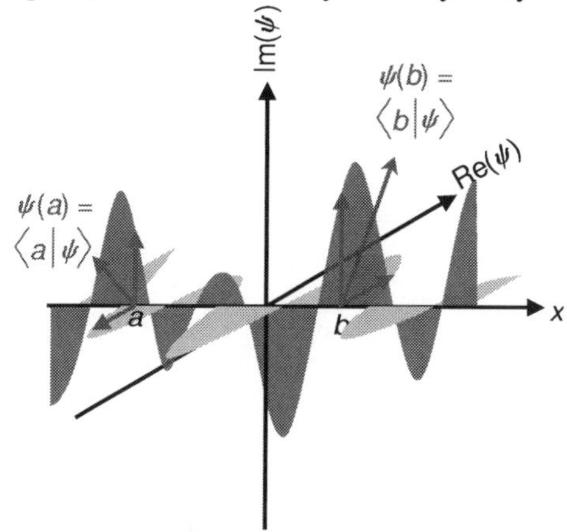

Continuous components $\psi(x)$ of a complex vector $|\psi\rangle = \int dx\ \psi(x)|x\rangle$, which belongs to an *uncountably infinite*-dimensional Hilbert space; there are infinitely many x values and basis vectors $|x\rangle$.
Components of complex vectors plotted against index number; discrete k and continuous x. Two particular components out of infinitely many are highlighted.

The Hilbert space of a spin−0 point particle is spanned by a "position basis" $\{\ |\mathbf{r}\rangle\ \}$, where the label $\mathbf{r}$ extends over the set of all points in position space. Since there are uncountably infinitely many vector components in the basis, this is an uncountably infinite-dimensional Hilbert space. The dimensions of the Hilbert space (usually infinite) and position space (usually 1, 2 or 3) are not to be conflated.

Starting from any ket $|\Psi\rangle$ in this Hilbert space, we can *define* a complex scalar function of $\mathbf{r}$, known as a wavefunction:

$$\Psi(\mathbf{r}) \stackrel{\text{def}}{=} \langle \mathbf{r}|\Psi\rangle$$

On the left side, $\Psi(\mathbf{r})$ is a function mapping any point in space to a complex number; on the right side, $|\Psi\rangle = \int d^3\mathbf{r}\ \Psi(\mathbf{r})\ |\mathbf{r}\rangle$ is a ket.

It is then customary to define linear operators acting on wavefunctions in terms of linear operators acting on kets, by

$$A\Psi(\mathbf{r}) \stackrel{\text{def}}{=} \langle \mathbf{r}|A|\Psi\rangle.$$

For instance, the momentum operator $\mathbf{p}$ has the following form,

$$\mathbf{p}\Psi(\mathbf{r}) \stackrel{\text{def}}{=} \langle\mathbf{r}|\mathbf{p}|\Psi\rangle = -i\hbar\nabla\Psi(\mathbf{r}).$$

One occasionally encounters a sloppy expression like

$$\nabla|\Psi\rangle,$$

though this is something of a (common) abuse of notation. The differential operator must be understood to be an abstract operator, acting on kets, that has the effect of differentiating wavefunctions once the expression is projected into the position basis,

$$\nabla\langle\mathbf{r}|\Psi\rangle,$$

even though, in the momentum basis, the operator amounts to a mere multiplication operator (by $i\hbar\boldsymbol{p}$).

22.2.2 Overlap of states

In quantum mechanics the expression $\langle\varphi|\psi\rangle$ is typically interpreted as the probability amplitude for the state ψ to collapse into the state φ. Mathematically, this means the coefficient for the projection of ψ onto φ. It is also described as the projection of state ψ onto state φ.

22.2.3 Changing basis for a spin-1/2 particle

A stationary spin-½ particle has a two-dimensional Hilbert space. One orthonormal basis is:

$$|\uparrow_z\rangle,\ |\downarrow_z\rangle$$

where $|\uparrow_z\rangle$ is the state with a definite value of the spin operator *Sz* equal to +1/2 and $|\downarrow_z\rangle$ is the state with a definite value of the spin operator *Sz* equal to −1/2.

Since these are a basis, *any* quantum state of the particle can be expressed as a linear combination (i.e., quantum superposition) of these two states:

$$|\psi\rangle = a_\psi|\uparrow_z\rangle + b_\psi|\downarrow_z\rangle$$

where $a\psi$, $b\psi$ are complex numbers.

A *different* basis for the same Hilbert space is:

$$|\uparrow_x\rangle,\ |\downarrow_x\rangle$$

defined in terms of *Sx* rather than *Sz*.

Again, *any* state of the particle can be expressed as a linear combination of these two:

$$|\psi\rangle = c_\psi|\uparrow_x\rangle + d_\psi|\downarrow_x\rangle$$

In vector form, you might write

$$|\psi\rangle \doteq \begin{pmatrix} a_\psi \\ b_\psi \end{pmatrix}, \quad \text{OR} \quad |\psi\rangle \doteq \begin{pmatrix} c_\psi \\ d_\psi \end{pmatrix}$$

depending on which basis you are using. In other words, the "coordinates" of a vector depend on the basis used.

There is a mathematical relationship between $a\psi$, $b\psi$, $c\psi$, $d\psi$; see change of basis.

22.2.4 Misleading uses

There are a few conventions and abuses of notation that are generally accepted by the physics community, but which might confuse the non-initiated.

It is common among physicists to use the same symbol for *labels* and *constants* in the same equation. It supposedly becomes easier to identify that the constant is related to the labeled object, and is claimed that the divergent nature of each will eliminate any ambiguity and no further differentiation is required. For example, $\hat{\alpha}\,|\alpha\rangle = \alpha|\alpha\rangle$, where the symbol α is used **simultaneously** as the *name of the operator* $\hat{\alpha}$, its *eigenvector* $|\alpha\rangle$ and the associated *eigenvalue* α.

Something similar occurs in component notation of vectors. While Ψ (uppercase) is traditionally associated with wavefunctions, ψ (lowercase) may be used to denote a *label*, a *wave function* or *complex constant* in the same context, usually differentiated only by a subscript.

The main abuses are including operations inside the vector labels. This is usually done for a fast notation of scaling vectors. E.g. if the vector $|\alpha\rangle$ is scaled by $\sqrt{2}$, it might be denoted by $|\alpha/\sqrt{2}\rangle$, which makes no sense since α is a label, not a function or a number, so you can't perform operations on it.

This is especially common when denoting vectors as tensor products, where part of the labels are moved **outside** the designed slot. E.g. $|\alpha\rangle = |\alpha/\sqrt{2}\rangle_1 \otimes |\alpha/\sqrt{2}\rangle_2$. Here part of the labeling that should state that all three vectors are different was moved outside the kets, as subscripts 1 and 2. And a further abuse occurs, since α is meant to refer to the norm of the first vector – which is a *label* is denoting a *value*.

22.3 Linear operators

See also: Linear operator

22.3.1 Linear operators acting on kets

A linear operator is a map that inputs a ket and outputs a ket. (In order to be called "linear", it is required to have certain properties.) In other words, if $\boldsymbol{A}$ is a linear operator and $|\psi\rangle$ is a ket, then $\boldsymbol{A}|\psi\rangle$ is another ket.

In an N-dimensional Hilbert space, $|\psi\rangle$ can be written as an N×1 column vector, and then $\boldsymbol{A}$ is an N×N matrix with complex entries. The ket $\boldsymbol{A}|\psi\rangle$ can be computed by normal matrix multiplication.

Linear operators are ubiquitous in the theory of quantum mechanics. For example, observable physical quantities are represented by self-adjoint operators, such as energy or momentum, whereas transformative processes are represented by unitary linear operators such as rotation or the progression of time.

22.3.2 Linear operators acting on bras

Operators can also be viewed as acting on bras *from the right hand side*. Specifically, if $\boldsymbol{A}$ is a linear operator and $\langle\varphi|$ is a bra, then $\langle\varphi|\boldsymbol{A}$ is another bra defined by the rule

$$\Big(\langle\phi|A\Big)\,|\psi\rangle = \langle\phi|\,\Big(A|\psi\rangle\Big),$$

(in other words, a function composition). This expression is commonly written as (cf. energy inner product)

$$\langle\phi|A|\psi\rangle$$

In an N-dimensional Hilbert space, $\langle\varphi|$ can be written as a 1×N row vector, and $\boldsymbol{A}$ (as in the previous section) is an N×N matrix. Then the bra $\langle\varphi|\boldsymbol{A}$ can be computed by normal matrix multiplication.

If the same state vector appears on both bra and ket side,

$$\langle\psi|A|\psi\rangle,$$

then this expression gives the expectation value, or mean or average value, of the observable represented by operator $\boldsymbol{A}$ for the physical system in the state $|\psi\rangle$.

22.3.3 Outer products

A convenient way to define linear operators on H is given by the outer product: if $\langle\varphi|$ is a bra and $|\psi\rangle$ is a ket, the outer product

$$|\phi\rangle\langle\psi|$$

denotes the rank-one operator with the rule

$$(|\phi\rangle\langle\psi|)(x) = \langle\psi, x\rangle\phi$$

For a finite-dimensional vector space, the outer product can be understood as simple matrix multiplication:

$$|\phi\rangle\,\langle\psi| \doteq \begin{pmatrix}\phi_1\\ \phi_2\\ \vdots\\ \phi_N\end{pmatrix}\begin{pmatrix}\psi_1^* & \psi_2^* & \cdots & \psi_N^*\end{pmatrix} = \begin{pmatrix}\phi_1\psi_1^* & \phi_1\psi_2^* & \cdots & \phi_1\psi_N^*\\ \phi_2\psi_1^* & \phi_2\psi_2^* & \cdots & \phi_2\psi_N^*\\ \vdots & \vdots & \ddots & \vdots\\ \phi_N\psi_1^* & \phi_N\psi_2^* & \cdots & \phi_N\psi_N^*\end{pmatrix}$$

The outer product is an N×N matrix, as expected for a linear operator.

One of the uses of the outer product is to construct projection operators. Given a ket $|\psi\rangle$ of norm 1, the orthogonal projection onto the subspace spanned by $|\psi\rangle$ is

$$|\psi\rangle\langle\psi|.$$

22.3.4 Hermitian conjugate operator

Main article: Hermitian conjugate

Just as kets and bras can be transformed into each other (making $|\psi\rangle$ into $\langle\psi|$), the element from the dual space corresponding to $A|\psi\rangle$ is $\langle\psi|A^\dagger$, where $A^\dagger$ denotes the Hermitian conjugate (or adjoint) of the operator A. In other words,

$$|\phi\rangle = A|\psi\rangle \text{ if and only if } \qquad \langle\phi| = \langle\psi|A^\dagger\ .$$

If A is expressed as an $N{\times}N$ matrix, then $A^\dagger$ is its conjugate transpose.

Self-adjoint operators, where $A = A^\dagger$, play an important role in quantum mechanics; for example, an observable is always described by a self-adjoint operator. If A is a self-adjoint operator, then $\langle\psi|A|\psi\rangle$ is always a real number (not complex). This implies that expectation values of observables are real.

22.4 Properties

bra-ket notation was designed to facilitate the formal manipulation of linear-algebraic expressions. Some of the properties that allow this manipulation are listed herein. In what follows, c_1 and c_2 denote arbitrary complex numbers, c^* denotes the complex conjugate of c, A and B denote arbitrary linear operators, and these properties are to hold for any choice of bras and kets.

22.4.1 Linearity

- Since bras are linear functionals,

$$\langle\phi|\Big(c_1|\psi_1\rangle + c_2|\psi_2\rangle\Big) = c_1\langle\phi|\psi_1\rangle + c_2\langle\phi|\psi_2\rangle.$$

- By the definition of addition and scalar multiplication of linear functionals in the dual space,[10]

$$\Big(c_1\langle\phi_1| + c_2\langle\phi_2|\Big)\ |\psi\rangle = c_1\langle\phi_1|\psi\rangle + c_2\langle\phi_2|\psi\rangle.$$

22.4.2 Associativity

Given any expression involving complex numbers, bras, kets, inner products, outer products, and/or linear operators (but not addition), written in bra-ket notation, the parenthetical groupings do not matter (i.e., the associative property holds). For example:

$$\langle\psi|(A|\phi\rangle) = (\langle\psi|A)|\phi\rangle \stackrel{\text{def}}{=} \langle\psi|A|\phi\rangle$$

$$(A|\psi\rangle)\langle\phi| = A(|\psi\rangle\langle\phi|) \stackrel{\text{def}}{=} A|\psi\rangle\langle\phi|$$

and so forth. The expressions on the right (with no parentheses whatsoever) are allowed to be written unambiguously *because* of the equalities on the left. Note that the associative property does *not* hold for expressions that include non-linear operators, such as the antilinear time reversal operator in physics.

22.4.3 Hermitian conjugation

bra-ket notation makes it particularly easy to compute the Hermitian conjugate (also called *dagger*, and denoted †) of expressions. The formal rules are:

- The Hermitian conjugate of a bra is the corresponding ket, and vice versa.

- The Hermitian conjugate of a complex number is its complex conjugate.
- The Hermitian conjugate of the Hermitian conjugate of anything (linear operators, bras, kets, numbers) is itself—i.e.,

$$(x^\dagger)^\dagger = x.$$

- Given any combination of complex numbers, bras, kets, inner products, outer products, and/or linear operators, written in bra-ket notation, its Hermitian conjugate can be computed by reversing the order of the components, and taking the Hermitian conjugate of each.

These rules are sufficient to formally write the Hermitian conjugate of any such expression; some examples are as follows:

- Kets:

$$(c_1|\psi_1\rangle + c_2|\psi_2\rangle)^\dagger = c_1^*\langle\psi_1| + c_2^*\langle\psi_2| \, .$$

- Inner products:

$$\langle\phi|\psi\rangle^* = \langle\psi|\phi\rangle \, .$$

- Matrix elements:

$$\langle\phi|A|\psi\rangle^* = \langle\psi|A^\dagger|\phi\rangle$$
$$\langle\phi|A^\dagger B^\dagger|\psi\rangle^* = \langle\psi|BA|\phi\rangle \, .$$

- Outer products:

$$((c_1|\phi_1\rangle\langle\psi_1|) + (c_2|\phi_2\rangle\langle\psi_2|))^\dagger = (c_1^*|\psi_1\rangle\langle\phi_1|) + (c_2^*|\psi_2\rangle\langle\phi_2|) \, .$$

22.5 Composite bras and kets

Two Hilbert spaces V and W may form a third space $V \otimes W$ by a tensor product. In quantum mechanics, this is used for describing composite systems. If a system is composed of two subsystems described in V and W respectively, then the Hilbert space of the entire system is the tensor product of the two spaces. (The exception to this is if the subsystems are actually identical particles. In that case, the situation is a little more complicated.)

If $|\psi\rangle$ is a ket in V and $|\varphi\rangle$ is a ket in W, the direct product of the two kets is a ket in $V \otimes W$. This is written in various notations:

$$|\psi\rangle|\phi\rangle \, , \quad |\psi\rangle \otimes |\phi\rangle \, , \quad |\psi\phi\rangle \, , \quad |\psi,\phi\rangle \, .$$

See quantum entanglement and the EPR paradox for applications of this product.

22.6 The unit operator

Consider a complete orthonormal system (*basis*), $\{e_i \mid i \in \mathbb{N}\}$, for a Hilbert space *H*, with respect to the norm from an inner product $\langle \cdot, \cdot \rangle$. From basic functional analysis we know that any ket $|\psi\rangle$ can also be written as

$$|\psi\rangle = \sum_{i \in \mathbb{N}} \langle e_i|\psi\rangle |e_i\rangle,$$

with $\langle \cdot|\cdot \rangle$ the inner product on the Hilbert space.

From the commutativity of kets with (complex) scalars now follows that

$$\sum_{i \in \mathbb{N}} |e_i\rangle\langle e_i| = \hat{1}$$

must be the identity operator, which sends each vector to itself. This can be inserted in any expression without affecting its value, for example

$$\langle v|w\rangle = \langle v|\sum_{i \in \mathbb{N}} |e_i\rangle\langle e_i|w\rangle = \langle v|\sum_{i \in \mathbb{N}} |e_i\rangle\langle e_i|\sum_{j \in \mathbb{N}} |e_j\rangle\langle e_j|w\rangle = \langle v|e_i\rangle\langle e_i|e_j\rangle\langle e_j|w\rangle$$

where, in the last identity, the Einstein summation convention has been used.

In quantum mechanics, it often occurs that little or no information about the inner product $\langle \psi|\phi\rangle$ of two arbitrary (state) kets is present, while it is still possible to say something about the expansion coefficients $\langle \psi|e_i\rangle = \langle e_i|\psi\rangle^*$ and $\langle e_i|\phi\rangle$ of those vectors with respect to a specific (orthonormalized) basis. In this case, it is particularly useful to insert the unit operator into the bracket one time or more.

For more information, see Resolution of the identity, $1 = \int dx\, |x\rangle\langle x| = \int dp\, |p\rangle\langle p|$, where $|p\rangle = \int dx\, e^{ixp/\hbar}|x\rangle/\sqrt{2\pi\hbar}$; since $\langle x'|x\rangle = \delta(x - x')$, plane waves follow, $\langle x|p\rangle = \exp(ixp/\hbar)/\sqrt{2\pi\hbar}$.

22.7 Notation used by mathematicians

The object physicists are considering when using the "bra-ket" notation is a Hilbert space (a complete inner product space).

Let $\mathcal{H}$ be a Hilbert space and $h \in \mathcal{H}$ is a vector in $\mathcal{H}$. What physicists would denote as $|h\rangle$ is the vector itself. That is

$$|h\rangle \in \mathcal{H}$$

Let $\mathcal{H}^*$ be the dual space of $\mathcal{H}$. This is the space of linear functionals on $\mathcal{H}$. The isomorphism $\Phi : \mathcal{H} \to \mathcal{H}^*$ is defined by $\Phi(h) = \phi_h$ where for all $g \in \mathcal{H}$ we have

$$\phi_h(g) = \mathrm{IP}(h, g) = (h, g) = \langle h, g\rangle = \langle h|g\rangle$$

where $\mathrm{IP}(\cdot,\cdot), (\cdot,\cdot), \langle \cdot,\cdot\rangle$ and $\langle \cdot|\cdot\rangle$ are just different notations for expressing an inner product between two elements in a Hilbert space (or for the first three, in *any* inner product space). Notational confusion arises when identifying ϕ_h and g with $\langle h|$ and $|g\rangle$ respectively. This is because of literal symbolic substitutions. Let $\phi_h = H = \langle h|$ and let $g = G = |g\rangle$. This gives

$$\phi_h(g) = H(g) = H(G) = \langle h|(G) = \langle h|(|g\rangle).$$

One ignores the parentheses and removes the double bars. Some properties of this notation are convenient since we are dealing with linear operators and composition acts like a ring multiplication.

Moreover, mathematicians usually write the dual entity not at the first place, as the physicists do, but at the second one, and they don't use the *-symbol, but an overline (which the physicists reserve for averages and Dirac conjugation) to denote conjugate-complex numbers, i.e. for scalar products mathematicians usually write

$$(\phi, \psi) = \int \phi(x) \cdot \overline{\psi(x)} \, \mathrm{d}x \,,$$

whereas physicists would write for the same quantity

$$\langle \psi | \phi \rangle = \int \mathrm{d}x \, \psi^*(x) \cdot \phi(x) \,.$$

22.8 See also

- Angular momentum diagrams (quantum mechanics)
- N-slit interferometric equation
- Quantum state
- Inner product

22.9 References and notes

[1] PAM Dirac (1939). "A new notation for quantum mechanics". *Mathematical Proceedings of the Cambridge Philosophical Society* **35** (3): 416–418. Bibcode:1939PCPS...35..416D. doi:10.1017/S0305004100021162.

[2] H. Grassmann (1862). *Extension Theory*. History of Mathematics Sources. American Mathematical Society, London Mathematical Society, 2000 translation by Lloyd C. Kannenberg.

[3] Cajori, Florian (1929). *A History Of Mathematical Notations Volume II*. Open Court Publishing. p. 134. ISBN 978-0-486-67766-8.

[4] McMahon, D. (2006). *Quantum Mechanics Demystified*. McGraw-Hill. ISBN 0-07-145546-9.

[5] Sakurai (1994). *Modern Quantum Mechanics* (Revised ed.). Addison-Wesley. p. 20. ISBN 0-201-53929-2.

[6] Carfì, David (April 2003). "Dirac-orthogonality in the space of tempered distributions". *Journal of Computational and Applied Mathematics* **153** (1–2): 99–107. Bibcode:2003JCoAM.153...99C. doi:10.1016/S0377-0427(02)00634-9.

[7] Carfì, David (April 2003). "Some properties of a new product in the space of tempered distributions". *Journal of Computational and Applied Mathematics* **153** (1–2): 109–118. Bibcode:2003JCoAM.153..109C. doi:10.1016/S0377-0427(02)00635-0.

[8] Carfì, David (2007). "TOPOLOGICAL CHARACTERIZATIONS OF S-LINEARITY". *AAPP-PHYSICAL, MATHEMATICAL AND NATURAL SCIENCES* **85** (2): 1–16. doi:10.1478/C1A0702005.

[9] Carfì, David (2005). "S-DIAGONALIZABLE OPERATORS IN QUANTUM MECHANICS". *Glasnik Matematicki* **40** (2): 261–301. doi:10.3336/gm.40.2.08.

[10] Lecture notes by Robert Littlejohn, eqns 12 and 13

22.10 Further reading

- Feynman, Leighton and Sands (1965). *The Feynman Lectures on Physics Vol. III*. Addison-Wesley. ISBN 0-201-02115-3.

22.11 External links

- Richard Fitzpatrick, “Quantum Mechanics: A graduate level course”, The University of Texas at Austin.
  - 1. Ket space
  - 2. Bra space
  - 3. Operators
  - 4. The outer product
  - 5. Eigenvalues and eigenvectors
- Robert Littlejohn, Lecture notes on “The Mathematical Formalism of Quantum mechanics”, including bra-ket notation.

Chapter 23

Lagrangian (field theory)

Lagrangian field theory is a formalism in classical field theory. It is the field theoretic analogue of Lagrangian mechanics. Lagrangian mechanics is used for discrete particles each with a finite number of degrees of freedom. Lagrangian field theory applies to continua and fields, which have an infinite number of degrees of freedom.

This article uses $\mathcal{L}$ for the Lagrangian density, and L for the Lagrangian.

The Lagrangian mechanics formalism was generalized further to handle field theory. In field theory, the independent variable is replaced by an event in spacetime (x, y, z, t), or more generally still by a point s on a manifold. The dependent variables (q) are replaced by the value of a field at that point in spacetime $\varphi(x, y, z, t)$ so that the equations of motion are obtained by means of an action principle, written as:

$$\frac{\delta \mathcal{S}}{\delta \varphi_i} = 0,$$

where the *action*, $\mathcal{S}$, is a functional of the dependent variables $\varphi i(s)$ with their derivatives and s itself

$$\mathcal{S}\left[\varphi_i\right] = \int \mathcal{L}\left(\varphi_i(s), \frac{\partial \varphi_i(s)}{\partial s^\alpha}, s^\alpha\right) \mathrm{d}^n s$$

and where $s = \{ s^\alpha \}$ denotes the set of n independent variables of the system, indexed by $\alpha = 1, 2, 3,..., n$. Notice L is used in the case of one independent variable (t) and $\mathcal{L}$ is used in the case of multiple independent variables (usually four: x, y, z, t).

23.1 Definitions

In Lagrangian field theory, the Lagrangian as a function of generalized coordinates is replaced by a Lagrangian density, a function of the fields in the system and their derivatives, and possibly the space and time coordinates themselves. In field theory, the independent variable t is replaced by an event in spacetime (x, y, z, t) or still more generally by a point s on a manifold.

Often, a "Lagrangian density" is simply referred to as a "Lagrangian".

23.1.1 Scalar fields

For one scalar field φ, the Lagrangian density will take the form:[nb 1][1]

$$\mathcal{L}(\phi, \nabla\phi, \partial\phi/\partial t, \mathbf{x}, t)$$

For many scalar fields

$$\mathcal{L}(\phi_1, \nabla\phi_1, \partial\phi_1/\partial t, \ldots, \phi_2, \nabla\phi_2, \partial\phi_2/\partial t, \ldots, \mathbf{x}, t)$$

23.1.2 Vector fields, tensor fields, spinor fields

The above can be generalized for vector fields, tensor fields, and spinor fields. In physics fermions are described by spinor fields and bosons by tensor fields.

23.1.3 Action

The time integral of the Lagrangian is called the action denoted by *S*. In field theory, a distinction is occasionally made between the **Lagrangian** *L*, of which the time integral is the action

$$\mathcal{S} = \int L \, \mathrm{d}t \,,$$

and the **Lagrangian density** $\mathcal{L}$, which one integrates over all spacetime to get the action:

$$\mathcal{S}[\phi] = \int \mathcal{L}(\phi, \nabla\phi, \partial\phi/\partial t, \mathbf{x}, t) \, \mathrm{d}^3\mathbf{x}\mathrm{d}t.$$

The spatial volume integral of the Lagrangian density is the Lagrangian, in 3d

$$L = \int \mathcal{L} \, d^3x \,.$$

Quantum field theories in particle physics, such as quantum electrodynamics, are usually described in terms of $\mathcal{L}$, and the terms in this form of the Lagrangian translate quickly to the rules used in evaluating Feynman diagrams.

Notice that, in the presence of gravity or when using general curvilinear coordinates, the Lagrangian density $\mathcal{L}$ will include a factor of $\sqrt{g}$ or its equivalent to ensure that it is a scalar density so that the integral will be invariant.

23.1.4 Mathematical formalism

Suppose we have an *n*-dimensional manifold, *M*, and a target manifold, *T*. Let $\mathcal{C}$ be the configuration space of smooth functions from *M* to *T*.

In field theory, *M* is the spacetime manifold and the target space is the set of values the fields can take at any given point. For example, if there are m real-valued scalar fields, ϕ_1, ..., ϕm, then the target manifold is $\mathbb{R}^m$. If the field is a real vector field, then the target manifold is isomorphic to $\mathbb{R}^n$. There is actually a much more elegant way using tangent bundles over *M*, but we will just stick to this version.

Consider a functional,

$$\mathcal{S} : \mathcal{C} \to \mathbb{R}$$

called the action. Physical considerations require it be a mapping to $\mathbb{R}$ (the set of all real numbers), not $\mathbb{C}$ (the set of all complex numbers).

In order for the action to be local, we need additional restrictions on the action. If $\varphi \in \mathcal{C}$, we assume $\mathcal{S}[\varphi]$ is the integral over *M* of a function of φ, its derivatives and the position called the **Lagrangian**, $\mathcal{L}(\varphi, \partial\varphi, \partial\partial\varphi, ..., x)$. In other words,

$$\forall\varphi \in \mathcal{C}, \ \ \mathcal{S}[\varphi] \equiv \int_M \mathrm{d}^n x \mathcal{L}\big(\varphi(x), \partial\varphi(x), \partial\partial\varphi(x), ..., x\big).$$

It is assumed below, in addition, that the Lagrangian depends on only the field value and its first derivative but not the higher derivatives.

Given boundary conditions, basically a specification of the value of φ at the boundary if *M* is compact or some limit on φ as $x \to \infty$ (this will help in doing integration by parts), the subspace of $\mathcal{C}$ consisting of functions, φ, such that all functional derivatives of *S* at φ are zero and φ satisfies the given boundary conditions is the subspace of on shell solutions.

The solution is given by the Euler–Lagrange equations (thanks to the boundary conditions),

$$\frac{\delta\mathcal{S}}{\delta\varphi} = -\partial_\mu \left(\frac{\partial\mathcal{L}}{\partial(\partial_\mu\varphi)} \right) + \frac{\partial\mathcal{L}}{\partial\varphi} = 0.$$

The left hand side is the functional derivative of the action with respect to φ.

23.2 Examples

To go with the section on test particles above, here are the equations for the fields in which they move. The equations below pertain to the fields in which the test particles described above move and allow the calculation of those fields. The equations below will not give you the equations of motion of a test particle in the field but will instead give you the potential (field) induced by quantities such as mass or charge density at any point $(\mathbf{x}, t)$. For example, in the case of Newtonian gravity, the Lagrangian density integrated over spacetime gives you an equation which, if solved, would yield $\Phi(\mathbf{x}, t)$. This $\Phi(\mathbf{x}, t)$, when substituted back in equation (**1**), the Lagrangian equation for the test particle in a Newtonian gravitational field, provides the information needed to calculate the acceleration of the particle.

23.2.1 Newtonian gravity

The Lagrangian (density) is $\mathcal{L}$ in J·m^{-3}. The interaction term $m\Phi$ is replaced by a term involving a continuous mass density ρ in kg·m^{-3}. This is necessary because using a point source for a field would result in mathematical difficulties. The resulting Lagrangian for the classical gravitational field is:

$$\mathcal{L}(\mathbf{x}, t) = -\rho(\mathbf{x}, t)\Phi(\mathbf{x}, t) - \frac{1}{8\pi G}(\nabla\Phi(\mathbf{x}, t))^2$$

where *G* in m^3·kg^{-1}·s^{-2} is the gravitational constant. Variation of the integral with respect to Φ gives:

$$\delta\mathcal{L}(\mathbf{x}, t) = -\rho(\mathbf{x}, t)\delta\Phi(\mathbf{x}, t) - \frac{2}{8\pi G}(\nabla\Phi(\mathbf{x}, t)) \cdot (\nabla\delta\Phi(\mathbf{x}, t)).$$

Integrate by parts and discard the total integral. Then divide out by $\delta\Phi$ to get:

$$0 = -\rho(\mathbf{x}, t) + \frac{1}{4\pi G}\nabla \cdot \nabla\Phi(\mathbf{x}, t)$$

and thus

$$4\pi G\rho(\mathbf{x},t) = \nabla^2\Phi(\mathbf{x},t)$$

which yields Gauss's law for gravity.

23.2.2 Einstein Gravity

Main article: Einstein–Hilbert action

The Lagrange density for general relativity in the presence of matter fields is

$$\mathcal{L}_{\text{GR}} = \mathcal{L}_{\text{EH}} + \mathcal{L}_{\text{matter}} = \frac{c^4}{16\pi G}(R - 2\Lambda) + \mathcal{L}_{\text{matter}}$$

R is the curvature scalar, which is the Ricci tensor contracted with the metric tensor, and the Ricci tensor is the Riemann tensor contracted with a Kronecker delta. The integral of $\mathcal{L}_{\text{EH}}$ is known as the Einstein-Hilbert action. The Riemann tensor is the tidal force tensor, and is constructed out of Christoffel symbols and derivatives of Christoffel symbols, which are the gravitational force field. Plugging this Lagrangian into the Euler-Lagrange equation and taking the metric tensor $g_{\mu\nu}$ as the field, we obtain the Einstein field equations

$$R_{\mu\nu} - \frac{1}{2}Rg_{\mu\nu} + g_{\mu\nu}\Lambda = \frac{8\pi G}{c^4}T_{\mu\nu}$$

The last tensor is the energy momentum tensor and is defined by

$$T_{\mu\nu} \equiv \frac{-2}{\sqrt{-g}}\frac{\delta(\mathcal{L}_{\text{matter}}\sqrt{-g})}{\delta g^{\mu\nu}} = -2\frac{\delta\mathcal{L}_{\text{matter}}}{\delta g^{\mu\nu}} + g_{\mu\nu}\mathcal{L}_{\text{matter}}.$$

g is the determinant of the metric tensor when regarded as a matrix. Λ is the Cosmological constant. Generally, in general relativity, the integration measure of the action of Lagrange density is $\sqrt{-g}d^4x$. This makes the integral coordinate independent, as the root of the metric determinant is equivalent to the Jacobian determinant. The minus sign is a consequence of the metric signature (the determinant by itself is negative).[2]

23.2.3 Electromagnetism in special relativity

The interaction terms

$$-q\phi(\mathbf{x}(t),t) + q\dot{\mathbf{x}}(t)\cdot\mathbf{A}(\mathbf{x}(t),t)$$

are replaced by terms involving a continuous charge density ρ in A·s·m^{-3} and current density $\mathbf{j}$ in A·m^{-2}. The resulting Lagrangian for the electromagnetic field is:

$$\mathcal{L}(\mathbf{x},t) = -\rho(\mathbf{x},t)\phi(\mathbf{x},t) + \mathbf{j}(\mathbf{x},t)\cdot\mathbf{A}(\mathbf{x},t) + \frac{\epsilon_0}{2}E^2(\mathbf{x},t) - \frac{1}{2\mu_0}B^2(\mathbf{x},t).$$

Varying this with respect to ϕ, we get

$$0 = -\rho(\mathbf{x}, t) + \epsilon_0 \nabla \cdot \mathbf{E}(\mathbf{x}, t)$$

which yields Gauss' law.

Varying instead with respect to $\mathbf{A}$, we get

$$0 = \mathbf{j}(\mathbf{x}, t) + \epsilon_0 \dot{\mathbf{E}}(\mathbf{x}, t) - \frac{1}{\mu_0} \nabla \times \mathbf{B}(\mathbf{x}, t)$$

which yields Ampère's law.

Using tensor notation, we can write all this more compactly. The term $-\rho\phi(\mathbf{x}, t) + \mathbf{j} \cdot \mathbf{A}$ is actually the inner product of two four-vectors. We package the charge density into the current 4-vector and the potential into the potential 4-vector. These two new vectors are

$$j^\mu = (\rho, \mathbf{j}) \quad \text{and} \quad A_\mu = (-\phi, \mathbf{A})$$

We can then write the interaction term as

$$-\rho\phi + \mathbf{j} \cdot \mathbf{A} = j^\mu A_\mu$$

Additionally, we can package the E and B fields into what is known as the electromagnetic tensor $F_{\mu\nu}$. We define this tensor as

$$F_{\mu\nu} = \partial_\mu A_\nu - \partial_\nu A_\mu$$

The term we are looking out for turns out to be

$$\frac{\epsilon_0}{2} E^2 - \frac{1}{2\mu_0} B^2 = -\frac{1}{4\mu_0} F_{\mu\nu} F^{\mu\nu} = -\frac{1}{4\mu_0} F_{\mu\nu} F_{\rho\sigma} \eta^{\mu\rho} \eta^{\nu\sigma}$$

We have made use of the Minkowski metric to raise the indices on the EMF tensor. In this notation, Maxwell's equations are

$$\partial_\mu F^{\mu\nu} = -\mu_0 j^\nu \quad \text{and} \quad \epsilon^{\mu\nu\lambda\sigma} \partial_\nu F_{\lambda\sigma} = 0$$

where ε is the Levi-Civita tensor. So the Lagrange density for electromagnetism in special relativity written in terms of Lorentz vectors and tensors is

$$\mathcal{L}(x) = j^\mu(x) A_\mu(x) - \frac{1}{4\mu_0} F_{\mu\nu}(x) F^{\mu\nu}(x)$$

In this notation it is apparent that classical electromagnetism is a Lorentz-invariant theory. By the equivalence principle, it becomes simple to extend the notion of electromagnetism to curved spacetime.[3][4]

23.2.4 Electromagnetism in general relativity

The Lagrange density of electromagnetism in general relativity also contains the Einstein-Hilbert action from above. The pure electromagnetic Lagrangian is precisely a matter Lagrangian $\mathcal{L}_{\text{matter}}$. The Lagrangian is

$$\begin{aligned}\mathcal{L}(x) &= j^\mu(x)A_\mu(x) - \frac{1}{4\mu_0}F_{\mu\nu}(x)F_{\rho\sigma}(x)g^{\mu\rho}(x)g^{\nu\sigma}(x) + \frac{c^4}{16\pi G}R(x)\\ &= \mathcal{L}_{\text{Maxwell}} + \mathcal{L}_{\text{Einstein-Hilbert}}.\end{aligned}$$

This Lagrangian is obtained by simply replacing the Minkowski metric in the above flat Lagrangian with a more general (possibly curved) metric $g_{\mu\nu}(x)$. We can generate the Einstein Field Equations in the presence of an EM field using this lagrangian. The energy-momentum tensor is

$$T^{\mu\nu}(x) = \frac{2}{\sqrt{-g(x)}}\frac{\delta}{\delta g_{\mu\nu}(x)}\mathcal{S}_{\text{Maxwell}} = \frac{1}{\mu_0}\left(F^\mu_{\ \lambda}(x)F^{\nu\lambda}(x) - \frac{1}{4}g^{\mu\nu}(x)F_{\rho\sigma}(x)F^{\rho\sigma}(x)\right)$$

It can be shown that this energy momentum tensor is traceless, i.e. that

$$T = g_{\mu\nu}T^{\mu\nu} = 0$$

If we take the trace of both sides of the Einstein Field Equations, we obtain

$$R = -\frac{8\pi G}{c^4}T$$

So the tracelessness of the energy momentum tensor implies that the curvature scalar in an electromagnetic field vanishes. The Einstein equations are then

$$R^{\mu\nu} = \frac{8\pi G}{c^4}\frac{1}{\mu_0}\left(F^\mu_{\ \lambda}(x)F^{\nu\lambda}(x) - \frac{1}{4}g^{\mu\nu}(x)F_{\rho\sigma}(x)F^{\rho\sigma}(x)\right)$$

Additionally, Maxwell's equations are

$$D_\mu F^{\mu\nu} = -\mu_0 j^\nu$$

where D_μ is the covariant derivative. For free space, we can set the current tensor equal to zero, $j^\mu = 0$. Solving both Einstein and Maxwell's equations around a spherically symmetric mass distribution in free space leads to the Reissner-Nordstrom charged black hole, with the defining line element (written in natural units and with charge Q):[5]

$$ds^2 = \left(1 - \frac{2M}{r} + \frac{Q^2}{r^2}\right)dt^2 - \left(1 - \frac{2M}{r} + \frac{Q^2}{r^2}\right)^{-1}dr^2 - r^2 d\Omega^2$$

23.2.5 Electromagnetism using differential forms

Using differential forms, the electromagnetic action S in vacuum on a (pseudo-) Riemannian manifold $\mathcal{M}$ can be written (using natural units, $c = \varepsilon_0 = 1$) as

$$\mathcal{S}[\mathbf{A}] = \int_{\mathcal{M}} \left(-\frac{1}{2}\mathbf{F} \wedge \star\mathbf{F} + \mathbf{A} \wedge \star\mathbf{J} \right).$$

Here, **A** stands for the electromagnetic potential 1-form, **J** is the current 1-form, **F** is the field strength 2-form and the star denotes the Hodge star operator. This is exactly the same Lagrangian as in the section above, except that the treatment here is coordinate-free; expanding the integrand into a basis yields the identical, lengthy expression. Note that with forms, an additional integration measure is not necessary because forms have coordinate differentials built in. Variation of the action leads to

$$\mathrm{d}\star\mathbf{F} = \mathbf{J}.$$

These are Maxwell's equations for the electromagnetic potential. Substituting **F** = d**A** immediately yields the equation for the fields,

$$\mathrm{d}\mathbf{F} = 0$$

because **F** is an exact form.

23.2.6 Dirac Lagrangian

The Lagrangian density for a Dirac field is:[6]

$$\mathcal{L} = i\hbar c\bar{\psi}\partial\!\!\!/\,\psi - mc^2\bar{\psi}\psi$$

where ψ is a Dirac spinor (annihilation operator), $\bar{\psi}=\psi^\dagger\gamma^0$ is its Dirac adjoint (creation operator), and $\partial\!\!\!/$ is Feynman slash notation for $\gamma^\sigma\partial_\sigma$.

23.2.7 Quantum electrodynamic Lagrangian

The Lagrangian density for QED is:

$$\mathcal{L}_{\mathrm{QED}} = i\hbar c\bar{\psi}D\!\!\!\!/\,\psi - mc^2\bar{\psi}\psi - \frac{1}{4\mu_0}F_{\mu\nu}F^{\mu\nu}$$

where $F^{\mu\nu}$ is the electromagnetic tensor, D is the gauge covariant derivative, and $D\!\!\!\!/$ is Feynman notation for $\gamma^\sigma D_\sigma$.

23.2.8 Quantum chromodynamic Lagrangian

The Lagrangian density for quantum chromodynamics is:[7][8][9]

$$\mathcal{L}_{\mathrm{QCD}} = \sum_n \left(i\hbar c\bar{\psi}_n D\!\!\!\!/\,\psi_n - m_n c^2\bar{\psi}_n\psi_n \right) - \frac{1}{4}G^\alpha{}_{\mu\nu}G_\alpha{}^{\mu\nu}$$

where D is the QCD gauge covariant derivative, n = 1, 2, ...6 counts the quark types, and $G^\alpha{}_{\mu\nu}$ is the gluon field strength tensor.

23.3 See also

23.4 Footnotes

[1] It is a standard abuse of notation to abbreviate all the derivatives and coordinates in the Lagrangian density as follows:

$$\mathcal{L}(\phi, \partial_\mu \phi, x_\mu)$$

see four gradient. The μ is an index which takes values 0 (for the time coordinate), and 1, 2, 3 (for the spatial coordinates), so strictly only one derivative or coordinate would be present. In general, all the spatial and time derivatives will appear in the Lagrangian density, for example in Cartesian coordinates, the Lagrangian density has the full form:

$$\mathcal{L}\left(\phi, \frac{\partial \phi}{\partial x}, \frac{\partial \phi}{\partial y}, \frac{\partial \phi}{\partial z}, \frac{\partial \phi}{\partial t}, x, y, z, t\right)$$

Here we write the same thing, but using ∇ to abbreviate all spatial derivatives as a vector.

23.5 Notes

[1] Mandl F., Shaw G., *Quantum Field Theory*, chapter 2

[2] Zee, A. (2013). *Einstein gravity in a nutshell.* Princeton: Princeton University Press. pp. 344–390. ISBN 9780691145587.

[3] Zee, A. (2013). *Einstein gravity in a nutshell.* Princeton: Princeton University Press. pp. 244–253. ISBN 9780691145587.

[4] Mexico, Kevin Cahill, University of New (2013). *Physical mathematics* (Repr. ed.). Cambridge: Cambridge University Press. ISBN 9781107005211.

[5] Zee, A. (2013). *Einstein gravity in a nutshell.* Princeton: Princeton University Press. pp. 381–383, 477–478. ISBN 9780691145587.

[6] Itzykson-Zuber, eq. 3-152

[7] http://www.fuw.edu.pl/~{}dobaczew/maub-42w/node9.html

[8] http://smallsystems.isn-oldenburg.de/Docs/THEO3/publications/semiclassical.qcd.prep.pdf

[9] http://www-zeus.physik.uni-bonn.de/~{}brock/teaching/jets_ws0405/seminar09/sluka_quark_gluon_jets.pdf

Chapter 24

Principle of least action

This article discusses the history of the principle of least action. For the application, please refer to action (physics).

In non-relativistic physics, the **principle of least action** – or, more accurately, the **principle of stationary action** – is a variational principle that, when applied to the action of a mechanical system, can be used to obtain the equations of motion for that system by stating a system follows the path where the average difference between the kinetic energy and potential energy is minimized or maximized over any time period. It is called stable if minimized. In relativity, a different average must be minimized or maximized. The principle can be used to derive Newtonian, Lagrangian, and Hamiltonian equations of motion. It was historically called "least" because its solution requires finding the path that has the least change from nearby paths.[1] Its classical mechanics and electromagnetic expressions are a consequence of quantum mechanics, but the stationary action method helped in the development of quantum mechanics. [2]

The principle remains central in modern physics and mathematics, being applied in the theory of relativity, quantum mechanics and quantum field theory, and a focus of modern mathematical investigation in Morse theory. Maupertuis' principle and Hamilton's principle exemplify the principle of stationary action.

The action principle is preceded by earlier ideas in surveying and optics. Rope stretchers in ancient Egypt stretched corded ropes to measure the distance between two points. Ptolemy, in his *Geography* (Bk 1, Ch 2), emphasized that one must correct for "deviations from a straight course". In ancient Greece, Euclid wrote in his *Catoptrica* that, for the path of light reflecting from a mirror, the angle of incidence equals the angle of reflection. Hero of Alexandria later showed that this path was the shortest length and least time.[3]

Scholars often credit Pierre Louis Maupertuis for formulating the principle of least action because he wrote about it in 1744[4] and 1746.[5] However, Leonhard Euler discussed the principle in 1744,[6] and evidence shows that Gottfried Leibniz preceded both by 39 years.[7][8][9]

In 1932, Paul Dirac discerned the quantum mechanical underpinning of the principle in the quantum interference of amplitudes:For macroscopic systems, the dominant contribution to the apparent path is the classical path (the stationary, action-extremizing one), while any other path is possible in the quantum realm.

24.1 General statement

The starting point is the *action*, denoted $\mathcal{S}$ (calligraphic S), of a physical system. It is defined as the integral of the Lagrangian L between two instants of time t_1 and t_2 - technically a functional of the N generalized coordinates $\mathbf{q} = (q_1, q_2 \ldots qN)$ which define the configuration of the system:

$$\mathcal{S}[\mathbf{q}(t)] = \int_{t_1}^{t_2} L(\mathbf{q}(t), \mathbf{q}(t), t)dt$$

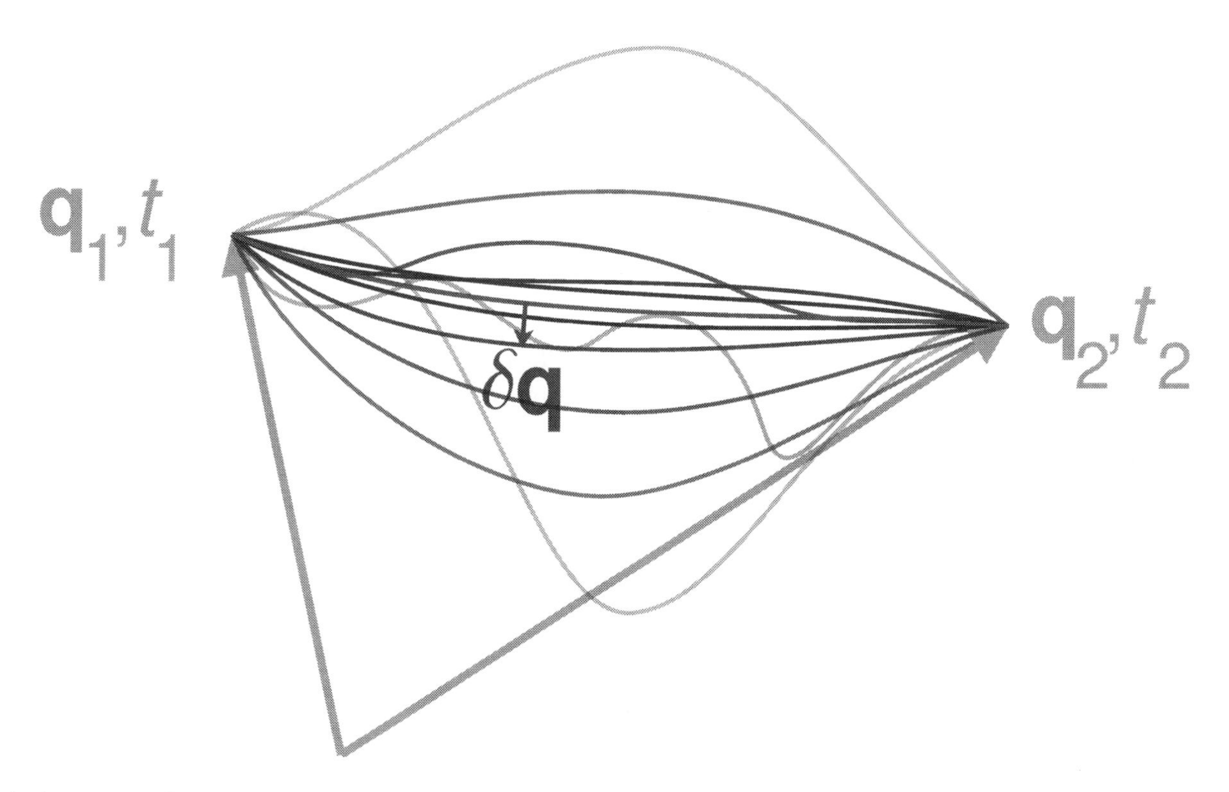

As the system evolves, ***q*** *traces a path through configuration space (only some are shown). The path taken by the system (red) has a stationary action (δS = 0) under small changes in the configuration of the system (δ**q**).*[10]

where the dot denotes the time derivative, and t is time.

Mathematically the principle is[11][12][13]

$$\delta \mathcal{S} = 0$$

where δ (Greek lowercase delta) means a *small* change. In words this reads:[10]

> *The path taken by the system between times t_1 and t_2 is the one for which the* ***action*** *is* ***stationary (no change)*** *to* ***first order****.*

In applications the statement and definition of action are taken together:[14]

$$\delta \int_{t_1}^{t_2} L(\mathbf{q}, \mathbf{q}, t) dt = 0$$

The action and Lagrangian both contain the dynamics of the system for all times. The term "path" simply refers to a curve traced out by the system in terms of the coordinates in the configuration space, i.e. the curve $\mathbf{q}(t)$, parameterized by time (see also parametric equation for this concept).

24.2 Origins, statements, and controversy

24.2.1 Fermat

Main article: Fermat's principle

In the 1600s, Pierre de Fermat postulated that "*light travels between two given points along the path of shortest time*," which is known as the **principle of least time** or **Fermat's principle**.[13]

24.2.2 Maupertuis

Main article: Maupertuis principle

Credit for the formulation of the **principle of least action** is commonly given to Pierre Louis Maupertuis, who felt that "Nature is thrifty in all its actions", and applied the principle broadly:

> The laws of movement and of rest deduced from this principle being precisely the same as those observed in nature, we can admire the application of it to all phenomena. The movement of animals, the vegetative growth of plants ... are only its consequences; and the spectacle of the universe becomes so much the grander, so much more beautiful, the worthier of its Author, when one knows that a small number of laws, most wisely established, suffice for all movements.
> —Pierre Louis Maupertuis[15]

This notion of Maupertuis, although somewhat deterministic today, does capture much of the essence of mechanics.

In application to physics, Maupertuis suggested that the quantity to be minimized was the product of the duration (time) of movement within a system by the "vis viva",

which is the integral of twice what we now call the kinetic energy T of the system.

24.2.3 Euler

Leonhard Euler gave a formulation of the action principle in 1744, in very recognizable terms, in the *Additamentum 2* to his *Methodus Inveniendi Lineas Curvas Maximi Minive Proprietate Gaudentes*. Beginning with the second paragraph:

As Euler states, $\int Mvds$ is the integral of the momentum over distance travelled, which, in modern notation, equals the reduced action

Thus, Euler made an equivalent and (apparently) independent statement of the variational principle in the same year as Maupertuis, albeit slightly later. Curiously, Euler did not claim any priority, as the following episode shows.

24.2.4 Disputed priority

Maupertuis' priority was disputed in 1751 by the mathematician Samuel König, who claimed that it had been invented by Gottfried Leibniz in 1707. Although similar to many of Leibniz's arguments, the principle itself has not been documented in Leibniz's works. König himself showed a *copy* of a 1707 letter from Leibniz to Jacob Hermann with the principle, but the *original* letter has been lost. In contentious proceedings, König was accused of forgery,[7] and even the King of Prussia entered the debate, defending Maupertuis (the head of his Academy), while Voltaire defended König.

Euler, rather than claiming priority, was a staunch defender of Maupertuis, and Euler himself prosecuted König for forgery before the Berlin Academy on 13 April 1752.[7] The claims of forgery were re-examined 150 years later, and archival work by C.I. Gerhardt in 1898[8] and W. Kabitz in 1913[9] uncovered other copies of the letter, and three others cited by König, in the Bernoulli archives.

24.3 Further development

Euler continued to write on the topic; in his *Reflexions sur quelques loix generales de la nature* (1748), he called the quantity "effort". His expression corresponds to what we would now call potential energy, so that his statement of least action in statics is equivalent to the principle that a system of bodies at rest will adopt a configuration that minimizes total potential energy.

24.3.1 Lagrange and Hamilton

Main article: Hamilton's principle

Much of the calculus of variations was stated by Joseph-Louis Lagrange in 1760[17][18] and he proceeded to apply this to problems in dynamics. In *Méchanique Analytique* (1788) Lagrange derived the general equations of motion of a mechanical body.[19] William Rowan Hamilton in 1834 and 1835[20] applied the variational principle to the classical Lagrangian function

$$L = T - V$$

to obtain the Euler–Lagrange equations in their present form.

24.3.2 Jacobi and Morse

In 1842, Carl Gustav Jacobi tackled the problem of whether the variational principle always found minima as opposed to other stationary points (maxima or stationary saddle points); most of his work focused on geodesics on two-dimensional surfaces.[21] The first clear general statements were given by Marston Morse in the 1920s and 1930s,[22] leading to what is now known as Morse theory. For example, Morse showed that the number of conjugate points in a trajectory equalled the number of negative eigenvalues in the second variation of the Lagrangian.

24.3.3 Gauss and Hertz

Other extremal principles of classical mechanics have been formulated, such as Gauss's principle of least constraint and its corollary, Hertz's principle of least curvature.

24.4 Disputes about possible teleological aspects

The mathematical equivalence of the differential equations of motion and their integral counterpart has important philosophical implications. The differential equations are statements about quantities localized to a single point in space or single moment of time. For example, Newton's second law

$$\mathbf{F} = m\mathbf{a}$$

states that the *instantaneous* force **F** applied to a mass *m* produces an acceleration **a** at the same *instant*. By contrast, the action principle is not localized to a point; rather, it involves integrals over an interval of time and (for fields) an extended region of space. Moreover, in the usual formulation of classical action principles, the initial and final states of the system are fixed, e.g.,

> *Given that the particle begins at position x_1 at time t_1 and ends at position x_2 at time t_2, the physical trajectory that connects these two endpoints is an extremum of the action integral.*

In particular, the fixing of the *final* state has been interpreted as giving the action principle a teleological character which has been controversial historically. However, according to W. Yourgrau and S. Mandelstam, *the teleological approach... presupposes that the variational principles themselves have mathematical characteristics which they* de facto *do not possess*[23] In addition, some critics maintain this apparent teleology occurs because of the way in which the question was asked. By specifying some but not all aspects of both the initial and final conditions (the positions but not the velocities) we are making some inferences about the initial conditions from the final conditions, and it is this "backward" inference that can be seen as a teleological explanation. Teleology can also be overcome if we consider the classical description as a limiting case of the quantum formalism of path integration, in which stationary paths are obtained as a result of interference of amplitudes along all possible paths.

The short story *Story of Your Life* by the speculative fiction writer Ted Chiang contains visual depictions of Fermat's Principle along with a discussion of its teleological dimension. Keith Devlin's *The Math Instinct* contains a chapter, "Elvis the Welsh Corgi Who Can Do Calculus" that discusses the calculus "embedded" in some animals as they solve the "least time" problem in actual situations.

24.5 More Fundamental Than Newton's 2nd Law

According to Richard Feynman, the principle of least action is mathematically more specific than Newton's 2nd law and more fundamental in theoretical physics because it explains a wider range of physical law. You can derive Newton's 2nd law from least action, but the converse is not true without also applying Newton's 1st and 3rd laws and disallowing non-conservative forces like friction. By being more specific and thereby explaining only conservative forces, the principle of least action is able to solve problems Newton's 2nd law can't, but the converse is not true. The principle of least action can be used to derive the conservation of momentum and energy if its symmetry in space and time are assumed.[24] It correctly does not allow non-conservative potential fields, but Newton's 2nd law allows for them by allowing for non-conservative momentums and forces (such as friction) which are not fundamental forces.[25] The mathematical basis for the difference is that Newton's 2nd law (correctly stated as $F=dp/dt$ instead of $F=ma$) allows for momentums $p(t)=q(t)+C$ where $q(t)$ are conserved momentums allowed by least action and C is a constant that can be non-zero in Newton's 2nd law but not in least action. The constant allows for non-conservative momentums and therefore non-conservative forces and potentials in Newton's 2nd law. Newton's 2nd law explains conservation of energy and momentum and can be used to show equivalency with least action when forces are properly conserved, e.g. when forces are summed to zero in accordance with Newton's 1st and 3rd laws and when accounting for heat generated by friction. Derivations of Lagrangian and Hamiltonian methods do not begin with Newton's 2nd law, but with a more modern mathematical formulation of it that requires forces to be conservative.

24.6 See also

- Action (physics)
- Analytical mechanics
- Calculus of variations
- Hamiltonian mechanics
- Lagrangian mechanics
- Occam's razor
- Path of least resistance

24.7 Notes and references

[1] Chapter 19 of Volume II, Feynman R, Leighton R, and Sands M. *The Feynman Lectures on Physics* . 3 volumes 1964, 1966. Library of Congress Catalog Card No. 63-20717. ISBN 0-201-02115-3 (1970 paperback three-volume set); ISBN 0-201-50064-7 (1989 commemorative hardcover three-volume set); ISBN 0-8053-9045-6 (2006 the definitive edition (2nd printing); hardcover)

[2] "The Character of Physical Law" Richard Feynman

[3] Kline, Morris (1972). *Mathematical Thought from Ancient to Modern Times.* New York: Oxford University Press. pp. 167–68. ISBN 0-19-501496-0.

[4] P.L.M. de Maupertuis, *Accord de différentes lois de la nature qui avaient jusqu'ici paru incompatibles.* (1744) Mém. As. Sc. Paris p. 417. (English translation)

[5] P.L.M. de Maupertuis, *Le lois de mouvement et du repos, déduites d'un principe de métaphysique.* (1746) Mém. Ac. Berlin, p. 267.(English translation)

[6] Leonhard Euler, *Methodus Inveniendi Lineas Curvas Maximi Minive Proprietate Gaudentes.* (1744) Bousquet, Lausanne & Geneva. 320 pages. Reprinted in *Leonhardi Euleri Opera Omnia: Series I vol 24.* (1952) C. Cartheodory (ed.) Orell Fuessli, Zurich. scanned copy of complete text at *The Euler Archive*, Dartmouth.

[7] J J O'Connor and E F Robertson, "The Berlin Academy and forgery", (2003), at *The MacTutor History of Mathematics archive.*

[8] Gerhardt CI. (1898) "Über die vier Briefe von Leibniz, die Samuel König in dem Appel au public, Leide MDCCLIII, veröffentlicht hat", *Sitzungsberichte der Königlich Preussischen Akademie der Wissenschaften*, **I**, 419-427.

[9] Kabitz W. (1913) "Über eine in Gotha aufgefundene Abschrift des von S. König in seinem Streite mit Maupertuis und der Akademie veröffentlichten, seinerzeit für unecht erklärten Leibnizbriefes", *Sitzungsberichte der Königlich Preussischen Akademie der Wissenschaften*, **II**, 632-638.

[10] R. Penrose (2007). *The Road to Reality.* Vintage books. p. 474. ISBN 0-679-77631-1.

[11] Encyclopaedia of Physics (2nd Edition), R.G. Lerner, G.L. Trigg, VHC publishers, 1991, ISBN (Verlagsgesellschaft) 3-527-26954-1, ISBN (VHC Inc.) 0-89573-752-3

[12] McGraw Hill Encyclopaedia of Physics (2nd Edition), C.B. Parker, 1994, ISBN 0-07-051400-3

[13] Analytical Mechanics, L.N. Hand, J.D. Finch, Cambridge University Press, 2008, ISBN 978-0-521-57572-0

[14] Classical Mechanics, T.W.B. Kibble, European Physics Series, McGraw-Hill (UK), 1973, ISBN 0-07-084018-0

[15] Chris Davis. *Idle theory* (1998)

[16] Euler, Additamentum II (external link), ibid. (English translation)

[17] D. J. Struik, ed. (1969). *A Source Book in Mathematics, 1200-1800.* Cambridge, Mass: MIT Press. pp. 406-413

[18] Kline, Morris (1972). *Mathematical Thought from Ancient to Modern Times.* New York: Oxford University Press. ISBN 0-19-501496-0. pp. 582-589

[19] Lagrange, Joseph-Louis (1788). *Mécanique Analytique*. p. 226

[20] W. R. Hamilton, "On a General Method in Dynamics", *Philosophical Transaction of the Royal Society* Part I (1834) p.247-308; Part II (1835) p. 95-144. (*From the collection Sir William Rowan Hamilton (1805-1865): Mathematical Papers edited by David R. Wilkins, School of Mathematics, Trinity College, Dublin 2, Ireland. (2000); also reviewed as On a General Method in Dynamics*)

[21] G.C.J. Jacobi, *Vorlesungen über Dynamik, gehalten an der Universität Königsberg im Wintersemester 1842-1843.* A. Clebsch (ed.) (1866); Reimer; Berlin. 290 pages, available online Œuvres complètes volume **8** at Gallica-Math from the Gallica Bibliothèque nationale de France.

[22] Marston Morse (1934). "The Calculus of Variations in the Large", *American Mathematical Society Colloquium Publication* **18**; New York.

[23] Stöltzner, Michael (1994). *Inside Versus Outside: Action Principles and Teleology*. Springer. pp. 33–62. ISBN 978-3-642-48649-4.

[24] "The Character of Physical Law" Richard Feynman

[25] "The Principle of Least Action" Richard Feynman

24.8 External links

- Interactive explanation of the principle of least action
- Interactive applet to construct trajectories using principle of least action
- Georgi Yordanov Georgiev 2012 , A quantitative measure, mechanism and attractor for self-organization in networked complex systems, in Lecture Notes in Computer Science (LNCS 7166), F.A. Kuipers and P.E. Heegaard (Eds.): IFIP International Federation for Information Processing, Proceedings of the Sixth International Workshop on Self-Organizing Systems (IWSOS 2012), pp. 90–95, Springer-Verlag (2012).
- Georgi Yordanov Georgiev and Iskren Yordanov Georgiev 2002 , The least action and the metric of an organized system, in Open Systems and Information Dynamics, 9(4), p. 371-380 (2002)

Chapter 25

Gauge theory

For a more accessible and less technical introduction to this topic, see Introduction to gauge theory.

In physics, a **gauge theory** is a type of field theory in which the Lagrangian is invariant under a continuous group of local transformations.

The term *gauge* refers to redundant degrees of freedom in the Lagrangian. The transformations between possible gauges, called *gauge transformations*, form a Lie group—referred to as the *symmetry group* or the *gauge group* of the theory. Associated with any Lie group is the Lie algebra of group generators. For each group generator there necessarily arises a corresponding vector field called the *gauge field*. Gauge fields are included in the Lagrangian to ensure its invariance under the local group transformations (called *gauge invariance*). When such a theory is quantized, the quanta of the gauge fields are called *gauge bosons*. If the symmetry group is non-commutative, the gauge theory is referred to as *non-abelian*, the usual example being the Yang–Mills theory.

Many powerful theories in physics are described by Lagrangians that are invariant under some symmetry transformation groups. When they are invariant under a transformation identically performed at *every* point in the space in which the physical processes occur, they are said to have a global symmetry. The requirement of local symmetry, the cornerstone of gauge theories, is a stricter constraint. In fact, a global symmetry is just a local symmetry whose group's parameters are fixed in space-time.

Gauge theories are important as the successful field theories explaining the dynamics of elementary particles. Quantum electrodynamics is an abelian gauge theory with the symmetry group U(1) and has one gauge field, the electromagnetic four-potential, with the photon being the gauge boson. The Standard Model is a non-abelian gauge theory with the symmetry group U(1)×SU(2)×SU(3) and has a total of twelve gauge bosons: the photon, three weak bosons and eight gluons.

Gauge theories are also important in explaining gravitation in the theory of general relativity. Its case is somewhat unique in that the gauge field is a tensor, the Lanczos tensor. Theories of quantum gravity, beginning with gauge gravitation theory, also postulate the existence of a gauge boson known as the graviton. Gauge symmetries can be viewed as analogues of the principle of general covariance of general relativity in which the coordinate system can be chosen freely under arbitrary diffeomorphisms of spacetime. Both gauge invariance and diffeomorphism invariance reflect a redundancy in the description of the system. An alternative theory of gravitation, gauge theory gravity, replaces the principle of general covariance with a true gauge principle with new gauge fields.

Historically, these ideas were first stated in the context of classical electromagnetism and later in general relativity. However, the modern importance of gauge symmetries appeared first in the relativistic quantum mechanics of electrons – quantum electrodynamics, elaborated on below. Today, gauge theories are useful in condensed matter, nuclear and high energy physics among other subfields.

25.1 History and importance

The earliest field theory having a gauge symmetry was Maxwell's formulation of electrodynamics in 1864. The importance of this symmetry remained unnoticed in the earliest formulations. Similarly unnoticed, Hilbert had derived the Einstein field equations by postulating the invariance of the action under a general coordinate transformation. Later Hermann Weyl, in an attempt to unify general relativity and electromagnetism, conjectured that *Eichinvarianz* or invariance under the change of scale (or "gauge") might also be a local symmetry of general relativity. After the development of quantum mechanics, Weyl, Vladimir Fock and Fritz London modified gauge by replacing the scale factor with a complex quantity and turned the scale transformation into a change of phase, which is a U(1) gauge symmetry. This explained the electromagnetic field effect on the wave function of a charged quantum mechanical particle. This was the first widely recognised gauge theory, popularised by Pauli in the 1940s.[1]

In 1954, attempting to resolve some of the great confusion in elementary particle physics, Chen Ning Yang and Robert Mills introduced **non-abelian gauge theories** as models to understand the strong interaction holding together nucleons in atomic nuclei. (Ronald Shaw, working under Abdus Salam, independently introduced the same notion in his doctoral thesis.) Generalizing the gauge invariance of electromagnetism, they attempted to construct a theory based on the action of the (non-abelian) SU(2) symmetry group on the isospin doublet of protons and neutrons. This is similar to the action of the U(1) group on the spinor fields of quantum electrodynamics. In particle physics the emphasis was on using **quantized gauge theories**.

This idea later found application in the quantum field theory of the weak force, and its unification with electromagnetism in the electroweak theory. Gauge theories became even more attractive when it was realized that non-abelian gauge theories reproduced a feature called asymptotic freedom. Asymptotic freedom was believed to be an important characteristic of strong interactions. This motivated searching for a strong force gauge theory. This theory, now known as quantum chromodynamics, is a gauge theory with the action of the SU(3) group on the color triplet of quarks. The Standard Model unifies the description of electromagnetism, weak interactions and strong interactions in the language of gauge theory.

In the 1970s, Sir Michael Atiyah began studying the mathematics of solutions to the classical Yang–Mills equations. In 1983, Atiyah's student Simon Donaldson built on this work to show that the differentiable classification of smooth 4-manifolds is very different from their classification up to homeomorphism. Michael Freedman used Donaldson's work to exhibit exotic $\mathbf{R}^4$s, that is, exotic differentiable structures on Euclidean 4-dimensional space. This led to an increasing interest in gauge theory for its own sake, independent of its successes in fundamental physics. In 1994, Edward Witten and Nathan Seiberg invented gauge-theoretic techniques based on supersymmetry that enabled the calculation of certain topological invariants (the Seiberg–Witten invariants). These contributions to mathematics from gauge theory have led to a renewed interest in this area.

The importance of gauge theories in physics is exemplified in the tremendous success of the mathematical formalism in providing a unified framework to describe the quantum field theories of electromagnetism, the weak force and the strong force. This theory, known as the Standard Model, accurately describes experimental predictions regarding three of the four fundamental forces of nature, and is a gauge theory with the gauge group SU(3) × SU(2) × U(1). Modern theories like string theory, as well as general relativity, are, in one way or another, gauge theories.

See Pickering[2] *for more about the history of gauge and quantum field theories.*

25.2 Description

25.2.1 Global and local symmetries

In physics, the mathematical description of any physical situation usually contains excess degrees of freedom; the same physical situation is equally well described by many equivalent mathematical configurations. For instance, in Newtonian dynamics, if two configurations are related by a Galilean transformation (an inertial change of reference frame) they represent the same physical situation. These transformations form a group of "symmetries" of the theory, and a physical situation corresponds not to an individual mathematical configuration but to a class of configurations related to one another by this symmetry group.

This idea can be generalized to include local as well as global symmetries, analogous to much more abstract "changes of coordinates" in a situation where there is no preferred "inertial" coordinate system that covers the entire physical system. A gauge theory is a mathematical model that has symmetries of this kind, together with a set of techniques for making physical predictions consistent with the symmetries of the model.

25.2.2 Example of global symmetry

When a quantity occurring in the mathematical configuration is not just a number but has some geometrical significance, such as a velocity or an axis of rotation, its representation as numbers arranged in a vector or matrix is also changed by a coordinate transformation. For instance, if one description of a pattern of fluid flow states that the fluid velocity in the neighborhood of (x=1, y=0) is 1 m/s in the positive x direction, then a description of the same situation in which the coordinate system has been rotated clockwise by 90 degrees states that the fluid velocity in the neighborhood of (x=0, y=1) is 1 m/s in the positive y direction. The coordinate transformation has affected both the coordinate system used to identify the *location* of the measurement and the basis in which its *value* is expressed. As long as this transformation is performed globally (affecting the coordinate basis in the same way at every point), the effect on values that represent the *rate of change* of some quantity along some path in space and time as it passes through point P is the same as the effect on values that are truly local to P.

25.2.3 Use of fiber bundles to describe local symmetries

In order to adequately describe physical situations in more complex theories, it is often necessary to introduce a "coordinate basis" for some of the objects of the theory that do not have this simple relationship to the coordinates used to label points in space and time. (In mathematical terms, the theory involves a fiber bundle in which the fiber at each point of the base space consists of possible coordinate bases for use when describing the values of objects at that point.) In order to spell out a mathematical configuration, one must choose a particular coordinate basis at each point (a *local section* of the fiber bundle) and express the values of the objects of the theory (usually "fields" in the physicist's sense) using this basis. Two such mathematical configurations are equivalent (describe the same physical situation) if they are related by a transformation of this abstract coordinate basis (a change of local section, or *gauge transformation*).

In most gauge theories, the set of possible transformations of the abstract gauge basis at an individual point in space and time is a finite-dimensional Lie group. The simplest such group is U(1), which appears in the modern formulation of quantum electrodynamics (QED) via its use of complex numbers. QED is generally regarded as the first, and simplest, physical gauge theory. The set of possible gauge transformations of the entire configuration of a given gauge theory also forms a group, the *gauge group* of the theory. An element of the gauge group can be parameterized by a smoothly varying function from the points of spacetime to the (finite-dimensional) Lie group, such that the value of the function and its derivatives at each point represents the action of the gauge transformation on the fiber over that point.

A gauge transformation with constant parameter at every point in space and time is analogous to a rigid rotation of the geometric coordinate system; it represents a global symmetry of the gauge representation. As in the case of a rigid rotation, this gauge transformation affects expressions that represent the rate of change along a path of some gauge-dependent quantity in the same way as those that represent a truly local quantity. A gauge transformation whose parameter is *not* a constant function is referred to as a local symmetry; its effect on expressions that involve a derivative is qualitatively different from that on expressions that don't. (This is analogous to a non-inertial change of reference frame, which can produce a Coriolis effect.)

25.2.4 Gauge fields

The "gauge covariant" version of a gauge theory accounts for this effect by introducing a gauge field (in mathematical language, an Ehresmann connection) and formulating all rates of change in terms of the covariant derivative with respect to this connection. The gauge field becomes an essential part of the description of a mathematical configuration. A configuration in which the gauge field can be eliminated by a gauge transformation has the property that its field strength (in mathematical language, its curvature) is zero everywhere; a gauge theory is *not* limited to these configurations. In other words, the distinguishing characteristic of a gauge theory is that the gauge field does not merely compensate for a poor choice of coordinate system; there is generally no gauge transformation that makes the gauge field vanish.

When analyzing the dynamics of a gauge theory, the gauge field must be treated as a dynamical variable, similarly to other objects in the description of a physical situation. In addition to its interaction with other objects via the covariant derivative, the gauge field typically contributes energy in the form of a "self-energy" term. One can obtain the equations for the gauge theory by:

- starting from a naïve ansatz without the gauge field (in which the derivatives appear in a "bare" form);
- listing those global symmetries of the theory that can be characterized by a continuous parameter (generally an abstract equivalent of a rotation angle);
- computing the correction terms that result from allowing the symmetry parameter to vary from place to place; and
- reinterpreting these correction terms as couplings to one or more gauge fields, and giving these fields appropriate self-energy terms and dynamical behavior.

This is the sense in which a gauge theory "extends" a global symmetry to a local symmetry, and closely resembles the historical development of the gauge theory of gravity known as general relativity.

25.2.5 Physical experiments

Gauge theories are used to model the results of physical experiments, essentially by:

- limiting the universe of possible configurations to those consistent with the information used to set up the experiment, and then
- computing the probability distribution of the possible outcomes that the experiment is designed to measure.

The mathematical descriptions of the "setup information" and the "possible measurement outcomes" (loosely speaking, the "boundary conditions" of the experiment) are generally not expressible without reference to a particular coordinate system, including a choice of gauge. (If nothing else, one assumes that the experiment has been adequately isolated from "external" influence, which is itself a gauge-dependent statement.) Mishandling gauge dependence in boundary conditions is a frequent source of anomalies in gauge theory calculations, and gauge theories can be broadly classified by their approaches to anomaly avoidance.

25.2.6 Continuum theories

The two gauge theories mentioned above (continuum electrodynamics and general relativity) are examples of continuum field theories. The techniques of calculation in a continuum theory implicitly assume that:

- given a completely fixed choice of gauge, the boundary conditions of an individual configuration can in principle be completely described;
- given a completely fixed gauge and a complete set of boundary conditions, the principle of least action determines a unique mathematical configuration (and therefore a unique physical situation) consistent with these bounds;
- the likelihood of possible measurement outcomes can be determined by:
    - establishing a probability distribution over all physical situations determined by boundary conditions that are consistent with the setup information,
    - establishing a probability distribution of measurement outcomes for each possible physical situation, and

  - convolving these two probability distributions to get a distribution of possible measurement outcomes consistent with the setup information; and

- fixing the gauge introduces no anomalies in the calculation, due either to gauge dependence in describing partial information about boundary conditions or to incompleteness of the theory.

These assumptions are close enough to be valid across a wide range of energy scales and experimental conditions, to allow these theories to make accurate predictions about almost all of the phenomena encountered in daily life, from light, heat, and electricity to eclipses and spaceflight. They fail only at the smallest and largest scales (due to omissions in the theories themselves) and when the mathematical techniques themselves break down (most notably in the case of turbulence and other chaotic phenomena).

25.2.7 Quantum field theories

Other than these classical continuum field theories, the most widely known gauge theories are quantum field theories, including quantum electrodynamics and the Standard Model of elementary particle physics. The starting point of a quantum field theory is much like that of its continuum analog: a gauge-covariant action integral that characterizes "allowable" physical situations according to the principle of least action. However, continuum and quantum theories differ significantly in how they handle the excess degrees of freedom represented by gauge transformations. Continuum theories, and most pedagogical treatments of the simplest quantum field theories, use a gauge fixing prescription to reduce the orbit of mathematical configurations that represent a given physical situation to a smaller orbit related by a smaller gauge group (the global symmetry group, or perhaps even the trivial group).

More sophisticated quantum field theories, in particular those that involve a non-abelian gauge group, break the gauge symmetry within the techniques of perturbation theory by introducing additional fields (the Faddeev–Popov ghosts) and counterterms motivated by anomaly cancellation, in an approach known as BRST quantization. While these concerns are in one sense highly technical, they are also closely related to the nature of measurement, the limits on knowledge of a physical situation, and the interactions between incompletely specified experimental conditions and incompletely understood physical theory . The mathematical techniques that have been developed in order to make gauge theories tractable have found many other applications, from solid-state physics and crystallography to low-dimensional topology.

25.3 Classical gauge theory

25.3.1 Classical electromagnetism

Historically, the first example of gauge symmetry discovered was classical electromagnetism. In electrostatics, one can either discuss the electric field, **E**, or its corresponding electric potential, *V*. Knowledge of one makes it possible to find the other, except that potentials differing by a constant, $V \to V + C$, correspond to the same electric field. This is because the electric field relates to *changes* in the potential from one point in space to another, and the constant *C* would cancel out when subtracting to find the change in potential. In terms of vector calculus, the electric field is the gradient of the potential, $\mathbf{E} = -\nabla V$. Generalizing from static electricity to electromagnetism, we have a second potential, the vector potential **A**, with

$$\mathbf{E} = -\nabla V - \frac{\partial \mathbf{A}}{\partial t}$$
$$\mathbf{B} = \nabla \times \mathbf{A}$$

The general gauge transformations now become not just $V \to V + C$ but

$$\mathbf{A} \to \mathbf{A} + \nabla f$$
$$V \to V - \frac{\partial f}{\partial t}$$

where f is any function that depends on position and time. The fields remain the same under the gauge transformation, and therefore Maxwell's equations are still satisfied. That is, Maxwell's equations have a gauge symmetry.

25.3.2 An example: Scalar O(n) gauge theory

The remainder of this section requires some familiarity with classical or quantum field theory, and the use of Lagrangians.

Definitions in this section: gauge group, gauge field, interaction Lagrangian, gauge boson.

The following illustrates how local gauge invariance can be "motivated" heuristically starting from global symmetry properties, and how it leads to an interaction between originally non-interacting fields.

Consider a set of n non-interacting real scalar fields, with equal masses m. This system is described by an action that is the sum of the (usual) action for each scalar field φ_i

$$\mathcal{S} = \int \mathrm{d}^4x \sum_{i=1}^{n} \left[\frac{1}{2}\partial_\mu\varphi_i\partial^\mu\varphi_i - \frac{1}{2}m^2\varphi_i^2\right]$$

The Lagrangian (density) can be compactly written as

$$\mathcal{L} = \frac{1}{2}(\partial_\mu\Phi)^T\partial^\mu\Phi - \frac{1}{2}m^2\Phi^T\Phi$$

by introducing a vector of fields

$$\Phi = (\varphi_1, \varphi_2, \ldots, \varphi_n)^T$$

The term ∂_μ is Einstein notation for the partial derivative of Φ in each of the four dimensions. It is now transparent that the Lagrangian is invariant under the transformation

$$\Phi \mapsto \Phi' = G\Phi$$

whenever G is a *constant* matrix belonging to the n-by-n orthogonal group O(n). This is seen to preserve the Lagrangian, since the derivative of Φ transforms identically to Φ and both quantities appear inside dot products in the Lagrangian (orthogonal transformations preserve the dot product).

$$(\partial_\mu\Phi) \mapsto (\partial_\mu\Phi)' = G\partial_\mu\Phi$$

This characterizes the *global* symmetry of this particular Lagrangian, and the symmetry group is often called the **gauge group**; the mathematical term is **structure group**, especially in the theory of G-structures. Incidentally, Noether's theorem implies that invariance under this group of transformations leads to the conservation of the *currents*

$$J_\mu^a = i\partial_\mu\Phi^T T^a\Phi$$

where the T^a matrices are generators of the SO(n) group. There is one conserved current for every generator.

Now, demanding that this Lagrangian should have *local* O(n)-invariance requires that the G matrices (which were earlier constant) should be allowed to become functions of the space-time coordinates x.

Unfortunately, the G matrices do not "pass through" the derivatives, when $G = G(x)$,

$$\partial_\mu(G\Phi) \neq G(\partial_\mu \Phi)$$

The failure of the derivative to commute with "G" introduces an additional term (in keeping with the product rule), which spoils the invariance of the Lagrangian. In order to rectify this we define a new derivative operator such that the derivative of Φ again transforms identically with Φ

$$(D_\mu \Phi)' = G D_\mu \Phi$$

This new "derivative" is called a (gauge) covariant derivative and takes the form

$$D_\mu = \partial_\mu + igA_\mu$$

Where g is called the coupling constant; a quantity defining the strength of an interaction. After a simple calculation we can see that the **gauge field** $A(x)$ must transform as follows

$$A'_\mu = GA_\mu G^{-1} + \frac{i}{g}(\partial_\mu G)G^{-1}$$

The gauge field is an element of the Lie algebra, and can therefore be expanded as

$$A_\mu = \sum_a A_\mu^a T^a$$

There are therefore as many gauge fields as there are generators of the Lie algebra.

Finally, we now have a *locally gauge invariant* Lagrangian

$$\mathcal{L}_{\text{loc}} = \frac{1}{2}(D_\mu \Phi)^T D^\mu \Phi - \frac{1}{2}m^2 \Phi^T \Phi$$

Pauli uses the term *gauge transformation of the first type* to mean the transformation of Φ , while the compensating transformation in A is called a *gauge transformation of the second type*.

The difference between this Lagrangian and the original *globally gauge-invariant* Lagrangian is seen to be the **interaction Lagrangian**

$$\mathcal{L}_{\text{int}} = i\frac{g}{2}\Phi^T A_\mu^T \partial^\mu \Phi + i\frac{g}{2}(\partial_\mu \Phi)^T A^\mu \Phi - \frac{g^2}{2}(A_\mu \Phi)^T A^\mu \Phi$$

This term introduces interactions between the n scalar fields just as a consequence of the demand for local gauge invariance. However, to make this interaction physical and not completely arbitrary, the mediator $A(x)$ needs to propagate in space. That is dealt with in the next section by adding yet another term, $\mathcal{L}_{\text{gf}}$, to the Lagrangian. In the quantized version of the obtained classical field theory, the quanta of the gauge field $A(x)$ are called gauge bosons. The interpretation of the interaction Lagrangian in quantum field theory is of scalar bosons interacting by the exchange of these gauge bosons.

25.3.3 The Yang–Mills Lagrangian for the gauge field

Main article: Yang–Mills theory

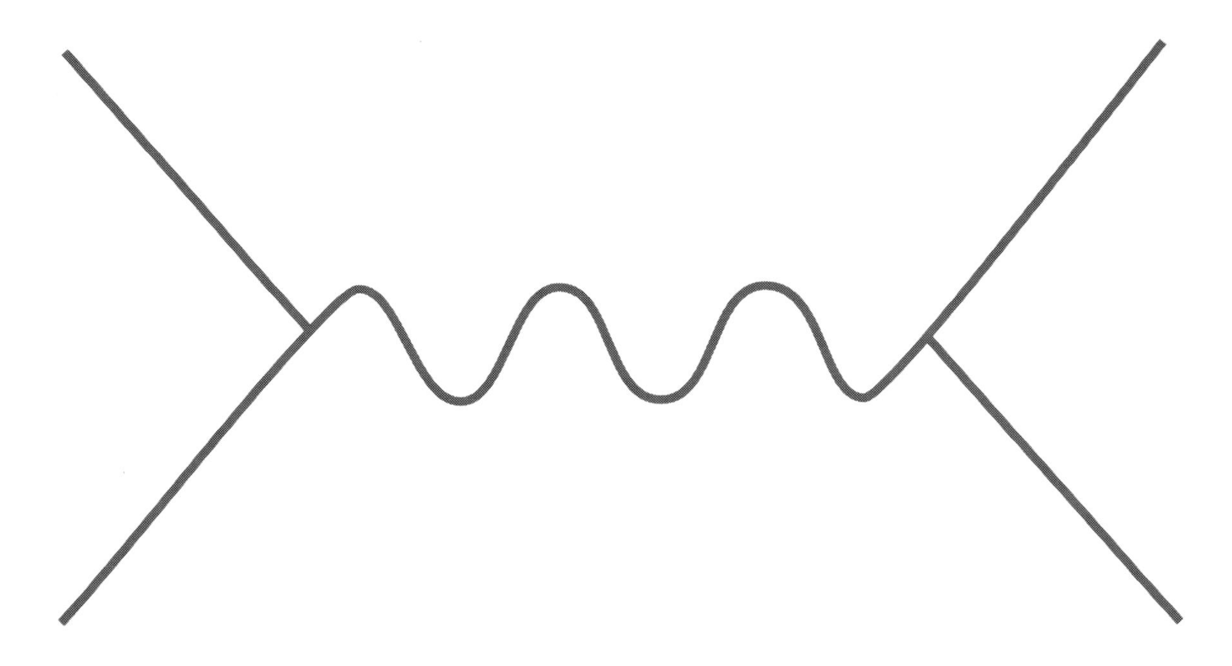

Feynman diagram of scalar bosons interacting via a gauge boson

The picture of a classical gauge theory developed in the previous section is almost complete, except for the fact that to define the covariant derivatives *D*, one needs to know the value of the gauge field $A(x)$ at all space-time points. Instead of manually specifying the values of this field, it can be given as the solution to a field equation. Further requiring that the Lagrangian that generates this field equation is locally gauge invariant as well, one possible form for the gauge field Lagrangian is (conventionally) written as

$$\mathcal{L}_{\text{gf}} = -\frac{1}{2}\operatorname{Tr}(F^{\mu\nu}F_{\mu\nu})$$

with

$$F_{\mu\nu} = \frac{1}{ig}[D_\mu, D_\nu]$$

and the trace being taken over the vector space of the fields. This is called the **Yang–Mills action**. Other gauge invariant actions also exist (e.g., nonlinear electrodynamics, Born–Infeld action, Chern–Simons model, theta term, etc.).

Note that in this Lagrangian term there is no field whose transformation counterweighs the one of A . Invariance of this term under gauge transformations is a particular case of *a priori* classical (geometrical) symmetry. This symmetry must be restricted in order to perform quantization, the procedure being denominated gauge fixing, but even after restriction, gauge transformations may be possible.[3]

The complete Lagrangian for the gauge theory is now

$$\mathcal{L} = \mathcal{L}_{\text{loc}} + \mathcal{L}_{\text{gf}} = \mathcal{L}_{\text{global}} + \mathcal{L}_{\text{int}} + \mathcal{L}_{\text{gf}}$$

25.3.4 An example: Electrodynamics

As a simple application of the formalism developed in the previous sections, consider the case of electrodynamics, with only the electron field. The bare-bones action that generates the electron field's Dirac equation is

$$\mathcal{S} = \int \bar{\psi}(i\hbar c\,\gamma^\mu \partial_\mu - mc^2)\psi\, \mathrm{d}^4x$$

The global symmetry for this system is

$$\psi \mapsto e^{i\theta}\psi$$

The gauge group here is U(1), just rotations of the phase angle of the field, with the particular rotation determined by the constant θ.

"Localising" this symmetry implies the replacement of θ by $\theta(x)$. An appropriate covariant derivative is then

$$D_\mu = \partial_\mu - i\frac{e}{\hbar}A_\mu$$

Identifying the "charge" *e* (not to be confused with the mathematical constant e in the symmetry description) with the usual electric charge (this is the origin of the usage of the term in gauge theories), and the gauge field *A*(*x*) with the four-vector potential of electromagnetic field results in an interaction Lagrangian

$$\mathcal{L}_{\text{int}} = \frac{e}{\hbar}\bar{\psi}(x)\gamma^\mu\psi(x)A_\mu(x) = J^\mu(x)A_\mu(x)$$

where $J^\mu(x)$ is the usual four vector electric current density. The gauge principle is therefore seen to naturally introduce the so-called minimal coupling of the electromagnetic field to the electron field.

Adding a Lagrangian for the gauge field $A_\mu(x)$ in terms of the field strength tensor exactly as in electrodynamics, one obtains the Lagrangian used as the starting point in quantum electrodynamics.

$$\mathcal{L}_{\text{QED}} = \bar{\psi}(i\hbar c\,\gamma^\mu D_\mu - mc^2)\psi - \frac{1}{4\mu_0}F_{\mu\nu}F^{\mu\nu}$$

See also: Dirac equation, Maxwell's equations, Quantum electrodynamics

25.4 Mathematical formalism

Gauge theories are usually discussed in the language of differential geometry. Mathematically, a *gauge* is just a choice of a (local) section of some principal bundle. A **gauge transformation** is just a transformation between two such sections.

Although gauge theory is dominated by the study of connections (primarily because it's mainly studied by high-energy physicists), the idea of a connection is not central to gauge theory in general. In fact, a result in general gauge theory shows that affine representations (i.e., affine modules) of the gauge transformations can be classified as sections of a jet bundle satisfying certain properties. There are representations that transform covariantly pointwise (called by physicists gauge transformations of the first kind), representations that transform as a connection form (called by physicists gauge transformations of the second kind, an affine representation)—and other more general representations, such as the B field in BF theory. There are more general nonlinear representations (realizations), but these are extremely complicated. Still, nonlinear sigma models transform nonlinearly, so there are applications.

If there is a principal bundle *P* whose base space is space or spacetime and structure group is a Lie group, then the sections of *P* form a principal homogeneous space of the group of gauge transformations.

Connections (gauge connection) define this principal bundle, yielding a covariant derivative ∇ in each associated vector bundle. If a local frame is chosen (a local basis of sections), then this covariant derivative is represented by the connection form A, a Lie algebra-valued 1-form, which is called the **gauge potential** in physics. This is evidently not an intrinsic but a frame-dependent quantity. The curvature form F, a Lie algebra-valued 2-form that is an intrinsic quantity, is constructed from a connection form by

$$\mathbf{F} = \mathrm{d}\mathbf{A} + \mathbf{A} \wedge \mathbf{A}$$

where d stands for the exterior derivative and $\wedge$ stands for the wedge product. ($\mathbf{A}$ is an element of the vector space spanned by the generators T^a , and so the components of $\mathbf{A}$ do not commute with one another. Hence the wedge product $\mathbf{A} \wedge \mathbf{A}$ does not vanish.)

Infinitesimal gauge transformations form a Lie algebra, which is characterized by a smooth Lie-algebra-valued scalar, ε. Under such an infinitesimal gauge transformation,

$$\delta_\varepsilon \mathbf{A} = [\varepsilon, \mathbf{A}] - \mathrm{d}\varepsilon$$

where $[\cdot, \cdot]$ is the Lie bracket.

One nice thing is that if $\delta_\varepsilon X = \varepsilon X$, then $\delta_\varepsilon DX = \varepsilon DX$ where D is the covariant derivative

$$DX \stackrel{\text{def}}{=} \mathrm{d}X + \mathbf{A}X$$

Also, $\delta_\varepsilon \mathbf{F} = \varepsilon \mathbf{F}$, which means $\mathbf{F}$ transforms covariantly.

Not all gauge transformations can be generated by infinitesimal gauge transformations in general. An example is when the base manifold is a compact manifold without boundary such that the homotopy class of mappings from that manifold to the Lie group is nontrivial. See instanton for an example.

The *Yang–Mills action* is now given by

$$\frac{1}{4g^2} \int \mathrm{Tr}[*F \wedge F]$$

where * stands for the Hodge dual and the integral is defined as in differential geometry.

A quantity which is **gauge-invariant** (i.e., invariant under gauge transformations) is the Wilson loop, which is defined over any closed path, γ, as follows:

$$\chi^{(\rho)} \left(\mathcal{P} \left\{ e^{\int_\gamma A} \right\} \right)$$

where χ is the character of a complex representation ρ and $\mathcal{P}$ represents the path-ordered operator.

25.5 Quantization of gauge theories

Main article: Quantum gauge theory

Gauge theories may be quantized by specialization of methods which are applicable to any quantum field theory. However, because of the subtleties imposed by the gauge constraints (see section on Mathematical formalism, above) there are many technical problems to be solved which do not arise in other field theories. At the same time, the richer structure of gauge theories allows simplification of some computations: for example Ward identities connect different renormalization constants.

25.5.1 Methods and aims

The first gauge theory quantized was quantum electrodynamics (QED). The first methods developed for this involved gauge fixing and then applying canonical quantization. The Gupta–Bleuler method was also developed to handle this problem. Non-abelian gauge theories are now handled by a variety of means. Methods for quantization are covered in the article on quantization.

The main point to quantization is to be able to compute quantum amplitudes for various processes allowed by the theory. Technically, they reduce to the computations of certain correlation functions in the vacuum state. This involves a renormalization of the theory.

When the running coupling of the theory is small enough, then all required quantities may be computed in perturbation theory. Quantization schemes intended to simplify such computations (such as canonical quantization) may be called **perturbative quantization schemes**. At present some of these methods lead to the most precise experimental tests of gauge theories.

However, in most gauge theories, there are many interesting questions which are non-perturbative. Quantization schemes suited to these problems (such as lattice gauge theory) may be called **non-perturbative quantization schemes**. Precise computations in such schemes often require supercomputing, and are therefore less well-developed currently than other schemes.

25.5.2 Anomalies

Some of the symmetries of the classical theory are then seen not to hold in the quantum theory; a phenomenon called an **anomaly**. Among the most well known are:

- The scale anomaly, which gives rise to a *running coupling constant*. In QED this gives rise to the phenomenon of the Landau pole. In Quantum Chromodynamics (QCD) this leads to asymptotic freedom.
- The chiral anomaly in either chiral or vector field theories with fermions. This has close connection with topology through the notion of instantons. In QCD this anomaly causes the decay of a pion to two photons.
- The gauge anomaly, which must cancel in any consistent physical theory. In the electroweak theory this cancellation requires an equal number of quarks and leptons.

25.6 Pure gauge

A pure gauge is the set of field configurations obtained by a gauge transformation on the null-field configuration, i.e., a gauge-transform of zero. So it is a particular "gauge orbit" in the field configuration's space.

Thus, in the abelian case, where $A_\mu(x) \to A'_\mu(x) = A_\mu(x) + \partial_\mu f(x)$, the pure gauge is just the set of field configurations $A'_\mu(x) = \partial_\mu f(x)$ for all $f(x)$.

25.7 See also

25.8 References

[1] Wolfgang Pauli (1941) "Relativistic Field Theories of Elementary Particles," *Rev. Mod. Phys.* **13**: 203–32.

[2] Pickering, A. (1984). *Constructing Quarks*. University of Chicago Press. ISBN 0-226-66799-5.

[3] Sakurai, *Advanced Quantum Mechanics*, sect 1–4

25.9 Bibliography

General readers

- Schumm, Bruce (2004) *Deep Down Things*. Johns Hopkins University Press. Esp. chpt. 8. A serious attempt by a physicist to explain gauge theory and the Standard Model with little formal mathematics.

Texts

- Bromley, D.A. (2000). *Gauge Theory of Weak Interactions*. Springer. ISBN 3-540-67672-4.
- Cheng, T.-P.; Li, L.-F. (1983). *Gauge Theory of Elementary Particle Physics*. Oxford University Press. ISBN 0-19-851961-3.
- Frampton, P. (2008). *Gauge Field Theories* (3rd ed.). Wiley-VCH.
- Kane, G.L. (1987). *Modern Elementary Particle Physics*. Perseus Books. ISBN 0-201-11749-5.

Articles

- Becchi, C. (1997). "Introduction to Gauge Theories". p. 5211. arXiv:hep-ph/9705211. Bibcode:1997hep.ph....5211B.
- Gross, D. (1992). "Gauge theory – Past, Present and Future" (PDF). Retrieved 2009-04-23.
- Jackson, J.D. (2002). "From Lorenz to Coulomb and other explicit gauge transformations". *Am.J.Phys* **70** (9): 917–928. arXiv:physics/0204034. Bibcode:2002AmJPh..70..917J. doi:10.1119/1.1491265.
- Svetlichny, George (1999). "Preparation for Gauge Theory". p. 2027. arXiv:math-ph/9902027. Bibcode:1999math.ph...2(

25.10 External links

- Hazewinkel, Michiel, ed. (2001), "Gauge transformation", *Encyclopedia of Mathematics*, Springer, ISBN 978-1-55608-010-4
- Yang–Mills equations on DispersiveWiki
- Gauge theories on Scholarpedia

Chapter 26

Weak isospin

In particle physics, **weak isospin** is a quantum number relating to the weak interaction, and parallels the idea of isospin under the strong interaction. Weak isospin is usually given the symbol T or I with the third component written as Tz, T_3, I_z or I_3.[1] Weak isospin is a complement of the weak hypercharge, which unifies weak interactions with electromagnetic interactions.

The **weak isospin conservation law** relates the conservation of T_3; all weak interactions must preserve T_3. It is also conserved by the other interactions and is therefore a conserved quantity in general. For this reason T_3 is more important than T and often the term "weak isospin" refers to the "3rd component of weak isospin".

26.1 Relation with chirality

Fermions with negative chirality (also called left-handed fermions) have $T = ½$ and can be grouped into doublets with $T_3 = ±½$ that behave the same way under the weak interaction. For example, up-type quarks (u, c, t) have $T_3 = +½$ and always transform into down-type quarks (d, s, b), which have $T_3 = −½$, and vice versa. On the other hand, a quark never decays weakly into a quark of the same T_3. Something similar happens with left-handed leptons, which exist as doublets containing a charged lepton (e−, μ−, τ−) with $T_3 = −½$ and a neutrino (ν
e, ν
μ, ν
τ) with $T_3 = ½$.

Fermions with positive chirality (also called right-handed fermions) have $T = 0$ and form singlets that do not undergo weak interactions.

Electric charge, Q, is related to weak isospin, T_3, and weak hypercharge, YW, by

$$Q = T_3 + \frac{Y_W}{2}.$$

26.2 Weak isospin and the W bosons

The symmetry associated with spin is SU(2). This requires gauge bosons to transform between weak isospin charges: bosons W+, W− and W0. This implies that W bosons have a $T = 1$, with three different values of T_3.

- W+ boson ($T_3 = +1$) is emitted in transitions $\{(T_3 = +½) \rightarrow (T_3 = -½)\}$,
- W− boson ($T_3 = -1$) is emitted in transitions $\{(T_3 = -½) \rightarrow (T_3 = +½)\}$.

- W0 boson ($T_3 = 0$) would be emitted in reactions where T_3 does not change. However, under electroweak unification, the W0 boson mixes with the weak hypercharge gauge boson B, resulting in the observed Z0 boson and the photon of Quantum Electrodynamics.

26.3 See also

- Field theoretical formulation of standard model
- Weak hypercharge

26.4 References

[1] Ambiguities: I is also used as sign for the 'normal' isospin, same for the third component I_3 aka I_z. T is also used as the sign for Topness. This article uses T and T_3.

Chapter 27

Weak hypercharge

The **weak hypercharge** in particle physics is a quantum number relating the electric charge and the third component of weak isospin. It is conserved (only terms that are overall weak-hypercharge neutral are allowed in the Lagrangian) and is similar to the Gell-Mann–Nishijima formula for the hypercharge of strong interactions (which is not conserved in weak interactions). It is frequently denoted *YW* and corresponds to the gauge symmetry U(1).[1]

27.1 Definition

Weak hypercharge is the generator of the U(1) component of the electroweak gauge group, SU(2)xU(1) and its associated quantum field *B* mixes with the W^3 electroweak quantum field to produce the observed Z gauge boson and the photon of quantum electrodynamics.

Weak hypercharge, usually written as *YW*, satisfies the equality:

$$Q = T_3 + \frac{Y_{\rm W}}{2}$$

where Q is the electrical charge (in elementary charge units) and T_3 is the third component of weak isospin. Rearranging, the weak hypercharge can be explicitly defined as:

$$Y_{\rm W} = 2(Q - T_3)$$

Note: sometimes weak hypercharge is scaled so that

$$Y_{\rm W} = Q - T_3$$

although this is a minority usage.[2]

Hypercharge assignments in the Standard Model are determined up to a twofold ambiguity by demanding cancellation of all anomalies.

27.2 Baryon and lepton number

Weak hypercharge is related to baryon number minus lepton number via:

$X + 2Y_W = 5(B - L)$

where X is a GUT-associated conserved quantum number. Since weak hypercharge is always conserved this implies that baryon number minus lepton number is also always conserved, within the Standard Model and most extensions.

27.2.1 Neutron decay

n → p + e− + ν
e

Hence neutron decay conserves baryon number B and lepton number L separately, so also the difference $B - L$ is conserved.

27.2.2 Proton decay

Proton decay is a prediction of many grand unification theories.

p+ → e+ + π0 → e+ + 2γ

Hence proton decay conserves $B - L$, even though it violates both lepton number and baryon number conservation.

27.3 See also

- Standard Model (mathematical formulation)

27.4 Notes

[1] J. F. Donoghue, E. Golowich, B. R. Holstein (1994). *Dynamics of the standard model*. Cambridge University Press. p. 52. ISBN 0-521-47652-6.

[2] M. R. Anderson (2003). *The mathematical theory of cosmic strings*. CRC Press. p. 12. ISBN 0-7503-0160-0.

Chapter 28

Color charge

Color charge is a property of quarks and gluons that is related to the particles' strong interactions in the theory of quantum chromodynamics (QCD). The color charge of quarks and gluons is completely unrelated to visual perception of color,[1] because it is a property that has almost no manifestation at distances above the size of an atomic nucleus. The term *color* was chosen because the charge responsible for the strong force between particles can be analogized to the three primary colors of human vision: red, green, and blue.[2] Another color scheme is "red, yellow, and blue",[3] using paint, rather than light as the perceptible analogy.

Particles have corresponding antiparticles. A particle with red, green, or blue charge has a corresponding antiparticle in which the color charge must be the anticolor of red, green, and blue, respectively, for the color charge to be conserved in particle-antiparticle creation and annihilation. Particle physicists call these antired, antigreen, and antiblue. All three colors mixed together, or any one of these colors and its complement (or negative), is "colorless" or "white" and has a net color charge of zero. Free particles have a color charge of zero: baryons are composed of three quarks, but the individual quarks can have red, green, or blue charges, or negatives; mesons are made from a quark and antiquark, the quark can be any color, and the antiquark will have the negative of that color. This color charge differs from electromagnetic charges since electromagnetic charges have only one kind of value. Positive and negative electrical charges are the same kind of charge as they only differ by the sign.

Shortly after the existence of quarks was first proposed in 1964, Oscar W. Greenberg introduced the notion of color charge to explain how quarks could coexist inside some hadrons in otherwise identical quantum states without violating the Pauli exclusion principle. The theory of quantum chromodynamics has been under development since the 1970s and constitutes an important component of the Standard Model of particle physics.

28.1 Red, green, and blue

In QCD, a quark's color can take one of three values or charges, red, green, and blue. An antiquark can take one of three anticolors, called antired, antigreen, and antiblue (represented as cyan, magenta and yellow, respectively). Gluons are mixtures of two colors, such as red and antigreen, which constitutes their color charge. QCD considers eight gluons of the possible nine color–anticolor combinations to be unique; see *eight gluon colors* for an explanation.

The following illustrates the coupling constants for color-charged particles:

- The quark colors (red, green, blue) combine to be colorless
- The quark anticolors (antired, antigreen, antiblue) also combine to be colorless
- A hadron with 3 quarks (red, green, blue) before a color change
- Blue quark emits a blue-antigreen gluon

- Green quark has absorbed the blue-antigreen gluon and is now blue; color remains conserved
- An animation of the interaction inside a neutron. The gluons are represented as circles with the color charge in the center and the anti-color charge on the outside.

28.1.1 Field lines from color charges

Main article: Field (physics)

Analogous to an electric field and electric charges, the strong force acting between color charges can be depicted using field lines. However, the color field lines do not arc outwards from one charge to another as much, because they are pulled together tightly by gluons (within 1 fm).[4] This effect confines quarks within hadrons.

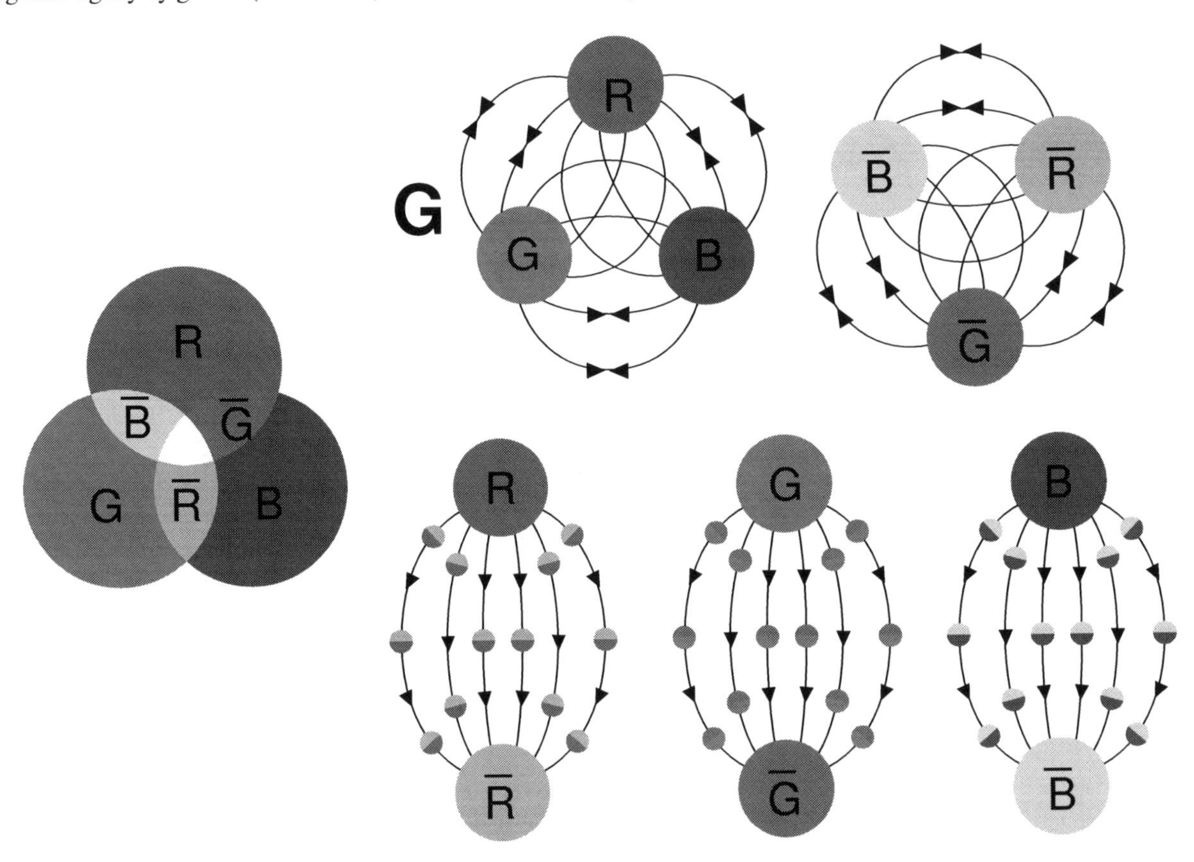

*Fields due to color charges, as in quarks (**G** is the gluon field strength tensor). These are "colorless" combinations. **Top:** Color charge has "ternary neutral states" as well as binary neutrality (analogous to electric charge). **Bottom:** Quark/antiquark combinations.*[5][6]

28.2 Coupling constant and charge

In a quantum field theory, a coupling constant and a charge are different but related notions. The coupling constant sets the magnitude of the force of interaction; for example, in quantum electrodynamics, the fine-structure constant is a coupling constant. The charge in a gauge theory has to do with the way a particle transforms under the gauge symmetry; i.e., its representation under the gauge group. For example, the electron has charge -1 and the positron has charge $+1$, implying that the gauge transformation has opposite effects on them in some sense. Specifically, if a local gauge transformation $\phi(x)$ is applied in electrodynamics, then one finds (using tensor index notation):

$$A_\mu \to A_\mu + \partial_\mu \phi(x)\,,\ \psi \to \exp[iQ\phi(x)]\psi \text{ and } \overline{\psi} \to \exp[-iQ\phi(x)]\overline{\psi}$$

where A_μ is the photon field, and ψ is the electron field with $Q=-1$ (a bar over ψ denotes its antiparticle — the positron). Since QCD is a non-abelian theory, the representations, and hence the color charges, are more complicated. They are dealt with in the next section.

28.3 Quark and gluon fields and color charges

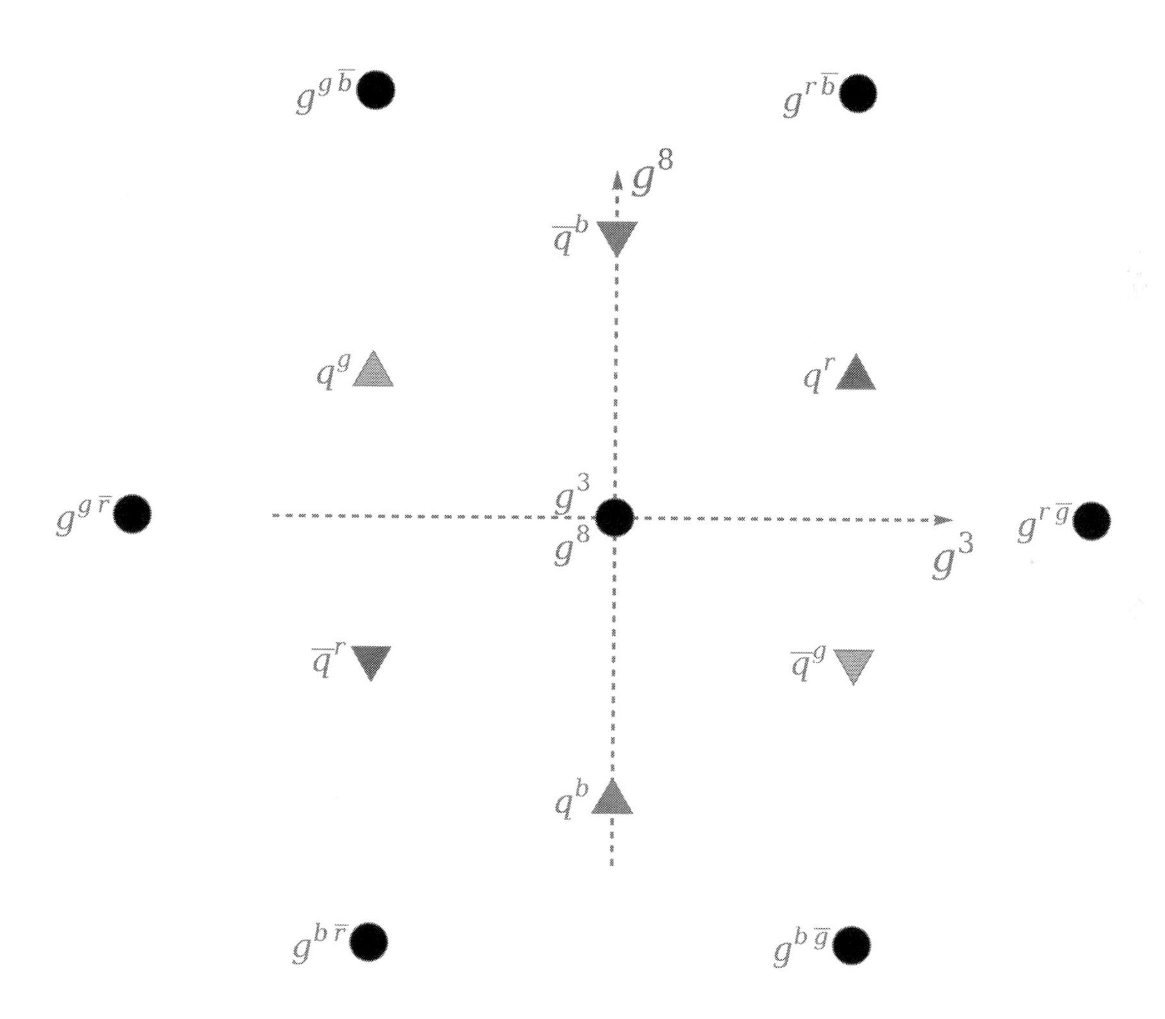

The pattern of strong charges for the three colors of quark, three antiquarks, and eight gluons (with two of zero charge overlapping).

In QCD the gauge group is the non-abelian group SU(3). The *running coupling* is usually denoted by α_s. Each flavor of quark belongs to the fundamental representation (**3**) and contains a triplet of fields together denoted by ψ. The antiquark field belongs to the complex conjugate representation (**3***) and also contains a triplet of fields. We can write

$$\psi = \begin{pmatrix} \psi_1 \\ \psi_2 \\ \psi_3 \end{pmatrix} \text{ and } \overline{\psi} = \begin{pmatrix} \overline{\psi}_1^* \\ \overline{\psi}_2^* \\ \overline{\psi}_3^* \end{pmatrix}.$$

The gluon contains an octet of fields (see gluon field), and belongs to the adjoint representation (**8**), and can be written using the Gell-Mann matrices as

$$\mathbf{A}_\mu = A_\mu^a \lambda_a.$$

(there is an implied summation over a = 1, 2, ... 8). All other particles belong to the trivial representation (**1**) of color SU(3). The **color charge** of each of these fields is fully specified by the representations. Quarks have a color charge of red, green or blue and antiquarks have a color charge of antired, antigreen or antiblue. Gluons have a combination of two color charges (one of red, green or blue and one of antired, antigreen and antiblue) in a superposition of states which are given by the Gell-Mann matrices. All other particles have zero color charge. Mathematically speaking, the color charge of a particle is the value of a certain quadratic Casimir operator in the representation of the particle.

In the simple language introduced previously, the three indices "1", "2" and "3" in the quark triplet above are usually identified with the three colors. The colorful language misses the following point. A gauge transformation in color SU(3) can be written as $\psi \rightarrow U\,\psi$, where U is a 3 × 3 matrix which belongs to the group SU(3). Thus, after gauge transformation, the new colors are linear combinations of the old colors. In short, the simplified language introduced before is not gauge invariant.

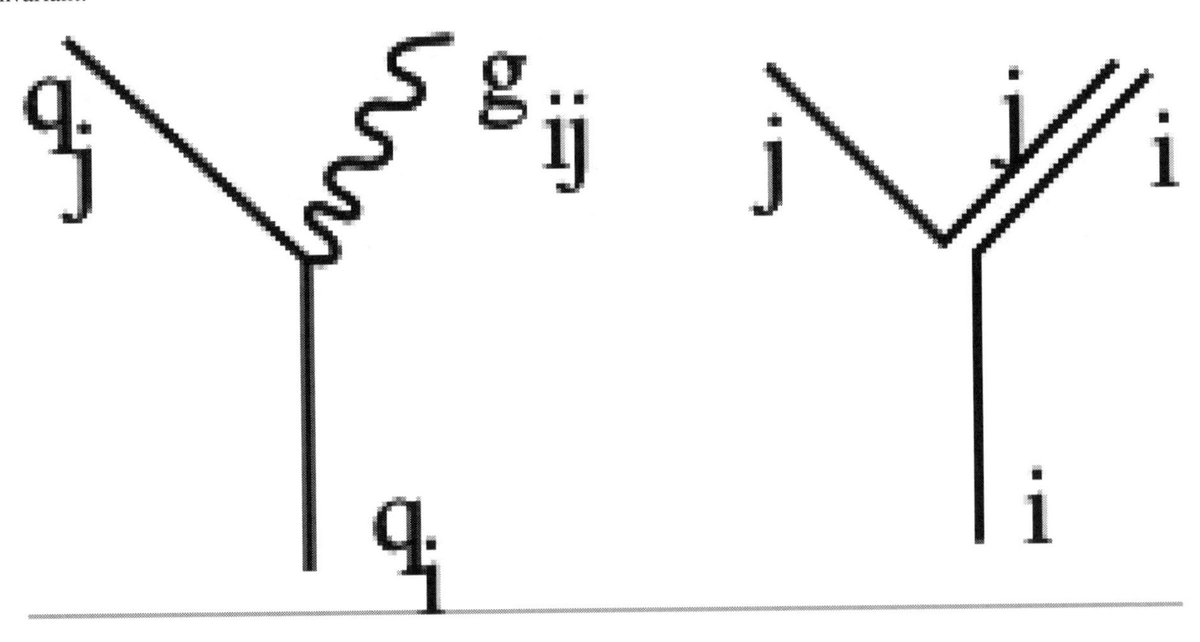

Color-line representation of QCD vertex

Color charge is conserved, but the book-keeping involved in this is more complicated than just adding up the charges, as is done in quantum electrodynamics. One simple way of doing this is to look at the interaction vertex in QCD and replace it by a color-line representation. The meaning is the following. Let ψ_i represent the i-th component of a quark field (loosely called the i-th color). The *color* of a gluon is similarly given by **A** which corresponds to the particular Gell-Mann matrix it is associated with. This matrix has indices i and j. These are the *color labels* on the gluon. At the interaction vertex one has $qi \rightarrow gi\,j + qj$. The **color-line** representation tracks these indices. Color charge conservation means that the ends of these color-lines must be either in the initial or final state, equivalently, that no lines break in the middle of a diagram.

Since gluons carry color charge, two gluons can also interact. A typical interaction vertex (called the three gluon vertex) for gluons involves g + g → g. This is shown here, along with its color-line representation. The color-line diagrams can be

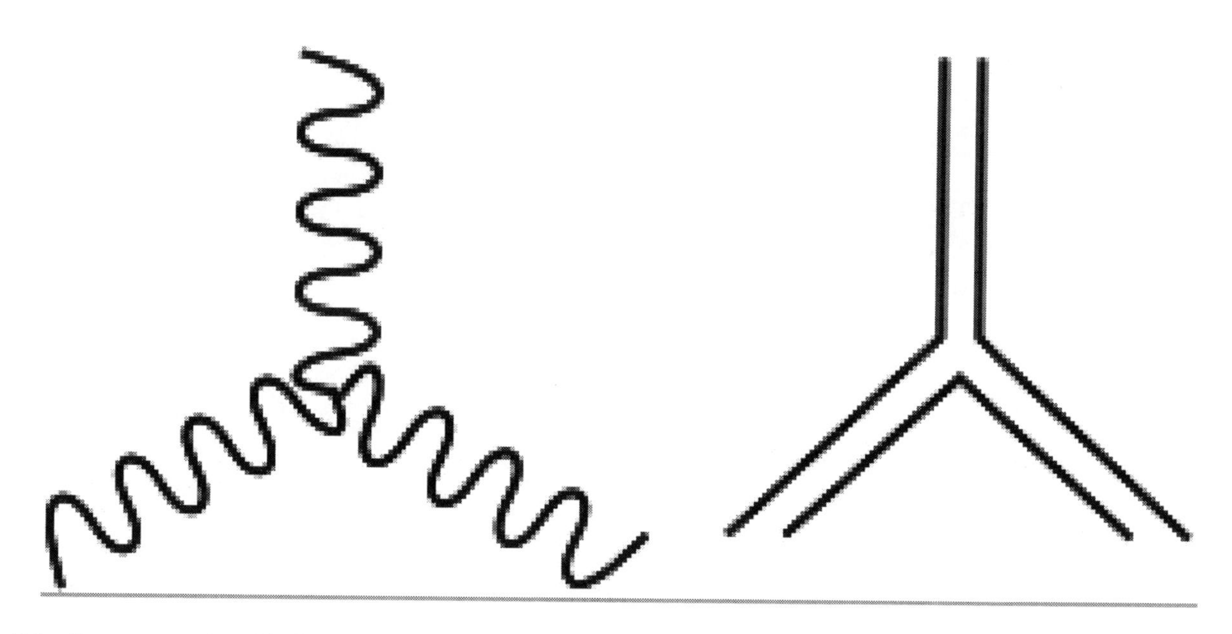

Color-line representation of 3-gluon vertex

restated in terms of conservation laws of color; however, as noted before, this is not a gauge invariant language. Note that in a typical non-abelian gauge theory the gauge boson carries the charge of the theory, and hence has interactions of this kind; for example, the W boson in the electroweak theory. In the electroweak theory, the W also carries electric charge, and hence interacts with a photon.

28.4 See also

- Color confinement
- Gluon field strength tensor

28.5 References

[1] Feynman, Richard (1985), *QED: The Strange Theory of Light and Matter*, Princeton University Press, p. 136, ISBN 0-691-08388-6, The idiot physicists, unable to come up with any wonderful Greek words anymore, call this type of polarization by the unfortunate name of 'color,' which has nothing to do with color in the normal sense.

[2] Close (2007)

[3] R. Penrose (2005). *The Road to Reality*. Vintage books. p. 648. ISBN 978-00994-40680.

[4] R. Resnick, R. Eisberg (1985), *Quantum Physics of Atoms, Molecules, Solids, Nuclei and Particles* (2nd ed.), John Wiley & Sons, p. 684, ISBN 978-0-471-87373-0

[5] Parker, C.B. (1994), *McGraw Hill Encyclopaedia of Physics* (2nd ed.), Mc Graw Hill, ISBN 0-07-051400-3

[6] M. Mansfield, C. O'Sullivan (2011), *Understanding Physics* (4th ed.), John Wiley & Sons, ISBN 978-0-47-0746370

28.6 Further reading

- Georgi, Howard (1999), *Lie algebras in particle physics*, Perseus Books Group, ISBN 0-7382-0233-9.

- Griffiths, David J. (1987), *Introduction to Elementary Particles*, New York: John Wiley & Sons, ISBN 0-471-60386-4.
- Christman, J. Richard (2001), "Colour and Charm" (PDF), *Project PHYSNET document MISN-0-283*.
- Hawking, Stephen (1998), *A Brief History of Time*, Bantam Dell Publishing Group, ISBN 978-0-553-10953-5.
- Close, Frank (2007), *The New Cosmic Onion*, Taylor & Francis, ISBN 1-58488-798-2.

Chapter 29

Weinberg angle

The **Weinberg angle** or **weak mixing angle** is a parameter in the Weinberg–Salam theory of the electroweak interaction, and is usually denoted as θW. It is the angle by which spontaneous symmetry breaking rotates the original W0 and B^0 vector boson plane, producing as a result the Z0 boson, and the photon.

$$\begin{pmatrix}\gamma \\ Z^0\end{pmatrix} = \begin{pmatrix}\cos\theta_W & \sin\theta_W \\ -\sin\theta_W & \cos\theta_W\end{pmatrix}\begin{pmatrix}B^0 \\ W^0\end{pmatrix}$$

It also gives the relationship between the masses of the W and Z bosons (denoted as *m*W and *m*Z):

$$m_Z = \frac{m_W}{\cos\theta_W}$$

The angle can be expressed in terms of the $SU(2)_L$ and $U(1)_Y$ coupling constants (g and g', respectively):

$$\cos\theta_W = \frac{g}{\sqrt{g^2+g'^2}} \text{ and } \sin\theta_W = \frac{g'}{\sqrt{g^2+g'^2}}$$

As the value of the mixing angle is currently determined empirically, it has been mathematically defined as:[1]

$$\cos\theta_W = \frac{m_W}{m_Z}$$

The value of θW varies as a function of the momentum transfer, *Q*, at which it is measured. This variation, or 'running', is a key prediction of the electroweak theory. The most precise measurements have been carried out in electron-positron collider experiments at a value of *Q* = 91.2 GeV/c, corresponding to the mass of the Z boson, *m*Z.

In practice the quantity $\sin^2\theta$W is more frequently used. The 2004 best estimate of $\sin^2\theta$W, at *Q* = 91.2 GeV/c, in the MS scheme is 0.23120 ± 0.00015. Atomic parity violation experiments yield values for $\sin^2\theta$W at smaller values of *Q*, below 0.01 GeV/c, but with much lower precision. In 2005 results were published from a study of parity violation in Møller scattering in which a value of $\sin^2\theta$W = 0.2397 ± 0.0013 was obtained at *Q* = 0.16 GeV/c, establishing experimentally the 'running' of the weak mixing angle. These values correspond to a Weinberg angle of ~30°.

Note, however, that the specific value of the angle is *not* a prediction of the standard model: it is an open, unfixed parameter. At this time, there is no generally accepted theory that explains why the measured value is what it is.

29.1 See also

- Cabibbo angle

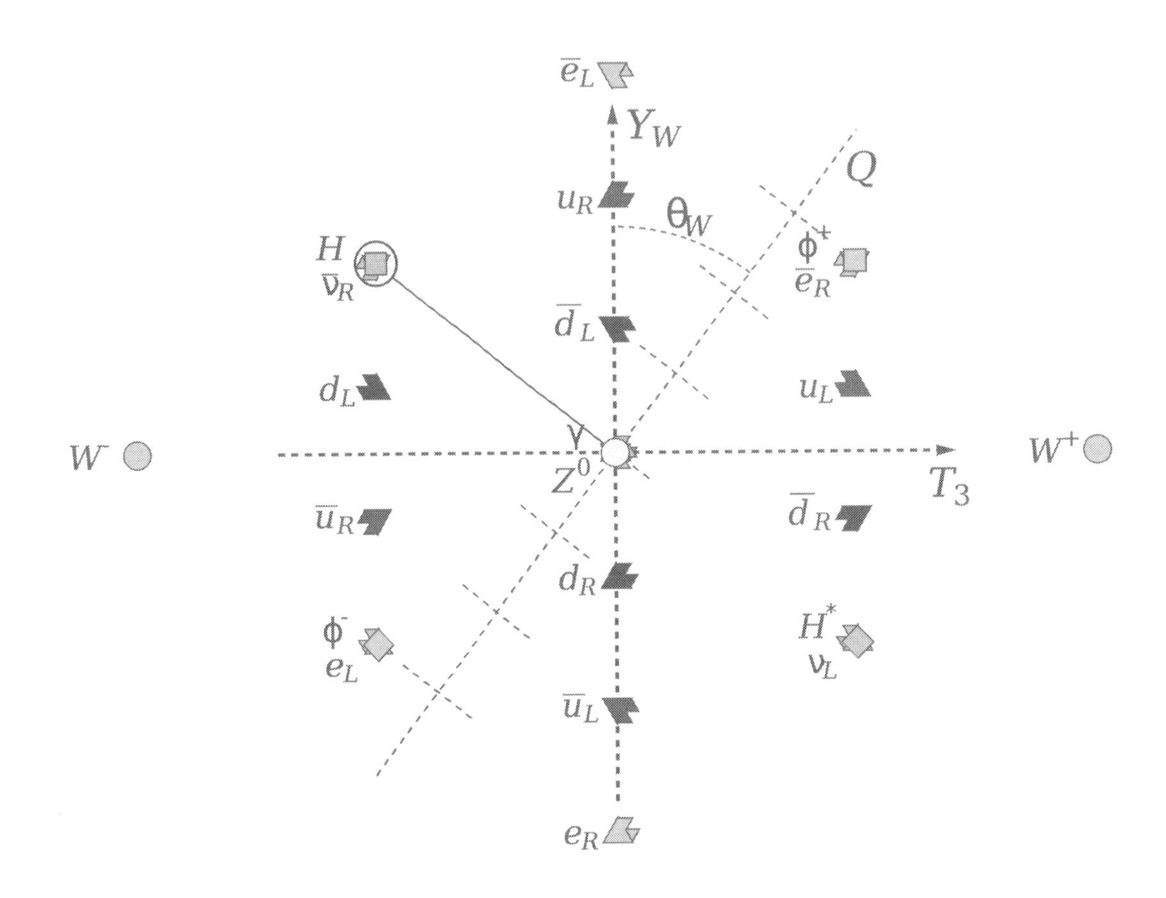

The pattern of weak isospin, T_3, and weak hypercharge, YW, of the known elementary particles, showing electric charge, Q, along the Weinberg angle. The neutral Higgs field (circled) breaks the electroweak symmetry and interacts with other particles to give them mass. Three components of the Higgs field become part of the massive W and Z bosons.

29.2 References

[1] L. B. Okun (1982). *Leptons and Quarks*. North-Holland Physics Publishing. p. 214. ISBN 0-444-86924-7.

- C. Amsler *et al.* (Particle Data Group) (2008). "Review of Particle Physics – Electroweak model and constraints on new physics" (PDF). *Physics Letters B* **667** (1): 1. Bibcode:2008PhLB..667....1P. doi:10.1016/j.physletb.2008.07.018.
- E158: A Precision Measurement of the Weak Mixing Angle in Møller Scattering
- Q-weak: A Precision Test of the Standard Model and Determination of the Weak Charges of the Quarks through Parity-Violating Electron Scattering

Chapter 30

Electroweak interaction

In particle physics, the **electroweak interaction** is the unified description of two of the four known fundamental interactions of nature: electromagnetism and the weak interaction. Although these two forces appear very different at everyday low energies, the theory models them as two different aspects of the same force. Above the unification energy, on the order of 100 GeV, they would merge into a single **electroweak force**. Thus, if the universe is hot enough (approximately 10^{15} K, a temperature exceeded until shortly after the Big Bang), then the electromagnetic force and weak force merge into a combined electroweak force. During the electroweak epoch, the electroweak force separated from the strong force. During the quark epoch, the electroweak force split into the electromagnetic and weak force.

For contributions to the unification of the weak and electromagnetic interaction between elementary particles, Sheldon Glashow, Abdus Salam, and Steven Weinberg were awarded the Nobel Prize in Physics in 1979.[1][2] The existence of the electroweak interactions was experimentally established in two stages, the first being the discovery of neutral currents in neutrino scattering by the Gargamelle collaboration in 1973, and the second in 1983 by the UA1 and the UA2 collaborations that involved the discovery of the W and Z gauge bosons in proton–antiproton collisions at the converted Super Proton Synchrotron. In 1999, Gerardus 't Hooft and Martinus Veltman were awarded the Nobel prize for showing that the electroweak theory is renormalizable.

30.1 Formulation

Mathematically, the unification is accomplished under an $SU(2) \times U(1)$ gauge group. The corresponding gauge bosons are the **three** W bosons of weak isospin from SU(2) (W+, W0, and W−), and the B0 boson of weak hypercharge from U(1), respectively, all of which are massless.

In the Standard Model, the W± and Z0 bosons, and the photon, are produced by the spontaneous symmetry breaking of the **electroweak symmetry** from $SU(2) \times U(1)Y$ to $U(1)_{\text{em}}$, caused by the Higgs mechanism (see also Higgs boson).[3][4][5][6] $U(1)Y$ and $U(1)_{\text{em}}$ are different copies of $U(1)$; the generator of $U(1)_{\text{em}}$ is given by $Q = Y/2 + I_3$, where Y is the generator of $U(1)Y$ (called the weak hypercharge), and I_3 is one of the $SU(2)$ generators (a component of weak isospin).

The spontaneous symmetry breaking causes the W0 and B0 bosons to coalesce together into two different bosons – the Z0 boson, and the photon (γ) as follows:

$$\begin{pmatrix} \gamma \\ Z^0 \end{pmatrix} = \begin{pmatrix} \cos\theta_W & \sin\theta_W \\ -\sin\theta_W & \cos\theta_W \end{pmatrix} \begin{pmatrix} B^0 \\ W^0 \end{pmatrix}$$

Where θW is the *weak mixing angle*. The axes representing the particles have essentially just been rotated, in the (W0, B0) plane, by the angle θW. This also introduces a discrepancy between the mass of the Z0 and the mass of the W± particles (denoted as MZ and MW, respectively);

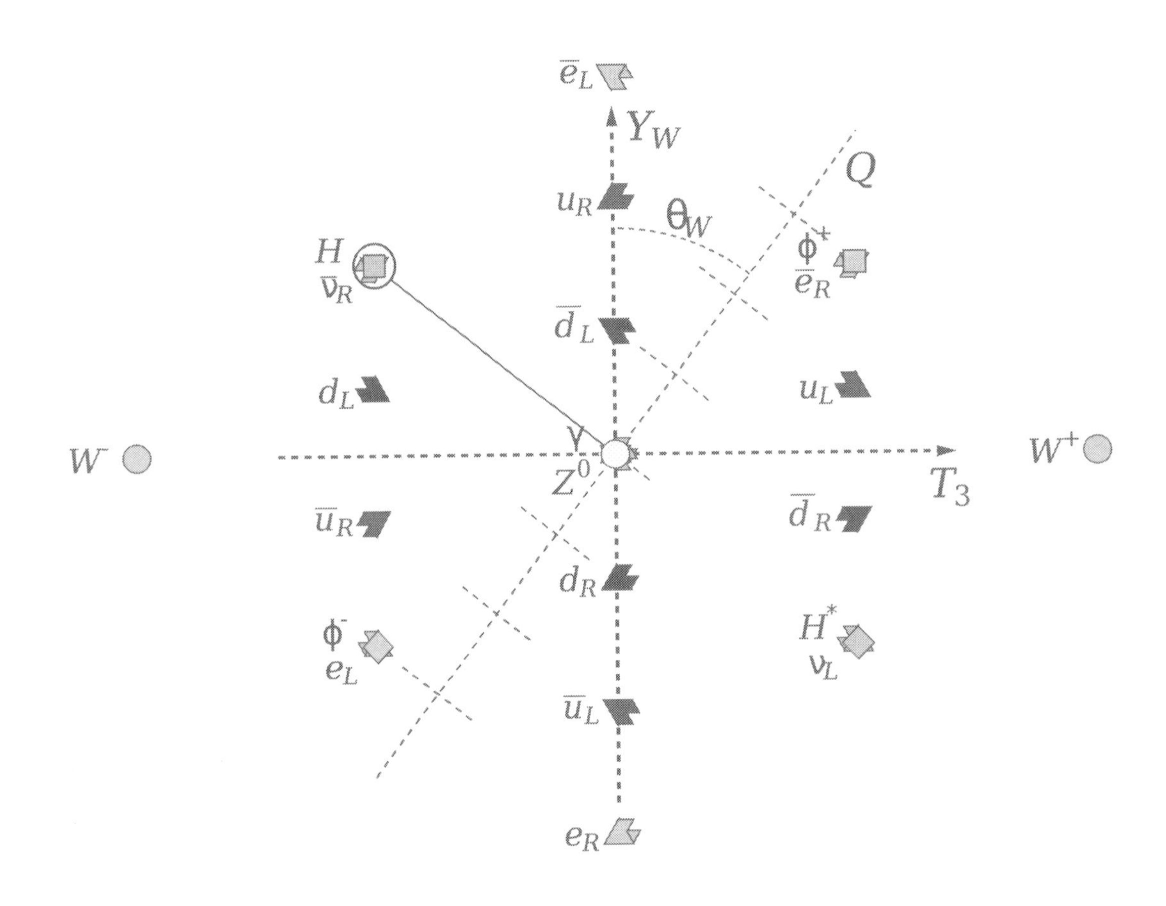

The pattern of weak isospin, T_3, and weak hypercharge, YW, of the known elementary particles, showing electric charge, Q, along the weak mixing angle. The neutral Higgs field (circled) breaks the electroweak symmetry and interacts with other particles to give them mass. Three components of the Higgs field become part of the massive W and Z bosons.

$$M_Z = \frac{M_W}{\cos\theta_W}$$

The distinction between electromagnetism and the weak force arises because there is a (nontrivial) linear combination of Y and I_3 that vanishes for the Higgs boson (it is an eigenstate of both Y and I_3, so the coefficients may be taken as $-I_3$ and Y): $U(1)_{\mathrm{em}}$ is defined to be the group generated by this linear combination, and is unbroken because it does not interact with the Higgs.

30.2 Lagrangian

30.2.1 Before electroweak symmetry breaking

The Lagrangian for the electroweak interactions is divided into four parts before electroweak symmetry breaking

$$\mathcal{L}_{EW} = \mathcal{L}_g + \mathcal{L}_f + \mathcal{L}_h + \mathcal{L}_y.$$

The $\mathcal{L}_g$ term describes the interaction between the three W particles and the B particle.

$$\mathcal{L}_g = -\frac{1}{4}W_a^{\mu\nu}W_{\mu\nu}^a - \frac{1}{4}B^{\mu\nu}B_{\mu\nu}$$

where $W^{a\mu\nu}$ ($a = 1,2,3$) and $B^{\mu\nu}$ are the field strength tensors for the weak isospin and weak hypercharge fields.

$\mathcal{L}_f$ is the kinetic term for the Standard Model fermions. The interaction of the gauge bosons and the fermions are through the gauge covariant derivative.

$$\mathcal{L}_f = \overline{Q}_i i\not{D} Q_i + \overline{u}_i i\not{D} u_i + \overline{d}_i i\not{D} d_i + \overline{L}_i i\not{D} L_i + \overline{e}_i i\not{D} e_i$$

where the subscript i runs over the three generations of fermions, Q , u , and d are the left-handed doublet, right-handed singlet up, and right handed singlet down quark fields, and L and e are the left-handed doublet and right-handed singlet electron fields.

The *h* term describes the Higgs field F.

$$\mathcal{L}_h = |D_\mu h|^2 - \lambda\left(|h|^2 - \frac{v^2}{2}\right)^2$$

The *y* term gives the Yukawa interaction that generates the fermion masses after the Higgs acquires a vacuum expectation value.

$$\mathcal{L}_y = -y_{u\,ij}\epsilon^{ab}\,h_b^\dagger\,\overline{Q}_{ia}u_j^c - y_{d\,ij}\,h\,\overline{Q}_i d_j^c - y_{e\,ij}\,h\,\overline{L}_i e_j^c + h.c.$$

30.2.2 After electroweak symmetry breaking

The Lagrangian reorganizes itself after the Higgs boson acquires a vacuum expectation value. Due to its complexity, this Lagrangian is best described by breaking it up into several parts as follows.

$$\mathcal{L}_{EW} = \mathcal{L}_K + \mathcal{L}_N + \mathcal{L}_C + \mathcal{L}_H + \mathcal{L}_{HV} + \mathcal{L}_{WWV} + \mathcal{L}_{WWVV} + \mathcal{L}_Y$$

The kinetic term $\mathcal{L}_K$ contains all the quadratic terms of the Lagrangian, which include the dynamic terms (the partial derivatives) and the mass terms (conspicuously absent from the Lagrangian before symmetry breaking)

$$\mathcal{L}_K = \sum_f \overline{f}(i\not{\partial} - m_f)f - \frac{1}{4}A_{\mu\nu}A^{\mu\nu} - \frac{1}{2}W_{\mu\nu}^+W^{-\mu\nu} + m_W^2 W_\mu^+ W^{-\mu}$$
$$-\frac{1}{4}Z_{\mu\nu}Z^{\mu\nu} + \frac{1}{2}m_Z^2 Z_\mu Z^\mu + \frac{1}{2}(\partial^\mu H)(\partial_\mu H) - \frac{1}{2}m_H^2 H^2$$

where the sum runs over all the fermions of the theory (quarks and leptons), and the fields $A_{\mu\nu}$, $Z_{\mu\nu}$, $W_{\mu\nu}^-$, and $W_{\mu\nu}^+ \equiv (W_{\mu\nu}^-)^\dagger$ are given as

$X_{\mu\nu} = \partial_\mu X_\nu - \partial_\nu X_\mu + g f^{abc} X^b_\mu X^c_\nu$, (replace X by the relevant field, and f^{abc} with the structure constants for the gauge group).

The neutral current $\mathcal{L}_N$ and charged current $\mathcal{L}_C$ components of the Lagrangian contain the interactions between the fermions and gauge bosons.

$$\mathcal{L}_N = eJ^{em}_\mu A^\mu + \frac{g}{\cos\theta_W}(J^3_\mu - \sin^2\theta_W J^{em}_\mu)Z^\mu$$

where the electromagnetic current J^{em}_μ and the neutral weak current J^3_μ are

$$J^{em}_\mu = \sum_f q_f \bar{f}\gamma_\mu f$$

and

$$J^3_\mu = \sum_f I^3_f \bar{f}\gamma_\mu \frac{1-\gamma^5}{2} f$$

q_f and I^3_f are the fermions' electric charges and weak isospin.

The charged current part of the Lagrangian is given by

$$\mathcal{L}_C = -\frac{g}{\sqrt{2}}\left[\overline{u}_i\gamma^\mu \frac{1-\gamma^5}{2} M^{CKM}_{ij} d_j + \overline{\nu}_i\gamma^\mu \frac{1-\gamma^5}{2} e_i\right] W^+_\mu + h.c.$$

$\mathcal{L}_H$ contains the Higgs three-point and four-point self interaction terms.

$$\mathcal{L}_H = -\frac{g m_H^2}{4 m_W} H^3 - \frac{g^2 m_H^2}{32 m_W^2} H^4$$

$\mathcal{L}_{HV}$ contains the Higgs interactions with gauge vector bosons.

$$\mathcal{L}_{HV} = \left(g m_W H + \frac{g^2}{4}H^2\right)\left(W^+_\mu W^{-\mu} + \frac{1}{2\cos^2\theta_W} Z_\mu Z^\mu\right)$$

$\mathcal{L}_{WWV}$ contains the gauge three-point self interactions.

$$\mathcal{L}_{WWV} = -ig[(W^+_{\mu\nu}W^{-\mu} - W^{+\mu}W^-_{\mu\nu})(A^\nu \sin\theta_W - Z^\nu\cos\theta_W) + W^-_\nu W^+_\mu(A^{\mu\nu}\sin\theta_W - Z^{\mu\nu}\cos\theta_W)]$$

$\mathcal{L}_{WWVV}$ contains the gauge four-point self interactions

$$\mathcal{L}_{WWVV} = -\frac{g^2}{4}\Big\{[2W^+_\mu W^{-\mu} + (A_\mu \sin\theta_W - Z_\mu\cos\theta_W)^2]^2 - [W^+_\mu W^-_\nu + W^+_\nu W^-_\mu + (A_\mu\sin\theta_W - Z_\mu\cos\theta_W)(A_\nu\sin\theta_W - Z_\nu\cos\theta_W)]^2\Big\}$$

and $\mathcal{L}_Y$ contains the Yukawa interactions between the fermions and the Higgs field.

$$\mathcal{L}_Y = -\sum_f \frac{g m_f}{2 m_W}\bar{f} f H$$

Note the $\frac{1-\gamma^5}{2}$ factors in the weak couplings: these factors project out the left handed components of the spinor fields. This is why electroweak theory (after symmetry breaking) is commonly said to be a chiral theory.

30.3 See also

- Fundamental forces
- Formulation of the standard model
- Weinberg angle
- Unitarity gauge

30.4 References

[1] S. Bais (2005). *The Equations: Icons of knowledge*. p. 84. ISBN 0-674-01967-9.

[2] "The Nobel Prize in Physics 1979". The Nobel Foundation. Retrieved 2008-12-16.

[3] F. Englert, R. Brout (1964). "Broken Symmetry and the Mass of Gauge Vector Mesons". *Physical Review Letters* **13** (9): 321–323. Bibcode:1964PhRvL..13..321E. doi:10.1103/PhysRevLett.13.321.

[4] P.W. Higgs (1964). "Broken Symmetries and the Masses of Gauge Bosons". *Physical Review Letters* **13** (16): 508–509. Bibcode:1964PhRvL..13..508H. doi:10.1103/PhysRevLett.13.508.

[5] G.S. Guralnik, C.R. Hagen, T.W.B. Kibble (1964). "Global Conservation Laws and Massless Particles". *Physical Review Letters* **13** (20): 585–587. Bibcode:1964PhRvL..13..585G. doi:10.1103/PhysRevLett.13.585.

[6] G.S. Guralnik (2009). "The History of the Guralnik, Hagen and Kibble development of the Theory of Spontaneous Symmetry Breaking and Gauge Particles". *International Journal of Modern Physics A* **24** (14): 2601–2627. arXiv:0907.3466. Bibcode:2009IJMPA..24.2601G. doi:10.1142/S0217751X09045431.

30.4.1 General readers

- B. A. Schumm (2004). *Deep Down Things: The Breathtaking Beauty of Particle Physics*. Johns Hopkins University Press. ISBN 0-8018-7971-X. Conveys much of the Standard Model with no formal mathematics. Very thorough on the weak interaction.

30.4.2 Texts

- D. J. Griffiths (1987). *Introduction to Elementary Particles*. John Wiley & Sons. ISBN 0-471-60386-4.
- W. Greiner, B. Müller (2000). *Gauge Theory of Weak Interactions*. Springer. ISBN 3-540-67672-4.
- G. L. Kane (1987). *Modern Elementary Particle Physics*. Perseus Books. ISBN 0-201-11749-5.

30.4.3 Articles

- E. S. Abers, B. W. Lee (1973). "Gauge theories". *Physics Reports* **9**: 1–141. Bibcode:1973PhR.....9....1A. doi:10.1016/0370-1573(73)90027-6.
- Y. Hayato et al. (1999). "Search for Proton Decay through $p \to \nu K^+$ in a Large Water Cherenkov Detector". *Physical Review Letters* **83** (8): 1529. arXiv:hep-ex/9904020. Bibcode:1999PhRvL..83.1529H. doi:10.1103/PhysRevLett.83.1529.
- J. Hucks (1991). "Global structure of the standard model, anomalies, and charge quantization". *Physical Review D* **43** (8): 2709–2717. Bibcode:1991PhRvD..43.2709H. doi:10.1103/PhysRevD.43.2709.
- S. F. Novaes (2000). "Standard Model: An Introduction". arXiv:hep-ph/0001283 [hep-ph].
- D. P. Roy (1999). "Basic Constituents of Matter and their Interactions — A Progress Report". arXiv:hep-ph/9912523 [hep-ph].

Chapter 31

Wave function

Not to be confused with Wave equation.

A **wave function** in quantum mechanics describes the quantum state of an isolated system of one or more particles. There is *one* wave function containing all the information about the entire system, not a separate wave function for each particle in the system. Its interpretation is that of a probability amplitude. Quantities associated with measurements, such as the average momentum of a particle, can be derived from the wave function. It is a central entity in quantum mechanics and is important in all modern theories, like quantum field theory incorporating quantum mechanics, while its interpretation may differ. The most common symbols for a wave function are the Greek letters ψ or Ψ (lower-case and capital psi).

For a given system, once a representation corresponding to a maximal set of commuting observables and a suitable coordinate system is chosen, the wave function is a complex-valued function of the system's degrees of freedom corresponding to the chosen representation and coordinate system, continuous as well as discrete. Such a set of observables, by a postulate of quantum mechanics, are Hermitian linear operators on the space of states representing a set of **physical observables**, like position, momentum and spin that can, in principle, be simultaneously measured with arbitrary precision. Wave functions can be added together and multiplied by complex numbers to form new wave functions, and hence are elements of a vector space. This is the superposition principle of quantum mechanics. This vector space is endowed with an inner product such that it is a complete metric topological space with respect to the metric induced by the inner product. In this way the set of wave functions for a system form a function space that is a Hilbert space. The inner product is a measure of the overlap between physical states and is used in the foundational probabilistic interpretation of quantum mechanics, the Born rule, relating transition probabilities to inner products. The actual space depends on the system's degrees of freedom (hence on the chosen representation and coordinate system) and the exact form of the Hamiltonian entering the equation governing the dynamical behavior. In the non-relativistic case, disregarding spin, this is the Schrödinger equation.

The Schrödinger equation determines the allowed wave functions for the system and how they evolve over time. A wave function behaves qualitatively like other waves, such as water waves or waves on a string, because the Schrödinger equation is mathematically a type of wave equation. This explains the name "wave function", and gives rise to wave–particle duality. The wave of the wave function, however, is not a wave in physical space; it is a wave in an abstract mathematical "space", and in this respect it differs fundamentally from water waves or waves on a string.[1][2][3][4][5][6][7]

For a given system, the choice of which relevant degrees of freedom to use are not unique, and correspondingly the domain of the wave function is not unique. It may be taken to be a function of all the position coordinates of the particles over *position space*, or the momenta of all the particles over *momentum space*, the two are related by a Fourier transform. These descriptions are the most important, but they are not the only possibilities. Just like in classical mechanics, canonical transformations may be used in the description of a quantum system. Some particles, like electrons and photons, have nonzero spin, and the wave function must include this fundamental property as an intrinsic discrete degree of freedom. In general, for a particle with *half-integer* spin the wave function is a spinor, for a particle with *integer* spin the wave function is a tensor. Particles with spin zero are called scalar particles, those with spin 1 vector particles, and more generally for higher integer spin, tensor particles. The terminology derives from how the wave functions transform under a rotation of the coordinate system. No *elementary* particle with spin $^3/_2$ or higher is known, except for the hypothesized spin 2 graviton. Other discrete variables can be included, such as isospin. When a system has internal degrees of freedom, the

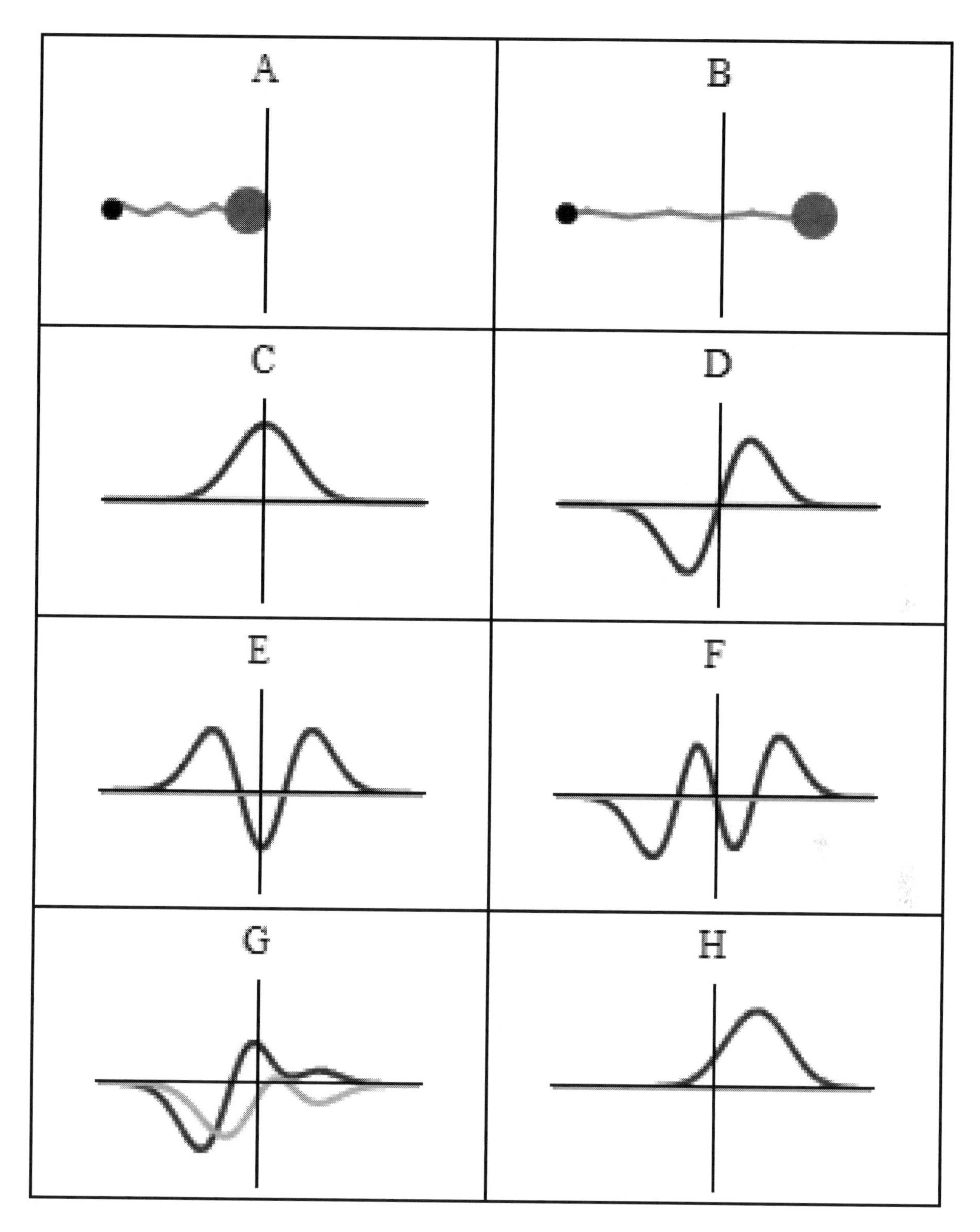

Comparison of classical and quantum harmonic oscillator conceptions for a single spinless particle. *The two processes differ greatly. The classical process (A–B) is represented as the motion of a particle along a trajectory. The quantum process (C–H) has no such trajectory. Rather, it is represented as a wave. Panels (C–F) show four different standing wave solutions of the Schrödinger equation. Panels (G–H) further show two different wave functions that are solutions of the Schrödinger equation but not standing waves.*

wave function at each point in the continuous degrees of freedom (e.g. a point in space) assigns a complex number for *each* possible value of the discrete degrees of freedom (e.g. z-component of spin). These values are often displayed in a column matrix (e.g. a 2×1 column vector for a non-relativistic electron with spin $1/2$).

In the Copenhagen interpretation, an interpretation of quantum mechanics, the squared modulus of the wave function, $|\psi|^2$, is a real number interpreted as the probability density of measuring a particle as being at a given place at a given time or having a definite momentum, and possibly having definite values for discrete degrees of freedom. The integral of this quantity, over all the system's degrees of freedom, must be 1 in accordance with the probability interpretation, this general requirement a wave function must satisfy is called the *normalization condition*. Since the wave function is complex valued, only its relative phase and relative magnitude can be measured. Its value does not in isolation tell anything about the magnitudes or directions of measurable observables; one has to apply quantum operators, whose eigenvalues correspond to sets of possible results of measurements, to the wave function ψ and calculate the statistical distributions for measurable quantities.

The unit of measurement for ψ depends on the system, and can be found by dimensional analysis of the normalization condition for the system. For one particle in three dimensions, its units are $[\text{length}]^{-3/2}$, because an integral of $|\psi|^2$ over a region of three-dimensional space is a dimensionless probability.[8]

31.1 Historical background

In 1905 Einstein postulated the proportionality between the frequency of a photon and its energy, $E = hf$,[9] and in 1916 the corresponding relation between photon momentum and wavelength, $\lambda = h/p$.[10] In 1923, De Broglie was the first to suggest that the relation $\lambda = h/p$, now called the De Broglie relation, holds for *massive* particles, the chief clue being Lorentz invariance,[11] and this can be viewed as the starting point for the modern development of quantum mechanics. The equations represent wave–particle duality for both massless and massive particles.

In the 1920s and 1930s, quantum mechanics was developed using calculus and linear algebra. Those who used the techniques of calculus included Louis de Broglie, Erwin Schrödinger, and others, developing "wave mechanics". Those who applied the methods of linear algebra included Werner Heisenberg, Max Born, and others, developing "matrix mechanics". Schrödinger subsequently showed that the two approaches were equivalent.[12]

In 1926, Schrödinger published the famous wave equation now named after him, indeed the Schrödinger equation, based on classical energy conservation using quantum operators and the de Broglie relations such that the solutions of the equation are the wave functions for the quantum system.[13] However, no one was clear on how to *interpret it*.[14] At first, Schrödinger and others thought that wave functions represent particles that are spread out with most of the particle being where the wave function is large.[15] This was shown to be incompatible with how elastic scattering of a wave packet representing a particle off a target appears; it spreads out in all directions.[16] While a scattered particle may scatter in any direction, it does not break up and take off in all directions. In 1926, Born provided the perspective of probability amplitude.[16][17][18] This relates calculations of quantum mechanics directly to probabilistic experimental observations. It is accepted as part of the Copenhagen interpretation of quantum mechanics. There are many other interpretations of quantum mechanics. In 1927, Hartree and Fock made the first step in an attempt to solve the N-body wave function, and developed the *self-consistency cycle*: an iterative algorithm to approximate the solution. Now it is also known as the Hartree–Fock method.[19] The Slater determinant and permanent (of a matrix) was part of the method, provided by John C. Slater.

Schrödinger did encounter an equation for the wave function that satisfied relativistic energy conservation *before* he published the non-relativistic one, but discarded it as it predicted negative probabilities and negative energies. In 1927, Klein, Gordon and Fock also found it, but incorporated the electromagnetic interaction and proved that it was Lorentz invariant. De Broglie also arrived at the same equation in 1928. This relativistic wave equation is now most commonly known as the Klein–Gordon equation.[20]

In 1927, Pauli phenomenologically found a non-relativistic equation to describe spin-1/2 particles in electromagnetic fields, now called the Pauli equation.[21] Pauli found the wave function was not described by a single complex function of space and time, but needed two complex numbers, which respectively correspond to the spin $+1/2$ and $-1/2$ states of the fermion. Soon after in 1928, Dirac found an equation from the first successful unification of special relativity and quantum mechanics applied to the electron, now called the Dirac equation. In this, the wave function is a *spinor* represented by four complex-valued components:[19] two for the electron and two for the electron's antiparticle, the positron. In the non-relativistic limit, the Dirac wave function resembles the Pauli wave function for the electron. Later, other relativistic wave equations were found.

31.1.1 Wave functions and wave equations in modern theories

All these wave equations are of enduring importance. The Schrödinger equation and the Pauli equation are under many circumstances excellent approximations of the relativistic variants. They are considerably easier to solve in practical problems than the relativistic equations. The Klein-Gordon equation and the Dirac equation, while being relativistic, do not represent full reconciliation of quantum mechanics and special relativity. The branch of quantum mechanics where these equations are studied the same way as the Schrödinger equation, often called relativistic quantum mechanics, while very successful, has its limitations (see e.g. Lamb shift) and conceptual problems (see e.g. Dirac sea).

Relativity makes it inevitable that the number of particles in a system is not constant. For full reconciliation, quantum field theory is needed.[22] In this theory, the wave equations and the wave functions have their place, but in a somewhat different guise. The main objects of interest are not the wave functions, but rather operators, so called *field operators* (or just fields where "operator" is understood) on the Hilbert space of states (to be described next section). It turns out that the original relativistic wave equations and their solutions are still needed to build the Hilbert space. Moreover, the *free fields operators*, i.e. when interactions are assumed not to exist, turn out to (formally) satisfy the same equation as do the fields (wave functions) in many cases.

Thus the Klein-Gordon equation (spin 0) and the Dirac equation (spin $1/2$) in this guise remain in the theory. Higher spin analogues include the Proca equation (spin 1), Rarita–Schwinger equation (spin $3/2$), and, more generally, the Bargmann–Wigner equations. For *massless* free fields two examples are the free field Maxwell equation (spin 1) and the free field Einstein equation (spin 2) for the field operators.[23] All of them are essentially a direct consequence of the requirement of Lorentz invariance. Their solutions must transform under Lorentz transformation in a prescribed way, i.e. under a particular representation of the Lorentz group and that together with few other reasonable demands, e.g. the *cluster decomposition principle*,[24] with implications for causality is enough to fix the equations.

It should be emphasized that this applies to free field equations; interactions are not included. It should also be noted that the equations and their solutions, though needed for the theories, are not the central objects of study.

31.2 Wave functions and function spaces

The concept of Function spaces enters naturally in the discussion about wave functions. A function space is a set of functions, usually with some defining requirements on the functions, together with a topology on that set. The latter will sparsely be used here, it is only needed to obtain a precise definition of what it means for a subset of a function space to be closed. A wave function is an element of a function space partly characterized by the following concrete and abstract descriptions.

- The Schrödinger equation is linear. This means that the solutions to it, wave functions, can be added and multiplied by scalars to form a new solution.
- The superposition principle of quantum mechanics. If Ψ and Φ are two states in the abstract space of **states** of a quantum mechanical system, then $a\Psi + b\Phi$ is a valid state as well.

The first item says that the set of solutions to the Schrödinger equation is a vector space. The second item says that the set of allowable states is a vector space. This similarity is of course not accidental. Not all properties of the respective spaces have been given so far. There are also a distinctions between the spaces to keep in mind.

- Basic states are characterized by a set of quantum numbers. This is a set of eigenvalues of a maximal set of commuting observables. A choice of such a set may be called a choice of **representation**. It is a postulate of quantum mechanics that a physically observable quantity of a system, like position, momentum and spin, is represented by a linear Hermitian operator on the state space. The possible outcomes of measurement of the quantity are

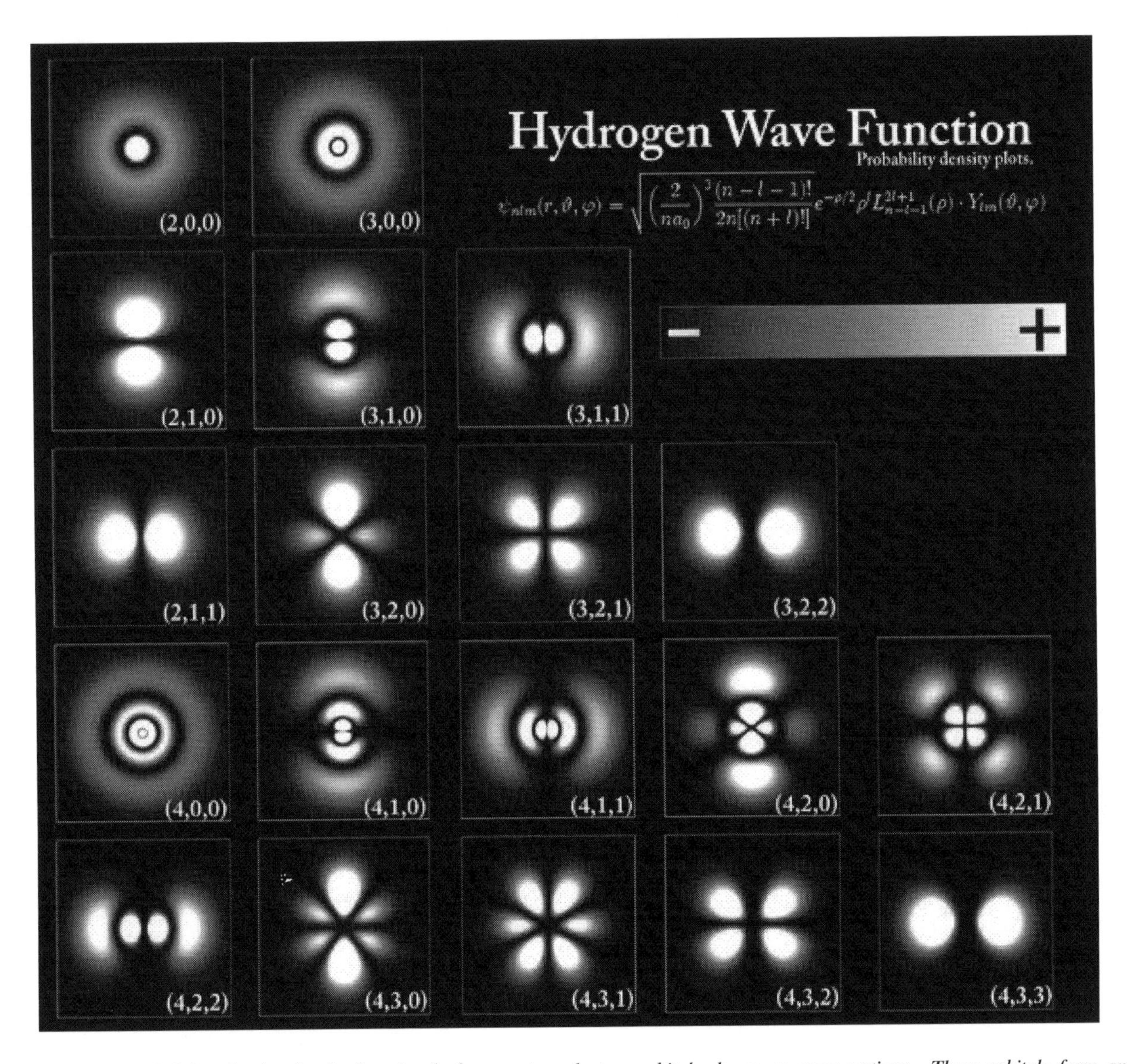

The electron probability density for the first few hydrogen atom electron orbitals shown as cross-sections. These orbitals form an orthonormal basis for the wave function of the electron. Different orbitals are depicted with different scale.

the eigenvalues of the operator.[15] Maximality refers to that no more algebraically independent linear Hermitian operator can be added to the set that commutes with the ones already present. The physical interpretation is that such a set represents what can – in theory – be simultaneously be measured with arbitrary position at a given time. The set is non-unique. It may for a one-particle system, for example, be position and spin z-projection, (x, Sz), or it may be momentum and spin y-projection, (p, Sy). At a deeper level, most observables, perhaps all, arise as generators of symmetries.[15][25][nb 1]

- Once a representation is chosen, there is still arbitrariness. It remains to choose a coordinate system. This may, for example, correspond to a choice of x, y- and z-axis, or a choice of **curvlinear coordinates** as exemplified by the spherical coordinates used for the atomic wave functions illustrated below. This final choice also fixes a basis in abstract Hilbert space. The basic states are labeled by the quantum numbers corresponding to the maximal set of commuting observables and an appropriate coordinate system.[nb 2]
- Wave functions corresponding to a state are accordingly not unique. This has been exemplified already with momentum and position space wave functions describing the same abstract state. This non-uniqueness reflects the non-uniqueness in the choice of a maximal set of commuting observables.

- The abstract states are "abstract" only in that an arbitrary choice necessary for a particular *explicit* description of it is not given. This is the same as saying that no choice of maximal set of commuting observables has been given. This is analogous to a vector space without a specified basis.
- The wave functions of position and momenta, respectively, can be seen as a choice of representation yielding two different, but entirely equivalent, explicit descriptions of the same state for a system with no discrete degrees of freedom.
- Corresponding to the two examples in the first item, to a particular state there corresponds two wave functions, $\Psi(x, Sz)$ and $\Psi(p, Sy)$, both describing the same state. For each choice of maximal commuting sets of observables for the abstract state space, there is a corresponding representation that is associated to a function space of wave functions.
- Each choice of representation should be thought of as specifying a unique function space in which wave functions corresponding to that choice of representation lives. This distinction is best kept, even if one could argue that two such function spaces are mathematically equal, e.g. being the set of square integrable functions. One can then think of the function spaces as two distinct copies of that set.
- Between all these different function spaces and the abstract state space, there are one-to-one correspondences (here disregarding normalization and unobservable phase factors), the common denominator here being a particular abstract state. The relationship between the momentum and position space wave functions, for instance, describing the same state is the Fourier transform.

To make this concrete, in the figure to the right, the 19 sub-images are images of wave functions in position space (their norm squared). The wave functions each represent the abstract state characterized by the triple of quantum numbers (n, l, m), in the lower right of each image. These are the principal quantum number, the orbital angular momentum quantum number and the magnetic quantum number. Together with one spin-projection quantum number of the electron, this is a complete set of observables.

The figure can serve to illustrate some further properties of the function spaces of wave functions.

- In this case, the wave functions are square integrable. One can initially take the function space as the space of square integrable functions, usually denoted L^2.
- The displayed functions are solutions to the Schrödinger equation. Obviously, not every function in L^2 satisfies the Schrödinger equation for the hydrogen atom. The function space is thus a subspace of L^2.
- The displayed functions form part of a basis for the function space. To each triple (n, l, m), there corresponds a basis wave function. If spin is taken into account, there are two basis functions for each triple. The function space thus has a countable basis.
- The basis functions are mutually orthonormal. For this concept to have a meaning, there must exist an inner product. The function space is thus an inner product space. The inner product between two states intuitively measures the "overlap" between the states. The physical interpretation is that the norm squared is proportional to the transition probability between the states. That is.

  $$P(\Psi \to \Phi_i) = |(\Psi, \Phi_i)|^2$$

  where the i is an index composed of quantum numbers corresponding to a representation and the probabilities are the probabilities of finding the state Ψ in the definite state represented by Φi upon measurement of the physical observables corresponding to the representation, for instance, i could be the quadruple (n, l, m, Sz). This is the Born rule,[16] and is one of the fundamental postulates of quantum mechanics.

These observations encapsulate the essence of the function spaces of which wave functions are elements. Mathematically, this is expressed (in one spatial dimension, disregarding here unimportant issues of normalization) for a particle with no internal degrees of freedom as

$$\Psi = I\Psi = \int \Phi_x(\Phi_x, \Psi)dx = \int \Psi(x)\Phi_x dx = \int \Phi_p(\Phi_p, \Psi)dp = \int \Psi(p)\Phi_p dp,$$

where Ψ is any "abstract" state, Φ*x* is an eigenfunction of the position operator representing a particle localized at *x*, (·,·) represents the inner product, Φ*p* is an eigenfunction of the momentum operator representing a particle with precise momentum *p*, *I* is the **identity operator** and the integrals (first and third) represent the completeness of momentum and position eigenstates, Ψ(*x*) is the coordinate space wave function and Ψ(*p*) is the wave function in momentum space. In Dirac notation, the above equation reads

$$|\Psi\rangle = I|\Psi\rangle = \int |x\rangle\langle x|\Psi\rangle dx = \int \Psi(x)|x\rangle dx = \int |p\rangle\langle p|\Psi\rangle dp = \int \Psi(p)|p\rangle dp.$$

The description is not yet complete. There is a further technical requirement on the function space, that of completeness, that allows one to take limits of sequences in the function space, and be ensured that, if the limit exists, it is an element of the function space. A complete inner product space is called a Hilbert space. The property of completeness is crucial in advanced treatments and applications of quantum mechanics. It will not be very important in the subsequent discussion of wave functions, and technical details and links may be found in footnotes like the one that follows.[nb 3] The space L^2 is a Hilbert space, with inner product presented later. The function space of the example of the figure is a subspace of L^2. A subspace of a Hilbert space is a Hilbert space if it is closed. It is here that the topology of the function space enters into its description.

It is also important to note, in order to avoid confusion, that not all functions to be discussed are elements of some Hilbert space, say L^2. The most glaring example is the set of functions $e^{2\pi ipx/h}$. These are solutions of the Schrödinger equation for a free particle, but are not normalizable, hence not in L^2. But they are nonetheless fundamental for the description. One can, using them, express functions that *are* normalizable using wave packets. They are, in a sense to be made precise later, a basis (but not a Hilbert space basis) in which wave functions of interest can be expressed. There is also the artifact "normalization to a delta function" that is frequently employed for notational convenience, see further down. The delta functions themselves aren't square integrable either.

31.2.1 Physical requirements

The above description of the function space containing the wave functions is mostly mathematically motivated. The function spaces are, due to completeness, very *large* in a certain sense. Not all functions are realistic descriptions of any physical system. For instance, in the function space L^2 one can find the function that takes on the value 0 for all rational numbers and -*i* for the irrationals in the interval [0, 1]. This *is* square integrable,[nb 4] but can hardly represent a physical state.

The following constraints on the wave function are sometimes explicitly formulated for the calculations and physical interpretation to make sense:[26][27]

- The wave function must be square integrable. This is motivated by the Copenhagen interpretation of the wave function as a probability amplitude.
- It must everywhere be everywhere continuous and everywhere continuously differentiable. This is motivated by the appearance of the Schrödinger equation.

It is possible to relax these conditions somewhat for special purposes.[nb 5] If these requirements are not met, it is not possible to interpret the wave function as a probability amplitude.[28]

This does not alter the structure of the Hilbert space that these particular wave functions inhabit, but it should be pointed out that the subspace of the square-integrable functions L^2, which is a Hilbert space, satisfying the second requirement *is not closed* in L^2, hence not a Hilbert space in itself.[nb 6] The functions that does not meet the requirements are still needed for both technical and practical reasons.[nb 7][nb 8]

31.3 Definition (one spinless particle in 1d)

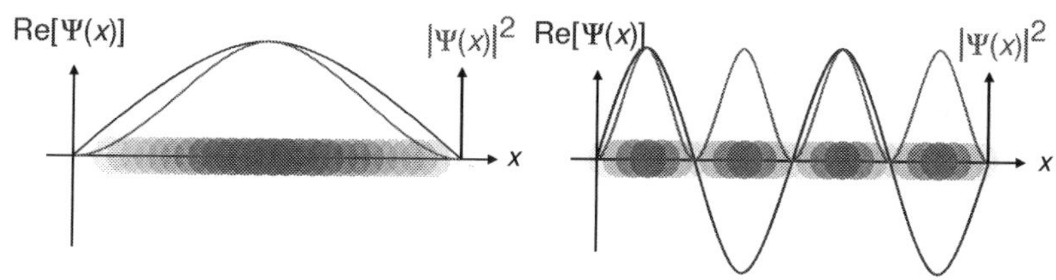

Standing waves for a particle in a box, examples of stationary states.

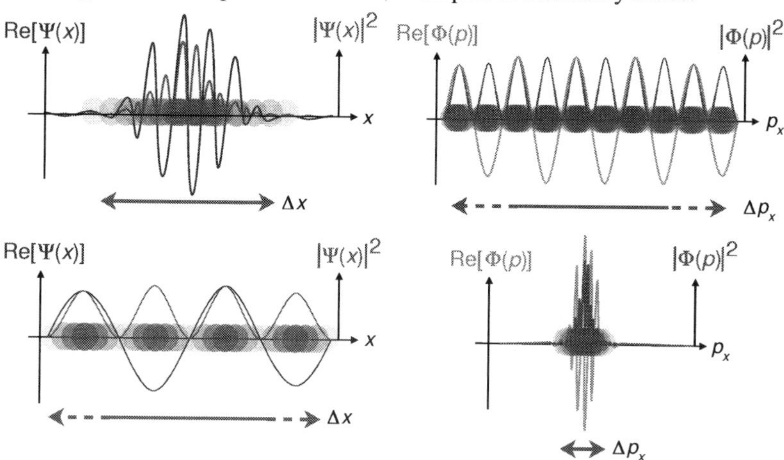

Travelling waves of a free particle.
The real parts of position wave function $\Psi(x)$ and momentum wave function $\Phi(p)$, and corresponding probability densities $|\Psi(x)|^2$ and $|\Phi(p)|^2$, for one spin-0 particle in one x or p dimension. The colour opacity of the particles corresponds to the probability density (*not* the wave function) of finding the particle at position x or momentum p.

For now, consider the simple case of a single particle, without spin, in one spatial dimension. More general cases are discussed below.

31.3.1 Position-space wave function

The state of such a particle is completely described by its wave function,

$$\Psi(x,t)\,,$$

where x is position and t is time. This is a complex-valued function of two real variables x and t.

If interpreted as a probability amplitude, the square modulus of the wave function, the positive real number

$$|\Psi(x,t)|^2 = \Psi(x,t)^*\Psi(x,t) = \rho(x,t),$$

is interpreted as the probability density that the particle is at x. The asterisk indicates the complex conjugate. If the particle's position is measured, its location cannot be determined from the wave function, but is described by a probability distribution. The probability that its position x will be in the interval $a \le x \le b$ is the integral of the density over this interval:

$$P_{a\le x\le b}(t) = \int_a^b dx\,|\Psi(x,t)|^2$$

where t is the time at which the particle was measured. This leads to the **normalization condition**:

$$\int_{-\infty}^{\infty} dx\, |\Psi(x,t)|^2 = 1\,,$$

because if the particle is measured, there is 100% probability that it will be *somewhere.*

Since the Schrödinger equation is linear, if any number of wave functions Ψn for $n = 1, 2, \ldots$ are solutions of the equation, then so is their sum, and their scalar multiples by complex numbers an. Taking scalar multiplication and addition together is known as a linear combination:

$$\sum_n a_n \Psi_n(x,t) = a_1\Psi_1(x,t) + a_2\Psi_2(x,t) + \cdots$$

This is the superposition principle. Multiplying a wave function Ψ by any nonzero constant complex number c to obtain $c\Psi$ does not change any information about the quantum system, because c cancels in the Schrödinger equation for $c\Psi$.

31.3.2 Momentum-space wave function

The particle also has a wave function in momentum space:

$$\Phi(p,t)$$

where p is the momentum in one dimension, which can be any value from $-\infty$ to $+\infty$, and t is time.

All the previous remarks on superposition, normalization, etc. apply similarly. In particular, if the particle's momentum is measured, the result is not deterministic, but is described by a probability distribution:

$$P_{a\le p\le b}(t) = \int_a^b dp\, |\Phi(p,t)|^2\,,$$

and the normalization condition is:

$$\int_{-\infty}^{\infty} dp\, |\Phi\,(p,t)|^2 = 1\,.$$

31.3.3 Relation between wave functions

The position-space and momentum-space wave functions are Fourier transforms of each other, therefore both contain the same information, and either one alone is sufficient to calculate any property of the particle. As elements of **abstract physical Hilbert space**, whose elements are the possible states of the system under consideration, they represent the same object, but they are not equal when viewed as square-integrable functions. (A function and its Fourier transform are not equal.) For one dimension,[29]

$$\Phi(p,t) = \frac{1}{\sqrt{2\pi\hbar}} \int_{-\infty}^{\infty} dx\, e^{-ipx/\hbar}\Psi(x,t) \quad \rightleftharpoons \quad \Psi(x,t) = \frac{1}{\sqrt{2\pi\hbar}} \int_{-\infty}^{\infty} dp\, e^{ipx/\hbar}\Phi(p,t)$$

In practice, the position-space wave function is used much more often than the momentum-space wave function. The potential entering the Schrödinger equation determines in which basis the description is easiest. For the harmonic oscillator, x and p enter symmetrically, so there it doesn't matter which description one uses.

31.4 Definitions (other cases)

Following are the general forms of the wave function for systems in higher dimensions and more particles, as well as including other degrees of freedom than position coordinates or momentum components.

The position-space wave function of a single particle in three spatial dimensions is similar to the case of one spatial dimension above:

$$\Psi(\mathbf{r},t)$$

where **r** is the position vector in three-dimensional space, and t is time. As always $\Psi(\mathbf{r}, t)$ is a complex number, for this case a complex-valued function of four real variables.

If there are many particles, in general there is only one wave function, not a separate wave function for each particle. The fact that *one* wave function describes *many* particles is what makes quantum entanglement and the EPR paradox possible. The position-space wave function for N particles is written:[19]

$$\Psi(\mathbf{r}_1,\mathbf{r}_2\cdots\mathbf{r}_N,t)$$

where ***ri*** is the position of the ith particle in three-dimensional space, and t is time. Altogether, this is a complex-valued function of $3N + 1$ real variables.

In quantum mechanics there is a fundamental distinction between *identical particles* and *distinguishable* particles. For example, any two electrons are identical and fundamentally indistinguishable from each other; the laws of physics make it impossible to "stamp an identification number" on a certain electron to keep track of it.[29] This translates to a requirement on the wave function for a system of identical particles:

$$\Psi\left(\ldots\mathbf{r}_a,\ldots,\mathbf{r}_b,\ldots\right)=\pm\Psi\left(\ldots\mathbf{r}_b,\ldots,\mathbf{r}_a,\ldots\right)$$

where the + sign occurs if the particles are *all bosons* and – sign if they are *all fermions*. In other words, the wave function is either totally symmetric in the positions of bosons, or totally antisymmetric in the positions of fermions.[30] The physical interchange of particles corresponds to mathematically switching arguments in the wave function. The antisymmetry feature of fermionic wave functions leads to the Pauli principle. Generally, bosonic and fermionic symmetry requirements are the manifestation of particle statistics and are present in other quantum state formalisms.

For N *distinguishable* particles (no two being identical, i.e. no two having the same set of quantum numbers), there is no requirement for the wave function to be either symmetric or antisymmetric.

For a collection of particles, some identical with coordinates $\mathbf{r}_1$, $\mathbf{r}_2$, ... and others distinguishable $\mathbf{x}_1$, $\mathbf{x}_2$, ... (not identical with each other, and not identical to the aforementioned identical particles), the wave function is symmetric or antisymmetric in the identical particle coordinates ***ri*** only:

$$\Psi\left(\ldots\mathbf{r}_a,\ldots,\mathbf{r}_b,\ldots,\mathbf{x}_1,\mathbf{x}_2,\ldots\right)=\pm\Psi\left(\ldots\mathbf{r}_b,\ldots,\mathbf{r}_a,\ldots,\mathbf{x}_1,\mathbf{x}_2,\ldots\right)$$

Again, there is no symmetry requirement for the distinguishable particle coordinates ***xi***.

For a particle with spin, the wave function can be written in "position–spin space" as:

$$\Psi(\mathbf{r},t,s_z)$$

which is a complex-valued function of position **r** in three-dimensional space, time t, and s_z, the spin projection quantum number along the z axis. (The z axis is an arbitrary choice; other axes can be used instead if the wave function is transformed appropriately, see below.) The sz parameter, unlike **r** and t, is a *discrete variable*. For example, for a spin-1/2 particle, s_z can only be +1/2 or −1/2, and not any other value. (In general, for spin s, sz can be $s, s-1, \dots, -s+1, -s$.)

Often, the complex values of the wave function for all the spin numbers are arranged into a column vector, in which there are as many entries in the column vector as there are allowed values of sz. In this case, the spin dependence is placed in indexing the entries and the wave function is a complex vector-valued function of space and time only:

$$\Psi(\mathbf{r},t) = \begin{bmatrix} \Psi(\mathbf{r},t,s) \\ \Psi(\mathbf{r},t,s-1) \\ \vdots \\ \Psi(\mathbf{r},t,-(s-1)) \\ \Psi(\mathbf{r},t,-s) \end{bmatrix}$$

The wave function for N particles each with spin is the complex-valued function:

$$\Psi(\mathbf{r}_1, \mathbf{r}_2 \cdots \mathbf{r}_N, s_{z\,1}, s_{z\,2} \cdots s_{z\,N}, t)$$

Concerning the general case of N particles with spin in 3d, if Ψ is interpreted as a probability amplitude, the probability density is:

$$\rho\left(\mathbf{r}_1 \cdots \mathbf{r}_N, s_{z\,1} \cdots s_{z\,N}, t\right) = \left|\Psi\left(\mathbf{r}_1 \cdots \mathbf{r}_N, s_{z\,1} \cdots s_{z\,N}, t\right)\right|^2$$

and the probability that particle 1 is in region R_1 with spin $sz_1 = m_1$ *and* particle 2 is in region R_2 with spin $sz_2 = m_2$ etc. at time t is the integral of the probability density over these regions and spins:

$$P_{\mathbf{r}_1 \in R_1, s_{z\,1}=m_1, \dots, \mathbf{r}_N \in R_N, s_{z\,N}=m_N}(t) = \int_{R_1} d^3\mathbf{r}_1 \int_{R_2} d^3\mathbf{r}_2 \cdots \int_{R_N} d^3\mathbf{r}_N \left|\Psi\left(\mathbf{r}_1 \cdots \mathbf{r}_N, m_1 \cdots m_N, t\right)\right|^2$$

The multidimensional Fourier transforms of the position or position–spin space wave functions yields momentum or momentum–spin space wave functions.

31.4.1 Decompositions into products

For systems in time-independent potentials, the wave function can always be written as a function of the degrees of freedom multiplied by a time-dependent phase factor, the form of which is given by the Schrödinger equation. For the case of N particles position-spin space,

$$\Psi(\mathbf{r}_1, \mathbf{r}_2, \dots, \mathbf{r}_N, t, s_{z1}, s_{z2}, \dots, s_{zN}) = e^{-iEt/\hbar}\psi(\mathbf{r}_1, \mathbf{r}_2, \dots, \mathbf{r}_N, s_{z1}, s_{z2}, \dots, s_{zN}),$$

where E is the energy eigenvalue of the system corresponding to the eigenstate Ψ. Wave functions of this form are called stationary states.

In some situations, the wave function for a particle with spin factors into a product of a space function ψ and a spin function ξ, where each are complex-valued functions, and the time dependence can be placed in either function:

$$\Psi(\mathbf{r}, t, s_z) = \psi(\mathbf{r}, t)\xi(s_z) = \phi(\mathbf{r})\zeta(s_z, t).$$

The dynamics of each factor can be studied in isolation. This factorization is always possible when the orbital and spin angular momenta of the particle are separable in the Hamiltonian operator, that is, the Hamiltonian can be split into an orbital term and a spin term.[31] It is not possible for those interactions where an external field or any space-dependent quantity couples to the spin; examples include a particle in a magnetic field, and spin-orbit coupling. For the time-independent case this reduces to

$$\Psi(\mathbf{r},t,s_z) = e^{-iEt/\hbar}\psi(\mathbf{r})\xi(s_z)\,,$$

where again E is the energy eigenvalue of the system corresponding to the eigenstate Ψ. This extends to the case of N particles:

$$\Psi(\mathbf{r},t,s_z) = \psi(\mathbf{r}_1,\mathbf{r}_2,\ldots,\mathbf{r}_N,t)\xi(s_{z1},s_{z2},\ldots,s_{zN}) = \phi(\mathbf{r}_1,\mathbf{r}_2,\ldots,\mathbf{r}_N)\zeta(s_{z1},s_{z2},\ldots,s_{zN},t)\,.$$

and for the case of identical particles, each factor has to have the correct antisymmetry or symmetry, to make the overall wave function antisymmetric for fermions or symmetric for bosons.

31.5 Inner product

31.5.1 Position-space inner products

The **inner product** of two wave functions Ψ_1 and Ψ_2 is useful and important for a number of reasons given below. For the case of one spinless particle in 1d, it can be defined as the complex number (at time t)[nb 9]

$$\langle\Psi_1,\Psi_2\rangle = \int\limits_{-\infty}^{\infty} dx\,\Psi_1^*(x,t)\Psi_2(x,t).$$

More generally, the formulae for the inner products are integrals over all coordinates or momenta and sums over all spin quantum numbers. That is, for one spinless particle in 3d the inner product of two wave functions can be defined as the complex number:

$$\langle\Psi_1,\Psi_2\rangle = \int\limits_{\text{all space}} d^3\mathbf{r}\,\Psi_1^*(\mathbf{r},t)\Psi_2(\mathbf{r},t)\,,$$

while for many spinless particles in 3d:

$$\langle\Psi_1,\Psi_2\rangle = \int\limits_{\text{all space}} d^3\mathbf{r}_1 \int\limits_{\text{all space}} d^3\mathbf{r}_2\cdots \int\limits_{\text{all space}} d^3\mathbf{r}_N\,\Psi_1^*(\mathbf{r}_1\cdots\mathbf{r}_N,t)\Psi_2(\mathbf{r}_1\cdots\mathbf{r}_N,t)$$

(altogether, this is N three-dimensional volume integrals with differential volume elements d^3**r***i*, also written "dVi" or "*dxi dyi dzi*"). For one particle with spin in 3d:

$$\langle\Psi_1,\Psi_2\rangle = \sum_{\text{all } s_z}\int\limits_{\text{all space}} d^3\mathbf{r}\Psi_1^*(\mathbf{r},t,s_z)\Psi_2(\mathbf{r},t,s_z)\,,$$

and for the general case of N particles with spin in 3d:

$$\langle \Psi_1, \Psi_2 \rangle = \sum_{s_{z\,N}} \cdots \sum_{s_{z\,2}} \sum_{s_{z\,1}} \int_{\text{all space}} d^3\mathbf{r}_1 \int_{\text{all space}} d^3\mathbf{r}_2 \cdots \int_{\text{all space}} d^3\mathbf{r}_N \Psi_1^* \left(\mathbf{r}_1 \cdots \mathbf{r}_N, s_{z\,1} \cdots s_{z\,N}, t\right) \Psi_2 \left(\mathbf{r}_1 \cdots \mathbf{r}_N, s_{z\,1} \cdots s_{z\,N}, t\right)$$

(altogether, *N* three-dimensional volume integrals followed by *N* sums over the spins).

In the Copenhagen interpretation, the modulus squared of the inner product (a complex number) gives a real number

$$|\langle \Psi_1, \Psi_2 \rangle|^2 = P\left(\Psi_2 \rightarrow \Psi_1\right) ,$$

which is interpreted as the probability of the wave function Ψ_2 "collapsing" to the new wave function Ψ_1 upon measurement of an observable, whose eigenvalues are the possible results of the measurement, with Ψ_1 being an eigenvector of the resulting eigenvalue.

Although the inner product of two wave functions is a complex number, the inner product of a wave function Ψ with itself,

$$\langle \Psi, \Psi \rangle = \|\Psi\|^2 ,$$

is *always* a positive real number. The number $\|\Psi\|$ (not $\|\Psi\|^2$) is called the **norm** of the wave function Ψ, and is not the same as the modulus $|\Psi|$.

A wave function is normalized if:

$$\langle \Psi, \Psi \rangle = 1 .$$

If Ψ is not normalized, then dividing by its norm gives the normalized function $\Psi/\|\Psi\|$.

Two wave functions Ψ_1 and Ψ_2 are orthogonal if their inner product is zero:

$$\langle \Psi_1, \Psi_2 \rangle = 0 .$$

A set of wave functions Ψ_1, Ψ_2, ... are orthonormal if they are each normalized and are all orthogonal to each other:

$$\langle \Psi_m, \Psi_n \rangle = \delta_{mn} ,$$

where m and n each take values 1, 2, ..., and δmn is the Kronecker delta (+1 for $m = n$ and 0 for $m \neq n$). Orthonormality of wave functions is instructive to consider since this guarantees linear independence of the functions. (However, the wave functions do not have to be orthonormal and can still be linearly independent, but the inner product of Ψm and Ψn is more complicated than the mere δmn).

Returning to the superposition above:

$$\Psi = \sum_n a_n \psi_n$$

if the basis wave functions ψn are orthonormal, then the coefficients have a particularly simple form:

$$a_n = \langle \psi_n, \Psi \rangle$$

If the basis wave functions were not orthonormal, then the coefficients would be different.

31.5.2 Momentum-space inner products

Analogous to the position case, the inner product of two wave functions $\Phi_1(p, t)$ and $\Phi_2(p, t)$ can be defined as:

$$\langle \Phi_1, \Phi_2 \rangle = \int_{-\infty}^{\infty} dp \, \Phi_1^*(p,t)\Phi_2(p,t) \, ,$$

and similarly for more particles in higher dimensions.

One particular solution to the time-independent Schrödinger equation is

$$\Psi_p(x) = e^{ipx/\hbar},$$

a plane wave, which can be used in the description of a particle with momentum exactly p, since it is an eigenfunction of the momentum operator. These functions are not normalizable to unity (they aren't square-integrable), so they are not really elements of physical Hilbert space. The set

$$\{\Psi_p(x,t), -\infty \leq p \leq \infty\}$$

forms what is called the **momentum basis**. This "basis" is not a basis in the usual mathematical sense. For one thing, since the functions aren't normalizable, they are instead **normalized to a delta function**,

$$\langle \Psi_p, \Psi_{p'} \rangle = \delta(p - p').$$

For another thing, though they are linearly independent, there are too many of them (they form an uncountable set) for a basis for physical Hilbert space. They can still be used to express all functions in it using Fourier transforms as described above.

31.6 Units of the wave function

Although wave functions are complex numbers, both the real and imaginary parts each have the same units (the imaginary unit i is a pure number without physical units). The units of ψ depend on the number of particles N the wave function describes, and the number of spatial or momentum dimensions n of the system.

When integrating $|\psi|^2$ over all the coordinates, the volume element $d^n\mathbf{r}_1 d^n\mathbf{r}_2 ... d^n\mathbf{r}N$ has units of $[\text{length}]^{Nn}$. Since the normalization conditions require the integral to be the unitless number 1, $|\psi|^2$ must have units of $[\text{length}]^{-Nn}$, thus the units of $|\psi|$ and hence ψ are $[\text{length}]^{-Nn/2}$. Likewise, in momentum space, length is replaced by momentum, and the units are $[\text{momentum}]^{-Nn/2}$. These results are true for particles of any spin, since for particles with spin, the summations are over dimensionless spin quantum numbers.

31.7 More on wave functions and abstract state space

Main article: Quantum state

As has been demonstrated, the set of all possible normalizable wave functions for a system with a particular choice of basis constitute a Hilbert space. This vector space is in general infinite-dimensional. Due to the multiple possible choices of basis, these Hilbert spaces are not unique. One therefore talks about an abstract Hilbert space, **state space**, where the choice of basis is left undetermined. The choice of basis corresponds to a choice of a maximal set of quantum numbers, each quantum number corresponding to an observable. Two observables corresponding to quantum numbers in the maximal set must commute, therefore, the basis isn't entirely arbitrary, but nonetheless, there are always several choices.

Specifically, each state is represented as an abstract vector in state space[32]

$$|\Psi\rangle$$

where $|\Psi\rangle$ is a "ket" (a vector) written in Dirac's bra–ket notation.[33] Kets that differ by multiplication by a scalar represent the same state. A **ray** in Hilbert space is a set of normalized vectors differing by a complex number of modulus 1. If $|\psi\rangle$ and $|\phi\rangle$ are two states in the vector space, and *a* and *b* are two complex numbers, then the linear combination

$$|\Psi\rangle = a|\psi\rangle + b|\phi\rangle$$

is also in the same vector space. The state space is postulated to have an inner product, denoted by

$$\langle\Psi_1|\Psi_2\rangle,$$

that is (usually, this differs) linear in the first argument and antilinear in the second argument. The dual vectors are denoted as "bras", $\langle\Psi|$. These are linear functionals, elements of the dual space to the state space. The inner product, once chosen, can be used to define a unique map from state space to its dual, see Riesz representation theorem. this map is antilinear. One has

$$\langle\Psi| = a^*\langle\psi| + b^*\langle\phi| \leftrightarrow a|\psi\rangle + b|\phi\rangle = |\Psi\rangle,$$

where the asterisk denotes the complex conjugate. For this reason one has under this map

$$\langle\Phi|\Psi\rangle = \langle\Phi|(|\Psi\rangle),$$

and one may, as a practical consequence, at least notation-wise in this formalism, ignore that bra's are dual vectors.

The state vector for the system evolves in time according to the Schrödinger equation, or other dynamical pictures of quantum mechanics- In bra-ket notation this reads,

$$i\hbar\frac{d}{dt}|\Psi\rangle = \hat{H}|\Psi\rangle$$

Abstract state space is also, by definition, required to be a Hilbert space. The only requirement missing for this in the description so far is completeness. See the quantum state article for more explanation of the Hilbert space formalism and its consequences to quantum physics.

The connection to the Hilbert spaces of wave functions is made as follows. If (a, b, ... l, m, ...) is a maximal set of quantum numbers, denote the state corresponding to *fixed choices* of these quantum numbers by

$$|a, b, \ldots, l, m, \ldots\rangle.$$

The wave function corresponding to an arbitrary state $|\Psi\rangle$ is denoted

$$\langle a, b, \ldots, l, m, \ldots|\Psi\rangle,$$

for a concrete example,

$$\Psi(x) = \langle x|\Psi\rangle.$$

There are several advantages to understanding wave functions as representing elements of an abstract vector space:

- All the powerful tools of linear algebra can be used to manipulate and understand wave functions. For example:
  - Linear algebra explains how a vector space can be given a basis, and then any vector in the vector space can be expressed in this basis. This explains the relationship between a wave function in position space and a wave function in momentum space, and suggests that there are other possibilities too.
  - Bra–ket notation can be used to manipulate wave functions.
- The idea that quantum states are vectors in an abstract vector space (technically, a complex projective Hilbert space) is completely general in all aspects of quantum mechanics and quantum field theory, whereas the idea that quantum states are complex-valued "wave" functions of space is only true in certain situations.

Following is a summary of the bra–ket formalism applied to wave functions, with general discrete or continuous bases.

31.7.1 Discrete and continuous bases

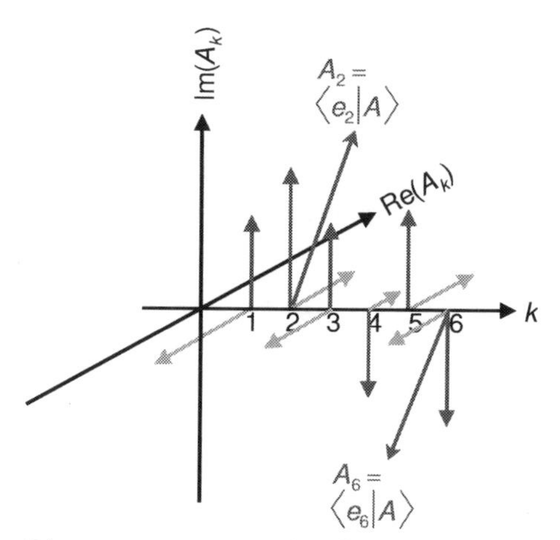

Discrete components Ak of a complex vector $|A\rangle = \sum k\ Ak|ek\rangle$, which belongs to a *countably infinite*-dimensional Hilbert space; there are countably infinitely many k values and basis vectors $|ek\rangle$.

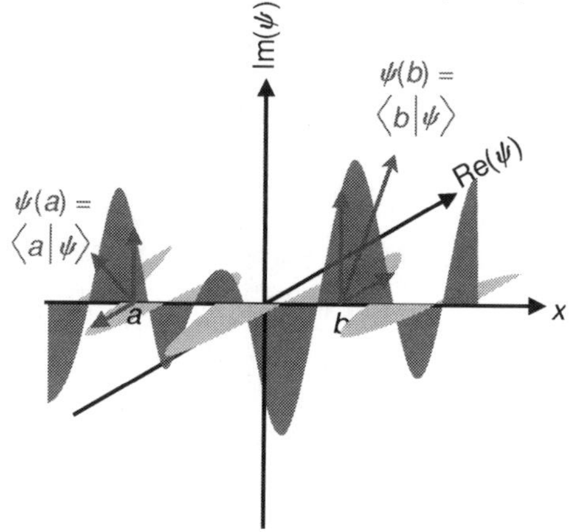

Continuous components $\psi(x)$ of a complex vector $|\psi\rangle = \int dx\ \psi(x)|x\rangle$, which belongs to an *uncountably infinite*-dimensional Hilbert space; there are uncountably infinitely many x values and basis vectors $|x\rangle$.

Components of complex vectors plotted against index number; discrete k and continuous x. Two probability amplitudes out of infinitely many are highlighted.

A Hilbert space with a discrete basis $|\varepsilon_i\rangle$ for $i = 1, 2...n$ is orthonormal if the inner product of all pairs of basis kets are given by the Kronecker delta:

$$\langle \varepsilon_i | \varepsilon_j \rangle = \delta_{ij} \,.$$

Orthonormal bases are convenient to work with because the inner product of two vectors have simple expressions. A wave function $|\Psi\rangle$ expressed in this discrete basis of the Hilbert space, and the corresponding bra in the dual space, are respectively given by:

$$|\Psi\rangle = \sum_{i=1}^{n} c_i |\varepsilon_i\rangle = \begin{bmatrix} c_1 \\ \vdots \\ c_n \end{bmatrix} \qquad \langle \Psi| = |\Psi\rangle^\dagger = \sum_{i=1}^{n} c_i^* \langle \varepsilon_i| = \begin{bmatrix} c_1^* & \cdots & c_n^* \end{bmatrix},$$

where the complex numbers

$$c_i = \langle \varepsilon_i | \Psi \rangle$$

are the components of the vector. The column vector is a useful way to list the numbers, and operations on the entire vector can be done according to matrix addition and multiplication. The entire vector $|\Psi\rangle$ is independent of the basis, but the components depend on the basis. If a change of basis is made, the components of the vector must also change to compensate.

A Hilbert space with a continuous basis { $|\varepsilon\rangle$ } is orthonormal if the inner product of all pairs of basis kets are given by the Dirac delta function:

$$\langle \varepsilon | \varepsilon' \rangle = \delta(\varepsilon - \varepsilon') \,.$$

As with the discrete bases, a symbol ε is used in the basis states, two common notations are $|\varepsilon\rangle$ and sometimes $|\Psi_\varepsilon\rangle$. A particular basis ket may be subscripted $|\varepsilon_0\rangle \equiv |\Psi_{\varepsilon_0}\rangle$ or primed $|\varepsilon'\rangle \equiv |\Psi_{\varepsilon'}\rangle$, or simply given another symbol in place of ε.

While discrete basis vectors are summed over a discrete index, continuous basis vectors are integrated over a continuous index (a variable of a function). In what follows, all integrals are with respect to the real-valued basis variable ε (not complex-valued), over the required range. Usually this is just the real line or subsets of it. The state $|\Psi\rangle$ in the continuous basis of the Hilbert space, with the corresponding bra in the dual space, are respectively given by:[34]

$$|\Psi\rangle = \int d\varepsilon |\varepsilon\rangle \Psi(\varepsilon) \,, \quad \langle \Psi| = \int d\varepsilon \langle \varepsilon | \Psi(\varepsilon)^* \,,$$

where the components are the complex-valued functions

$$\Psi(\varepsilon) = \langle \varepsilon | \Psi \rangle$$

of a real variable ε.

31.7.2 Completeness conditions

The **completeness conditions** (also called **closure relations**) are

$$\sum_{i=1}^{n}|\varepsilon_i\rangle\langle\varepsilon_i| = 1\,, \quad \int d\varepsilon\,|\varepsilon\rangle\langle\varepsilon| = 1$$

for discrete and continuous orthonormal bases, respectively. An orthonormal set of kets form bases if and only if they satisfy these relations.[34] In each case, the equality to unity means this is an identity operator; its action on any state leaves it unchanged. Multiplying any state on the right of these gives the representation of the state $|\Psi\rangle$ in the basis. The inner product of a first state $|\Psi_1\rangle$ with a second $|\Psi_2\rangle$ can also be obtained by multiplying $|\Psi_1\rangle$ on the left and $|\Psi_2\rangle$ on the right of the relevant completeness condition.

31.7.3 Inner product

Physically, the nature of the inner product is dependent on the basis in use, because the basis is chosen to reflect the quantum state of the system.

If $|\Psi_1\rangle$ is a state in the above basis with components c_1, c_2, ..., cn and $|\Psi_2\rangle$ is another state in the same basis with components z_1, z_2, ..., zn, the inner product is the complex number:

$$\langle\Psi_1|\Psi_2\rangle = \left(\sum_i z_i^*\langle\varepsilon_i|\right)\left(\sum_j c_j|\varepsilon_j\rangle\right) = \sum_{ij} z_i^* c_j\langle\varepsilon_i|\varepsilon_j\rangle = \sum_i z_i^* c_i\,.$$

If $|\Psi_1\rangle$ is a state in the above continuous basis with components $\Psi_1(\varepsilon')$, and $|\Psi_2\rangle$ is another state in the same basis with components $\Psi_2(\varepsilon)$, the inner product is the complex number:

$$\langle\Psi_1|\Psi_2\rangle = \left(\int d\varepsilon'\Psi_1(\varepsilon')^*\langle\varepsilon'|\right)\left(\int d\varepsilon\Psi_2(\varepsilon)|\varepsilon\rangle\right) = \int d\varepsilon'\int d\varepsilon\Psi_1(\varepsilon')^*\Psi_2(\varepsilon)\langle\varepsilon'|\varepsilon\rangle = \int d\varepsilon\Psi_1(\varepsilon)^*\Psi_2(\varepsilon)\,.$$

where the integrals are taken over all ε and ε'.

The square of the **norm (magnitude)** of the state vector $|\Psi\rangle$ is given by the inner product of $|\Psi\rangle$ with itself, a real number:

$$\|\Psi\|^2 = \langle\Psi|\Psi\rangle = \sum_{j=1}^{n}|c_j|^2\,, \quad \|\Psi\|^2 = \langle\Psi|\Psi\rangle = \int d\varepsilon\,|\Psi(\varepsilon)|^2$$

for the discrete and continuous bases, respectively. Each say the projection of a complex probability amplitude onto itself is real. If $|\Psi\rangle$ is normalized, these expressions would be each separately equal to 1. If the state is not normalized, then dividing by its magnitude normalizes the state:

$$|\Psi_N\rangle = \frac{1}{\|\Psi\|}|\Psi\rangle$$

31.7.4 Normalized components and probabilities

In the literature, the following results are often presented with normalized wavefunctions. Here, we keep the normalization factors to show where they appear if the wavefunction is not already normalized.

For the discrete basis, projecting the normalized state $|\Psi N\rangle$ onto a particular state the system may collapse to, $|\varepsilon q\rangle$, gives the complex number;

$$\langle\varepsilon_q|\Psi_N\rangle = \langle\varepsilon_q|\frac{1}{\|\Psi\|}\left(\sum_{i=1}^{n} c_i|\varepsilon_i\rangle\right) = \frac{c_q}{\|\Psi\|},$$

so the modulus squared of this gives a real number;

$$P(\varepsilon_q) = |\langle\varepsilon_q|\Psi_N\rangle|^2 = \frac{|c_q|^2}{\|\Psi\|^2},$$

In the Copenhagen interpretation, this is the probability of state $|\varepsilon q\rangle$ occurring.

In the continuous basis, the projection of the normalized state onto some particular basis $|\varepsilon'\rangle$ is a complex-valued function;

$$\langle\varepsilon'|\Psi_N\rangle = \langle\varepsilon'|\left(\frac{1}{\|\Psi\|}\int d\varepsilon|\varepsilon\rangle\Psi(\varepsilon)\right) = \frac{1}{\|\Psi\|}\int d\varepsilon\langle\varepsilon'|\varepsilon\rangle\Psi(\varepsilon) = \frac{1}{\|\Psi\|}\int d\varepsilon\delta(\varepsilon'-\varepsilon)\Psi(\varepsilon) = \frac{\Psi(\varepsilon')}{\|\Psi\|},$$

so the squared modulus is a real-valued function

$$\rho(\varepsilon') = |\langle\varepsilon'|\Psi_N\rangle|^2 = \frac{|\Psi(\varepsilon')|^2}{\|\Psi\|^2}$$

In the Copenhagen interpretation, this function is the *probability density function* of measuring the observable ε', so integrating this with respect to ε' between $a \le \varepsilon' \le b$ gives:

$$P_{a\le\varepsilon\le b} = \frac{1}{\|\Psi\|^2}\int_a^b d\varepsilon'|\Psi(\varepsilon')|^2 = \frac{1}{\|\Psi\|^2}\int_a^b d\varepsilon'|\langle\varepsilon'|\Psi\rangle|^2,$$

the probability of finding the system with ε' between $\varepsilon' = a$ and $\varepsilon' = b$.

31.7.5 Wave function collapse

The physical meaning of the components of $|\Psi\rangle$ is given by the *wave function collapse postulate*, also known as wave function collapse. If the observable(s) ε (momentum and/or spin, position and/or spin, etc.) corresponding to states $|\varepsilon i\rangle$ has distinct and definite values, λi, and a measurement of that variable is performed on a system in the state $|\Psi\rangle$ then the probability of measuring λi is $|\langle\varepsilon i|\Psi\rangle|^2$. If the measurement yields λi, the system "collapses" to the state $|\varepsilon i\rangle$ irreversibly and instantaneously.

31.7.6 Time dependence

Main article: Dynamical pictures (quantum mechanics)

In the Schrödinger picture, the states evolve in time, so the time dependence is placed in $|\Psi\rangle$ according to[35]

$$|\Psi(t)\rangle = \sum_i |\varepsilon_i\rangle\langle\varepsilon_i|\Psi(t)\rangle = \sum_i c_i(t)|\varepsilon_i\rangle$$

for discrete bases, or

$$|\Psi(t)\rangle = \int d\varepsilon \, |\varepsilon\rangle\langle\varepsilon|\Psi(t)\rangle = \int d\varepsilon \, \Psi(\varepsilon, t)|\varepsilon\rangle$$

for continuous bases. However, in the Heisenberg picture the states |Ψ⟩ are constant in time and time dependence is placed in the Heisenberg operators, so |Ψ⟩ is not written as |Ψ(t)⟩. The Heisenberg picture wave function is a snapshot of a Schrödinger picture wave function, representing the whole spacetime history of the system. In the interaction picture (also called Dirac picture), the time dependence is placed in both the states and operators, the subdivision depending on the interaction term in the Hamiltonian, and can be viewed as intermediate between the Heisenberg and Schrödinger pictures. It is useful primarily in computing S-matrix elements.[36]

31.7.7 Tensor product

Further information: Bra-ket notation § Composite bras and kets

It is useful to introduce another operation with the physical interpretation of forming composite states from a collection of other states. This is the tensor product. Given two systems described by states |Ψ⟩ and |Φ⟩, the tensor product of the states forms the composite state denoted by |Ψ⟩⊗|Φ⟩ or simply without any operation symbol |Ψ⟩|Φ⟩, and the new system includes both of the original systems together. The tensor product state |Ψ⟩|Φ⟩ lives in a new space; the tensor product of the original Hilbert spaces. The bases spanning this space are the tensor products of the original bases. The product is not commutative in general, so |Ψ⟩|Φ⟩ ≠ |Φ⟩|Ψ⟩. If |Ψ⟩ has components *ci* and |Φ⟩ has components *zj*, each in a discrete orthonormal basis |ε*k*⟩, then:

$$|\Psi\rangle|\Phi\rangle = \left(\sum_i c_i|\varepsilon_i\rangle\right)\left(\sum_j z_j|\varepsilon_j\rangle\right) = \sum_{i,j} c_i z_j |\varepsilon_i\rangle|\varepsilon_j\rangle$$

and the notation can be simplified by abbreviating |*A*⟩ = |Ψ⟩|Φ⟩, *Aij* = *cizj*, and |*Eij*⟩ = |ε*i*⟩|ε*j*⟩, so that

$$|A\rangle = \sum_{i,j} A_{ij}|E_{ij}\rangle$$

The same procedure follows for continuous bases using integration. This can also be extended to any number of states, however taking tensor products for fermions and bosons is complicated by the symmetry requirements, see identical particles for general results.

31.8 Position representations

This section applies mostly to non-relativistic quantum mechanics. In relativistic quantum mechanics, eigenstates of the position operator are problematic due to a relativistic extension of Heisenberg's uncertainty principle. In relativistic quantum field theory, they are not used at all to label physical states. Associated to a particle perfectly localized to a point in space is an infinite uncertainty in energy. This leads to pair production in the relativistic regime. Thus such a particle automatically has companions, leading to a breakdown of the description.

31.8.1 State space for one spin-0 particle in 1d

For a spinless particle in one spatial dimension (the *x*-axis or real line), the state |Ψ⟩ can be expanded in terms of a continuum of basis states; |*x*⟩, also written |Ψ*x*⟩, corresponding to the set of all position coordinates *x*. The completeness condition for this basis is

$$1 = \int_{-\infty}^{\infty} dx\, |x\rangle\langle x|$$

and the orthogonality relation is

$$\langle x'|x\rangle = \delta(x' - x)$$

The state $|\Psi\rangle$ is expressed by:

$$|\Psi\rangle = \left(\int_{-\infty}^{\infty} dx\, |x\rangle\langle x| \right) |\Psi\rangle = \int_{-\infty}^{\infty} dx\, |x\rangle\langle x|\Psi\rangle = \int_{-\infty}^{\infty} dx\, \Psi(x)|x\rangle$$

in which the "wave function" described as a function is a component of the complex state vector.

$$\Psi(x) = \langle x|\Psi\rangle$$

The inner product as stated at the beginning of this article is:

$$\langle \Psi_1|\Psi_2\rangle = \langle \Psi_1| \left(\int_{-\infty}^{\infty} dx\, |x\rangle\langle x| \right) |\Psi_2\rangle = \int_{-\infty}^{\infty} dx\, \langle \Psi_1|x\rangle\langle x|\Psi_2\rangle = \int_{-\infty}^{\infty} dx\, \Psi_1(x)^* \Psi_2(x)\,.$$

If the particle is confined to a region *R* (a subset of the *x*-axis), the integrals in the inner product and completeness condition would be integrals over *R*.

31.8.2 State space (other cases)

The previous example can be extended to more particles in higher dimensions, and include spin.

For one spinless particle in 3d, the basis states are $|\mathbf{r}\rangle$ and any state vector $|\Psi\rangle$ in this space is expressed in terms of the basis vectors as $|\mathbf{r}\rangle$:

$$|\Psi\rangle = \int_{\text{all space}} d^3\mathbf{r}|\mathbf{r}\rangle\langle\mathbf{r}|\Psi\rangle$$

with components:

$$\langle\mathbf{r}|\Psi\rangle = \Psi(\mathbf{r})$$

For *N* spinless particles in 3d, the basis states are $|\mathbf{r}_1, ..., \mathbf{r}N\rangle$. This is the tensor product of the one-particle position bases $|\mathbf{r}_1\rangle$, $|\mathbf{r}_2\rangle$, ..., $|\mathbf{r}N\rangle$, each of which spans the separate one-particle Hilbert spaces, so $|\mathbf{r}_1, ..., \mathbf{r}N\rangle$ are the basis states for the tensor product of the one-particle Hilbert spaces (the Hilbert space for the composite many particle system). Any state vector $|\Psi\rangle$ in this space is

$$|\Psi\rangle = \int_{\text{all space}} d^3\mathbf{r}_N \cdots \int_{\text{all space}} d^3\mathbf{r}_2 \int_{\text{all space}} d^3\mathbf{r}_1 |\mathbf{r}_1, \mathbf{r}_2, \ldots, \mathbf{r}_N\rangle\langle\mathbf{r}_1, \mathbf{r}_2, \ldots, \mathbf{r}_N|\Psi\rangle$$

with components:

$$\langle\mathbf{r}_1, \mathbf{r}_2, \ldots, \mathbf{r}_N|\Psi\rangle = \Psi(\mathbf{r}_1, \mathbf{r}_2, \ldots, \mathbf{r}_N)$$

For one particle with spin in 3d, the basis states are $|\mathbf{r}, s_z\rangle$, the tensor product of the position basis $|\mathbf{r}\rangle$ and spin basis $|s_z\rangle$, which exists in a new space from the spin space and position space alone. Any state $|\Psi\rangle$ in this space is:

$$|\Psi\rangle = \sum_{s_z} \int_{\text{all space}} d^3\mathbf{r} |\mathbf{r}, s_z\rangle\langle\mathbf{r}, s_z|\Psi\rangle$$

with components:

$$\langle\mathbf{r}, s_z|\Psi\rangle = \Psi(\mathbf{r}, s_z)$$

For N particles with spin in 3d, the basis states are $|\mathbf{r}_1, \ldots, \mathbf{r}_N, s_{z\,1}, \ldots, s_{z\,N}\rangle$, the tensor product of the position basis $|\mathbf{r}_1, \ldots, \mathbf{r}_N\rangle$ and spin basis $|s_{z\,1}, \ldots, s_{z\,N}\rangle$, which exists in a new space from the spin space and position space alone. Any state in this space is:

$$|\Psi\rangle = \sum_{s_{z\,1},\ldots,s_{z\,N}} \int_{\text{all space}} d^3\mathbf{r}_N \cdots \int_{\text{all space}} d^3\mathbf{r}_1 \, |\mathbf{r}_1, \ldots, \mathbf{r}_N, s_{z\,1}, \ldots, s_{z\,N}\rangle\langle\mathbf{r}_1, \ldots, \mathbf{r}_N, s_{z\,1}, \ldots, s_{z\,N}|\Psi\rangle$$

with components:

$$\langle\mathbf{r}_1, \ldots, \mathbf{r}_N, s_{z\,1}, \ldots, s_{z\,N}|\Psi\rangle = \Psi(\mathbf{r}_1, \ldots, \mathbf{r}_N, s_{z\,1}, \ldots, s_{z\,N})$$

If the particles are restricted to regions of position space, then the integrals in the completeness relations are taken over those regions, rather than the entire coordinate space. For the general case of many particles with spin in 3d, if particle 1 is in region R_1, particle 2 is in region R_2, and so on, the state in this position–spin representation is:

$$|\Psi\rangle = \sum_{s_{z\,1},\ldots,s_{z\,N}} \int_{R_N} d^3\mathbf{r}_N \cdots \int_{R_1} d^3\mathbf{r}_1 \, \Psi(\mathbf{r}_1, \ldots, \mathbf{r}_N, s_{z\,1}, \ldots, s_{z\,N})|\mathbf{r}_1, \ldots, \mathbf{r}_N, s_{z\,1}, \ldots, s_{z\,N}\rangle$$

The orthogonality relation for this basis is:

$$\langle\mathbf{x}_1, \ldots, \mathbf{x}_N, m_1, \ldots, m_N|\mathbf{r}_1, \ldots, \mathbf{r}_N, s_{z\,1}, \ldots, s_{z\,N}\rangle = \delta_{m_1\,s_{z\,1}} \cdots \delta_{m_N\,s_{z\,N}}\delta(\mathbf{x}_1 - \mathbf{r}_1)\cdots\delta(\mathbf{x}_N - \mathbf{r}_N)$$

and the inner product of $|\Psi_1\rangle$ and $|\Psi_2\rangle$ is:

$$\langle\Psi_1|\Psi_2\rangle = \sum_{s_{z\,1},\ldots,s_{z\,N}} \int_{R_N} d^3\mathbf{r}_N \cdots \int_{R_1} d^3\mathbf{r}_1 \Psi_1(\mathbf{r}_1, \ldots, \mathbf{r}_N, s_{z\,1}, \ldots, s_{z\,N})^*\Psi_2(\mathbf{r}_1, \ldots, \mathbf{r}_N, s_{z\,1}, \ldots, s_{z\,N}).$$

Momentum space wave functions are similar, using the momentum vectors of the particles as continuous bases, namely $|\mathbf{p}\rangle$, $|\mathbf{p}_1, \mathbf{p}_2, \ldots, \mathbf{p}_N\rangle$, etc.

31.9 Ontology

Main article: Interpretations of quantum mechanics

Whether the wave function really exists, and what it represents, are major questions in the interpretation of quantum mechanics. Many famous physicists of a previous generation puzzled over this problem, such as Schrödinger, Einstein and Bohr. Some advocate formulations or variants of the Copenhagen interpretation (e.g. Bohr, Wigner and von Neumann) while others, such as Wheeler or Jaynes, take the more classical approach[37] and regard the wave function as representing information in the mind of the observer, i.e. a measure of our knowledge of reality. Some, including Schrödinger, Bohm and Everett and others, argued that the wave function must have an objective, physical existence. Einstein thought that a complete description of physical reality should refer directly to physical space and time, as distinct from the wave function, which refers to an abstract mathematical space.[38]

31.10 Examples

31.10.1 Free particle

Main article: Free particle

A free particle in 3d with wave vector **k** and angular frequency ω has a wave function

$$\Psi(\mathbf{r},t) = Ae^{i(\mathbf{k}\cdot\mathbf{r}-\omega t)}\,.$$

31.10.2 Particle in a box

Main article: Particle in a box

A particle is restricted to a 1D region between $x = 0$ and $x = L$; its wave function is:

$$\begin{aligned}\Psi(x,t) &= Ae^{i(kx-\omega t)}, & 0 \le x \le L\\ \Psi(x,t) &= 0, & x < 0, x > L\end{aligned}$$

To normalize the wave function we need to find the value of the arbitrary constant A; solved from

$$\int_{-\infty}^{\infty} dx\, |\Psi|^2 = 1.$$

From Ψ, we have $|\Psi|^2 = A^2$, so the integral becomes;

$$\int_{-\infty}^{0} dx \cdot 0 + \int_{0}^{L} dx\, A^2 + \int_{L}^{\infty} dx \cdot 0 = 1,$$

Solving this equation gives $A = 1/\sqrt{L}$, so the normalized wave function in the box is;

$$\Psi(x,t) = \frac{1}{\sqrt{L}}e^{i(kx-\omega t)}, \quad 0 \le x \le L\,.$$

31.10.3 One-dimensional quantum tunnelling

Main articles: Finite potential barrier and Quantum tunnelling
One of most prominent features of the wave mechanics is a possibility for a particle to reach a location with a prohibitive (in classical mechanics) force potential. In the one-dimensional case of particles with energy less than V_0 in the square potential

$$V(x) = \begin{cases} V_0 & |x| < a \\ 0 & \text{otherwise}, \end{cases}$$

the steady-state solutions to the wave equation have the form (for some constants k, κ)

$$\psi(x) = \begin{cases} A_{\mathrm{r}} \exp(ikx) + A_{\mathrm{l}} \exp(-ikx) & x < -a, \\ B_{\mathrm{r}} \exp(\kappa x) + B_{\mathrm{l}} \exp(-\kappa x) & |x| \le a, \\ C_{\mathrm{r}} \exp(ikx) + C_{\mathrm{l}} \exp(-ikx) & x > a. \end{cases}$$

Note that these wave functions are not normalized; see scattering theory for discussion.

The standard interpretation of this is as a stream of particles being fired at the step from the left (the direction of negative x): setting $A_r = 1$ corresponds to firing particles singly; the terms containing A_r and C_r signify motion to the right, while A_l and C_l – to the left. Under this beam interpretation, put $C_l = 0$ since no particles are coming from the right. By applying the continuity of wave functions and their derivatives at the boundaries, it is hence possible to determine the constants above.

31.10.4 Quantum Dots

In a semiconductor crystallite whose radius is smaller than the size of its exciton Bohr radius, the excitons are squeezed, leading to quantum confinement. The energy levels can then be modeled using the particle in a box model in which the energy of different states is dependent on the length of the box.

31.10.5 Other

Some examples of wave functions for specific applications include:

- Finite square well
- Delta potential
- Quantum harmonic oscillator
- Hydrogen atom and Hydrogen-like atom

31.11 See also

- Boson
- de Broglie–Bohm theory
- Double-slit experiment
- Faraday wave

- Fermion
- Schrödinger equation
- Wave function collapse
- Wave packet
- Phase space formulation of quantum mechanics, wave functions are replaced by quasi-probability distributions that place the position and momenta variables on equal footing.

31.12 Remarks

[1] For this statement to make sense, the observables need to be elements of a maximal commuting set. To see this, it is a simple matter to note that, for example, the momentum operator of the i'th particle in an n-particle system is *not* a generator of any symmetry in nature. On the other hand, the *total* angular momentum *is* a generator of a symmetry in nature; the translational symmetry.

[2] The resulting basis may or may not technically be a basis in the mathematical sense of Hilbert spaces. For instance, states of definite position and definite momentum are not square integrable. This may be overcome with the use of wave packets or by enclosing the system in a "box". See further remarks below.

[3] In technical terms, this is formulated the following way. The inner product yields a norm. This norm in turn induces a metric. If this metric is complete, then the aforementioned limits will be in the function space. The inner product space is then called complete. A complete inner product space is a Hilbert space. The abstract state space is always taken as a Hilbert space. The matching requirement for the function spaces is a natural one. The Hilbert space property of the abstract state space was originally extracted from the observation that the function spaces forming normalizable solutions to the Schrödinger equation are Hilbert spaces.

[4] As is explained in a later footnote, the integral must be taken to be the Lebesgue integral, the Riemann integral is not sufficient.

[5] One such relaxation is that the wave function must belong to the Sobolev space $W^{1,2}$. It means that it is differentiable in the sense of distributions, and its gradient is square-integrable. This relaxation is necessary for potentials that are not functions but are distributions, such as the Dirac delta function.

[6] It is easy to visualize a sequence of functions meeting the requirement that converges to a *discontinuous* function. For this, modify an example given in Inner product space#Examples. This element though *is* an element of L^2.

[7] For instance, in perturbation theory one may construct a sequence of functions approximating the true wave function. This sequence will be guaranteed to converge in a larger space, but without the assumption of a full-fledged Hilbert space, it will not be guaranteed that the convergence is to a function in the relevant space and hence solving the original problem.

[8] Some functions not being square-integrable, like the plane-wave free particle solutions are necessary for the description as outlined in a previous note and also further below.

[9] The functions are here assumed to be elements of L^2, the space of square integrable functions. The elements of this space are more precisely equivalence classes of square integrable functions, two functions declared equivalent if they differ on a set of Lebesgue measure 0. This is necessary to obtain an inner product (that is, $(\Psi, \Psi) = 0 \Rightarrow \Psi \equiv 0$) as opposed to a **semi-inner product**. The integral is taken to be the Lebesque integral. This is essential for completeness of the space, thus yielding a complete inner product space = Hilbert space.

31.13 Notes

[1] Born 1927, pp. 354–357

[2] Heisenberg 1958, p. 143

[3] Heisenberg, W. (1927/1985/2009). Heisenberg is translated by Camilleri 2009, p. 71, (from Bohr 1985, p. 142).

[4] Murdoch 1987, p. 43

[5] de Broglie 1960, p. 48

[6] Landau Lifshitz, p. 6

[7] Newton 2002, pp. 19–21

[8] Lerner & Trigg 1991, pp. 1223–1229

[9] Einstein 1905, pp. 132–148 (in German), Arons & Peppard 1965, p. 367 (in English)

[10] Einstein 1916, pp. 47–62 and a nearly identical version Einstein 1917, pp. 121–128 translated in ter Haar 1967, pp. 167–183.

[11] de Broglie 1923, pp. 507–510,548,630

[12] Hanle 1977, pp. 606–609

[13] Schrödinger 1926, pp. 1049–1070

[14] Tipler, Mosca & Freeman 2008

[15] Weinberg 2013

[16] Born 1926a, translated in Wheeler & Zurek 1983 at pages 52–55.

[17] Born 1926b, translated in Ludwig 1968, pp. 206–225. Also here.

[18] Young & Freedman 2008, p. 1333

[19] Atkins 1974

[20] Martin & Shaw 2008

[21] Pauli 1927, pp. 601–623.

[22] Weinberg (2002) takes the standpoint that quantum field theory appears the way it does because it is the *only* way to reconcile quantum mechanics with special relativity.

[23] Weinberg (2002) See especially chapter 5, where some of these results are derived.

[24] Weinberg 2002 Chapter 4.

[25] Weinberg 2002

[26] Eisberg & Resnick 1985

[27] Rae 2008

[28] Atkins 1974, p. 258

[29] Griffiths 2004

[30] Zettili 2009, p. 463

[31] Shankar 1994, p. 378–379

[32] Dirac 1982

[33] Dirac 1939

[34] (Peleg et al. 2010) pp. 64–65.

[35] (Peleg et al. 2010, pp. 68–69)

[36] Weinberg 2002 Chapter 3, Scattering matrix.

[37] Jaynes 2003

[38] Einstein 1998, p. 682

31.14 References

- Atkins, P. W. (1974). *Quanta: A Handbook of Concepts*. ISBN 0-19-855494-X.
- Arons, A. B.; Peppard, M. B. (1965). “Einstein’s proposal of the photon concept: A translation of the *Annalen der Physik* paper of 1905” (PDF). *American Journal of Physics* **33** (5): 367. Bibcode:1965AmJPh..33..367A. doi:10.1119/1.1971542.
- Bohr, N. (1985). J. Kalckar, ed. *Niels Bohr - Collected Works: Foundations of Quantum Physics I (1926 - 1932)* **6**. Amsterdam: North Holland. ISBN 9780444532893.
- Born, M. (1926a). “Zur Quantenmechanik der Stossvorgange”. *Z. f. Physik* **37**: 863–867. Bibcode:1926ZPhy...37..863B. doi:10.1007/bf01397477.
- Born, M. (1926b). “Quantenmechanik der Stossvorgange”. *Z. f. Physik* **38**: 803–827. Bibcode:1926ZPhy...38..803B. doi:10.1007/bf01397184.
- Born, M. (1927). “Physical aspects of quantum mechanics”. *Nature* **119**: 354–357. Bibcode:1927Natur.119..354B. doi:10.1038/119354a0.
- de Broglie, L. (1923). “Radiations—Ondes et quanta” [Radiation—Waves and quanta]. *Comptes Rendus* (in French) **177**: 507–510, 548, 630. Online copy (French) Online copy (English)
- de Broglie, L. (1960). *Non-linear Wave Mechanics: a Causal Interpretation*. Amsterdam: Elsevier.
- Camilleri, K. (2009). *Heisenberg and the Interpretation of Quantum Mechanics: the Physicist as Philosopher*. Cambridge UK: Cambridge University Press. ISBN 978-0-521-88484-6.
- Dirac, P. A. M. (1982). *The principles of quantum mechanics*. The international series on monographs on physics (4th ed.). Oxford University Press. ISBN 0 19 852011 5.
- Dirac, P. A. M. (1939). “A new notation for quantum mechanics”. *Mathematical Proceedings of the Cambridge Philosophical Society* **35** (3): 416–418. Bibcode:1939PCPS...35..416D. doi:10.1017/S0305004100021162.
- Einstein, A. (1905). "Über einen die Erzeugung und Verwandlung des Lichtes betreffenden heuristischen Gesichtspunkt”. *Annalen der Physik* (in German) **17** (6): 132–148. Bibcode:1905AnP...322..132E. doi:10.1002/andp.19053220607.
- Einstein, A. (1916). “Zur Quantentheorie der Strahlung”. *Mitteilungen der Physikalischen Gesellschaft Zürich* **18**: 47–62.
- Einstein, A. (1917). “Zur Quantentheorie der Strahlung”. *Physikalische Zeitschrift* (in German) **18**: 121–128. Bibcode:1917PhyZ...18..121E.
- Einstein, A. (1998). P. A. Schlipp, ed. *Albert Einstein: Philosopher-Scientist*. The Library of Living Philosophers **VII** (3rd ed.). La Salle Publishing Company, Illinois: Open Court. ISBN 0-87548-133-7.
- Eisberg, R.; Resnick, R. (1985). *Quantum Physics of Atoms, Molecules, Solids, Nuclei and Particles* (2nd ed.). John Wiley & Sons. ISBN 978-0-471-87373-0.
- Griffiths, D. J. (2004). *Introduction to Quantum Mechanics* (2nd ed.). Essex England: Pearson Education Ltd. ISBN 978-0131118928.

- Heisenberg, W. (1958). *Physics and Philosophy: the Revolution in Modern Science*. New York: Harper & Row.
- Hanle, P.A. (1977), "Erwin Schrodinger's Reaction to Louis de Broglie's Thesis on the Quantum Theory.", *Isis* **68** (4), doi:10.1086/351880
- Jaynes, E. T. (2003). G. Larry Bretthorst, ed. *Probability Theory: The Logic of Science*. Cambridge University Press. ISBN 978-0-521 59271-0.
- Landau, L.D.; Lifshitz, E. M. (1977). *Quantum Mechanics: Non-Relativistic Theory*. Vol. 3 (3rd ed.). Pergamon Press. ISBN 978-0-08-020940-1. Online copy
- Lerner, R.G.; Trigg, G.L. (1991). *Encyclopaedia of Physics* (2nd ed.). VHC Publishers. ISBN 0-89573-752-3.
- Ludwig, G. (1968). *Wave Mechanics*. Oxford UK: Pergamon Press. ISBN 0-08-203204-1. LCCN 66-30631.
- Murdoch, D. (1987). *Niels Bohr's Philosophy of Physics*. Cambridge UK: Cambridge University Press. ISBN 0-521-33320-2.
- Newton, R.G. (2002). *Quantum Physics: a Text for Graduate Student*. New York: Springer. ISBN 0-387-95473-2.
- Pauli, Wolfgang (1927). "Zur Quantenmechanik des magnetischen Elektrons". *Zeitschrift für Physik* (in German) **43**. Bibcode:1927ZPhy...43..601P. doi:10.1007/bf01397326.
- Peleg, Y.; Pnini, R.; Zaarur, E.; Hecht, E. (2010). *Quantum mechanics*. Schaum's outlines (2nd ed.). McGraw Hill. ISBN 978-0-07-162358-2.
- Rae, A.I.M. (2008). *Quantum Mechanics* **2** (5th ed.). Taylor & Francis Group. ISBN 1-5848-89705.
- Schrödinger, E. (1926). "An Undulatory Theory of the Mechanics of Atoms and Molecules" (PDF). *Physical Review* **28** (6): 1049–1070. Bibcode:1926PhRv...28.1049S. doi:10.1103/PhysRev.28.1049. Archived from the original (PDF) on 17 December 2008.
- Shankar, R. (1994). *Principles of Quantum Mechanics* (2nd ed.). ISBN 0306447908.
- Martin, B.R.; Shaw, G. (2008). *Particle Physics*. Manchester Physics Series (3rd ed.). John Wiley & Sons. ISBN 978-0-470-03294-7.
- ter Haar, D. (1967). *The Old Quantum Theory*. Pergamon Press. pp. 167–183. LCCN 66029628.
- Tipler, P. A.; Mosca, G.; Freeman (2008). *Physics for Scientists and Engineers – with Modern Physics* (6th ed.). ISBN 0-7167-8964-7.
- Weinberg, S. (2013), *Lectures in Quantum Mechanics*, Cambridge University Press, ISBN 978-1-107-02872-2
- Weinberg, S. (2002), *The Quantum Theory of Fields* **1**, Cambridge University Press, ISBN 0-521-55001-7
- Young, H. D.; Freedman, R. A. (2008). Pearson, ed. *Sears' and Zemansky's University Physics* (12th ed.). Addison-Wesley. ISBN 978-0-321-50130-1.
- Wheeler, J.A.; Zurek, W.H. (1983). *Quantum Theory and Measurement*. Princeton NJ: Princeton University Press.
- Zettili, N. (2009). *Quantum Mechanics: Concepts and Applications* (2nd ed.). ISBN 978-0-470-02679-3.

31.15 Further reading

- Yong-Ki Kim (September 2, 2000). "Practical Atomic Physics" (PDF). *National Institute of Standards and Technology* (Maryland): 1 (55 pages). Retrieved 2010-08-17.
- Polkinghorne, John (2002). *Quantum Theory, A Very Short Introduction*. Oxford University Press. ISBN 0-19-280252-6.

31.16 External links

- , , ,
- Normalization.
- Quantum Mechanics and Quantum Computation at BerkeleyX
- Einstein, *The quantum theory of radiation*

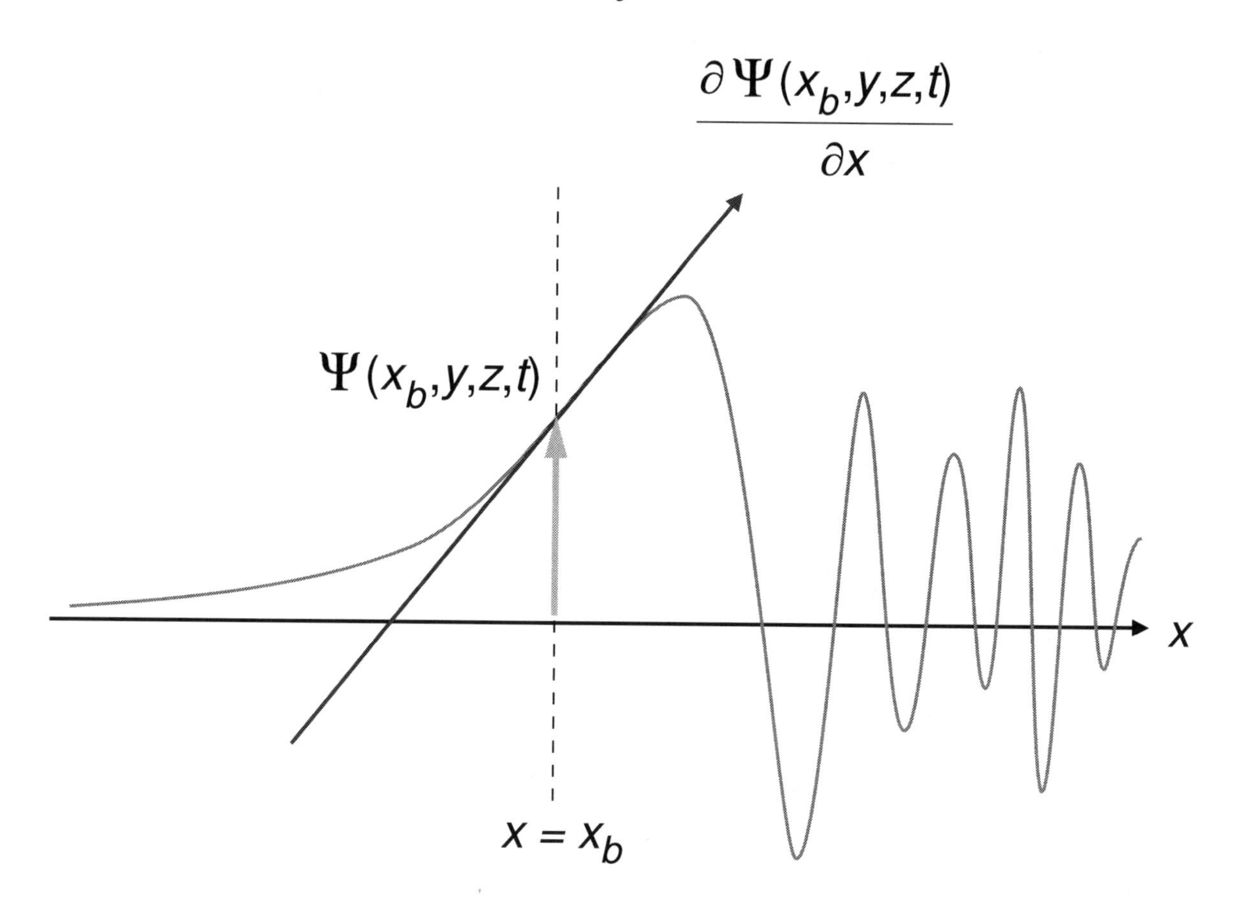
Continuously differentiable
$\frac{\partial \Psi(x_b,y,z,t)}{\partial x}$
$\Psi(x_b,y,z,t)$
x
$x = x_b$

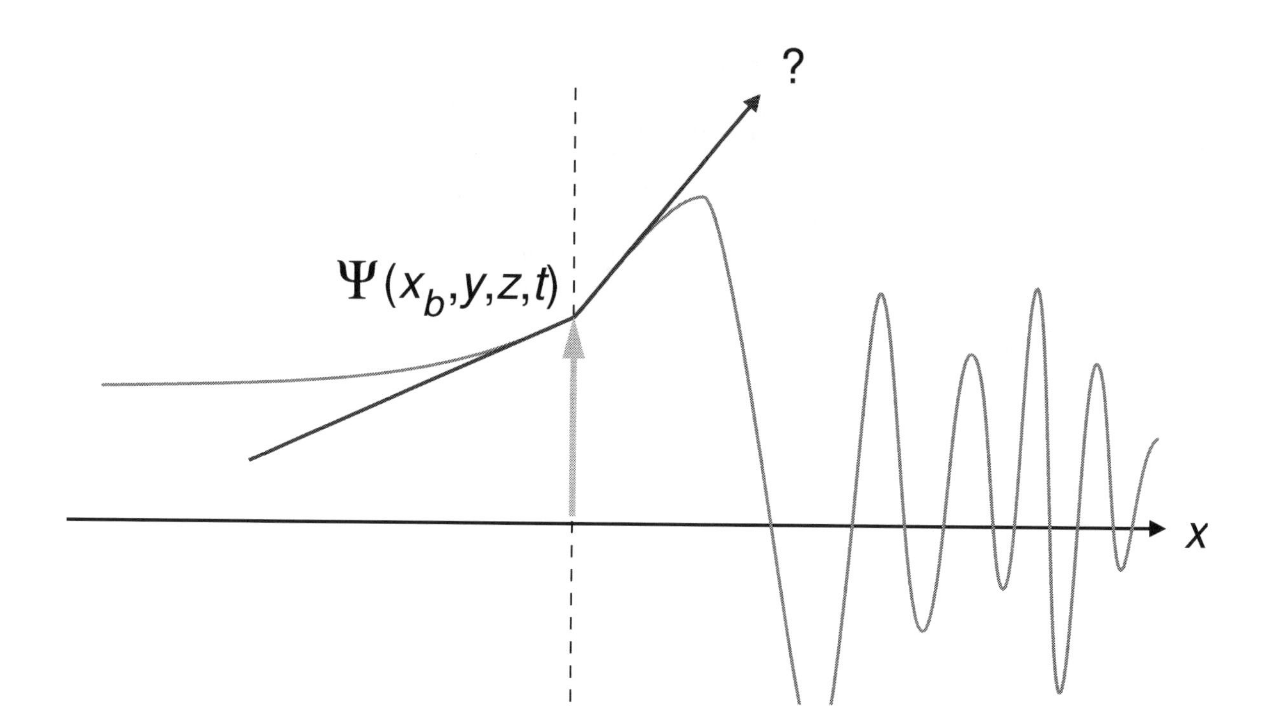
Discontinuous
?
$\Psi(x_b,y,z,t)$
x

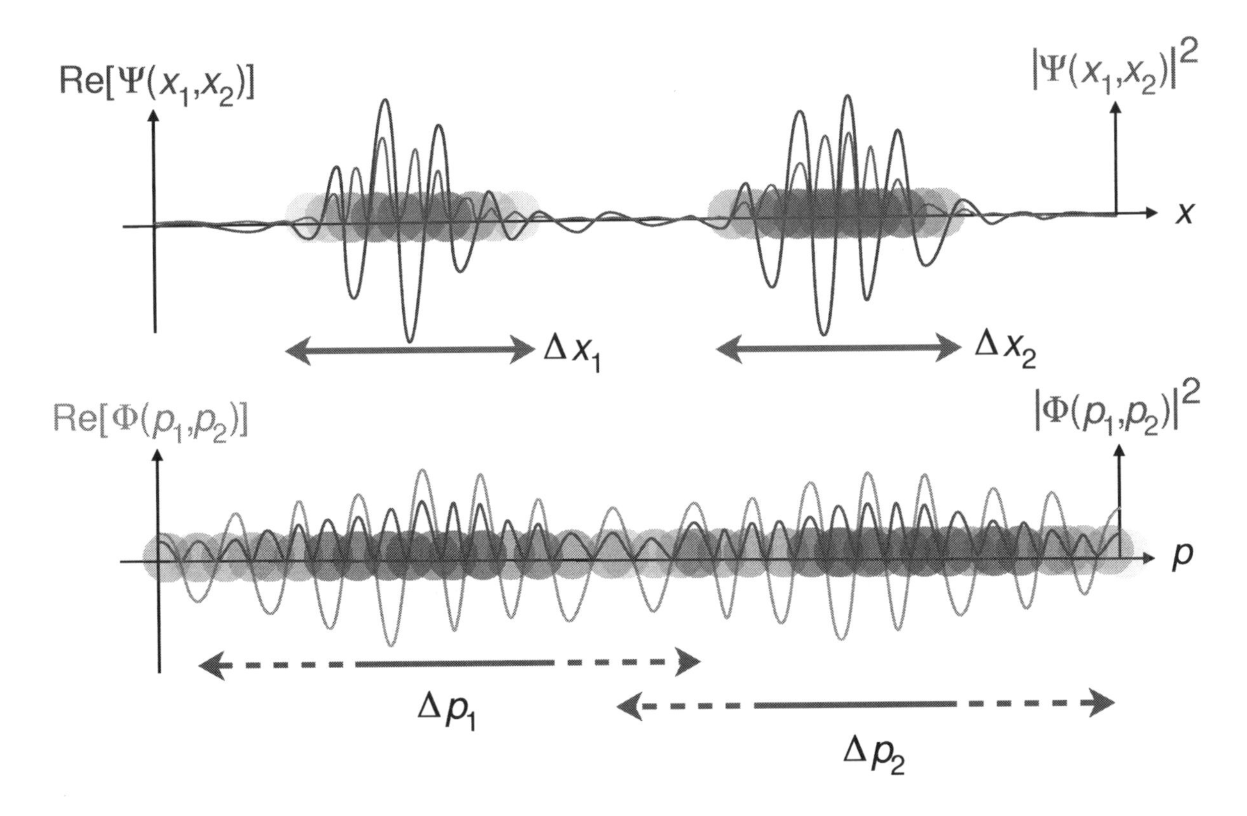

Traveling waves of two free particles, with two of three dimensions suppressed. Top is position space wave function, bottom is momentum space wave function, with corresponding probability densities.

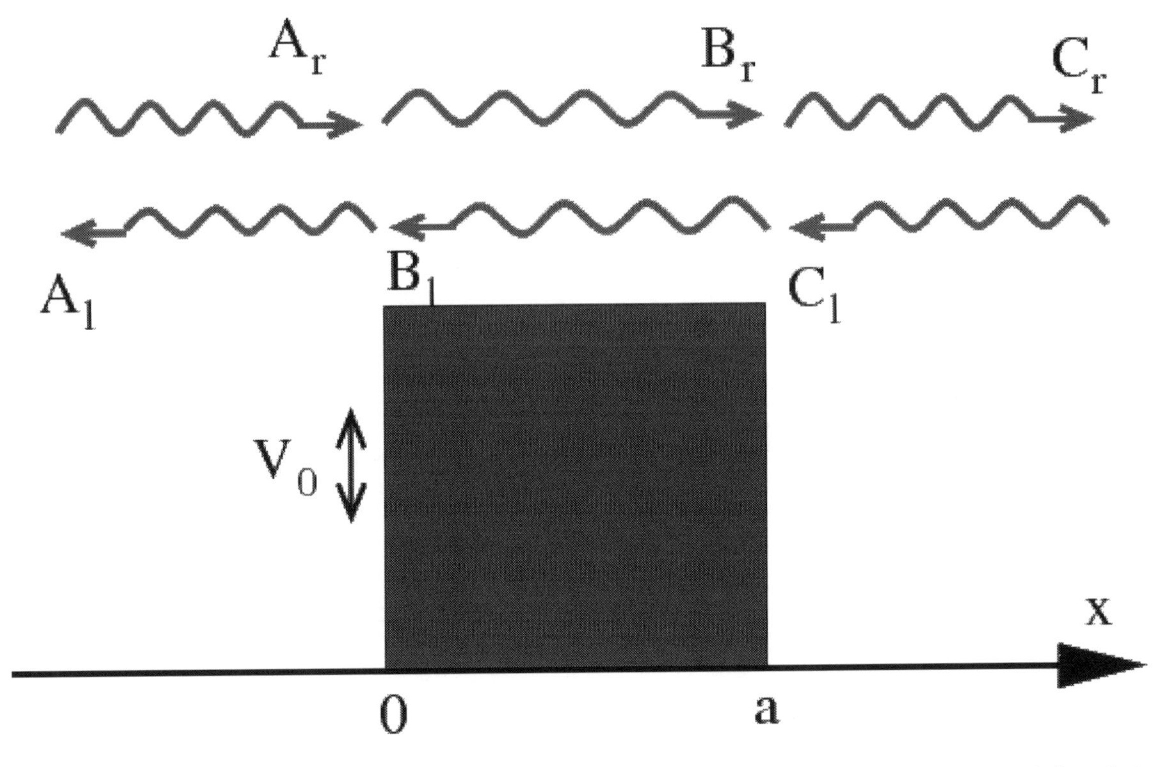

Scattering at a finite potential barrier of height V_0 The amplitudes and direction of left and right moving waves are indicated. In red, those waves used for the derivation of the reflection and transmission amplitude. $E > V_0$ for this illustration.

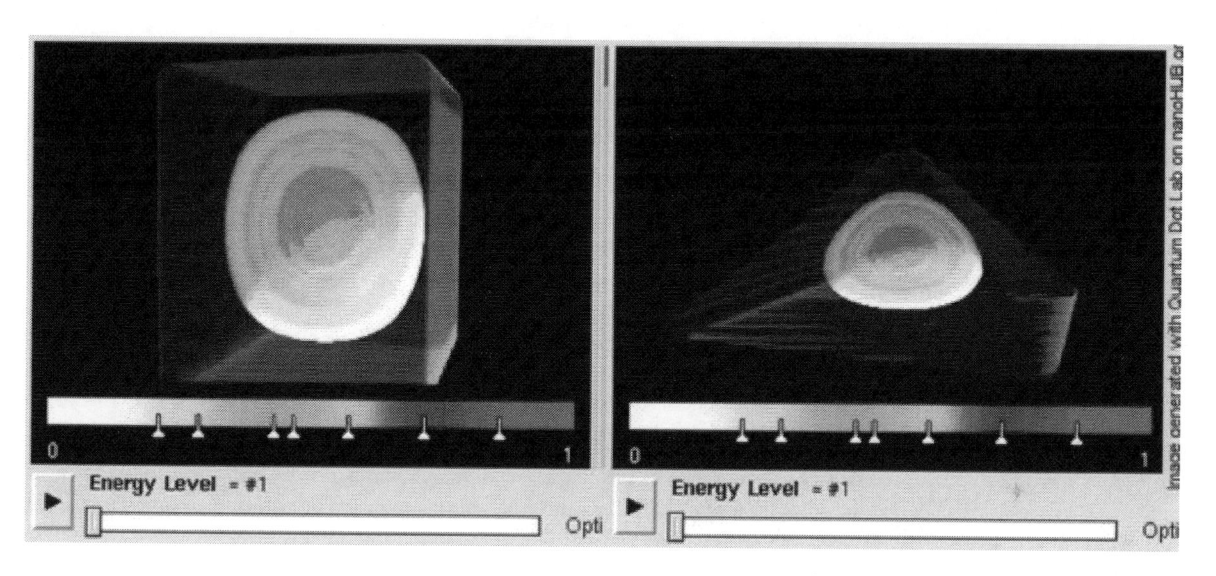

3D confined electron wave functions in a quantum dot. Here, rectangular and triangular-shaped quantum dots are shown. Energy states in rectangular dots are more s-type *and* p-type. *However, in a triangular dot the wave functions are mixed due to confinement symmetry. (Click for animation)*

Chapter 32

Quantum state

In quantum physics, **quantum state** refers to the state of a quantum system.

A quantum state can be either **pure** or **mixed**. A pure quantum state is represented by a vector, called a **state vector**, in a Hilbert space. For example, when dealing with the energy spectrum of the electron in a hydrogen atom, the relevant state vectors are identified by the principal quantum number, written $\{n\}$. For a more complicated case, consider Bohm's formulation of the EPR experiment, where the state vector

$$|\psi\rangle = \frac{1}{\sqrt{2}}\bigg(|\uparrow\downarrow\rangle - |\downarrow\uparrow\rangle\bigg)$$

involves superposition of joint spin states for two particles. Mathematically, a pure quantum state is represented by a state vector in a Hilbert space over complex numbers, which is a generalization of our more usual three-dimensional space.[1] If this Hilbert space is represented as a function space, then its elements are called wave functions.

A mixed quantum state corresponds to a probabilistic mixture of pure states; however, different distributions of pure states can generate equivalent (i.e., physically indistinguishable) mixed states. Mixed states are described by so-called density matrices. A pure state can also be recast as a density matrix; in this way, pure states can be represented as a subset of the more general mixed states.

For example, if the spin of an electron is measured in any direction, e.g. with a Stern–Gerlach experiment, there are two possible results: up or down. The Hilbert space for the electron's spin is therefore two-dimensional. A pure state here is represented by a two-dimensional complex vector (α, β) , with a length of one; that is, with

$$|\alpha|^2 + |\beta|^2 = 1,$$

where $|\alpha|$ and $|\beta|$ are the absolute values of α and β . A mixed state, in this case, is a 2×2 matrix that is Hermitian, positive-definite, and has trace 1.

Before a particular measurement is performed on a quantum system, the theory usually gives only a probability distribution for the outcome, and the form that this distribution takes is completely determined by the quantum state and the observable describing the measurement. These probability distributions arise for both mixed states and pure states: it is impossible in quantum mechanics (unlike classical mechanics) to prepare a state in which all properties of the system are fixed and certain. This is exemplified by the uncertainty principle, and reflects a core difference between classical and quantum physics. Even in quantum theory, however, for every observable there are some states that have an exact and determined value for that observable.[2][3]

32.1 Conceptual description

32.1.1 Pure states

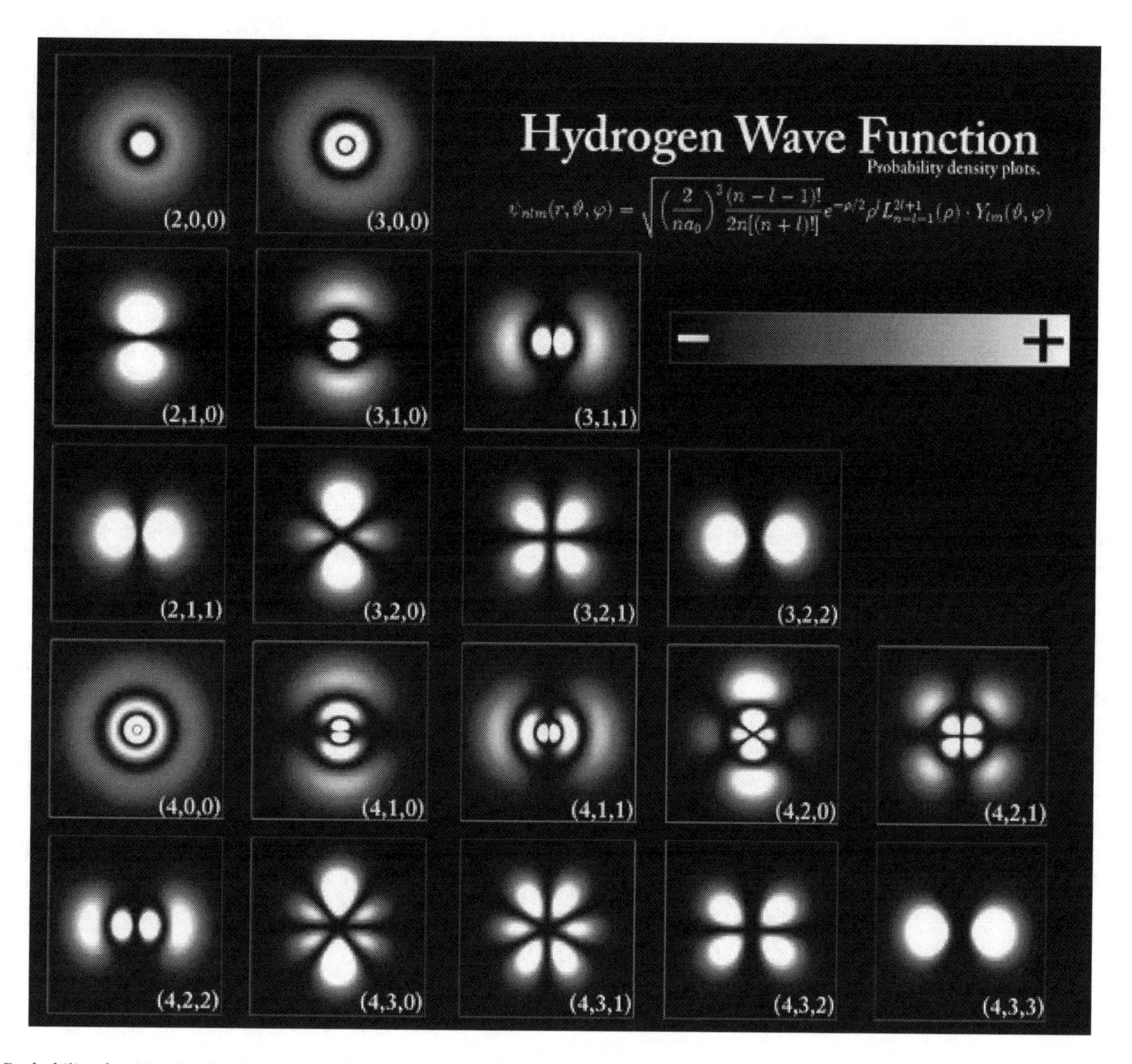

Probability densities for the electron of a hydrogen atom in different quantum states.

In the mathematical formulation of quantum mechanics, pure quantum states correspond to vectors in a Hilbert space, while each observable quantity (such as the energy or momentum of a particle) is associated with a mathematical operator. The operator serves as a linear function which acts on the states of the system. The eigenvalues of the operator correspond to the possible values of the observable, i.e. it is possible to observe a particle with a momentum of 1 kg·m/s if and only if one of the eigenvalues of the momentum operator is 1 kg·m/s. The corresponding eigenvector (which physicists call an **eigenstate**) with eigenvalue 1 kg·m/s would be a quantum state with a definite, well-defined value of momentum of 1 kg·m/s, with no quantum uncertainty. If its momentum were measured, the result is guaranteed to be 1 kg·m/s.

On the other hand, a system in a linear combination of multiple different eigenstates *does* in general have quantum uncertainty for the given observable. We can represent this linear combination of eigenstates as:

$$|\Psi(t)\rangle = \sum_n C_n(t)|\Phi_n\rangle$$

The coefficient which corresponds to a particular state in the linear combination is complex thus allowing interference effects between states. The coefficients are time dependent. How a quantum system changes in time is governed by the time evolution operator. The symbols | and ⟩ [4] surrounding the Ψ are part of bra–ket notation.

Statistical mixtures of states are different from a linear combination. A statistical mixture of states is a statistical ensemble of independent systems. Statistical mixtures represent the degree of knowledge whilst the uncertainty within quantum mechanics is fundamental. Mathematically, a statistical mixture is not a combination using complex coefficients, but rather a combination using real-valued, positive probabilities of different states Φ_n . A number P_n represents the probability of a randomly selected system being in the state Φ_n . Unlike the linear combination case each system is in a definite eigenstate.[5][6]

The expectation value $\langle A \rangle_\sigma$ of an observable *A* is a statistical mean of measured values of the observable. It is this mean, and the distribution of probabilities, that is predicted by physical theories.

There is no state which is simultaneously an eigenstate for *all* observables. For example, we cannot prepare a state such that both the position measurement *Q*(*t*) and the momentum measurement *P*(*t*) (at the same time *t*) are known exactly; at least one of them will have a range of possible values.[lower-alpha 1] This is the content of the Heisenberg uncertainty relation.

Moreover, in contrast to classical mechanics, it is unavoidable that *performing a measurement on the system generally changes its state* More precisely: After measuring an observable *A*, the system will be in an eigenstate of *A*; thus the state has changed, unless the system was already in that eigenstate. This expresses a kind of logical consistency: If we measure *A* twice in the same run of the experiment, the measurements being directly consecutive in time, then they will produce the same results. This has some strange consequences, however, as follows.

Consider two observables, *A* and *B*, where *A* corresponds to a measurement earlier in time than *B*.[7] Suppose that the system is in an eigenstate of *B* at the experiment's begin. If we measure only *B*, we will not notice statistical behaviour. If we measure first *A* and then *B* in the same run of the experiment, the system will transfer to an eigenstate of *A* after the first measurement, and we will generally notice that the results of *B* are statistical. Thus: *Quantum mechanical measurements influence one another*, and it is important in which order they are performed.

Another feature of quantum states becomes relevant if we consider a physical system that consists of multiple subsystems; for example, an experiment with two particles rather than one. Quantum physics allows for certain states, called *entangled states*, that show certain statistical correlations between measurements on the two particles which cannot be explained by classical theory. For details, see entanglement. These entangled states lead to experimentally testable properties (Bell's theorem) that allow us to distinguish between quantum theory and alternative classical (non-quantum) models.

32.1.2 Schrödinger picture vs. Heisenberg picture

One can take the observables to be dependent on time, while the state σ was fixed once at the beginning of the experiment. This approach is called the Heisenberg picture. (This approach was taken in the later part of the discussion above, with time-varying observables *P*(*t*), *Q*(*t*).) One can, equivalently, treat the observables as fixed, while the state of the system depends on time; that is known as the Schrödinger picture. (This approach was taken in the earlier part of the discussion above, with a time-varying state $|\Psi(t)\rangle = \sum_n C_n(t)|\Phi_n\rangle$.) Conceptually (and mathematically), the two approaches are equivalent; choosing one of them is a matter of convention.

Both viewpoints are used in quantum theory. While non-relativistic quantum mechanics is usually formulated in terms of the Schrödinger picture, the Heisenberg picture is often preferred in a relativistic context, that is, for quantum field theory. Compare with Dirac picture.[8]

32.2 Formalism in quantum physics

See also: Mathematical formulation of quantum mechanics

32.2.1 Pure states as rays in a Hilbert space

Quantum physics is most commonly formulated in terms of linear algebra, as follows. Any given system is identified with some finite- or infinite-dimensional Hilbert space. The pure states correspond to vectors of norm 1. Thus the set of all pure states corresponds to the unit sphere in the Hilbert space.

Multiplying a pure state by a scalar is physically inconsequential (as long as the state is considered by itself). If one vector is obtained from the other by multiplying by a scalar of unit magnitude, the two vectors are said to correspond to the same "ray" in Hilbert space[9] and also to the same point in the projective Hilbert space.

32.2.2 Bra–ket notation

Main article: Bra–ket notation

Calculations in quantum mechanics make frequent use of linear operators, inner products, dual spaces and Hermitian conjugation. In order to make such calculations flow smoothly, and to obviate the need (in some contexts) to fully understand the underlying linear algebra, Paul Dirac invented a notation to describe quantum states, known as *bra-ket notation*. Although the details of this are beyond the scope of this article (see the article bra–ket notation), some consequences of this are:

- The expression used to denote a state vector (which corresponds to a pure quantum state) takes the form $|\psi\rangle$ (where the " ψ " can be replaced by any other symbols, letters, numbers, or even words). This can be contrasted with the usual *mathematical* notation, where vectors are usually bold, lower-case letters, or letters with arrows on top.
- Instead of *vector*, the term *ket* is used synonymously.
- Each ket $|\psi\rangle$ is uniquely associated with a so-called *bra*, denoted $\langle\psi|$, which corresponds to the same physical quantum state. Technically, the bra is the adjoint of the ket. It is an element of the dual space, and related to the ket by the Riesz representation theorem. In a finite-dimensional space with a chosen basis, writing $|\psi\rangle$ as a column vector, $\langle\psi|$ is a row vector; to obtain it just take the transpose and entry-wise complex conjugate of $|\psi\rangle$.
- Inner products (also called *brackets*) are written so as to look like a bra and ket next to each other: $\langle\psi_1|\psi_2\rangle$. (The phrase "bra-ket" is supposed to resemble "bracket".)

32.2.3 Spin

Main article: Mathematical formulation of quantum mechanics § Spin

The angular momentum has the same dimension as the Planck constant and, at quantum scale, behaves as a *discrete* degree of freedom. Most particles possess a kind of intrinsic angular momentum that does not appear at all in classical mechanics and arises from Dirac's relativistic generalization of the theory. Mathematically it is described with spinors. In non-relativistic quantum mechanics the group representations of the Lie group SU(2) are used to describe this additional freedom. For a given particle, the choice of representation (and hence the range of possible values of the spin observable) is specified by a non-negative number S that, in units of Planck's reduced constant $\hbar$, is either an integer (0, 1, 2 ...) or a half-integer (1/2, 3/2, 5/2 ...). For a massive particle with spin S, its spin quantum number m always assumes one of the $2S + 1$ possible values in the set

$$\{-S, -S+1, \ldots +S-1, +S\}$$

As a consequence, the quantum state of a particle with spin is described by a vector-valued wave function with values in $\mathbf{C}^{2S+1}$. Equivalently, it is represented by a complex-valued function of four variables: one discrete quantum number variable (for the spin) is added to the usual three continuous variables (for the position in space).

32.2.4 Many-body states and particle statistics

Further information: Particle statistics

The quantum state of a system of *N* particles, each potentially with spin, is described by a complex-valued function with four variables per particle, e.g.

$$|\psi(\mathbf{r}_1, m_1; \ldots; \mathbf{r}_N, m_N)\rangle.$$

Here, the spin variables *mν* assume values from the set

$$\{-S_\nu, -S_\nu + 1, \ldots + S_\nu - 1, +S_\nu\}$$

where S_ν is the spin of *ν*th particle. $S_\nu = 0$ for a particle that does not exhibit spin.

The treatment of identical particles is very different for bosons (particles with integer spin) versus fermions (particles with half-integer spin). The above *N*-particle function must either be symmetrized (in the bosonic case) or anti-symmetrized (in the fermionic case) with respect to the particle numbers. If not all *N* particles are identical, but some of them are, then the function must be (anti)symmetrized separately over the variables corresponding to each group of identical variables, according to its statistics (bosonic or fermionic).

Electrons are fermions with *S* = 1/2, photons (quanta of light) are bosons with *S* = 1 (although in the vacuum they are massless and can't be described with Schrödingerian mechanics).

When symmetrization or anti-symmetrization is unnecessary, *N*-particle spaces of states can be obtained simply by tensor products of one-particle spaces, to which we will return later.

32.2.5 Basis states of one-particle systems

As with any Hilbert space, if a basis is chosen for the Hilbert space of a system, then any ket can be expanded as a linear combination of those basis elements. Symbolically, given basis kets $|k_i\rangle$, any ket $|\psi\rangle$ can be written

$$|\psi\rangle = \sum_i c_i |k_i\rangle$$

where *ci* are complex numbers. In physical terms, this is described by saying that $|\psi\rangle$ has been expressed as a *quantum superposition* of the states $|k_i\rangle$. If the basis kets are chosen to be orthonormal (as is often the case), then $c_i = \langle k_i|\psi\rangle$.

One property worth noting is that the *normalized* states $|\psi\rangle$ are characterized by

$$\sum_i |c_i|^2 = 1.$$

Expansions of this sort play an important role in measurement in quantum mechanics. In particular, if the $|k_i\rangle$ are eigenstates (with eigenvalues *ki*) of an observable, and that observable is measured on the normalized state $|\psi\rangle$, then the probability that the result of the measurement is *ki* is $|ci|^2$. (The normalization condition above mandates that the total sum of probabilities is equal to one.)

A particularly important example is the *position basis*, which is the basis consisting of eigenstates of the observable which corresponds to measuring position. If these eigenstates are nondegenerate (for example, if the system is a single, spinless particle), then any ket $|\psi\rangle$ is associated with a complex-valued function of three-dimensional space:

$$\psi(\mathbf{r}) \equiv \langle \mathbf{r}|\psi\rangle.$$

This function is called the **wavefunction** corresponding to $|\psi\rangle$.

32.2.6 Superposition of pure states

Main article: Quantum superposition

One aspect of quantum states, mentioned above, is that superpositions of them can be formed. If $|\alpha\rangle$ and $|\beta\rangle$ are two kets corresponding to quantum states, the ket

$$c_\alpha|\alpha\rangle + c_\beta|\beta\rangle$$

is a different quantum state (possibly not normalized). Note that *which* quantum state it is depends on both the amplitudes and phases (arguments) of c_α and c_β . In other words, for example, even though $|\psi\rangle$ and $e^{i\theta}|\psi\rangle$ (for real θ) correspond to the same physical quantum state, they are *not interchangeable*, since for example $|\phi\rangle + |\psi\rangle$ and $|\phi\rangle + e^{i\theta}|\psi\rangle$ do *not* (in general) correspond to the same physical state. However, $|\phi\rangle + |\psi\rangle$ and $e^{i\theta}(|\phi\rangle + |\psi\rangle)$ *do* correspond to the same physical state. This is sometimes described by saying that "global" phase factors are unphysical, but "relative" phase factors are physical and important.

One example of a quantum interference phenomenon that arises from superposition is the double-slit experiment. The photon state is a superposition of two different states, one of which corresponds to the photon having passed through the left slit, and the other corresponding to passage through the right slit. The relative phase of those two states has a value which depends on the distance from each of the two slits. Depending on what that phase is, the interference is constructive at some locations and destructive in others, creating the interference pattern. By the analogy with coherence in other wave phenomena, a superposed state can be referred to as a *coherent superposition*.

Another example of the importance of relative phase in quantum superposition is Rabi oscillations, where the relative phase of two states varies in time due to the Schrödinger equation. The resulting superposition ends up oscillating back and forth between two different states.

32.2.7 Mixed states

See also: Density matrix

A *pure quantum state* is a state which can be described by a single ket vector, as described above. A *mixed quantum state* is a statistical ensemble of pure states (see quantum statistical mechanics). Mixed states inevitably arise from pure states when, for a composite quantum system $H_1 \otimes H_2$ with an entangled state on it, the part H_2 is inaccessible to the observer. The state of the part H_1 is expressed then as the partial trace over H_2 .

A mixed state *cannot* be described as a ket vector. Instead, it is described by its associated *density matrix* (or *density operator*), usually denoted ϱ. Note that density matrices can describe both mixed *and* pure states, treating them on the same footing. Moreover, a mixed quantum state on a given quantum system described by a Hilbert space H can be always represented as the partial trace of a pure quantum state (called a purification) on a larger bipartite system $H \otimes K$ for a sufficiently large Hilbert space K .

The density matrix describing a mixed state is defined to be an operator of the form

$$\rho = \sum_s p_s|\psi_s\rangle\langle\psi_s|$$

where p_s is the fraction of the ensemble in each pure state $|\psi_s\rangle$. The density matrix can be thought of as a way of using the one-particle formalism to describe the behavior of many similar particles by giving a probability distribution (or ensemble) of states that these particles can be found in.

A simple criterion for checking whether a density matrix is describing a pure or mixed state is that the trace of ϱ^2 is equal to 1 if the state is pure, and less than 1 if the state is mixed.[10] Another, equivalent, criterion is that the von Neumann entropy is 0 for a pure state, and strictly positive for a mixed state.

The rules for measurement in quantum mechanics are particularly simple to state in terms of density matrices. For example, the ensemble average (expectation value) of a measurement corresponding to an observable A is given by

$$\langle A\rangle = \sum_s p_s \langle \psi_s | A | \psi_s \rangle = \sum_s \sum_i p_s a_i |\langle \alpha_i | \psi_s \rangle|^2 = \mathrm{tr}(\rho A)$$

where $|\alpha_i\rangle$, a_i are eigenkets and eigenvalues, respectively, for the operator A, and "tr" denotes trace. It is important to note that two types of averaging are occurring, one being a weighted quantum superposition over the basis kets $|\psi_s\rangle$ of the pure states, and the other being a statistical (said *incoherent*) average with the probabilities *ps* of those states.

According to Wigner,[11] the concept of mixture was put forward by Landau.[12][13]

32.3 Interpretation

Main article: Interpretations of quantum mechanics

Although theoretically, for a given quantum system, a state vector provides the full information about its evolution, it is not easy to understand what information about the "real world" it carries. Due to the uncertainty principle, a state, even if it has the value of one observable exactly defined (i.e. the observable has this state as an eigenstate), cannot exactly define values of *all* observables.

For state vectors (pure states), probability amplitudes offer a probabilistic interpretation. It can be generalized for all states (including mixed), for instance, as expectation values mentioned above.

32.4 Mathematical generalizations

States can be formulated in terms of observables, rather than as vectors in a vector space. These are positive normalized linear functionals on a C*-algebra, or sometimes other classes of algebras of observables. See State on a C*-algebra and Gelfand–Naimark–Segal construction for more details.

32.5 See also

- Atomic electron transition
- Bloch sphere
- Ground state
- Introduction to quantum mechanics
- No-cloning theorem
- Orthonormal basis
- PBR theorem
- Quantum harmonic oscillator
- Qubit
- State vector reduction, for historical reasons called a *wave function collapse*
- Stationary state
- W state

32.6 Notes

[1] To avoid misunderstandings: Here we mean that $Q(t)$ and $P(t)$ are measured in the same state, but *not* in the same run of the experiment.

32.7 References

[1] Griffiths, D.J.(2004), pp. 93–96.

[2] Griffiths, D.J.(2004), pp. 4–5.

[3] Ballentine, L. E. (1970), “The Statistical Interpretation of Quantum Mechanics”, *Reviews of Modern Physics* **42**: 358–381, Bibcode:1970RvMP...42..358B, doi:10.1103/RevModPhys.42.358

[4] Sometimes written ">"; see angle brackets.

[5] Statistical Mixture of States

[6] http://electron6.phys.utk.edu/qm1/modules/m6/statistical.htm

[7] For concreteness’ sake, suppose that $A = Q(t_1)$ and $B = P(t_2)$ in the above example, with $t_2 > t_1 > 0$.

[8] Gottfried, Kurt; Yan, Tung-Mow (2003). *Quantum Mechanics: Fundamentals* (2nd, illustrated ed.). Springer. p. 65. ISBN 9780387955766.

[9] Weinberg, Steven. “The Quantum Theory of Fields”, Vol. 1. Cambridge University Press, 1995 p. 50.

[10] Blum, *Density matrix theory and applications*, page 39. Note that this criterion works when the density matrix is normalized so that the trace of ρ is 1, as it is for the standard definition given in this section. Occasionally a density matrix will be normalized differently, in which case the criterion is $\mathrm{Tr}(\rho^2) = (\mathrm{Tr}\,\rho)^2$

[11] Eugene Wigner (1962). “Remarks on the mind-body question” (PDF). In I.J. Good. *The Scientist Speculates*. London: Heinemann. pp. 284–302. Footnote 13 on p.180

[12] Lev Landau (1927). “Das Dämpfungsproblem in der Wellenmechanik (The Damping Problem in Wave Mechanics)". *Zeitschrift für Physik* **45** (5–6): 430–441. Bibcode:1927ZPhy...45..430L. doi:10.1007/bf01343064. English translation reprinted in: D. Ter Haar, ed. (1965). *Collected papers of L.D. Landau*. Oxford: Pergamon Press. p.8–18

[13] Lev Landau; Evgeny Lifshitz (1965). *Quantum Mechanics — Non-Relativistic Theory* (PDF). Course of Theoretical Physics **3** (2nd ed.). London: Pergamon Press. p.38–41.

32.7.1 Books referred to

- Ballentine, Leslie (1998). *Quantum Mechanics: A Modern Development* (2nd, illustrated, reprint ed.). World Scientific. ISBN 9789810241056.
- Griffiths, David J. (2004), *Introduction to Quantum Mechanics (2nd ed.)*, Prentice Hall, ISBN 0-13-111892-7

32.8 Further reading

The concept of quantum states, in particular the content of the section Formalism in quantum physics above, is covered in most standard textbooks on quantum mechanics.

For a discussion of conceptual aspects and a comparison with classical states, see:

- Isham, Chris J (1995). *Lectures on Quantum Theory: Mathematical and Structural Foundations*. Imperial College Press. ISBN 978-1-86094-001-9.

For a more detailed coverage of mathematical aspects, see:

- Bratteli, Ola; Robinson, Derek W (1987). *Operator Algebras and Quantum Statistical Mechanics 1*. Springer. ISBN 978-3-540-17093-8. 2nd edition. In particular, see Sec. 2.3.

For a discussion of purifications of mixed quantum states, see Chapter 2 of John Preskill's lecture notes for Physics 219 at Caltech.

Chapter 33

Lorentz transformation

For the derivations, see Derivations of the Lorentz transformations.

In physics, the **Lorentz transformation** (or **transformations**) is named after the Dutch physicist Hendrik Lorentz. It was the result of attempts by Lorentz and others to explain how the speed of light was observed to be independent of the reference frame, and to understand the symmetries of the laws of electromagnetism. The Lorentz transformation is in accordance with special relativity, but was derived before special relativity.

The transformations describe how measurements related to events in space and time by two observers, in inertial frames moving at constant velocity with respect to each other, are related. They reflect the fact that observers moving at different velocities may measure different distances, elapsed times, and even different orderings of events. They supersede the Galilean transformation of Newtonian physics, which assumes an absolute space and time (see Galilean relativity). The Galilean transformation is a good approximation only at relative speeds much smaller than the speed of light.

The Lorentz transformation is a linear transformation. It may include a rotation of space; a rotation-free Lorentz transformation is called a **Lorentz boost**.

In Minkowski space, the Lorentz transformations preserve the spacetime interval between any two events. They describe only the transformations in which the spacetime event at the origin is left fixed, so they can be considered as a hyperbolic rotation of Minkowski space. The more general set of transformations that also includes translations is known as the Poincaré group.

33.1 History

Main article: History of Lorentz transformations

Many physicists, including Woldemar Voigt, George FitzGerald, Joseph Larmor, and Hendrik Lorentz himself had been discussing the physics implied by these equations since 1887.[1]

Early in 1889, Oliver Heaviside had shown from Maxwell's equations that the electric field surrounding a spherical distribution of charge should cease to have spherical symmetry once the charge is in motion relative to the ether. FitzGerald then conjectured that Heaviside's distortion result might be applied to a theory of intermolecular forces. Some months later, FitzGerald published the conjecture that bodies in motion are being contracted, in order to explain the baffling outcome of the 1887 ether-wind experiment of Michelson and Morley. In 1892, Lorentz independently presented the same idea in a more detailed manner, which was subsequently called FitzGerald–Lorentz contraction hypothesis.[2] Their explanation was widely known before 1905.[3]

Lorentz (1892–1904) and Larmor (1897–1900), who believed the luminiferous ether hypothesis, were also seeking the transformation under which Maxwell's equations are invariant when transformed from the ether to a moving frame. They extended the FitzGerald–Lorentz contraction hypothesis and found out that the time coordinate has to be modified as well ("local time"). Henri Poincaré gave a physical interpretation to local time (to first order in v/c) as the consequence of clock synchronization, under the assumption that the speed of light is constant in moving frames.[4] Larmor is credited to have been the first to understand the crucial time dilation property inherent in his equations.[5]

In 1905, Poincaré was the first to recognize that the transformation has the properties of a mathematical group, and named it after Lorentz.[6] Later in the same year Albert Einstein published what is now called special relativity, by deriving the Lorentz transformation under the assumptions of the principle of relativity and the constancy of the speed of light in any inertial reference frame, and by abandoning the mechanical aether.[7]

33.2 Derivation

Further information: Derivations of the Lorentz transformations

From Einstein's second postulate of relativity follows immediately

$$c^2(t_2-t_1)^2-(x_2-x_1)^2-(y_2-y_1)^2-(z_2-z_1)^2=0$$

in all reference frames for events connected by *light signals*. The quantity on the left is called the *spacetime interval*. The interval between *any two* is in fact invariant, as is shown here (where one can also find several more explicit derivations than presently given). The transformation sought after thus must possess the property that

$$c^2(t_2-t_1)^2-(x_2-x_1)^2-(y_2-y_1)^2-(z_2-z_1)^2=c^2(t_2'-t_1')^2-(x_2'-x_1')^2-(y_2'-y_1')^2-(z_2'-z_1')^2.$$

Now one observes that a *linear* solution to the simpler problem

$$c^2t^2-x^2-y^2-z^2=c^2t'^2-x'^2-y'^2-z'^2$$

will solve the general problem too. This is just a matter of look-up in the theory of classical groups that preserve bilinear forms of various signature. The Lorentz transformation is thus an element of the group O(3, 1) or, for those that prefer the other metric signature, O(1, 3). Connection between the matrix elements and physical quantities is made below.

33.3 Frames in standard configuration

Consider two observers O and O', each using their own Cartesian coordinate system to measure space and time intervals. O uses (t, x, y, z) and O' uses (t', x', y', z'). Assume further that the coordinate systems are oriented so that, in 3 dimensions, the x-axis and the x'-axis are collinear, the y-axis is parallel to the y'-axis, and the z-axis parallel to the z'-axis. The relative velocity between the two observers is v along the common x-axis; O measures O' to move at velocity v along the coincident xx' axes, while O' measures O to move at velocity $-v$ along the coincident xx' axes. Also assume that the origins of both coordinate systems are the same, that is, coincident times and positions. If all these hold, then the coordinate systems are said to be in **standard configuration**. The formulae below give the Lorentz transformations (boosts) for this configuration.

The inverse of a Lorentz transformation relates the coordinates the other way round; from the coordinates O' measures (t', x', y', z') to the coordinates O measures (t, x, y, z), so t, x, y, z are in terms of t', x', y', z'. The mathematical form is nearly identical to the original transformation; the only difference is the negation of the uniform relative velocity (from v to $-v$), and exchange of primed and unprimed quantities, because O' moves at velocity v relative to O, and equivalently, O moves at velocity $-v$ relative to O'. This symmetry makes it effortless to find the inverse transformation (carrying out the exchange and negation saves a lot of rote algebra), although more fundamentally; it highlights that all physical laws should remain unchanged under a Lorentz transformation.[8]

Below, the Lorentz transformations are called "boosts" in the stated directions.

33.3.1 Boost in the x-direction

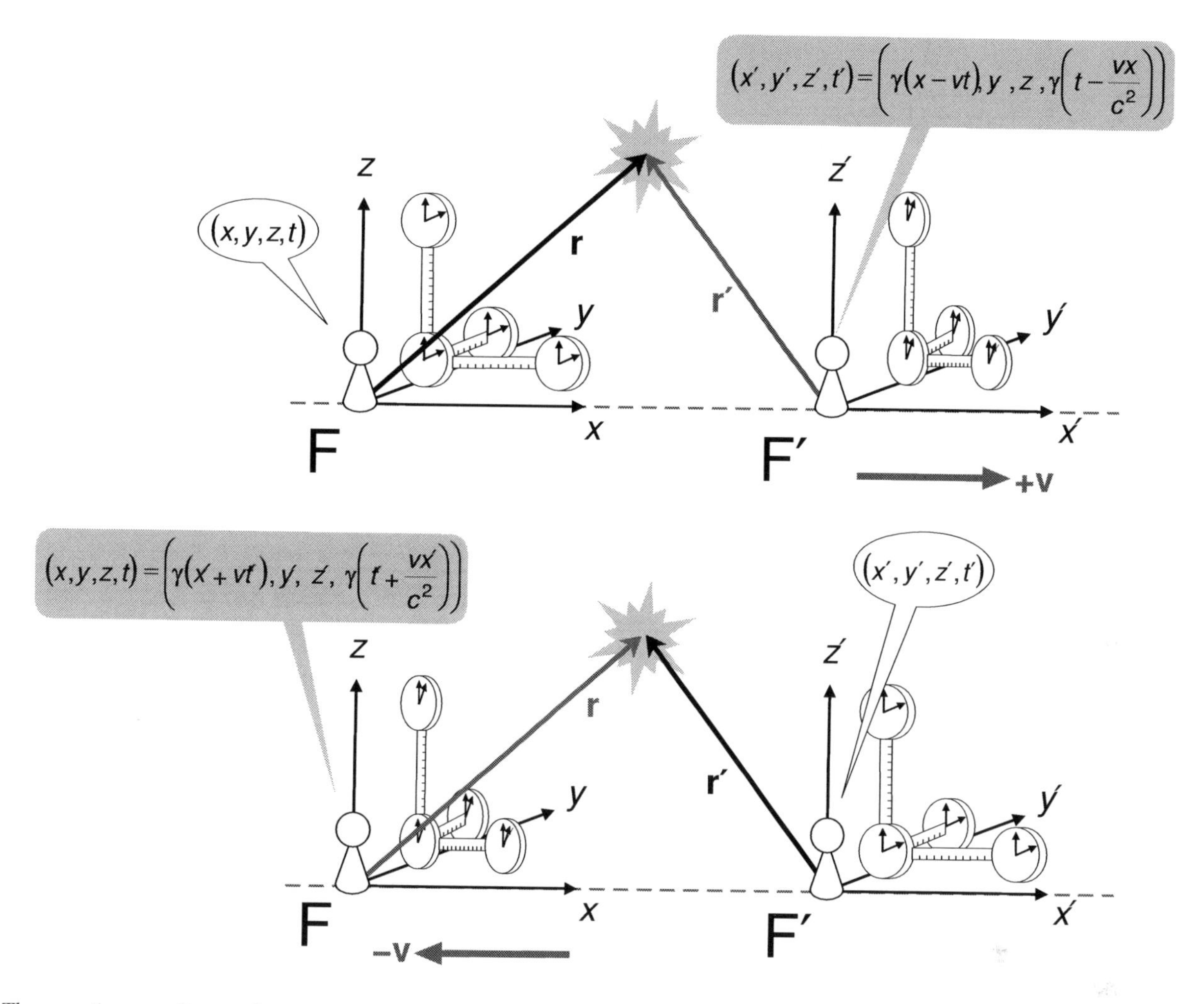

The spacetime coordinates of an event, as measured by each observer in their inertial reference frame (in standard configuration) are shown in the speech bubbles.
Top: frame ***F′*** *moves at velocity* v *along the* x*-axis of frame* ***F****.*
Bottom: frame ***F*** *moves at velocity* −v *along the* x′*-axis of frame* ***F′****.*[9]

These are the simplest forms. The Lorentz transformation for frames in standard configuration can be shown to be:[10]

$$t' = \gamma\left(t - \frac{vx}{c^2}\right)$$
$$x' = \gamma\left(x - vt\right)$$
$$y' = y$$
$$z' = z$$

where:

- v is the relative velocity between frames in the x-direction,
- c is the speed of light,
- $\gamma = \frac{1}{\sqrt{1-\beta^2}}$ is the Lorentz factor (Greek lowercase gamma),

- $\beta = \frac{v}{c}$ is the velocity coefficient (Greek lowercase beta), again for the x-direction.

The use of β and γ is standard throughout the literature.[11] For the remainder of the article – they will be also used throughout unless otherwise stated. Since the above is a linear system of equations (more technically a linear transformation), they can be written in matrix form:

$$\begin{bmatrix} ct' \\ x' \\ y' \\ z' \end{bmatrix} = \begin{bmatrix} \gamma & -\beta\gamma & 0 & 0 \\ -\beta\gamma & \gamma & 0 & 0 \\ 0 & 0 & 1 & 0 \\ 0 & 0 & 0 & 1 \end{bmatrix} \begin{bmatrix} ct \\ x \\ y \\ z \end{bmatrix},$$

According to the principle of relativity, there is no privileged frame of reference, so the inverse transformations frame F' to frame F must be given by simply negating v and exchanging primed and unprimed variables:

$$\begin{aligned} t &= \gamma\left(t' + \frac{vx'}{c^2}\right) \\ x &= \gamma\left(x' + vt'\right) \\ y &= y' \\ z &= z', \end{aligned}$$

where the value of γ remains unchanged. These equations are also obtained by algebraically solving the standard equations for the variables t, x, y, z.

33.3.2 Boost in the *y* or *z* directions

The above collection of equations apply only for a boost in the x-direction. The standard configuration works equally well in the y or z directions instead of x, and so the results are similar.

For the y-direction:

$$\begin{aligned} t' &= \gamma\left(t - \frac{vy}{c^2}\right) \\ x' &= x \\ y' &= \gamma\left(y - vt\right) \\ z' &= z \end{aligned}$$

summarized by

$$\begin{bmatrix} ct' \\ x' \\ y' \\ z' \end{bmatrix} = \begin{bmatrix} \gamma & 0 & -\beta\gamma & 0 \\ 0 & 1 & 0 & 0 \\ -\beta\gamma & 0 & \gamma & 0 \\ 0 & 0 & 0 & 1 \end{bmatrix} \begin{bmatrix} ct \\ x \\ y \\ z \end{bmatrix},$$

where v and so β are now in the y-direction. For the z-direction one obtains

$$\begin{bmatrix} ct' \\ x' \\ y' \\ z' \end{bmatrix} = \begin{bmatrix} \gamma & 0 & 0 & -\beta\gamma \\ 0 & 1 & 0 & 0 \\ 0 & 0 & 1 & 0 \\ -\beta\gamma & 0 & 0 & \gamma \end{bmatrix} \begin{bmatrix} ct \\ x \\ y \\ z \end{bmatrix}.$$

The Lorentz or boost matrix is usually denoted by $\mathbf{\Lambda}$ (Greek capital lambda). Above the transformations have been applied to the four-position $\mathbf{X}$,

$$\mathbf{X} = \begin{bmatrix} ct \\ x \\ y \\ z \end{bmatrix}, \quad \mathbf{X}' = \begin{bmatrix} ct' \\ x' \\ y' \\ z' \end{bmatrix},$$

The Lorentz transform for a boost in one of the above directions can be compactly written as a single matrix equation:

$$\mathbf{X}' = \mathbf{\Lambda}(v)\mathbf{X}.$$

33.3.3 Boost in any direction

Vector form

Further information: Euclidean vector and vector projection

For a boost in an arbitrary direction with velocity **v**, that is, O observes O' to move in direction **v** in the F coordinate frame, while O' observes O to move in direction −**v** in the F' coordinate frame, it is convenient to decompose the spatial vector **r** into components perpendicular and parallel to **v**:

$$\mathbf{r} = \mathbf{r}_\perp + \mathbf{r}_\|$$

so that

$$\mathbf{r}\cdot\mathbf{v} = \mathbf{r}_\perp\cdot\mathbf{v} + \mathbf{r}_\|\cdot\mathbf{v} = r_\| v$$

where • denotes the dot product (see also orthogonality for more information). Then, only time and the component **r**‖ in the direction of **v** are “warped” by the Lorentz factor:

$$t' = \gamma\left(t - \frac{\mathbf{r}\cdot\mathbf{v}}{c^2}\right)$$
$$\mathbf{r}' = \mathbf{r}_\perp + \gamma(\mathbf{r}_\| - \mathbf{v}t)$$
$$\gamma(\mathbf{v}) = \frac{1}{\sqrt{1-\frac{v^2}{c^2}}}$$

The parallel and perpendicular components can be eliminated, by substituting **r**⊥ = **r** − **r**‖ into **r**′:

$$\mathbf{r}' = \mathbf{r} + (\gamma - 1)\,\mathbf{r}_\| - \gamma\mathbf{v}t\,.$$

Since **r**‖ and **v** are parallel we have

$$\mathbf{r}_\| = r_\|\frac{\mathbf{v}}{v} = \left(\frac{\mathbf{r}\cdot\mathbf{v}}{v}\right)\frac{\mathbf{v}}{v}$$

where geometrically and algebraically:

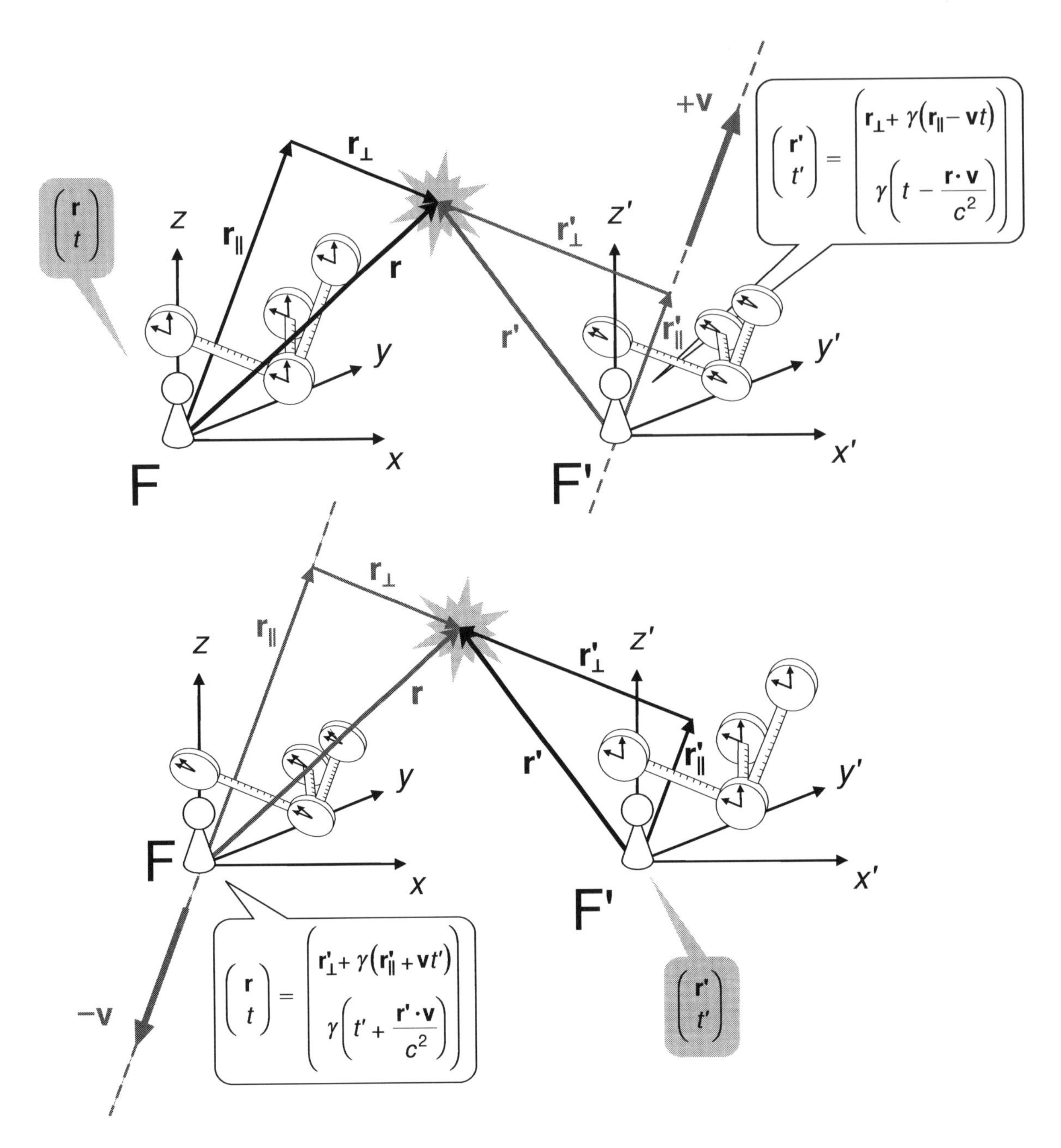

Boost in an arbitrary direction.

- $\mathbf{v}/v$ is a dimensionless unit vector pointing in the same direction as $\mathbf{r}\|$,
- $r\| = (\mathbf{r} \bullet \mathbf{v})/v$ is the projection of $\mathbf{r}$ into the direction of $\mathbf{v}$,

substituting for $\mathbf{r}\|$ and factoring $\mathbf{v}$ gives

$$\mathbf{r}' = \mathbf{r} + \left(\frac{\gamma - 1}{v^2}\mathbf{r} \cdot \mathbf{v} - \gamma t\right)\mathbf{v}\,.$$

This method, of eliminating parallel and perpendicular components, can be applied to any Lorentz transformation written in parallel-perpendicular form.

Matrix forms

These equations can be expressed in block matrix form as

$$\begin{bmatrix} ct' \\ \mathbf{r}' \end{bmatrix} = \begin{bmatrix} \gamma & -\gamma\boldsymbol{\beta}^{\mathrm{T}} \\ -\gamma\boldsymbol{\beta} & \mathbf{I} + (\gamma - 1)\boldsymbol{\beta}\boldsymbol{\beta}^{\mathrm{T}}/\beta^2 \end{bmatrix} \begin{bmatrix} ct \\ \mathbf{r} \end{bmatrix}$$

where $\mathbf{I}$ is the 3×3 identity matrix and $\boldsymbol{\beta} = \mathbf{v}/c$ is the relative velocity vector (in units of c) as a *column vector* – in cartesian and tensor index notation it is:

$$\boldsymbol{\beta} = \frac{\mathbf{v}}{c} \equiv \begin{bmatrix} \beta_x \\ \beta_y \\ \beta_z \end{bmatrix} = \frac{1}{c}\begin{bmatrix} v_x \\ v_y \\ v_z \end{bmatrix} \equiv \begin{bmatrix} \beta_1 \\ \beta_2 \\ \beta_3 \end{bmatrix} = \frac{1}{c}\begin{bmatrix} v_1 \\ v_2 \\ v_3 \end{bmatrix}$$

$\boldsymbol{\beta}^{\mathrm{T}} = \mathbf{v}^{\mathrm{T}}/c$ is the transpose – a row vector:

$$\boldsymbol{\beta}^{\mathrm{T}} = \frac{\mathbf{v}^{\mathrm{T}}}{c} \equiv \begin{bmatrix} \beta_x & \beta_y & \beta_z \end{bmatrix} = \frac{1}{c}\begin{bmatrix} v_x & v_y & v_z \end{bmatrix} \equiv \begin{bmatrix} \beta_1 & \beta_2 & \beta_3 \end{bmatrix} = \frac{1}{c}\begin{bmatrix} v_1 & v_2 & v_3 \end{bmatrix}$$

and β is the magnitude of $\boldsymbol{\beta}$:

$$\beta = |\boldsymbol{\beta}| = \sqrt{\beta_x^2 + \beta_y^2 + \beta_z^2}.$$

More explicitly stated:

$$\begin{bmatrix} ct' \\ x' \\ y' \\ z' \end{bmatrix} = \begin{bmatrix} \gamma & -\gamma\,\beta_x & -\gamma\,\beta_y & -\gamma\,\beta_z \\ -\gamma\,\beta_x & 1 + (\gamma - 1)\dfrac{\beta_x^2}{\beta^2} & (\gamma - 1)\dfrac{\beta_x\beta_y}{\beta^2} & (\gamma - 1)\dfrac{\beta_x\beta_z}{\beta^2} \\ -\gamma\,\beta_y & (\gamma - 1)\dfrac{\beta_y\beta_x}{\beta^2} & 1 + (\gamma - 1)\dfrac{\beta_y^2}{\beta^2} & (\gamma - 1)\dfrac{\beta_y\beta_z}{\beta^2} \\ -\gamma\,\beta_z & (\gamma - 1)\dfrac{\beta_z\beta_x}{\beta^2} & (\gamma - 1)\dfrac{\beta_z\beta_y}{\beta^2} & 1 + (\gamma - 1)\dfrac{\beta_z^2}{\beta^2} \end{bmatrix} \begin{bmatrix} ct \\ x \\ y \\ z \end{bmatrix}.$$

The transformation $\boldsymbol{\Lambda}$ can be written in the same form as before,

$$\mathbf{X}' = \boldsymbol{\Lambda}(\mathbf{v})\mathbf{X}.$$

which has the structure:[12]

$$\begin{bmatrix} ct' \\ x' \\ y' \\ z' \end{bmatrix} = \begin{bmatrix} \Lambda_{00} & \Lambda_{01} & \Lambda_{02} & \Lambda_{03} \\ \Lambda_{10} & \Lambda_{11} & \Lambda_{12} & \Lambda_{13} \\ \Lambda_{20} & \Lambda_{21} & \Lambda_{22} & \Lambda_{23} \\ \Lambda_{30} & \Lambda_{31} & \Lambda_{32} & \Lambda_{33} \end{bmatrix} \begin{bmatrix} ct \\ x \\ y \\ z \end{bmatrix}.$$

and the components deduced from above are:

$$\Lambda_{00} = \gamma,$$
$$\Lambda_{0i} = \Lambda_{i0} = -\gamma\beta_i,$$
$$\Lambda_{ij} = \Lambda_{ji} = (\gamma - 1)\frac{\beta_i\beta_j}{\beta^2} + \delta_{ij} = (\gamma - 1)\frac{v_i v_j}{v^2} + \delta_{ij},$$

where δij is the Kronecker delta, and by convention: Latin letters for indices take the values 1, 2, 3, for spatial components of a 4-vector (Greek indices take values 0, 1, 2, 3 for time and space components).

Note that this transformation is only the "boost," i.e., a transformation between two frames whose x, y, and z axis are parallel and whose spacetime origins coincide. The most general proper Lorentz transformation also contains a rotation of the three axes, because the composition of two boosts is not a pure boost but is a boost followed by a rotation. The rotation gives rise to Thomas precession. The boost is given by a symmetric matrix, but the general Lorentz transformation matrix need not be symmetric.

33.3.4 Composition of two boosts

The composition of two Lorentz boosts B(**u**) and B(**v**) of velocities **u** and **v** is given by:[13][14]

$$B(\mathbf{u})B(\mathbf{v}) = B\left(\mathbf{u} \oplus \mathbf{v}\right) \mathrm{Gyr}\left[\mathbf{u}, \mathbf{v}\right] = \mathrm{Gyr}\left[\mathbf{u}, \mathbf{v}\right] B\left(\mathbf{v} \oplus \mathbf{u}\right)$$

where

- B(**v**) is the 4×4 matrix that uses the components of **v**, i.e. v_1, v_2, v_3 in the entries of the matrix, or rather the components of **v**/c in the representation that is used above,
- $\mathbf{u} \oplus \mathbf{v}$ is the velocity-addition,
- Gyr[**u**,**v**] (capital G) is the rotation arising from the composition. If the 3×3 matrix form of the rotation applied to spatial coordinates is given by gyr[**u**,**v**], then the 4×4 matrix rotation applied to 4-coordinates is given by:[13]

$$\mathrm{Gyr}[\mathbf{u}, \mathbf{v}] = \begin{pmatrix} 1 & 0 \\ 0 & \mathrm{gyr}[\mathbf{u}, \mathbf{v}] \end{pmatrix}$$

- gyr (lower case g) is the gyrovector space abstraction of the *gyroscopic Thomas precession*, defined as an operator on a velocity **w** in terms of velocity addition:

$$\mathrm{gyr}[\mathbf{u}, \mathbf{v}]\mathbf{w} = \ominus(\mathbf{u} \oplus \mathbf{v}) \oplus (\mathbf{u} \oplus (\mathbf{v} \oplus \mathbf{w}))$$

for all **w**.

The composition of two Lorentz transformations L(**u**, U) and L(**v**, V) which include rotations U and V is given by:[15]

$$L(\mathbf{u}, U)L(\mathbf{v}, V) = L(\mathbf{u} \oplus U\mathbf{v}, \mathrm{gyr}[\mathbf{u}, U\mathbf{v}]UV)$$

33.4 Visualizing the transformations in Minkowski space

Main article: Minkowski space

Lorentz transformations can be depicted on the Minkowski light cone spacetime diagram.

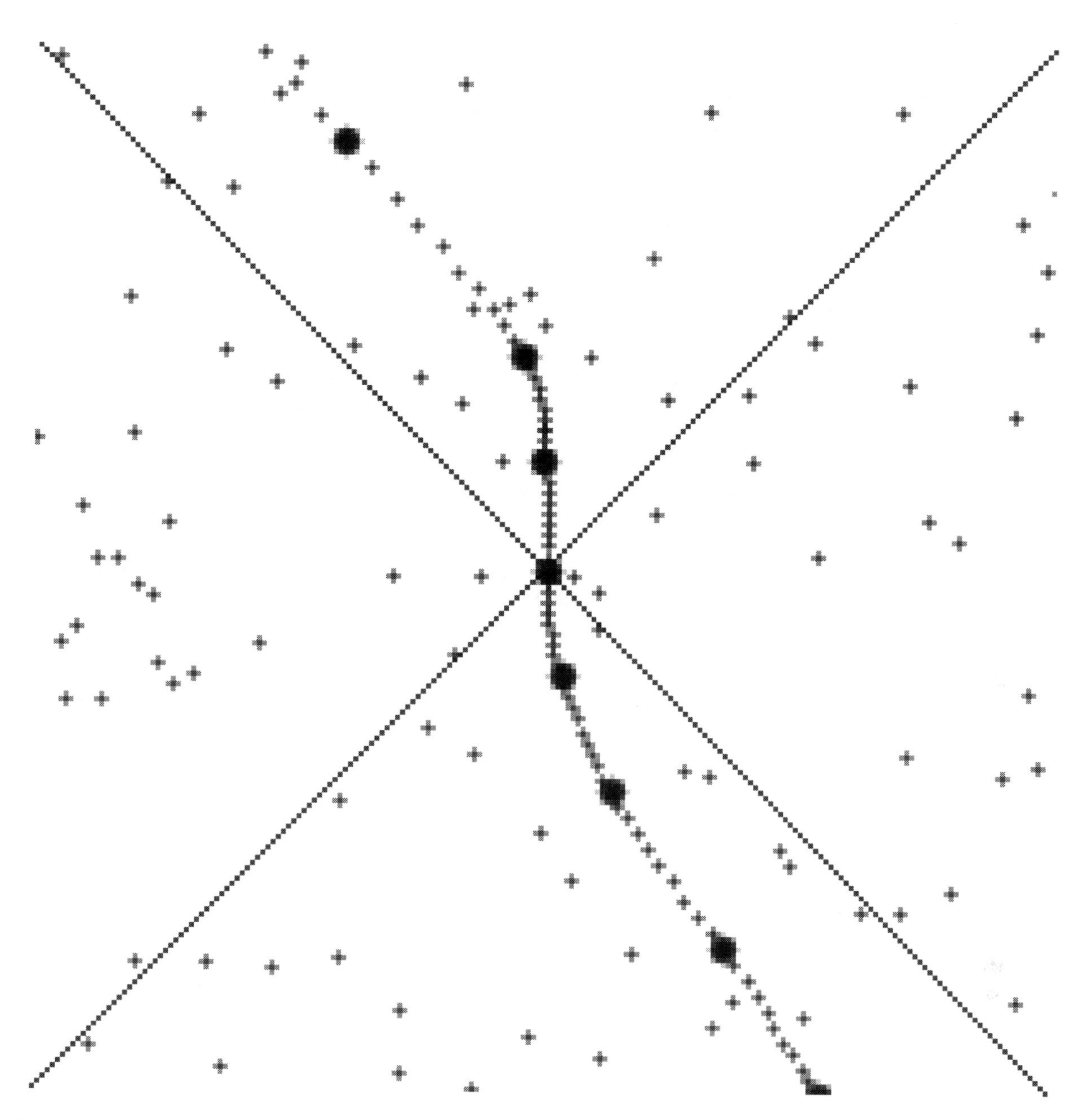

The momentarily co-moving inertial frames along the world line of a rapidly accelerating observer (center). The vertical direction indicates time, while the horizontal indicates distance, the dashed line is the spacetime trajectory ("world line") of the observer. The small dots are specific events in spacetime. If one imagines these events to be the flashing of a light, then the events that pass the two diagonal lines in the bottom half of the image (the past light cone of the observer in the origin) are the events visible to the observer. The slope of the world line (deviation from being vertical) gives the relative velocity to the observer. Note how the momentarily co-moving inertial frame changes when the observer accelerates.

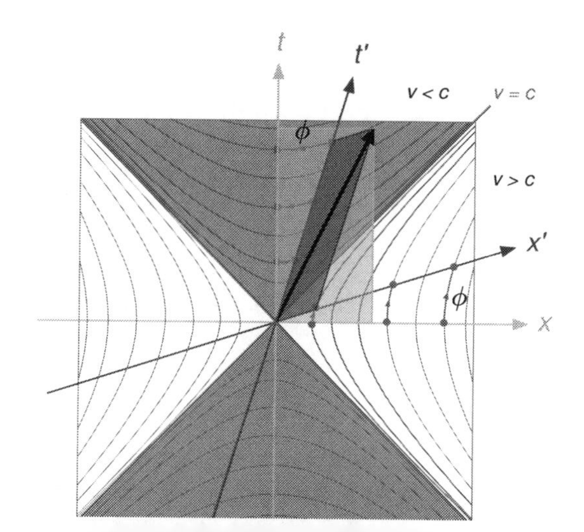

Particle travelling at constant velocity (straight worldline coincident with time t' axis).

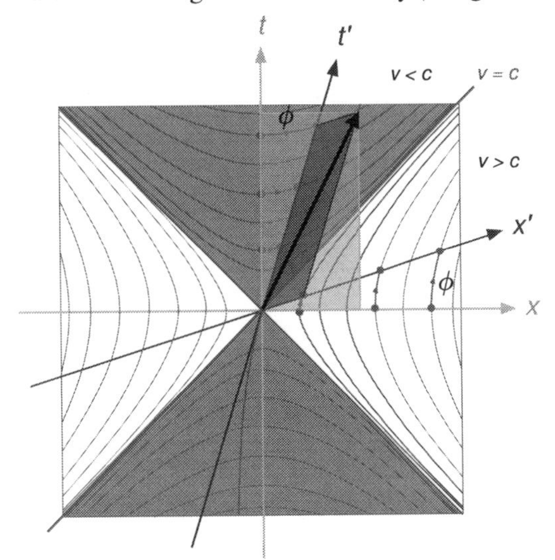

Accelerating particle (curved worldline).
Lorentz transformations on the Minkowski light cone spacetime diagram, for one space and one time dimension.

The yellow axes are the rest frame of an observer, the blue axes correspond to the frame of a moving observer

The red lines are world lines, a continuous sequence of events: straight for an object travelling at constant velocity, curved for an object accelerating. Worldlines of light form the boundary of the light cone.

The purple hyperbolae indicate this is a hyperbolic rotation, the hyperbolic angle φ is called rapidity (see below). The greater the relative speed between the reference frames, the more "warped" the axes become. The relative velocity cannot exceed c.

The black arrow is a displacement four-vector between two events (not necessarily on the same world line), showing that in a Lorentz boost; time dilation (fewer time intervals in moving frame) and length contraction (shorter lengths in moving frame) occur. The axes in the moving frame are orthogonal (even though they do not look so).

33.4.1 Rapidity

The Lorentz transformation can be cast into another useful form by defining a parameter ϕ called the rapidity (an instance of hyperbolic angle) such that

$$e^{\phi} = \gamma(1+\beta) = \gamma\left(1+\frac{v}{c}\right) = \sqrt{\frac{1+\frac{v}{c}}{1-\frac{v}{c}}},$$

and thus

$$e^{-\phi} = \gamma(1-\beta) = \gamma\left(1-\frac{v}{c}\right) = \sqrt{\frac{1-\frac{v}{c}}{1+\frac{v}{c}}}.$$

Equivalently:

$$\phi = \ln\left[\gamma(1+\beta)\right] = -\ln\left[\gamma(1-\beta)\right]$$

Then the Lorentz transformation in standard configuration is:

$$\begin{aligned}
ct - x &= e^{-\phi}(ct' - x') \\
ct + x &= e^{\phi}(ct' + x') \\
y &= y' \\
z &= z'.
\end{aligned}$$

Hyperbolic expressions

From the above expressions for e^{φ} and $e^{-\varphi}$

$$\gamma = \cosh\phi = \frac{e^{\phi} + e^{-\phi}}{2},$$

$$\beta\gamma = \sinh\phi = \frac{e^{\phi} - e^{-\phi}}{2},$$

and therefore,

$$\beta = \tanh\phi = \frac{e^{\phi} - e^{-\phi}}{e^{\phi} + e^{-\phi}}.$$

Hyperbolic rotation of coordinates

Substituting these expressions into the matrix form of the transformation, it is evident that

$$\begin{bmatrix} ct' \\ x' \\ y' \\ z' \end{bmatrix} = \begin{bmatrix} \cosh\phi & -\sinh\phi & 0 & 0 \\ -\sinh\phi & \cosh\phi & 0 & 0 \\ 0 & 0 & 1 & 0 \\ 0 & 0 & 0 & 1 \end{bmatrix} \begin{bmatrix} ct \\ x \\ y \\ z \end{bmatrix}.$$

Thus, the Lorentz transformation can be seen as a hyperbolic rotation of coordinates in Minkowski space, where the parameter ϕ represents the hyperbolic angle of rotation, often referred to as rapidity. This transformation is sometimes illustrated with a Minkowski diagram, as displayed above.

This 4-by-4 boost matrix can thus be written compactly as a matrix exponential,

$$\begin{bmatrix} \cosh\phi & -\sinh\phi & 0 & 0 \\ -\sinh\phi & \cosh\phi & 0 & 0 \\ 0 & 0 & 1 & 0 \\ 0 & 0 & 0 & 1 \end{bmatrix} = \exp\left(-\phi \begin{bmatrix} 0 & 1 & 0 & 0 \\ 1 & 0 & 0 & 0 \\ 0 & 0 & 0 & 0 \\ 0 & 0 & 0 & 0 \end{bmatrix} \right) \equiv \exp(-\phi K_x),$$

where the simpler Lie-algebraic hyperbolic rotation generator *Kx* is called a **boost generator**.

33.5 Transformation of other physical quantities

Main article: Representation theory of the Lorentz group
For the notation used, see Ricci calculus.

The transformation matrix is universal for all four-vectors, not just 4-dimensional spacetime coordinates. If $\mathbf{Z}$ is any four-vector, then:[12]

$$\mathbf{Z}' = \mathbf{\Lambda}(\mathbf{v})\mathbf{Z}.$$

or in tensor index notation:

$$Z^{\alpha'} = \Lambda^{\alpha'}{}_{\alpha} Z^{\alpha}.$$

in which the primed indices denote indices of $\mathbf{Z}$ in the primed frame.

More generally, the transformation of any tensor quantity $\boldsymbol{T}$ is given by:[16]

$$T^{\alpha'\beta'\cdots\zeta'}_{\theta'\iota'\cdots\kappa'} = \Lambda^{\alpha'}{}_{\mu}\Lambda^{\beta'}{}_{\nu}\cdots\Lambda^{\zeta'}{}_{\rho}\Lambda_{\theta'}{}^{\sigma}\Lambda_{\iota'}{}^{\upsilon}\cdots\Lambda_{\kappa'}{}^{\phi}T^{\mu\nu\cdots\rho}_{\sigma\upsilon\cdots\phi}$$

where $\Lambda_{\chi'}{}^{\psi}$ is the inverse matrix of $\Lambda^{\chi'}{}_{\psi}$.

33.6 Special relativity

The crucial insight of Einstein's clock-setting method is the idea that *time is relative*. In essence, each observer's frame of reference is associated with a unique set of clocks, the result being that time as measured for a location passes at different rates for different observers.[17] This was a direct result of the Lorentz transformations and is called time dilation. We can also clearly see from the Lorentz "local time" transformation that the concept of the relativity of simultaneity and of the relativity of length contraction are also consequences of that clock-setting hypothesis.[18]

33.6.1 Transformation of the electromagnetic field

For the transformation rules, see classical electromagnetism and special relativity.

Lorentz transformations can also be used to prove that magnetic and electric fields are simply different aspects of the same force — the electromagnetic force, as a consequence of relative motion between electric charges and observers.[19] The fact that the electromagnetic field shows relativistic effects becomes clear by carrying out a simple thought experiment:[20]

- Consider an observer measuring a charge at rest in a reference frame F. The observer will detect a static electric field. As the charge is stationary in this frame, there is no electric current, so the observer will not observe any magnetic field.
- Consider another observer in frame F′ moving at relative velocity $\mathbf{v}$ (relative to F and the charge). *This* observer will see a different electric field because the charge is moving at velocity $-\mathbf{v}$ in their rest frame. Further, in frame F′ the moving charge constitutes an electric current, and thus the observer in frame F′ will also see a magnetic field.

This shows that the Lorentz transformation also applies to electromagnetic field quantities when changing the frame of reference, given below in vector form.

33.6.2 The correspondence principle

For relative speeds much less than the speed of light, the Lorentz transformations reduce to the Galilean transformation in accordance with the correspondence principle.

The correspondence limit is usually stated mathematically as: as $v \to 0$, $c \to \infty$. In words: as velocity approaches 0, the speed of light (seems to) approach infinity. Hence, it is sometimes said that nonrelativistic physics is a physics of "instantaneous action at a distance".[17]

33.7 Spacetime interval

In a given coordinate system x^μ, if two events A and B are separated by

$$(\Delta t, \Delta x, \Delta y, \Delta z) = (t_B - t_A, x_B - x_A, y_B - y_A, z_B - z_A)\,,$$

the spacetime interval between them is given by

$$s^2 = -c^2(\Delta t)^2 + (\Delta x)^2 + (\Delta y)^2 + (\Delta z)^2\,.$$

This can be written in another form using the Minkowski metric. In this coordinate system,

$$\eta_{\mu\nu} = \begin{bmatrix} -1 & 0 & 0 & 0 \\ 0 & 1 & 0 & 0 \\ 0 & 0 & 1 & 0 \\ 0 & 0 & 0 & 1 \end{bmatrix}.$$

Then, we can write

$$s^2 = \begin{bmatrix} c\Delta t & \Delta x & \Delta y & \Delta z \end{bmatrix} \begin{bmatrix} -1 & 0 & 0 & 0 \\ 0 & 1 & 0 & 0 \\ 0 & 0 & 1 & 0 \\ 0 & 0 & 0 & 1 \end{bmatrix} \begin{bmatrix} c\Delta t \\ \Delta x \\ \Delta y \\ \Delta z \end{bmatrix}$$

or, using the Einstein summation convention,

$$s^2 = \eta_{\mu\nu} x^\mu x^\nu\,.$$

Now suppose that we make a coordinate transformation $x^\mu \rightarrow x'^\mu$. Then, the interval in this coordinate system is given by

$$s'^2 = \begin{bmatrix} c\Delta t' & \Delta x' & \Delta y' & \Delta z' \end{bmatrix} \begin{bmatrix} -1 & 0 & 0 & 0 \\ 0 & 1 & 0 & 0 \\ 0 & 0 & 1 & 0 \\ 0 & 0 & 0 & 1 \end{bmatrix} \begin{bmatrix} c\Delta t' \\ \Delta x' \\ \Delta y' \\ \Delta z' \end{bmatrix}$$

or

$$s'^2 = \eta_{\mu\nu} x'^\mu x'^\nu\,.$$

It is a result of special relativity that the interval is an invariant. That is, $s^2 = s'^2$, see invariance of interval. For this to hold, it can be shown[21] that it is necessary and sufficient for the coordinate transformation to be of the form

$$x'^\mu = x^\nu \Lambda^\mu_\nu + C^\mu\,,$$

where C^μ is a constant vector and $\Lambda^\mu{}_\nu$ a constant matrix, where we require that

$$\eta_{\mu\nu}\Lambda^\mu_\alpha\Lambda^\nu_\beta = \eta_{\alpha\beta} \ .$$

Such a transformation is called a *Poincaré transformation* or an *inhomogeneous Lorentz transformation*.[22] The C^a represents a spacetime translation. When $C^a = 0$, the transformation is called an *homogeneous Lorentz transformation*, or simply a *Lorentz transformation*.

Taking the determinant of

$$\eta_{\mu\nu}\Lambda^\mu{}_\alpha\Lambda^\nu{}_\beta = \eta_{\alpha\beta}$$

gives us

$$\det(\Lambda^a_b) = \pm 1 \ .$$

The cases are:

- **Proper Lorentz transformations** have $\det(\Lambda^\mu \nu) = +1$, and form a subgroup called the special orthogonal group SO(1,3).
- **Improper Lorentz transformations** are $\det(\Lambda^\mu \nu) = -1$, which do not form a subgroup, as the product of any two improper Lorentz transformations will be a proper Lorentz transformation.

From the above definition of Λ it can be shown that $(\Lambda^0{}_0)^2 \geq 1$, so either $\Lambda^0{}_0 \geq 1$ or $\Lambda^0{}_0 \leq -1$, called orthochronous and non-orthochronous respectively. An important subgroup of the proper Lorentz transformations are the **proper orthochronous Lorentz transformations** which consist purely of boosts and rotations. Any Lorentz transform can be written as a proper orthochronous, together with one or both of the two discrete transformations; space inversion P and time reversal T, whose non-zero elements are:

$$P^0_0 = 1, P^1_1 = P^2_2 = P^3_3 = -1$$

$$T^0_0 = -1, T^1_1 = T^2_2 = T^3_3 = 1$$

The set of Poincaré transformations satisfies the properties of a group and is called the Poincaré group. Under the Erlangen program, Minkowski space can be viewed as the geometry defined by the Poincaré group, which combines Lorentz transformations with translations. In a similar way, the set of all Lorentz transformations forms a group, called the Lorentz group.

A quantity invariant under Lorentz transformations is known as a Lorentz scalar.

33.8 See also

- Ricci calculus
- Electromagnetic field
- Galilean transformation
- Hyperbolic rotation
- Invariance mechanics
- Lorentz group

- Representation theory of the Lorentz group
- Principle of relativity
- Velocity-addition formula
- Algebra of physical space
- Relativistic aberration
- Prandtl–Glauert transformation
- Split-complex number

33.9 References

[1] O'Connor, John J.; Robertson, Edmund F., *A History of Special Relativity*

[2] Brown, Harvey R., *Michelson, FitzGerald and Lorentz: the Origins of Relativity Revisited*

[3] Rothman, Tony (2006), "Lost in Einstein's Shadow" (PDF), *American Scientist* **94** (2): 112f.

[4] Darrigol, Olivier (2005), "The Genesis of the theory of relativity" (PDF), *Séminaire Poincaré* **1**: 1–22, doi:10.1007/3-7643-7436-5_1

[5] Macrossan, Michael N. (1986), "A Note on Relativity Before Einstein", *Brit. Journal Philos. Science* **37**: 232–34, doi:10.1093/bjps/37.2.232

[6] The reference is within the following paper: Poincaré, Henri (1905), "On the Dynamics of the Electron", *Comptes rendus hebdomadaires des séances de l'Académie des sciences* **140**: 1504–1508

[7] Einstein, Albert (1905), "Zur Elektrodynamik bewegter Körper" (PDF), *Annalen der Physik* **322** (10): 891–921, Bibcode:1905AnP...322..891E, doi:10.1002/andp.19053221004. See also: English translation.

[8] A. Halpern (1988). *3000 Solved Problems in Physics*. Schaum Series. Mc Graw Hill. p. 688. ISBN 978-0-07-025734-4.

[9] University Physics – With Modern Physics (12th Edition), H.D. Young, R.A. Freedman (Original edition), Addison-Wesley (Pearson International), 1st Edition: 1949, 12th Edition: 2008, ISBN (10-) 0-321-50130-6, ISBN (13-) 978-0-321-50130-1

[10] Dynamics and Relativity, J.R. Forshaw, A.G. Smith, Manchester Physics Series, John Wiley & Sons Ltd, ISBN 978-0-470-01460-8

[11] Relativity DeMystified, D. McMahon, Mc Graw Hill (USA), 2006, ISBN 0-07-145545-0

[12] Gravitation, J.A. Wheeler, C. Misner, K.S. Thorne, W.H. Freeman & Co, 1973, ISBN 0-7167-0344-0

[13] Ungar, A. A. (1989). "The relativistic velocity composition paradox and the Thomas rotation". *Foundations of Physics* **19**: 1385–1396. Bibcode:1989FoPh...19.1385U. doi:10.1007/BF00732759.

[14] Ungar, A. A. (2000). "The relativistic composite-velocity reciprocity principle". *Foundations of Physics* (Springer) **30** (2): 331–342. CiteSeerX: 10.1.1.35.1131.

[15] Ungar, AA (1988). "Thomas rotation and the parameterization of the Lorentz transformation group". *Foundations of Physics Letters* (Kluwer Academic Publishers-Plenum Publishers) **1** (1): 55–89. doi:10.1007/BF00661317. ISSN 0894-9875. (subscription required (help)). eqn (55).

[16] M. Carroll, Sean (2004). *Spacetime and Geometry: An Introduction to General Relativity* (illustrated ed.). Addison Wesley. p. 22. ISBN 0-8053-8732-3.

[17] Einstein, Albert (1916). "Relativity: The Special and General Theory" (PDF). Retrieved 2012-01-23.

[18] Dynamics and Relativity, J.R. Forshaw, A.G. Smith, Wiley, 2009, ISBN 978 0 470 01460 8

[19] Electromagnetism (2nd Edition), I.S. Grant, W.R. Phillips, Manchester Physics, John Wiley & Sons, 2008, ISBN 9-780471-927129

[20] Introduction to Electrodynamics (3rd Edition), D.J. Griffiths, Pearson Education, Dorling Kindersley, 2007, ISBN 81-7758-293-3

[21] Weinberg, Steven (1972), *Gravitation and Cosmology*, New York, [NY.]: Wiley, ISBN 0-471-92567-5: (Section 2:1)

[22] Weinberg, Steven (1995), *The quantum theory of fields (3 vol.)*, Cambridge, [England] ; New York, [NY.]: Cambridge University Press, ISBN 0-521-55001-7 : volume 1.

33.10 Further reading

- Einstein, Albert (1961), *Relativity: The Special and the General Theory*, New York: Three Rivers Press (published 1995), ISBN 0-517-88441-0
- Ernst, A.; Hsu, J.-P. (2001), "First proposal of the universal speed of light by Voigt 1887" (PDF), *Chinese Journal of Physics* **39** (3): 211–230, Bibcode:2001ChJPh..39..211E
- Thornton, Stephen T.; Marion, Jerry B. (2004), *Classical dynamics of particles and systems* (5th ed.), Belmont, [CA.]: Brooks/Cole, pp. 546–579, ISBN 0-534-40896-6
- Voigt, Woldemar (1887), "Über das Doppler'sche princip", *Nachrichten von der Königlicher Gesellschaft den Wissenschaft zu Göttingen* **2**: 41–51

33.11 External links

- Derivation of the Lorentz transformations. This web page contains a more detailed derivation of the Lorentz transformation with special emphasis on group properties.
- The Paradox of Special Relativity. This webpage poses a problem, the solution of which is the Lorentz transformation, which is presented graphically in its next page.
- Relativity – a chapter from an online textbook
- Warp Special Relativity Simulator. A computer program demonstrating the Lorentz transformations on everyday objects.
- Animation clip on YouTube visualizing the Lorentz transformation.
- Lorentz Frames Animated *from John de Pillis.* Online Flash animations of Galilean and Lorentz frames, various paradoxes, EM wave phenomena, *etc.*

Chapter 34

Vector boson

In particle physics, a **vector boson** is a boson with the spin equal to 1. The vector bosons regarded as elementary particles in the Standard Model are the gauge bosons, which are the force carriers of fundamental interactions: the photon of electromagnetism, the W and Z bosons of the weak interaction, and the gluons of the strong interaction. Some composite particles are vector bosons, for instance any vector meson (quark and antiquark). During the 1970s and '80s, **intermediate vector bosons**—vector bosons of "intermediate" mass—drew much attention in particle physics.

34.1 Vector bosons and the Higgs

Certain vector bosons, i.e. the Z and W particles, play a prominent role in the recently confirmed (March 14, 2013)[1] Higgs Boson, as shown by the attached Feynman diagram.

34.2 Explanation

The name *vector boson* arises from quantum field theory. The component of such a particle's spin along any axis has the three eigenvalues $-\hbar$, 0, and $+\hbar$ (where $\hbar$ is the reduced Planck constant), meaning that any measurement of it can only yield one of these values. (This is, at least, true for massive vector bosons; the situation is a bit different for massless particles such as the photon, for reasons beyond the scope of this article. See Wigner's classification.[2]) The space of spin states therefore is a discrete degree of freedom consisting of three states, the same as the number of components of a vector in three-dimensional space. Quantum superpositions of these states can be taken such that they transform under rotations just like the spatial components of a rotating vector (the so named **3** representation of SU(2)). If the vector boson is taken to be the quantum of a field, the field is a vector field, hence the name.

34.3 See also

- Pseudovector meson

34.4 Notes

[1] http://www.livescience.com/27888-newfound-particle-is-higgs.html

[2] Weingard, Robert. "Some Comments Regarding Spin and Relativity"

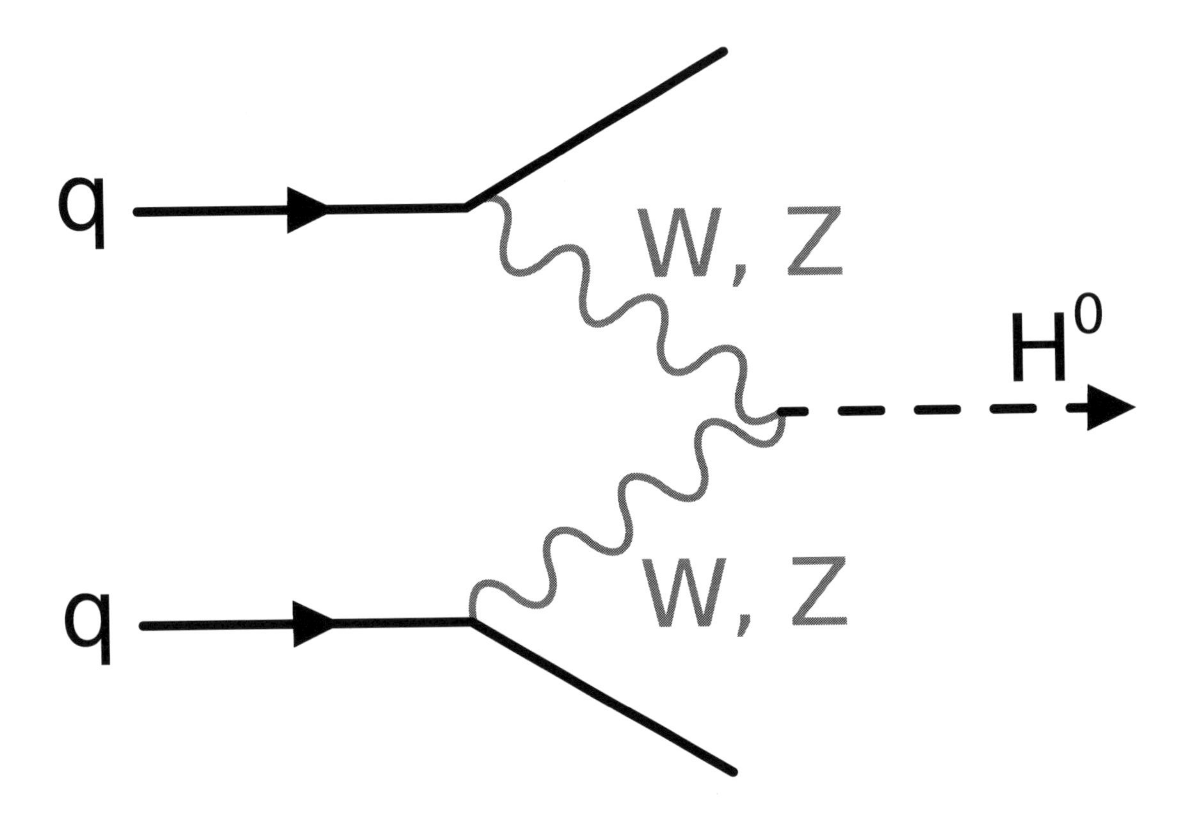

Feynman diagram of the fusion of two electroweak vector bosons to the scalar Higgs boson, which is a prominent process of the generation of Higgs bosons at particle accelerators.
(The symbol q means a quark particle, W and Z are the vector bosons of the electroweak interaction. H^0 is the Higgs boson.)

Chapter 35

Spinor

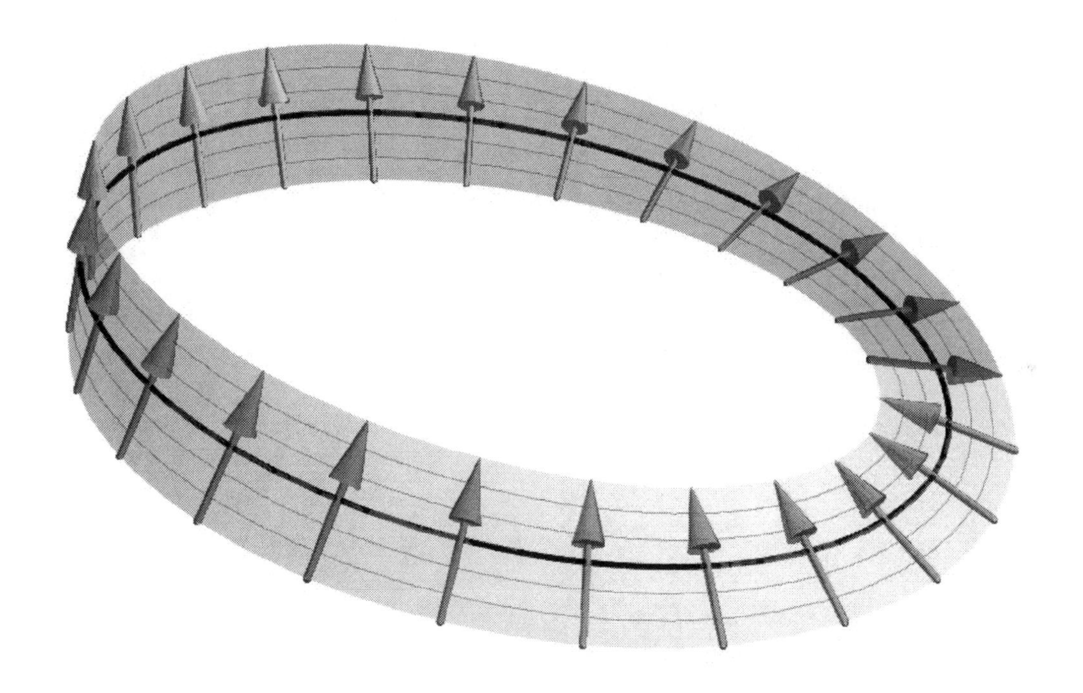

A spinor visualized as a vector pointing along the Möbius band, exhibiting a sign inversion when the circle (the "physical system") is rotated through a full turn of 360°.[nb 1]

In geometry and physics, **spinors** are elements of a (complex) vector space that can be associated with Euclidean space.[nb 2] Like geometric vectors and more general tensors, spinors transform linearly when the Euclidean space is subjected to a slight (infinitesimal) rotation.[nb 3] When a sequence of such small rotations is composed (integrated) to form an overall final rotation, however, the resulting spinor transformation depends on which sequence of small rotations was used, *unlike* for vectors and tensors. A spinor transforms to its negative when the space is rotated through a complete turn from 0° to 360° (see picture), and it is this property that characterizes spinors. It is also possible to associate a substantially similar notion of spinor to Minkowski space in which case the Lorentz transformations of special relativity play the role of rotations. Spinors were introduced in geometry by Élie Cartan in 1913.[1][2] In the 1920s physicists discovered that spinors are essential to describe the intrinsic angular momentum, or "spin", of the electron and other subatomic particles.[nb 4]

Spinors are characterized by the specific way in which they behave under rotations. They change in different ways depending not just on the overall final rotation, but the details of how that rotation was achieved (by a continuous path in the rotation group). There are two topologically distinguishable classes (homotopy classes) of paths through rotations that result in the same overall rotation, as famously illustrated by the belt trick puzzle (below). These two inequivalent classes yield spinor transformations of opposite sign. The spin group is the group of all rotations keeping track of the class.[nb 5] It doubly-covers the rotation group, since each rotation can be obtained in two inequivalent ways as the endpoint of a path. The space of spinors by definition is equipped with a (complex) linear representation of the spin group, meaning that elements of the spin group act as linear transformations on the space of spinors, in a way that genuinely depends on the homotopy class.[nb 6]

Although spinors can be defined purely as elements of a representation space of the spin group (or its Lie algebra of infinitesimal rotations), they are typically defined as elements of a vector space that carries a linear representation of the Clifford algebra. The Clifford algebra is an associative algebra that can be constructed from Euclidean space and its inner product in a basis independent way. Both the spin group and its Lie algebra are embedded inside the Clifford algebra in a natural way, and in applications the Clifford algebra is often the easiest to work with.[nb 7] After choosing an orthonormal basis of Euclidean space, a representation of the Clifford algebra is generated by gamma matrices, matrices that satisfy a set of canonical anti-commutation relations. The spinors are the column vectors on which these matrices act. In three Euclidean dimensions, for instance, the Pauli spin matrices are a set of gamma matrices,[nb 8] and the two-component complex column vectors on which these matrices act are spinors. However, the particular matrix representation of the Clifford algebra, and hence what precisely constitutes a "column vector" (or spinor), involves the choice of basis and gamma matrices in an essential way. As a representation of the spin group, this realization of spinors as (complex[nb 9]) column vectors will either be irreducible if the dimension is odd, or it will decompose into a pair of so-called "half-spin" or Weyl representations if the dimension is even.[nb 10]

35.1 Introduction

A gradual rotation can be visualized as a ribbon in space (the TNB frame of the ribbon defines a rotation continuously for each value of the arc length parameter). Two gradual rotations with different classes, one through 2π and one through

4π, are illustrated here in the belt trick puzzle. A solution of the puzzle is a (continuous) manipulation of the belt, fixing the endpoints, that untwists it. This is impossible with the 2π rotation, but possible with the 4π rotation. A solution, shown in the second animation, actually gives an explicit homotopy in the rotation group between the 4π rotation and the trivial (identity) rotation.

What characterizes spinors and distinguishes them from geometric vectors and other tensors is subtle. Consider applying a rotation to the coordinates of a system. No object in the system itself has moved, only the coordinates have, so there will always be a compensating change in those coordinate values when applied to any object of the system. Geometrical vectors, for example, have components that will undergo *the same* rotation as the coordinates. More broadly, any tensor associated with the system (for instance, the stress of some medium) also has coordinate descriptions that adjust to compensate for changes to the coordinate system itself. Spinors do not appear at this level of the description of a physical system, when one is concerned only with the properties of a single isolated rotation of the coordinates. Rather, spinors appear when we imagine that instead of a single rotation, the coordinate system is gradually (continuously) rotated between some initial and final configuration. For any of the familiar and intuitive ("tensorial") quantities associated with the system, the transformation law does not depend on the precise details of how the coordinates arrived at their final configuration. Spinors, on the other hand, are constructed in such a way that makes them *sensitive* to how the gradual rotation of the coordinates arrived there: they exhibit path-dependence. It turns out that, for any final configuration of the coordinates, there are actually two ("topologically") inequivalent *gradual* (continuous) rotations of the coordinate system that result in this same configuration. This ambiguity is called the homotopy class of the gradual rotation. The belt trick puzzle (shown) famously demonstrates two different rotations, one through an angle of 2π and the other through an angle of 4π, having the same final configurations but different classes. Spinors actually exhibit a sign-reversal that genuinely depends on this homotopy class. This distinguishes them from vectors and other tensors, none of which can feel the class.

Spinors can be exhibited as concrete objects using a choice of Cartesian coordinates. In three Euclidean dimensions, for instance, spinors can be constructed by making a choice of Pauli spin matrices corresponding to (angular momenta about) the three coordinate axes. These are 2×2 matrices with complex entries, and the two-component complex column vectors on which these matrices act by matrix multiplication are the spinors. In this case, the spin group is isomorphic to the group of 2×2 unitary matrices with determinant one, which naturally sits inside the matrix algebra. This group acts by conjugation on the real vector space spanned by the Pauli matrices themselves,[nb 11] realizing it as a group of rotations among them,[nb 12] but it also acts on the column vectors (that is, the spinors).

More generally, a Clifford algebra can be constructed from any vector space V equipped with a (nondegenerate) quadratic form, such as Euclidean space with its standard dot product or Minkowski space with its standard Lorentz metric. Given a suitably normalized basis of V, the Clifford algebra is generated by gamma matrices, matrices that satisfy a set of canonical anti-commutation relations, and the space of spinors is the space of column vectors with $2^{\lfloor \dim V/2 \rfloor}$ components on which those matrices act. Although the Clifford algebra can be defined abstractly in a coordinate-independent way, its particular realization as a specific algebra of matrices depends on which orthogonal axes the gamma matrices represent. So what precisely constitutes a "column vector" (or spinor) also depends on such arbitrary choices.[nb 13] The orthogonal Lie algebra (i.e., the infinitesimal "rotations") and the spin group associated to the quadratic form are both (canonically) contained in the Clifford algebra, so every Clifford algebra representation also defines a representation of the Lie algebra and the spin group.[nb 14] Depending on the dimension and metric signature, this realization of spinors as column vectors may be irreducible or it may decompose into a pair of so-called "half-spin" or Weyl representations.[nb 15]

35.2 Overview

There are essentially two frameworks for viewing the notion of a spinor.

One is representation theoretic. In this point of view, one knows beforehand that there are some representations of the Lie algebra of the orthogonal group that cannot be formed by the usual tensor constructions. These missing representations are then labeled the **spin representations**, and their constituents *spinors*. In this view, a spinor must belong to a representation of the double cover of the rotation group SO(n, **R**), or more generally of double cover of the generalized special orthogonal group $SO^+(p, q, \mathbf{R})$ on spaces with metric signature (p, q). These double covers are Lie groups, called the spin groups Spin(n) or Spin(p, q). All the properties of spinors, and their applications and derived objects, are manifested first in the spin group. Representations of the double covers of these groups yield projective representations of the groups themselves, which do not meet the full definition of a representation.

The other point of view is geometrical. One can explicitly construct the spinors, and then examine how they behave under the action of the relevant Lie groups. This latter approach has the advantage of providing a concrete and elementary description of what a spinor is. However, such a description becomes unwieldy when complicated properties of spinors, such as Fierz identities, are needed.

35.2.1 Clifford algebras

For more details on this topic, see Clifford algebra.

The language of Clifford algebras[3] (sometimes called geometric algebras) provides a complete picture of the spin representations of all the spin groups, and the various relationships between those representations, via the classification of Clifford algebras. It largely removes the need for *ad hoc* constructions.

In detail, let V be a finite-dimensional complex vector space with nondegenerate bilinear form g. The Clifford algebra $C\ell(V, g)$ is the algebra generated by V along with the anticommutation relation $xy + yx = 2g(x, y)$. It is an abstract version of the algebra generated by the gamma or Pauli matrices. If $V = \mathbf{C}^n$, with the standard form $g(x, y) = x^t y = x_1 y_1 + ... + x_n yn$ we denote the Clifford algebra by $C\ell n(\mathbf{C})$. Since by the choice of an orthonormal basis every complex vectorspace with non-degenerate form is isomorphic to this standard example, this notation is abused more generally if $\dim\mathbf{C}(V) = n$. If $n = 2k$ is even, $C\ell n(\mathbf{C})$ is isomorphic as an algebra (in a non-unique way) to the algebra $\mathrm{Mat}(2^k, \mathbf{C})$ of $2^k \times 2^k$ complex matrices (by the Artin-Wedderburn theorem and the easy to prove fact that the Clifford algebra is central simple). If $n = 2k + 1$ is odd, $C\ell_2k_{+1}(\mathbf{C})$ is isomorphic to the algebra $\mathrm{Mat}(2^k, \mathbf{C}) \oplus \mathrm{Mat}(2^k, \mathbf{C})$ of two copies of the $2^k \times 2^k$ complex matrices. Therefore, in either case $C\ell(V, g)$ has a unique (up to isomorphism) irreducible representation (also called simple Clifford module), commonly denoted by Δ, of dimension $2^{[n/2]}$. Since the Lie algebra $\mathbf{so}(V, g)$ is embedded as a Lie subalgebra in $C\ell(V, g)$ equipped with the Clifford algebra commutator as Lie bracket, the space Δ is also a Lie algebra representation of $\mathbf{so}(V, g)$ called a spin representation. If n is odd, this Lie algebra representation is irreducible. If n is even, it splits further into two irreducible representations $\Delta = \Delta_+ \oplus \Delta_-$ called the Weyl or *half-spin representations*.

Irreducible representations over the reals in the case when V is a real vector space are much more intricate, and the reader is referred to the Clifford algebra article for more details.

35.2.2 Spin groups

Spinors form a vector space, usually over the complex numbers, equipped with a linear group representation of the spin group that does not factor through a representation of the group of rotations (see diagram). The spin group is the group of rotations keeping track of the homotopy class. Spinors are needed to encode basic information about the topology of the group of rotations because that group is not simply connected, but the simply connected spin group is its double cover. So for every rotation there are two elements of the spin group that represent it. Geometric vectors and other tensors cannot feel the difference between these two elements, but they produce *opposite* signs when they affect any spinor under the representation. Thinking of the elements of the spin group as homotopy classes of one-parameter families of rotations, each rotation is represented by two distinct homotopy classes of paths to the identity. If a one-parameter family of rotations is visualized as a ribbon in space, with the arc length parameter of that ribbon being the parameter (its tangent, normal, binormal frame actually gives the rotation), then these two distinct homotopy classes are visualized in the two states of the belt trick puzzle (above). The space of spinors is an auxiliary vector space that can be constructed explicitly in coordinates, but ultimately only exists up to isomorphism in that there is no "natural" construction of them that does not rely on arbitrary choices such as coordinate systems. A notion of spinors can be associated, as such an auxiliary mathematical object, with any vector space equipped with a quadratic form such as Euclidean space with its standard dot product, or Minkowski space with its Lorentz metric. In the latter case, the "rotations" include the Lorentz boosts, but otherwise the theory is substantially similar.

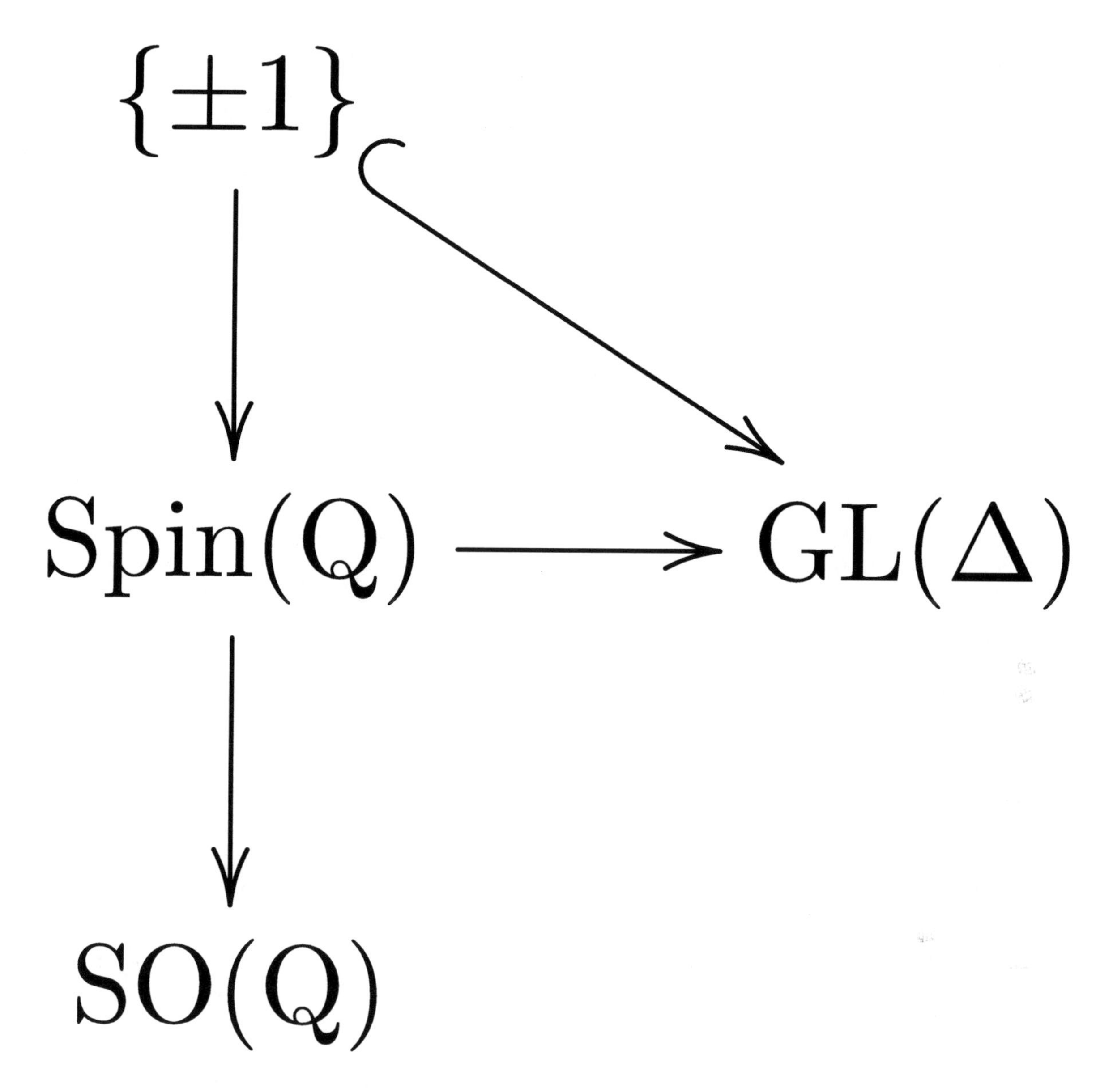

The spin representation Δ is a vector space equipped with a representation of the spin group that does not factor through a representation of the (special) orthogonal group.

35.2.3 Terminology in physics

The most typical type of spinor, the Dirac spinor,[4] is an element of the fundamental representation of $C\ell_{p+q}(\mathbf{C})$, the complexification of the Clifford algebra $C\ell_{p,q}(\mathbf{R})$, into which the spin group Spin(p, q) may be embedded. On a $2k$- or $2k$+1-dimensional space a Dirac spinor may be represented as a vector of 2^k complex numbers. (See Special unitary group.) In even dimensions, this representation is reducible when taken as a representation of Spin(p, q) and may be decomposed into two: the left-handed and right-handed **Weyl spinor**[5] representations. In addition, sometimes the non-complexified version of $C\ell_{p,q}(\mathbf{R})$ has a smaller real representation, the Majorana spinor representation.[6] If this happens in an even dimension, the Majorana spinor representation will sometimes decompose into two Majorana–Weyl spinor representations. Dirac and Weyl spinors are complex representations while Majorana spinors are real representations.

The Dirac, Lorentz, Weyl, and Majorana spinors are interrelated, and their relation can be elucidated on the basis of real geometric algebra.[7]

Massive particles, such as electrons, are described as Dirac spinors. The classical neutrino of the standard model of particle physics is an example of a Weyl spinor. However, because of observed neutrino oscillation, it is now believed that they are not Weyl spinors, but perhaps instead Majorana spinors.[8] It is not known whether (spin-1/2) Weyl spinors exist in nature. In 2015, an international team led by Princeton University scientists announced that they had found a quasiparticle that behaves as a Weyl fermion.[9]

35.2.4 Spinors in representation theory

Main article: Spin representation

One major mathematical application of the construction of spinors is to make possible the explicit construction of linear representations of the Lie algebras of the special orthogonal groups, and consequently spinor representations of the groups themselves. At a more profound level, spinors have been found to be at the heart of approaches to the Atiyah–Singer index theorem, and to provide constructions in particular for discrete series representations of semisimple groups.

The spin representations of the special orthogonal Lie algebras are distinguished from the tensor representations given by Weyl's construction by the weights. Whereas the weights of the tensor representations are integer linear combinations of the roots of the Lie algebra, those of the spin representations are half-integer linear combinations thereof. Explicit details can be found in the spin representation article.

35.2.5 Attempts at intuitive understanding

The spinor can be described, in simple terms, as "vectors of a space the transformations of which are related in a particular way to rotations in physical space".[10] Stated differently:[2]

> *Spinors [...] provide a linear representation of the group of rotations in a space with any number n of dimensions, each spinor having 2^ν components where $n = 2\nu + 1$ or 2ν.*

Several ways of illustrating everyday analogies have been formulated in terms of the plate trick, tangloids and other examples of orientation entanglement.

Nonetheless, the concept is generally considered notoriously difficult to understand, as illustrated by Michael Atiyah's statement that is recounted by Dirac's biographer Graham Farmelo:[11]

> *No one fully understands spinors. Their algebra is formally understood but their general significance is mysterious. In some sense they describe the "square root" of geometry and, just as understanding the square root of −1 took centuries, the same might be true of spinors.*

35.3 History

The most general mathematical form of spinors was discovered by Élie Cartan in 1913.[12] The word "spinor" was coined by Paul Ehrenfest in his work on quantum physics.[13]

Spinors were first applied to mathematical physics by Wolfgang Pauli in 1927, when he introduced his spin matrices.[14] The following year, Paul Dirac discovered the fully relativistic theory of electron spin by showing the connection between spinors and the Lorentz group.[15] By the 1930s, Dirac, Piet Hein and others at the Niels Bohr Institute (then known as the Institute for Theoretical Physics of the University of Copenhagen) created toys such as Tangloids to teach and model the calculus of spinors.

Spinor spaces were represented as left ideals of a matrix algebra in 1930, by G. Juvet[16] and by Fritz Sauter.[17][18] More specifically, instead of representing spinors as complex-valued 2D column vectors as Pauli had done, they represented them as complex-valued 2×2 matrices in which only the elements of the left column are non-zero. In this manner the spinor space became a minimal left ideal in Mat(2, **C**).[19][20]

In 1947 Marcel Riesz constructed spinor spaces as elements of a minimal left ideal of Clifford algebras. In 1966/1967, David Hestenes[21][22] replaced spinor spaces by the even subalgebra $C\ell^0{}_{1,3}(\mathbf{R})$ of the spacetime algebra $C\ell_{1,3}(\mathbf{R})$.[18][20] As of the 1980s, the theoretical physics group at Birkbeck College around David Bohm and Basil Hiley has been developing algebraic approaches to quantum theory that build on Sauter and Riesz' identification of spinors with minimal left ideals.

35.4 Examples

Some simple examples of spinors in low dimensions arise from considering the even-graded subalgebras of the Clifford algebra $C\ell p, q(\mathbf{R})$. This is an algebra built up from an orthonormal basis of $n = p + q$ mutually orthogonal vectors under addition and multiplication, p of which have norm +1 and q of which have norm −1, with the product rule for the basis vectors

$$e_i e_j = \begin{cases} +1 & i = j,\, i \in (1 \ldots p) \\ -1 & i = j,\, i \in (p+1 \ldots n) \\ -e_j e_i & i \neq j. \end{cases}$$

35.4.1 Two dimensions

The Clifford algebra $C\ell_{2,0}(\mathbf{R})$ is built up from a basis of one unit scalar, 1, two orthogonal unit vectors, σ_1 and σ_2, and one unit pseudoscalar $i = \sigma_1\sigma_2$. From the definitions above, it is evident that $(\sigma_1)^2 = (\sigma_2)^2 = 1$, and $(\sigma_1\sigma_2)(\sigma_1\sigma_2) = -\sigma_1\sigma_1\sigma_2\sigma_2 = -1$.

The even subalgebra $C\ell^0{}_{2,0}(\mathbf{R})$, spanned by *even-graded* basis elements of $C\ell_{2,0}(\mathbf{R})$, determines the space of spinors via its representations. It is made up of real linear combinations of 1 and $\sigma_1\sigma_2$. As a real algebra, $C\ell^0{}_{2,0}(\mathbf{R})$ is isomorphic to field of complex numbers **C**. As a result, it admits a conjugation operation (analogous to complex conjugation), sometimes called the *reverse* of a Clifford element, defined by

$$(a + b\sigma_1\sigma_2)^* = a + b\sigma_2\sigma_1$$

which, by the Clifford relations, can be written

$$(a + b\sigma_1\sigma_2)^* = a + b\sigma_2\sigma_1 = a - b\sigma_1\sigma_2$$

The action of an even Clifford element $\gamma \in C\ell^0{}_{2,0}(\mathbf{R})$ on vectors, regarded as 1-graded elements of $C\ell_{2,0}(\mathbf{R})$, is determined by mapping a general vector $u = a_1\sigma_1 + a_2\sigma_2$ to the vector

$$\gamma(u) = \gamma u \gamma^*$$

where γ^* is the conjugate of γ, and the product is Clifford multiplication. In this situation, a **spinor**[23] is an ordinary complex number. The action of γ on a spinor φ is given by ordinary complex multiplication:

$$\gamma(\phi) = \gamma\phi$$

An important feature of this definition is the distinction between ordinary vectors and spinors, manifested in how the even-graded elements act on each of them in different ways. In general, a quick check of the Clifford relations reveals that even-graded elements conjugate-commute with ordinary vectors:

$$\gamma(u) = \gamma u \gamma^* = \gamma^2 u$$

On the other hand, comparing with the action on spinors $\gamma(\varphi) = \gamma\varphi$, γ on ordinary vectors acts as the *square* of its action on spinors.

Consider, for example, the implication this has for plane rotations. Rotating a vector through an angle of θ corresponds to $\gamma^2 = \exp(\theta\ \sigma_1\sigma_2)$, so that the corresponding action on spinors is via $\gamma = \pm\ \exp(\theta\ \sigma_1\sigma_2/2)$. In general, because of logarithmic branching, it is impossible to choose a sign in a consistent way. Thus the representation of plane rotations on spinors is two-valued.

In applications of spinors in two dimensions, it is common to exploit the fact that the algebra of even-graded elements (that is just the ring of complex numbers) is identical to the space of spinors. So, by abuse of language, the two are often conflated. One may then talk about "the action of a spinor on a vector." In a general setting, such statements are meaningless. But in dimensions 2 and 3 (as applied, for example, to computer graphics) they make sense.

Examples

- The even-graded element

  $$\gamma = \tfrac{1}{\sqrt{2}}(1 - \sigma_1\sigma_2)$$

  corresponds to a vector rotation of 90° from σ_1 around towards σ_2, which can be checked by confirming that

  $$\tfrac{1}{2}(1 - \sigma_1\sigma_2)\,\{a_1\sigma_1 + a_2\sigma_2\}\,(1 - \sigma_2\sigma_1) = a_1\sigma_2 - a_2\sigma_1$$

  It corresponds to a spinor rotation of only 45°, however:

  $$\tfrac{1}{\sqrt{2}}(1 - \sigma_1\sigma_2)\,\{a_1 + a_2\sigma_1\sigma_2\} = \frac{a_1 + a_2}{\sqrt{2}} + \frac{-a_1 + a_2}{\sqrt{2}}\sigma_1\sigma_2$$

- Similarly the even-graded element $\gamma = -\sigma_1\sigma_2$ corresponds to a vector rotation of 180°:

  $$(-\sigma_1\sigma_2)\,\{a_1\sigma_1 + a_2\sigma_2\}\,(-\sigma_2\sigma_1) = -a_1\sigma_1 - a_2\sigma_2$$

  but a spinor rotation of only 90°:

  $$(-\sigma_1\sigma_2)\,\{a_1 + a_2\sigma_1\sigma_2\} = a_2 - a_1\sigma_1\sigma_2$$

- Continuing on further, the even-graded element $\gamma = -1$ corresponds to a vector rotation of 360°:

  $$(-1)\,\{a_1\sigma_1 + a_2\sigma_2\}\,(-1) = a_1\sigma_1 + a_2\sigma_2$$

  but a spinor rotation of 180°.

35.4.2 Three dimensions

Main articles Spinors in three dimensions, Quaternions and spatial rotation

The Clifford algebra $C\ell_{3,0}(\mathbf{R})$ is built up from a basis of one unit scalar, 1, three orthogonal unit vectors, σ_1, σ_2 and σ_3, the three unit bivectors $\sigma_1\sigma_2$, $\sigma_2\sigma_3$, $\sigma_3\sigma_1$ and the pseudoscalar $i = \sigma_1\sigma_2\sigma_3$. It is straightforward to show that $(\sigma_1)^2 = (\sigma_2)^2 = (\sigma_3)^2 = 1$, and $(\sigma_1\sigma_2)^2 = (\sigma_2\sigma_3)^2 = (\sigma_3\sigma_1)^2 = (\sigma_1\sigma_2\sigma_3)^2 = -1$.

The sub-algebra of even-graded elements is made up of scalar dilations,

$$u' = \rho^{(1/2)} u \rho^{(1/2)} = \rho u,$$

and vector rotations

$$u' = \gamma u \gamma^*,$$

where

$$\left.\begin{aligned} \gamma &= \cos(\theta/2) - \{a_1\sigma_2\sigma_3 + a_2\sigma_3\sigma_1 + a_3\sigma_1\sigma_2\}\sin(\theta/2) \\ &= \cos(\theta/2) - i\{a_1\sigma_1 + a_2\sigma_2 + a_3\sigma_3\}\sin(\theta/2) \\ &= \cos(\theta/2) - iv\sin(\theta/2) \end{aligned}\right\}$$

corresponds to a vector rotation through an angle θ about an axis defined by a unit vector $v = a_1\sigma_1 + a_2\sigma_2 + a_3\sigma_3$.

As a special case, it is easy to see that, if $v = \sigma_3$, this reproduces the $\sigma_1\sigma_2$ rotation considered in the previous section; and that such rotation leaves the coefficients of vectors in the σ_3 direction invariant, since

$$(\cos(\theta/2) - i\sigma_3\sin(\theta/2))\,\sigma_3\,(\cos(\theta/2) + i\sigma_3\sin(\theta/2)) = (\cos^2(\theta/2) + \sin^2(\theta/2))\,\sigma_3 = \sigma_3.$$

The bivectors $\sigma_2\sigma_3$, $\sigma_3\sigma_1$ and $\sigma_1\sigma_2$ are in fact Hamilton's quaternions **i**, **j** and **k**, discovered in 1843:

$$\begin{aligned} \mathbf{i} &= -\sigma_2\sigma_3 = -i\sigma_1 \\ \mathbf{j} &= -\sigma_3\sigma_1 = -i\sigma_2 \\ \mathbf{k} &= -\sigma_1\sigma_2 = -i\sigma_3. \end{aligned}$$

With the identification of the even-graded elements with the algebra **H** of quaternions, as in the case of two dimensions the only representation of the algebra of even-graded elements is on itself.[24] Thus the (real[25]) spinors in three-dimensions are quaternions, and the action of an even-graded element on a spinor is given by ordinary quaternionic multiplication.

Note that the expression (1) for a vector rotation through an angle θ, the angle appearing in γ was halved. Thus the spinor rotation $\gamma(\psi) = \gamma\psi$ (ordinary quaternionic multiplication) will rotate the spinor ψ through an angle one-half the measure of the angle of the corresponding vector rotation. Once again, the problem of lifting a vector rotation to a spinor rotation is two-valued: the expression (1) with $(180° + \theta/2)$ in place of $\theta/2$ will produce the same vector rotation, but the negative of the spinor rotation.

The spinor/quaternion representation of rotations in 3D is becoming increasingly prevalent in computer geometry and other applications, because of the notable brevity of the corresponding spin matrix, and the simplicity with which they can be multiplied together to calculate the combined effect of successive rotations about different axes.

35.5 Explicit constructions

A space of spinors can be constructed explicitly with concrete and abstract constructions. The equivalence of these constructions are a consequence of the uniqueness of the spinor representation of the complex Clifford algebra. For a complete example in dimension 3, see spinors in three dimensions.

35.5.1 Component spinors

Given a vector space V and a quadratic form g an explicit matrix representation of the Clifford algebra $C\ell(V, g)$ can be defined as follows. Choose an orthonormal basis $e^1 \ldots e^n$ for V i.e. $g(e^\mu e^\nu) = \eta^{\mu\nu}$ where $\eta^{\mu\mu} = \pm 1$ and $\eta^{\mu\nu} = 0$ for $\mu \neq$

v. Let $k = \lfloor n/2 \rfloor$. Fix a set of $2^k \times 2^k$ matrices $\gamma^1 \ldots \gamma^n$ such that $\gamma^\mu\gamma^\nu + \gamma^\nu\gamma^\mu = 2\eta^{\mu\nu}1$ (i.e. fix a convention for the gamma matrices). Then the assignment $e^\mu \to \gamma^\mu$ extends uniquely to an algebra homomorphism $C\ell(V, g) \to \mathrm{Mat}(2^k, \mathbf{C})$ by sending the monomial $e^{\mu_1} \ldots e^{\mu_k}$ in the Clifford algebra to the product $\gamma^{\mu_1} \ldots \gamma^{\mu_k}$ of matrices and extending linearly. The space $\Delta = \mathbf{C}^{2^k}$ on which the gamma matrices act is a now a space of spinors. One needs to construct such matrices explicitly, however. In dimension 3, defining the gamma matrices to be the Pauli sigma matrices gives rise to the familiar two component spinors used in non relativistic quantum mechanics. Likewise using the 4×4 Dirac gamma matrices gives rise to the 4 component Dirac spinors used in 3+1 dimensional relativistic quantum field theory. In general, in order to define gamma matrices of the required kind, one can use the Weyl–Brauer matrices.

In this construction the representation of the Clifford algebra $C\ell(V, g)$, the Lie algebra $\mathbf{so}(V, g)$, and the Spin group $\mathrm{Spin}(V, g)$, all depend on the choice of the orthonormal basis and the choice of the gamma matrices. This can cause confusion over conventions, but invariants like traces are independent of choices. In particular, all physically observable quantities must be independent of such choices. In this construction a spinor can be represented as a vector of 2^k complex numbers and is denoted with spinor indices (usually α, β, γ). In the physics literature, abstract spinor indices are often used to denote spinors even when an abstract spinor construction is used.

35.5.2 Abstract spinors

There are at least two different, but essentially equivalent, ways to define spinors abstractly. One approach seeks to identify the minimal ideals for the left action of $C\ell(V, g)$ on itself. These are subspaces of the Clifford algebra of the form $C\ell(V, g)\omega$, admitting the evident action of $C\ell(V, g)$ by left-multiplication: $c : x\omega \to cx\omega$. There are two variations on this theme: one can either find a primitive element ω that is a nilpotent element of the Clifford algebra, or one that is an idempotent. The construction via nilpotent elements is more fundamental in the sense that an idempotent may then be produced from it.[26] In this way, the spinor representations are identified with certain subspaces of the Clifford algebra itself. The second approach is to construct a vector space using a distinguished subspace of V, and then specify the action of the Clifford algebra *externally* to that vector space.

In either approach, the fundamental notion is that of an isotropic subspace W. Each construction depends on an initial freedom in choosing this subspace. In physical terms, this corresponds to the fact that there is no measurement protocol that can specify a basis of the spin space, even if a preferred basis of V is given.

As above, we let (V, g) be an n-dimensional complex vector space equipped with a nondegenerate bilinear form. If V is a real vector space, then we replace V by its complexification $V \otimes\mathbf{R}\, \mathbf{C}$ and let g denote the induced bilinear form on $V \otimes\mathbf{R}\, \mathbf{C}$. Let W be a maximal isotropic subspace, i.e. a maximal subspace of V such that $g|W = 0$. If $n = 2k$ is even, then let W^* be an isotropic subspace complementary to W. If $n = 2k + 1$ is odd, let W^* be a maximal isotropic subspace with $W \cap W^* = 0$, and let U be the orthogonal complement of $W \oplus W^*$. In both the even- and odd-dimensional cases W and W^* have dimension k. In the odd-dimensional case, U is one-dimensional, spanned by a unit vector u.

35.5.3 Minimal ideals

Since W' is isotropic, multiplication of elements of W' inside $C\ell(V, g)$ is skew. Hence vectors in W' anti-commute, and $C\ell(W', g|W') = C\ell(W', 0)$ is just the exterior algebra $\Lambda^* W'$. Consequently, the k-fold product of W' with itself, W'^k, is one-dimensional. Let ω be a generator of W'^k. In terms of a basis $w'_1,\ldots, w'_k$ of in W', one possibility is to set

$$\omega = w'_1 w'_2 \cdots w'_k.$$

Note that $\omega^2 = 0$ (i.e., ω is nilpotent of order 2), and moreover, $w'\omega = 0$ for all $w' \in W'$. The following facts can be proven easily:

1. If $n = 2k$, then the left ideal $\Delta = C\ell(V, g)\omega$ is a minimal left ideal. Furthermore, this splits into the two spin spaces $\Delta_+ = C\ell^{\mathrm{even}}\omega$ and $\Delta_- = C\ell^{\mathrm{odd}}\omega$ on restriction to the action of the even Clifford algebra.
2. If $n = 2k + 1$, then the action of the unit vector u on the left ideal $C\ell(V, g)\omega$ decomposes the space into a pair of isomorphic irreducible eigenspaces (both denoted by Δ), corresponding to the respective eigenvalues +1 and −1.

In detail, suppose for instance that n is even. Suppose that I is a non-zero left ideal contained in $C\ell(V, g)\omega$. We shall show that I must be equal to $C\ell(V, g)\omega$ by proving that it contains a nonzero scalar multiple of ω.

Fix a basis w_i of W and a complementary basis w_i' of W' so that

$$w_i w_j' + w_j' w_i = \delta_{ij}, \text{ and}$$
$$(w_i)^2 = 0, \ (w_i')^2 = 0.$$

Note that any element of I must have the form $\alpha\omega$, by virtue of our assumption that $I \subset C\ell(V, g)\,\omega$. Let $\alpha\omega \in I$ be any such element. Using the chosen basis, we may write

$$\alpha = \sum_{i_1<i_2<\cdots<i_p} a_{i_1\ldots i_p} w_{i_1}\cdots w_{i_p} + \sum_j B_j w_j'$$

where the $a_{i1\cdots ip}$ are scalars, and the B_j are auxiliary elements of the Clifford algebra. Observe now that the product

$$\alpha\omega = \sum_{i_1<i_2<\cdots<i_p} a_{i_1\ldots i_p} w_{i_1}\cdots w_{i_p}\omega.$$

Pick any nonzero monomial a in the expansion of α with maximal homogeneous degree in the elements w_i:

$$a = a_{i_1\ldots i_{max}} w_{i_1}\ldots w_{i_{max}}$$

then

$$w'_{i_{max}}\cdots w'_{i_1}\alpha\omega = a_{i_1\ldots i_{max}}\omega$$

is a nonzero scalar multiple of ω, as required.

Note that for n even, this computation also shows that

$$\Delta = C\ell(W)\omega = (\Lambda^* W)\omega$$

as a vector space. In the last equality we again used that W is isotropic. In physics terms, this shows that Δ is built up like a Fock space by creating spinors using anti-commuting creation operators in W acting on a vacuum ω.

35.5.4 Exterior algebra construction

The computations with the minimal ideal construction suggest that a spinor representation can also be defined directly using the exterior algebra $\Lambda^* W = \oplus j\, \Lambda^j W$ of the isotropic subspace W. Let $\Delta = \Lambda^* W$ denote the exterior algebra of W considered as vector space only. This will be the spin representation, and its elements will be referred to as spinors.[27]

The action of the Clifford algebra on Δ is defined first by giving the action of an element of V on Δ, and then showing that this action respects the Clifford relation and so extends to a homomorphism of the full Clifford algebra into the endomorphism ring End(Δ) by the universal property of Clifford algebras. The details differ slightly according to whether the dimension of V is even or odd.

When dim(V) is even, $V = W \oplus W'$ where W' is the chosen isotropic complement. Hence any $v \in V$ decomposes uniquely as $v = w + w'$ with $w \in W$ and $w' \in W'$. The action of v on a spinor is given by

$$c(v)w_1 \wedge \cdots \wedge w_n = (\epsilon(w) + i(w'))\,(w_1 \wedge \cdots \wedge w_n)$$

where $i(w')$ is interior product with w' using the non degenerate quadratic form to identify V with V^*, and ε(w) denotes the exterior product. It may be verified that

$$c(u)c(v) + c(v)c(u) = 2\, g(u,v),$$

and so c respects the Clifford relations and extends to a homomorphism from the Clifford algebra to End(Δ).

The spin representation Δ further decomposes into a pair of irreducible complex representations of the Spin group[28] (the half-spin representations, or Weyl spinors) via

$$\Delta_+ = \Lambda^{even}W,\ \Delta_- = \Lambda^{odd}W$$

When dim(V) is odd, $V = W \oplus U \oplus W'$, where U is spanned by a unit vector u orthogonal to W. The Clifford action c is defined as before on $W \oplus W'$, while the Clifford action of (multiples of) u is defined by

$$c(u)\alpha = \begin{cases} \alpha & \text{if } \alpha \in \Lambda^{even}W \\ -\alpha & \text{if } \alpha \in \Lambda^{odd}W \end{cases}$$

As before, one verifies that c respects the Clifford relations, and so induces a homomorphism.

35.5.5 Hermitian vector spaces and spinors

If the vector space V has extra structure that provides a decomposition of its complexification into two maximal isotropic subspaces, then the definition of spinors (by either method) becomes natural.

The main example is the case that the real vector space V is a hermitian vector space (V, h), i.e., V is equipped with a complex structure J that is an orthogonal transformation with respect to the inner product g on V. Then $V \otimes$**R C** splits in the $\pm i$ eigenspaces of J. These eigenspaces are isotropic for the complexification of g and can be identified with the complex vector space (V, J) and its complex conjugate (V, $-J$). Therefore for a hermitian vector space (V, h) the vector space Λ·
CV (as well as its complex conjugate Λ·
CV) is a spinor space for the underlying real euclidean vector space.

With the Clifford action as above but with contraction using the hermitian form, this construction gives a spinor space at every point of an almost Hermitian manifold and is the reason why every almost complex manifold (in particular every symplectic manifold) has a Spin^c structure. Likewise, every complex vector bundle on a manifold carries a Spin^c structure.[29]

35.6 Clebsch–Gordan decomposition

A number of Clebsch–Gordan decompositions are possible on the tensor product of one spin representation with another.[30] These decompositions express the tensor product in terms of the alternating representations of the orthogonal group.

For the real or complex case, the alternating representations are

- $\Gamma r = \Lambda^r V$, the representation of the orthogonal group on skew tensors of rank r.

In addition, for the real orthogonal groups, there are three characters (one-dimensional representations)

- $\sigma_+ : O(p, q) \to \{-1, +1\}$ given by $\sigma_+(R) = -1$, if R reverses the spatial orientation of V, $+1$, if R preserves the spatial orientation of V. (*The spatial character.*)
- $\sigma_- : O(p, q) \to \{-1, +1\}$ given by $\sigma_-(R) = -1$, if R reverses the temporal orientation of V, $+1$, if R preserves the temporal orientation of V. (*The temporal character.*)
- $\sigma = \sigma_+\sigma_-$. (*The orientation character.*)

The Clebsch–Gordan decomposition allows one to define, among other things:

- An action of spinors on vectors.
- A Hermitian metric on the complex representations of the real spin groups.
- A Dirac operator on each spin representation.

35.6.1 Even dimensions

If $n = 2k$ is even, then the tensor product of Δ with the contragredient representation decomposes as

$$\Delta \otimes \Delta^* \cong \bigoplus_{p=0}^{n} \Gamma_p \cong \bigoplus_{p=0}^{k-1} (\Gamma_p \oplus \sigma\Gamma_p) \oplus \Gamma_k$$

which can be seen explicitly by considering (in the Explicit construction) the action of the Clifford algebra on decomposable elements $\alpha\omega \otimes \beta\omega'$. The rightmost formulation follows from the transformation properties of the Hodge star operator. Note that on restriction to the even Clifford algebra, the paired summands $\Gamma p \oplus \sigma\Gamma p$ are isomorphic, but under the full Clifford algebra they are not.

There is a natural identification of Δ with its contragredient representation via the conjugation in the Clifford algebra:

$$(\alpha\omega)^* = \omega(\alpha^*).$$

So $\Delta \otimes \Delta$ also decomposes in the above manner. Furthermore, under the even Clifford algebra, the half-spin representations decompose

$$\begin{aligned} \Delta_+ \otimes \Delta_+^* \cong \Delta_- \otimes \Delta_-^* &\cong \bigoplus\nolimits_{p=0}^{k} \Gamma_{2p} \\ \Delta_+ \otimes \Delta_-^* \cong \Delta_- \otimes \Delta_+^* &\cong \bigoplus\nolimits_{p=0}^{k-1} \Gamma_{2p+1} \end{aligned}$$

For the complex representations of the real Clifford algebras, the associated reality structure on the complex Clifford algebra descends to the space of spinors (via the explicit construction in terms of minimal ideals, for instance). In this way, we obtain the complex conjugate Δ of the representation Δ, and the following isomorphism is seen to hold:

$$\bar{\Delta} \cong \sigma_-\Delta^*$$

In particular, note that the representation Δ of the orthochronous spin group is a unitary representation. In general, there are Clebsch–Gordan decompositions

$$\Delta \otimes \bar{\Delta} \cong \bigoplus_{p=0}^{k} (\sigma_-\Gamma_p \oplus \sigma_+\Gamma_p)\,.$$

In metric signature (p, q), the following isomorphisms hold for the conjugate half-spin representations

- If q is even, then $\bar{\Delta}_+ \cong \sigma_- \otimes \Delta_+^*$ and $\bar{\Delta}_- \cong \sigma_- \otimes \Delta_-^*$.
- If q is odd, then $\bar{\Delta}_+ \cong \sigma_- \otimes \Delta_-^*$ and $\bar{\Delta}_- \cong \sigma_- \otimes \Delta_+^*$.

Using these isomorphisms, one can deduce analogous decompositions for the tensor products of the half-spin representations $\Delta\pm \otimes \Delta\pm$.

35.6.2 Odd dimensions

If $n = 2k + 1$ is odd, then

$$\Delta \otimes \Delta^* \cong \bigoplus_{p=0}^{k} \Gamma_{2p}.$$

In the real case, once again the isomorphism holds

$$\bar{\Delta} \cong \sigma_- \Delta^*.$$

Hence there is a Clebsch–Gordan decomposition (again using the Hodge star to dualize) given by

$$\Delta \otimes \bar{\Delta} \cong \sigma_- \Gamma_0 \oplus \sigma_+ \Gamma_1 \oplus \cdots \oplus \sigma_\pm \Gamma_k$$

35.6.3 Consequences

There are many far-reaching consequences of the Clebsch–Gordan decompositions of the spinor spaces. The most fundamental of these pertain to Dirac's theory of the electron, among whose basic requirements are

- A manner of regarding the product of two spinors $\phi\psi$ as a scalar. In physical terms, a spinor should determine a probability amplitude for the quantum state.
- A manner of regarding the product $\psi\phi$ as a vector. This is an essential feature of Dirac's theory, which ties the spinor formalism to the geometry of physical space.
- A manner of regarding a spinor as acting upon a vector, by an expression such as $\psi v \psi$. In physical terms, this represents an electric current of Maxwell's electromagnetic theory, or more generally a probability current.

35.7 Summary in low dimensions

- In 1 dimension (a trivial example), the single spinor representation is formally Majorana, a real 1-dimensional representation that does not transform.
- In 2 Euclidean dimensions, the left-handed and the right-handed Weyl spinor are 1-component complex representations, i.e. complex numbers that get multiplied by $e^{\pm i\varphi/2}$ under a rotation by angle φ.
- In 3 Euclidean dimensions, the single spinor representation is 2-dimensional and quaternionic. The existence of spinors in 3 dimensions follows from the isomorphism of the groups $SU(2) \cong Spin(3)$ that allows us to define the action of Spin(3) on a complex 2-component column (a spinor); the generators of SU(2) can be written as Pauli matrices.

- In 4 Euclidean dimensions, the corresponding isomorphism is Spin(4) ≅ SU(2) × SU(2). There are two inequivalent quaternionic 2-component Weyl spinors and each of them transforms under one of the SU(2) factors only.
- In 5 Euclidean dimensions, the relevant isomorphism is Spin(5) ≅ USp(4) ≅ Sp(2) that implies that the single spinor representation is 4-dimensional and quaternionic.
- In 6 Euclidean dimensions, the isomorphism Spin(6) ≅ SU(4) guarantees that there are two 4-dimensional complex Weyl representations that are complex conjugates of one another.
- In 7 Euclidean dimensions, the single spinor representation is 8-dimensional and real; no isomorphisms to a Lie algebra from another series (A or C) exist from this dimension on.
- In 8 Euclidean dimensions, there are two Weyl–Majorana real 8-dimensional representations that are related to the 8-dimensional real vector representation by a special property of Spin(8) called triality.
- In $d + 8$ dimensions, the number of distinct irreducible spinor representations and their reality (whether they are real, pseudoreal, or complex) mimics the structure in d dimensions, but their dimensions are 16 times larger; this allows one to understand all remaining cases. See Bott periodicity.
- In spacetimes with p spatial and q time-like directions, the dimensions viewed as dimensions over the complex numbers coincide with the case of the $(p + q)$-dimensional Euclidean space, but the reality projections mimic the structure in $|p - q|$ Euclidean dimensions. For example, in 3 + 1 dimensions there are two non-equivalent Weyl complex (like in 2 dimensions) 2-component (like in 4 dimensions) spinors, which follows from the isomorphism SL(2, **C**) ≅ Spin(3,1).

35.8 See also

- Anyon
- Dirac equation in the algebra of physical space
- Einstein–Cartan theory
- Pure spinor
- Spin-½
- Spinor bundle
- Supercharge
- Twistor theory

35.9 Notes

[1] Spinors in three dimensions are points of a line bundle over a conic in the projective plane. In this picture, which is associated to spinors of a three-dimensional pseudo-Euclidean space of signature (1,2), the conic is an ordinary real conic (here the circle), the line bundle is the Möbius bundle, and the spin group is $SL_2(\mathbf{R})$. In Euclidean signature, the projective plane, conic and line bundle are over the complex instead, and this picture is just a real slice.

[2] Spinors can always be defined over the complex numbers. However, in some signatures there exist real spinors. Details can be found in spin representation.

[3] A formal definition of spinors at this level is that the space of spinors is a linear representation of the Lie algebra of infinitesimal rotations of a certain kind.

[4] More precisely, it is the fermions of spin-1/2 that are described by spinors, which is true both in the relativistic and non-relativistic theory. The wavefunction of the non-relativistic electron has values in 2 component spinors transforming under three-dimensional infinitesimal rotations. The relativistic Dirac equation for the electron is an equation for 4 component spinors transforming under infinitesimal Lorentz transformations for which a substantially similar theory of spinors exists.

[5] Formally, the spin group is the group of relative homotopy classes with fixed endpoints in the rotation group.

[6] More formally, the space of spinors can be defined as an (irreducible) representation of the spin group that does not factor through a representation of the rotation group (in general, the connected component of the identity of the orthogonal group).

[7] Geometric algebra is a name for the Clifford algebra in an applied setting.

[8] the Pauli matrices correspond to angular momenta operators about the three coordinate axes. This makes them slightly a-typical gamma matrices because in addition to their anti commutation relation they also satisfy commutation relations

[9] The metric signature relevant as well if we are concerned with real spinors. See spin representation.

[10] Whether the representation decomposes depends on whether they are regarded as representations of the spin group (or its Lie algebra), in which case it decomposes in even but not odd dimensions, or the Clifford algebra when it is the other way around. Other structures than this decomposition can also exist; precise criteria are covered at spin representation and Clifford algebra.

[11] This is the set of 2×2 complex traceless hermitian matrices.

[12] Except for a kernel of $\{\pm 1\}$ corresponding to the two different elements of the spin group that go to the same rotation.

[13] Although there are several more intrinsic constructions, the spin representations are not functorial in the quadratic form, so they cannot be built up naturally within the tensor algebra.

[14] So the ambiguity in identifying the spinors themselves persists from the point of view of the group theory, and still depends on choices.

[15] The Clifford algebra can be given an even/odd grading from the parity of the degree in the gammas, and the spin group and its Lie algebra both lie in the even part. Whether here by "representation" we mean representations of the spin group or the Clifford algebra will affect the determination of their reducibility. Other structures than this splitting can also exist; precise criteria are covered at spin representation and Clifford algebra.

35.10 References

[1] Cartan 1913.

[2] Quote from Elie Cartan: *The Theory of Spinors*, Hermann, Paris, 1966, first sentence of the Introduction section of the beginning of the book (before the page numbers start): "Spinors were first used under that name, by physicists, in the field of Quantum Mechanics. In their most general form, spinors were discovered in 1913 by the author of this work, in his investigations on the linear representations of simple groups*; they provide a linear representation of the group of rotations in a space with any number n of dimensions, each spinor having 2^ν components where $n = 2\nu + 1$ or 2ν ." The star (*) refers to Cartan 1913.

[3] Named after William Kingdon Clifford,

[4] Named after Paul Dirac.

[5] Named after Hermann Weyl.

[6] Named after Ettore Majorana.

[7] Matthew R. Francis, Arthur Kosowsky: *The Construction of Spinors in Geometric Algebra*, submitted 20 March 2004, version of 18 October 2004 arXiv:math-ph/0403040

[8]
- Wilczek, Frank (2009). "Majorana returns". *Nature Phys.* (Macmillan Publishers) **5** (9): 614–618. doi:10.1038/nphys1380. ISSN 1745-2473. (subscription required (help)).

[9] Xu, Yang-Su et al. (2015). "Discovery of a Weyl Fermion semimetal and topological Fermi arcs". *Science Magazine* (AAAS). doi:10.1126/science.aaa9297. ISSN 0036-8075. (subscription required (help)).

[10] Jean Hladik: *Spinors in Physics*, translated by J. M. Cole, Springer 1999, ISBN 978-0-387-98647-0, p. 3

[11] Graham Farmelo: *The Strangest Man. The Hidden Life of Paul Dirac, Quantum Genius*, Faber & Faber, 2009, ISBN 978-0-571-22286-5, p. 430

[12] Cartan 1913

[13] Tomonaga 1998, p. 129

[14] Pauli 1927.

[15] Dirac 1928.

[16] G. Juvet: *Opérateurs de Dirac et équations de Maxwell*, Commentarii Mathematici Helvelvetici, 2 (1930), pp. 225–235, doi:10.1007/BF01214461 (abstract in French language)

[17] F. Sauter: *Lösung der Diracschen Gleichungen ohne Spezialisierung der Diracschen Operatoren*, Zeitschrift für Physik, Volume 63, Numbers 11–12, 803–814, doi:10.1007/BF01339277 (abstract in German language)

[18] Pertti Lounesto: *Crumeyrolle's bivectors and spinors*, pp. 137–166, In: Rafał Abłamowicz, Pertti Lounesto (eds.): *Clifford algebras and spinor structures: A Special Volume Dedicated to the Memory of Albert Crumeyrolle (1919–1992)*, ISBN 0-7923-3366-7, 1995, p. 151

[19] The matrices of dimension $N \times N$ in which only the elements of the left column are non-zero form a *left ideal* in the $N \times N$ matrix algebra Mat(N, **C**) – multiplying such a matrix M from the left with any $N \times N$ matrix A gives the result AM that is again an $N \times N$ matrix in which only the elements of the left column are non-zero. Moreover, it can be shown that it is a *minimal left ideal*. See also: Pertti Lounesto: *Clifford algebras and spinors*, London Mathematical Society Lecture Notes Series 286, Cambridge University Press, Second Edition 2001, DOI 978-0-521-00551-7, p. 52

[20] Pertti Lounesto: *Clifford algebras and spinors*, London Mathematical Society Lecture Notes Series 286, Cambridge University Press, Second Edition 2001, DOI 978-0-521-00551-7, p. 148 f. and p. 327 f.

[21] D. Hestenes: *Space–Time Algebra*, Gordon and Breach, New York, 1966, 1987, 1992

[22] D. Hestenes: *Real spinor fields*, J. Math. Phys. 8 (1967), pp. 798–808

[23] These are the right-handed Weyl spinors in two dimensions. For the left-handed Weyl spinors, the representation is via $\gamma(\phi) = \gamma\phi$. The Majorana spinors are the common underlying real representation for the Weyl representations.

[24] Since, for a skew field, the kernel of the representation must be trivial. So inequivalent representations can only arise via an automorphism of the skew-field. In this case, there are a pair of equivalent representations: $\gamma(\phi) = \gamma\phi$, and its quaternionic conjugate $\gamma(\phi) = \phi\gamma$.

[25] The complex spinors are obtained as the representations of the tensor product **H** ⊗**R** **C** = $\mathrm{Mat}_2(\mathbf{C})$. These are considered in more detail in spinors in three dimensions.

[26] This construction is due to Cartan. The treatment here is based on Chevalley (1954).

[27] One source for this subsection is Fulton & Harris (1991).

[28] Via the even-graded Clifford algebra.

[29] Lawson & Michelsohn 1989, Appendix D.

[30] Brauer & Weyl 1935.

35.11 Further reading

- Brauer, Richard; Weyl, Hermann (1935), "Spinors in n dimensions", *American Journal of Mathematics* (The Johns Hopkins University Press) **57** (2): 425–449, doi:10.2307/2371218, JSTOR 2371218.
- Cartan, Élie (1913), "Les groupes projectifs qui ne laissent invariante aucune multiplicité plane" (PDF), *Bul. Soc. Math. France* **41**: 53–96.
- Cartan, Élie (1966), *The theory of spinors*, Paris, Hermann (reprinted 1981, Dover Publications), ISBN 978-0-486-64070-9
- Chevalley, Claude (1954), *The algebraic theory of spinors and Clifford algebras*, Columbia University Press (reprinted 1996, Springer), ISBN 978-3-540-57063-9.
- Dirac, Paul M. (1928), "The quantum theory of the electron", *Proceedings of the Royal Society of London* **A117**: 610–624, JSTOR 94981.
- Fulton, William; Harris, Joe (1991), *Representation theory. A first course*, Graduate Texts in Mathematics, Readings in Mathematics **129**, New York: Springer-Verlag, ISBN 0-387-97495-4, MR 1153249.
- Gilkey, Peter B. (1984), *Invariance Theory, the Heat Equation, and the Atiyah–Singer Index Theorem*, Publish or Perish, ISBN 0-914098-20-9.
- Harvey, F. Reese (1990), *Spinors and Calibrations*, Academic Press, ISBN 978-0-12-329650-4.
- Hazewinkel, Michiel, ed. (2001), "Spinor", *Encyclopedia of Mathematics*, Springer, ISBN 978-1-55608-010-4
- Hitchin, Nigel J. (1974), "Harmonic spinors", *Advances in Mathematics* **14**: 1–55, doi:10.1016/0001-8708(74)90021-8, MR 358873.
- Lawson, H. Blaine; Michelsohn, Marie-Louise (1989), *Spin Geometry*, Princeton University Press, ISBN 0-691-08542-0.
- Pauli, Wolfgang (1927), "Zur Quantenmechanik des magnetischen Elektrons", *Zeitschrift für Physik* **43** (9–10): 601–632, Bibcode:1927ZPhy...43..601P, doi:10.1007/BF01397326.
- Penrose, Roger; Rindler, W. (1988), *Spinors and Space–Time: Volume 2, Spinor and Twistor Methods in Space–Time Geometry*, Cambridge University Press, ISBN 0-521-34786-6.
- Tomonaga, Sin-Itiro (1998), "Lecture 7: The Quantity Which Is Neither Vector nor Tensor", *The story of spin*, University of Chicago Press, p. 129, ISBN 0-226-80794-0

Chapter 36

Abstract index notation

Not to be confused with tensor index notation.

Abstract index notation is a mathematical notation for tensors and spinors that uses indices to indicate their types, rather than their components in a particular basis. The indices are mere placeholders, not related to any fixed basis and, in particular, are non-numerical. Thus it should not be confused with the Ricci calculus. The notation was introduced by Roger Penrose as a way to use the formal aspects of the Einstein summation convention in order to compensate for the difficulty in describing contractions and covariant differentiation in modern abstract tensor notation, while preserving the explicit covariance of the expressions involved.

Let V be a vector space, and V^* its dual. Consider, for example, a rank-2 covariant tensor $h \in V^* \otimes V^*$. Then h can be identified with a bilinear form on V. In other words, it is a function of two arguments in V which can be represented as a pair of *slots*:

$$h = h(-, -).$$

Abstract index notation is merely a *labelling* of the slots by Latin letters, which have no significance apart from their designation as labels of the slots (i.e., they are non-numerical):

$$h = h_{ab}.$$

A contraction between two tensors is represented by the repetition of an index label, where one label is contravariant (an *upper index* corresponding to a tensor in V) and one label is covariant (a *lower index* corresponding to a tensor in V^*). Thus, for instance,

$$t_{ab}{}^{b}$$

is the trace of a tensor $t = t_{ab}{}^{c}$ over its last two slots. This manner of representing tensor contractions by repeated indices is formally similar to the Einstein summation convention. However, as the indices are non-numerical, it does not imply summation: rather it corresponds to the abstract basis-independent trace operation (or duality pairing) between tensor factors of type V and those of type V^*.

36.1 Abstract indices and tensor spaces

A general homogeneous tensor is an element of a tensor product of copies of V and V^*, such as

$$V \otimes V^* \otimes V^* \otimes V \otimes V^*.$$

Label each factor in this tensor product with a Latin letter in a raised position for each contravariant V factor, and in a lowered position for each covariant V^* position. In this way, write the product as

$$V^a V b V c V^d V e$$

or, simply

$$V^a{}_{bc}{}^d{}_e.$$

The last two expressions denote the same object as the first. Tensors of this type are denoted using similar notation, for example:

$$H^a{}_{bc}{}^d{}_e \in V^a{}_{bc}{}^d{}_e = V \otimes V^* \otimes V^* \otimes V \otimes V^*.$$

36.2 Contraction

In general, whenever one contravariant and one covariant factor occur in a tensor product of spaces, there is an associated *contraction* (or *trace*) map. For instance,

$$\mathrm{Tr}12 : V \otimes V^* \otimes V^* \otimes V \otimes V^* \to V^* \otimes V \otimes V^*$$

is the trace on the first two spaces of the tensor product.

$$\mathrm{Tr}15 : V \otimes V^* \otimes V^* \otimes V \otimes V^* \to V^* \otimes V^* \otimes V$$

is the trace on the first and last space.

These trace operations are signified on tensors by the repetition of an index. Thus the first trace map is given by

$$\mathrm{Tr}12 : H^a{}_{bc}{}^d{}_e \to H^a{}_{ac}{}^d{}_e$$

and the second by

$$\mathrm{Tr}15 : H^a{}_{bc}{}^d{}_e \to H^a{}_{bc}{}^d{}_a.$$

36.3 Braiding

To any tensor product on a single vector space, there are associated braiding maps. For example, the braiding map

$$\tau(12) : V \otimes V \to V \otimes V$$

interchanges the two tensor factors (so that its action on simple tensors is given by $\tau_{(12)}(v \otimes w) = w \otimes v$). In general, t
he
braiding maps are in one-to-one correspondence with elements of the symmetric group, acting by permuting the tensor factors. Here, we use τ_σ to denote the braiding map associated to the permutation σ (represented as a product of disjoint cyclic permutations).

Braiding maps are important in differential geometry, for instance, in order to express the Bianchi identity. Here let R denote the Riemann tensor, regarded as a tensor in $V^* \otimes V^* \otimes V^* \otimes V$. The first Bianchi identity then asserts tha
t

$$R + \tau_{(123)}R + \tau_{(132)}R = 0.$$

Abstract index notation handles braiding as follows. On a particular tensor product, an ordering of the abstract indices is fixed (usually this is a lexicographic ordering). The braid is then represented in notation by permuting the labels of the indices. Thus, for instance, with the Riemann tensor

$$R = R_{abc}{}^d \in V_{abc}{}^d = V^* \otimes V^* \otimes V^* \otimes V,$$

the Bianchi identity becomes

$$R_{abc}{}^d + R_{cab}{}^d + R_{bca}{}^d = 0.$$

36.4 See also

- Penrose graphical notation
- Einstein notation
- Index notation
- Tensor
- Antisymmetric tensor
- Raising and lowering indices
- Covariance and contravariance of vectors

36.5 References

- Roger Penrose, *The Road to Reality: A Complete Guide to the Laws of the Universe*, 2004, has a chapter explaining it.
- Roger Penrose and Wolfgang Rindler, *Spinors and space-time*, volume I, *two-spinor calculus and relativistic fields*

Made in the USA
San Bernardino, CA
23 October 2015